ACS SYMPOSIUM SERIES **599**

Fire and Polymers II

Materials and Tests for Hazard Prevention

Gordon L. Nelson, EDITOR
Florida Institute of Technology

Developed from a symposium sponsored
by the Division of Polymeric Materials:
Science and Engineering, Inc.
at the 208th National Meeting
of the American Chemical Society,
Washington, DC
August 21–26, 1994

American Chemical Society, Washington, DC 1995

Library of Congress Cataloging-in-Publication Data

Fire and polymers II: materials and tests for hazard prevention / Gordon L. Nelson, editor

    p.   cm.—(ACS symposium series; 599)

"Developed from a symposium sponsored by the Division of Polymeric Materials: Science and Engineering, Inc. at the 208th National Meeting of the American Chemical Society, Washington, D.C., August 21–26, 1994."

Includes bibliographical references and index.

ISBN 0–8412–3231–8

1. Polymers—Fires and fire prevention—Congresses.   I. Nelson, Gordon L., 1943–   .   II. American Chemical Society. Division of Polymeric Materials: Science and Engineering.   III. American Chemical Society. Meeting (208th: 1994: Washington, D.C.)   IV. Series.

TH9446.P65F568   1995
628.9'222—dc20

95–17788
CIP

This book is printed on acid-free, recycled paper.

Copyright © 1995

American Chemical Society

All Rights Reserved. The appearance of the code at the bottom of the first page of each chapter in this volume indicates the copyright owner's consent that reprographic copies of the chapter may be made for personal or internal use or for the personal or internal use of specific clients. This consent is given on the condition, however, that the copier pay the stated per-copy fee through the Copyright Clearance Center, Inc., 222 Rosewood Drive, Danvers, MA 01923, for copying beyond that permitted by Sections 107 or 108 of the U.S. Copyright Law. This consent does not extend to copying or transmission by any means—graphic or electronic—for any other purpose, such as for general distribution, for advertising or promotional purposes, for creating a new collective work, for resale, or for information storage and retrieval systems. The copying fee for each chapter is indicated in the code at the bottom of the first page of the chapter.

The citation of trade names and/or names of manufacturers in this publication is not to be construed as an endorsement or as approval by ACS of the commercial products or services referenced herein; nor should the mere reference herein to any drawing, specification, chemical process, or other data be regarded as a license or as a conveyance of any right or permission to the holder, reader, or any other person or corporation, to manufacture, reproduce, use, or sell any patented invention or copyrighted work that may in any way be related thereto. Registered names, trademarks, etc., used in this publication, even without specific indication thereof, are not to be considered unprotected by law.

PRINTED IN THE UNITED STATES OF AMERICA

1995 Advisory Board
ACS Symposium Series

Robert J. Alaimo
Procter & Gamble Pharmaceuticals

Mark Arnold
University of Iowa

David Baker
University of Tennessee

Arindam Bose
Pfizer Central Research

Robert F. Brady, Jr.
Naval Research Laboratory

Mary E. Castellion
ChemEdit Company

Margaret A. Cavanaugh
National Science Foundation

Arthur B. Ellis
University of Wisconsin at Madison

Gunda I. Georg
University of Kansas

Madeleine M. Joullie
University of Pennsylvania

Lawrence P. Klemann
Nabisco Foods Group

Douglas R. Lloyd
The University of Texas at Austin

Cynthia A. Maryanoff
R. W. Johnson Pharmaceutical Research Institute

Roger A. Minear
University of Illinois at Urbana–Champaign

Omkaram Nalamasu
AT&T Bell Laboratories

Vincent Pecoraro
University of Michigan

George W. Roberts
North Carolina State University

John R. Shapley
University of Illinois at Urbana–Champaign

Douglas A. Smith
Concurrent Technologies Corporation

L. Somasundaram
DuPont

Michael D. Taylor
Parke-Davis Pharmaceutical Research

William C. Walker
DuPont

Peter Willett
University of Sheffield (England)

Foreword

THE ACS SYMPOSIUM SERIES was first published in 1974 to provide a mechanism for publishing symposia quickly in book form. The purpose of this series is to publish comprehensive books developed from symposia, which are usually "snapshots in time" of the current research being done on a topic, plus some review material on the topic. For this reason, it is necessary that the papers be published as quickly as possible.

Before a symposium-based book is put under contract, the proposed table of contents is reviewed for appropriateness to the topic and for comprehensiveness of the collection. Some papers are excluded at this point, and others are added to round out the scope of the volume. In addition, a draft of each paper is peer-reviewed prior to final acceptance or rejection. This anonymous review process is supervised by the organizer(s) of the symposium, who become the editor(s) of the book. The authors then revise their papers according to the recommendations of both the reviewers and the editors, prepare camera-ready copy, and submit the final papers to the editors, who check that all necessary revisions have been made.

As a rule, only original research papers and original review papers are included in the volumes. Verbatim reproductions of previously published papers are not accepted.

Contents

Preface ... ix

1. **Fire and Polymers: An Overview** ... 1
   Gordon L. Nelson

STATE-OF-THE-ART PHOSPHORUS OR HALOGEN FLAME RETARDANTS

2. **Triarylphosphine Oxide Containing Nylon 6,6 Copolymers** ... 29
   I-Yuan Wan, J. E. McGrath, and Takashi Kashiwagi

3. **Copolycarbonates and Poly(arylates) Derived from Hydrolytically Stable Phosphine Oxide Comonomers** ... 41
   D. M. Knauss, J. E. McGrath, and Takashi Kashiwagi

4. **Aromatic Organic Phosphate Oligomers as Flame Retardants in Plastics** ... 56
   Rudolph D. Deanin and Mohammad Ali

5. **Chlorinated Flame Retardant Used in Combination with Other Flame Retardants** ... 65
   R. L. Markezich and D. G. Aschbacher

6. **Developments in Intumescent Fire-Retardant Systems: Ammonium Polyphosphate–Poly(ethyleneurea formaldehyde) Mixtures** ... 76
   G. Camino, M. P. Luda, and L. Costa

7. **Intumescent Systems for Flame Retarding of Polypropylene** ... 91
   Menachem Lewin and Makoto Endo

METALS AND COMPOUNDS AS FLAME RETARDANTS

8. **Reductive Coupling Promoted by Zerovalent Copper: A Potential New Method of Smoke Suppression for Vinyl Chloride Polymers** ... 118
   J. P. Jeng, S. A. Terranova, E. Bonaplata, K. Goldsmith, D. M. Williams, B. J. Wojciechowski, and W. H. Starnes, Jr.

9. **Effect of Some Tin and Sulfur Additives on the Thermal Degradation of Poly(methyl methacrylate)** 126
   Jayakody A. Chandrasiri and Charles A. Wilkie

10. **Effect of Zinc Chloride on the Thermal Stability of Styrene−Acrylonitrile Copolymers** 136
    Sang Yeol Oh, Eli M. Pearce, and T. K. Kwei

SURFACES AND CHAR

11. **Thermal Decomposition Chemistry of Poly(vinyl alcohol): Char Characterization and Reactions with Bismaleimides** 161
    Jeffrey W. Gilman, David L. VanderHart, and Takashi Kashiwagi

12. **New Types of Ecologically Safe Flame-Retardant Polymer Systems** 186
    G. E. Zaikov and S. M. Lomakin

13. **Some Practical and Theoretical Aspects of Melamine as a Flame Retardant** 199
    Edward D. Weil and Weiming Zhu

14. **Flammability Improvement of Polyurethanes by Incorporation of a Silicone Moiety into the Structure of Block Copolymers** 217
    Ramazan Benrashid and Gordon L. Nelson

15. **Surface Modification of Polymers To Achieve Flame Retardancy** 236
    Charles A. Wilkie, Xiaoxing Dong, and Masanori Suzuki

16. **Flammability Properties of Honeycomb Composites and Phenol−Formaldehyde Resins** 245
    Marc R. Nyden, James E. Brown, and S. M. Lomakin

17. **Synthesis and Characterization of Novel Carbon−Nitrogen Materials by Thermolysis of Monomers and Dimers of 4,5-Dicyanoimidazole** 256
    Eric C. Coad and Paul G. Rasmussen

18. **High-Temperature Copolymers from Inorganic−Organic Hybrid Polymer and Multi-ethynylbenzene** 267
    Teddy M. Keller

19. **Linear Siloxane−Acetylene Polymers as Precursors to High-Temperature Materials** 280
    David Y. Son and Teddy M. Keller

FIRE TOXICITY

20. **Further Development of the N-Gas Mathematical Model: An Approach for Predicting the Toxic Potency of Complex Combustion Mixtures** 293
    Barbara C. Levin, Emil Braun, Magdalena Navarro, and Maya Paabo

21. **Correlation of Atmospheric and Inhaled Blood Cyanide Levels in Miniature Pigs** 312
    F. W. Stemler, A. Kaminskis, T. M. Tezak-Reid, R. R. Stotts, T. S. Moran, H. H. Hurt, Jr., and N. W. Ahle

22. **Environmental Nitrogen Dioxide Exposure Hazards of Concern to the U.S. Army** 323
    M. A. Mayorga, A. J. Januszkiewicz, and B. E. Lehnert

23. **Application of the Naval Medical Research Institute Toxicology Detachment Neurobehavioral Screening Battery to Combustion Toxicology** 344
    G. D. Ritchie, J. Rossi III, and D. A. Macys

24. **Smoke Production from Advanced Composite Materials** 366
    D. J. Caldwell, K. J. Kuhlmann, and J. A. Roop

25. **Formation of Polybrominated Dibenzodioxins and Dibenzofurans in Laboratory Combustion Processes of Brominated Flame Retardants** 377
    Dieter Lenoir and Kathrin Kampke-Thiel

26. **Analysis of Soot Produced from the Combustion of Polymeric Materials** 393
    Kent J. Voorhees

TOOLS FOR FIRE SCIENCE

27. **The Computer Program Roomfire: A Compartment Fire Model Shell** 409
    Marc L. Janssens

28. **Upward Flame Spread on Composite Materials** 422
    T. J. Ohlemiller and T. G. Cleary

29. **Protocol for Ignitability, Lateral Flame Spread, and Heat Release Rate Using Lift Apparatus** 435
    Mark A. Dietenberger

30. Fire Properties of Materials for Model-Based Assessments for Hazards and Protection Needs 450
    A. Tewarson

31. Controlled-Atmosphere Cone Calorimeter 498
    M. Robert Christy, Ronald V. Petrella, and John J. Penkala

32. X-ray Photoelectron Spectroscopy (Electron Spectroscopy for Chemical Analysis) Studies in Flame Retardancy of Polymers 518
    Jianqi Wang

33. Thermal Analysis of Fire-Retardant Poly(vinyl chloride) Using Pyrolysis–Chemical Ionization Mass Spectrometry 536
    Sunit Shah, Vipul Davé, and Stanley C. Israel

SCIENCE-BASED REGULATION

34. Smoke Corrosivity: Technical Issues and Testing 553
    Marcelo M. Hirschler

35. "Green Products," A Challenge to Flame-Retardant Plastics: Recycling, Marking, Ecolabeling, and Product Take-Back 579
    Gordon L. Nelson

36. Tools Available To Predict Full-Scale Fire Performance of Furniture 593
    Marcelo M. Hirschler

37. Combustion Behavior of Upholstered Furniture Tested in Europe: Overview of Activities and a Project Description 609
    Björn Sundström

38. Fire-Safe Aircraft Cabin Materials 618
    Richard E. Lyon

Author Index 639

Affiliation Index 640

Subject Index 640

Preface

FIRE, A MAJOR SOCIAL ISSUE IN THE UNITED STATES and around the world, causes some $10 billion in property loss annually in the United States alone. Most of these fires involve the combustion of polymeric materials. Despite the magnitude of the issue in the United States—it has the highest rate of fire in the world—there are no peer-reviewed books on current fire science research topics.

Because fire and polymers are an importance social issue and because of the interest in and complexity of fire science, a symposium and workshop were organized in conjunction with the 208th ACS National Meeting in Washington, D.C. It was the first major symposium held on fire and polymers at an ACS National Meeting in more than five years. Leading experts were brought together to discuss new materials and new approaches to flame retardancy, combustion and toxicity, and testing and modeling. The workshop provided a basic grounding in fire chemistry, flame-retardant mechanisms, testing, toxicity, and regulation. A press conference held in conjunction with the symposium generated press clippings in more than 400 newspapers in the United States and numerous radio and television interviews of symposium participants. More press clippings were generated from this symposium than from any other symposium at the 208th ACS National Meeting.

This book is based upon the symposium and workshop. It is divided into six sections, each representing a significant segment of the latest research on fire and polymers. The book includes several key contributions not presented at the symposium. The contributors, major figures in the fire sciences, are drawn from the United States as well as Europe, Israel, and China.

Acknowledgments

I gratefully acknowledge the work of Marcelo M. Hirschler, GBH International; Barbara C. Levin, National Institute of Standards and Technology, Eli M. Pearce, Polytechnic University of New York; and Charles A. Wilkie, Marquette University. Their solicitation of papers ensured a broadly representative symposium. I thank the ACS Division of Polymeric Materials: Science and Engineering for support of travel for international speakers at the symposium. I also acknowledge Cheryl-Ann

Brown for her extensive assistance with various phases of this endeavor and Elaine N. Hooker for assistance with a number of figures.

GORDON L. NELSON
Dean, College of Science and Liberal Arts
Florida Institute of Technology
150 West University Boulevard
Melbourne, FL 32901–6988

February 6, 1995

Chapter 1

Fire and Polymers: An Overview

Gordon L. Nelson

Florida Institute of Technology, 150 West University Boulevard, Melbourne, FL 32901-6988

Our environment is largely one of polymers and all polymers burn whether natural or synthetic. The issue is not whether polymers burn but rather if a given polymer has a property profile appropriate to provide for an acceptable level of risk in a given application. How we make polymers flame retardant, what flame retardant means, and how codes achieve an acceptable level of risk are the subjects of this overview.

Fire is an ever present hazard in the built environment. In A.D. 64, during Nero's reign, ten of the fourteen districts of Rome burned in eight days. In 1666 the Great Fire of London destroyed 13,200 homes, 94 churches, and countless public buildings. In 1842, 4200 buildings were destroyed in Hamburg, Germany, by a fire which killed 100 and rendered 20% of the population homeless. The Chicago Fire of 1871 killed 766 and destroyed 17,500 buildings. San Francisco (1906), Tokyo (1923) and Yokohama were almost totally destroyed by fires after damaging earthquakes. In the San Francisco fire alone 28,000 buildings (10km^2) were destroyed and over 1000 lives lost. In a 1908 fire storm in Chelsea, Massachusetts, some 3500 buildings were lost. And lest we think fire is forgotten in the last years of the 20th century, the Oakland, California, hills fire leaped freeways and destroyed nearly 6000 buildings. When the lessons learned of construction, materials, building separation, and urban defense strategies are ignored, then fires of significant consequence remain possible.

Most fire deaths are not the consequence of large fires, however, but fires in single structures and involve 1 or 2 fatalities. In the U.S. some 6,000 people die from fire each year, some 30,000 people are injured in 2.5 million fires causing $10 billion in property damage. With improved building design and materials the fire death rate in the US has fallen from 9.1 per 100,000 in 1913 to 2.0 in 1988. Even though the rate has fallen it remains twice that of European countries and that fire death rate has stalled. The early 1980's brought smoke detectors to 80% of American homes. Unfortunately only 60% of those are operational now. A simple, cost effective strategy is defeated by indifference and a "fires only happen to someone else" attitude.

In fact we have a 40% chance that a fire big enough to cause a call to the fire department will occur in our household during our lifetime. One-third of all fire deaths involve residential furnishings as the first item ignited (smoking 27%, open flame 5%). The advent of cigarette resistant furniture is aimed at the former scenario. Indeed analysis of fire incidents has played an important role in pinpointing important scenarios and has allowed the development of defense strategies.[1]

Our environment is largely one of polymers and all polymers burn whether natural or synthetic. The issue is not whether polymers burn but rather if a given polymer has a proper profile appropriate to provide for an acceptable level of risk in a given application. How we make polymers flame retardant, what flame retardant means, and how codes achieve an acceptable level of risk are the subjects of this overview. The reader should note that hazard is the potential for harm, with the probability of the event (fire) equal to one and the probability of exposure (e.g. people) equal to one. Risk is the product of the probability of the event times the probability of exposure, times the potential for harm.

Flame Retardant Chemistry

Polymer combustion occurs in a continuous cycle (Figures 1-2). Heat generated in the flame is transferred back to the polymer surface producing volatile polymer fragments, or fuel. These fragments diffuse into the flame zone where they react with oxygen by free-radical chain processes. This in turn produces more heat and continues the cycle. Flame retardancy is achieved by interrupting this cycle.[3]

There are two ways to interrupt the cycle. One method, *solid phase inhibition*, involves changes in the polymer substrate. Systems that promote extensive polymer crosslinking at the surface, form a carbonaceous char upon heating. Char insulates the underlying polymer from the heat of the flame, preventing production of new fuel and further burning. Other systems evolve water during heating, cooling the surface and increasing the amount of energy needed to maintain the flame.

The second way of interrupting the flame cycle, *vapor phase inhibition*, involves changes in the flame chemistry. Reactive species are built into the polymer which are transformed into volatile free-radical inhibitors during burning. These materials diffuse into the flame and inhibit the branching radical reaction. (Figure 3) As a result, increased energy is required to maintain the flame and the cycle is interrupted. Of course, for many materials both solid and vapor phase inhibition are involved.

Polymers vary a great deal in their inherent flammability and can be divided roughly into three classes (Table I). The first group consists of relatively flame retardant structures containing either high halogen, or aromatic groups that confer high thermal stability as well as the ability to form char on burning. Second are the less flame retardant materials, many of which can be made more flame retardant by appropriate chemistry. The third class consists of quite flammable polymers which are more difficult to make flame retardant because they decompose readily, forming large quantities of fuel, but these can be made appropriately flame retardant for particular applications by the addition of additives.

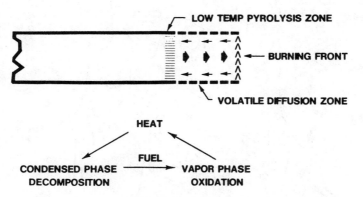

Figure 1. Polymer combustion process and cycle.

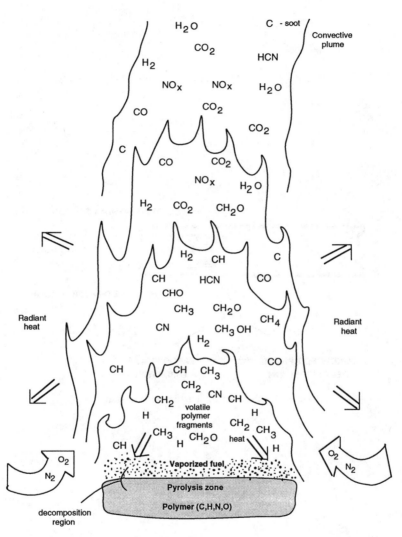

Figure 2. A typical flame involving organic fuel showing decomposition region where volatilized fuel decomposes before combustion. For many (most) fuels oxygen is not involved in fuel generation. For other materials like PTFE oxygen plays an important role in polymer decomposition. After reference 2

What Happens When Something Burns?

The simplest example of a combustion process is the burning of methane.

$$CH_4 + O_2 \longrightarrow CO_2 + H_2O$$

But even this simple combustion involves many free-radical production steps. Propagation steps that produce free radicals keep the burning going; chain-branching steps produce two free radicals, accelerating the reaction with explosive force; termination occurs when radicals are removed, quenching combustion. The flame-carrying radicals—H·, HO·, O·—occur in all flames, whether in methane or in polymers.

A Few Propagation Steps

$CH_4 + HO\cdot \longrightarrow CH_3\cdot + H_2O$
$CH_4 + H\cdot \longrightarrow CH_3\cdot + H_2$
$CH_3\cdot + O\cdot \longrightarrow CH_2O + H\cdot$
$CH_2O + HO\cdot \longrightarrow CHO\cdot + H_2O$
$CH_2O + H\cdot \longrightarrow CHO\cdot + H_2$
$CHO\cdot \longrightarrow CO + H\cdot$
$CO + HO\cdot \longrightarrow CO_2 + H\cdot$

Chain Branching

$H\cdot + O_2 \longrightarrow HO\cdot + O\cdot$

Termination Step

$H\cdot + R\cdot + M \longrightarrow RH + M^*$

where R is any organic radical and M is any surface. Heat is transferred to the surface, producing higher-energy M.

How Do Vapor-Phase Flame Retardants Affect Burning?

Flame retardancy involves taking out the active, flame-carrying radicals from the propagation and chain-branching steps. Such measures are particularly effective in the chain-branching step,

$$H\cdot + O_2 \longrightarrow HO\cdot + O\cdot$$

because they interfere with the explosive yield of large numbers of radicals. There are two ways to interfere with radical production: radical scavenging—replacing active reaction carriers by less-active halogen atoms—and radical recombination.

Radical Scavenging

$H\cdot + HBr \longrightarrow H_2 + Br\cdot$
$HO\cdot + HBr \longrightarrow H_2O + Br\cdot$

Radical Recombination

$HO\cdot + Na\cdot \longrightarrow NaOH$
$NaOH + H\cdot \longrightarrow H_2O + Na\cdot$

$$HO\cdot + H\cdot \xrightarrow{\text{catalyst}} H_2O$$

Figure 3. The combustion process and the role of flame retardants. Reference 3.

Table I. Polymers and Flammability Classification

Intrinsically Flame Retardant	Less Flame Retardant	Flammable
Polytetrafluoroethylene	Silicones	Polystyrene
Aromatic polyethersulfone	Polycarbonates	Polyacetal
All-aromatic polyimides	Polysulfone	Acrylics
All-aromatic polyamides	Wool	Polyethylene terephthalate
All-aromatic polyesters	Polysulfones	Polyolefins
All-aromatic polyethers		Cellulose (wood, cotton, paper)
Polyvinylidene dichloride		Polyurethanes

Reference 3

Intrinsically Flame Retardant Polymers. High temperature materials are intrinsically flame retardant. Some of these intrinsically flame retardant polymers are stable for a few minutes at extreme temperatures, 600-1000°C, while others can perform at 200-300 °C for long periods of time. There are three general types of structures: linear single-strand polymers such as aromatic polyimides and polyamides based on benzenoid systems, ladder polymers consisting of an uninterrupted sequence of cyclic aromatic or heterocyclic structures, and spiro polymers in which one carbon is common to two rings.

Behind each type of structure is the premise that polymers with high aromatic character and very strong connecting linkages between rings produce more char on heating, retaining most of the potential fuel of the original polymer as residue. A good example is polyphenylene, a crystalline, high-melting substance, with thermal degradation beginning at 500 to 550 °C and continuing to 900 °C with only 20-30% weight loss.

In practice, the choice of a polymer depends on cost and on the importance of flame retardancy in relation to its final use. Intrinsically flame retardant and high temperature materials are quite expensive. Less costly materials in the less-flame retardant or flammable classes can be made sufficiently flame retardant either by adding flame retardant chemicals or by modifying the polymer backbone.

Additive Approach. Flame retardant additives used with synthetic polymers include organic phosphorus compounds, organic halogen compounds, and combinations of organic halogen compounds with antimony oxide. Inorganic flame retardants include hydrated alumina, magnesium hydroxide, borates to mention only a few. Not all retardants function well in all polymers. Components may interact - components such as fillers, stabilizers, and processing aids must be considered. To be effective, the flame retardant must decompose near the decomposition temperature of the polymer in order to do the appropriate chemistry as the polymer decomposes, yet be stable at processing temperatures.

Most organic phosphorus compounds flame retard by solid phase inhibition, decomposing to form phosphorus acids and anhydrides to promote carbonaceous char. In other phosphorus compounds, reaction cooling is accomplished by endothermic reduction of phosphorus species by carbon. A few phosphorus compounds act as vapor phase inhibitors.

Most organic halogen compounds are vapor phase inhibitors and decompose to yield HBr or HCl which quench chain-branching free radical reactions in the flame. Also, in the solid state, some halogen acids catalyze char formation, particularly with polyolefins.

Combinations of antimony oxide with organic halogen compounds are even more effective vapor phase free-radical inhibitors than halogens alone. Antimony oxide reacts with the organic halogen compound producing antimony trihalide, which carries the halogen into the flame where it is released as hydrogen halide. The end product involving antimony is thought to be antimony oxide in finely divided form in the flame. Injecting fine particles into the reaction zone is known to reduce flame propagation rates.

Flame retardants in order of commercial importance are phosphorus compounds (including poly(vinyl chloride) plasticizers), halogen compounds (chlorine and bromine compounds), and combinations of halogen materials with antimony oxide. In addition, certain nitrogen and boron compounds as well as alkali metal salts and hydrates of metal oxides are important as flame retardants in specific polymers.

As reported in this volume increased understanding of specific degradation chemistry now allows one to use additives which will direct that chemistry to increase char or to alter vapor phase reactions. The use of metals, copper in PVC for example, is of particular interest in that regard and is discussed a chapter in this volume (Starnes, et al.).

Backbone Incorporation. Although the additive approach may be "simple," incorporating flame retardant chemical units directly into a polymer backbone may be more effective. The main advantages are the ability to bestow permanent flame retardancy and at the same time better maintain the original physical properties of the polymer. Monomeric additives may not be permanent and often change the polymer's physical characteristics. Flame retardant monomers used in backbone modification contain reactive functional groups to allow the monomers to be incorporated directly into the polymer chain. Many are halogen compounds, the source of the resulting improvement in flame retardancy. Generally 10-25% halogen is necessary to impart suitable flame retardancy in polymers, either by additive or backbone modification. The examples of phosphorus comonomers in Nylon 6,6 or polycarbonate are discussed in this volume.

Testing and Regulation

In many respects, to say a material is flame retardant is a misnomer. Fire is a sequence of events or phases (Figure 4). Depending upon the application different fire properties are important. These properties are:

1. Ease of ignition - how readily will a material ignite? to what kind of ignition source? a cigarette, a match, a large open flame?

2. Flame spread - how rapidly will fire spread across a polymer surface? horizontal, upward, downward, across a ceiling?

3. Rate of heat release - how much heat is released? how quickly?

4. Fire endurance - how rapidly will fire penetrate a wall, floor, or ceiling, or other barrier (fire penetration)?

5. Ease of extinction - how easily will the fire go out?

6. Smoke Release - how much smoke is released? how quickly?

7. Toxic gas evolution - how potent and how rapidly are toxic gases released? Are they irritating? Are they corrosive?

Fire starts with an ignition source and a first item to be ignited. For an electrical appliance one wants a material which will resist a small electrical arc or a small short duration flame. One does not want the appliance to be the source of fire. But one does not expect an appliance to survive a house fire. Ignition resistance is what is required.

Given ignition of an initial item then fire begins to spread and heat is released. For a wall covering, for example, flame spread should not be rapid. And indeed wall coverings and other interior finish are regulated for rate of flame spread.

As fire spreads from item to item and more heat is released and combustible gases rise, a point is reached when fire engulfs the upper part of the room (flashover). The upper part of the room exceeds 600 °C. If a door is open, fire moves down from the ceiling and begins to exit the room. The fire is fully developed.

Given a fully developed fire the issue then is the ability of barriers to contain the compartment fire. Can the assembly of materials resist penetration (fire endurance) to allow for evacuation and to protect adjacent property? Walls and floors/roofs are rated for their fire endurance.

The answer as to whether a material is flame retardant totally depends upon the application. A material suitable as an appliance enclosure because of its ignition resistance would be totally unsuitable as a seal in a wall or floor where the expectation might be the resistance to a room burnout on the other side of the wall of several hours duration.

It should not be surprising that there are hundreds of tests which are used in this country alone to access one or more aspects of the flammability of materials or assemblies. Tests by ASTM, UL, NFPA, the building codes, Federal standards, are only a portion of the test methods available. Some tests are very specific to applications, others provide data which can be used more broadly, for example in mathematical fire models. It is important for anyone interested in flame retardant materials to understand what tests measure and why they are being used.

Ignition. For ignition one might want to determine the temperature at which a material will ignite, the flux that will sustain ignition, or determine whether an ignition source such as a prescribed flame of specific duration will cause self-sustained ignition.

The Setchkin Apparatus (ASTM D1929) is used to heat a material in a furnace to determine the temperature at which a material will generate sufficient volatiles to ignite in the presence or absence of a pilot flame under specific conditions.

Thermal analysis (TGA) can determine decomposition temperature, and if interfaced to an infrared or GC/MS can determine decomposition products.

One can also determine other thermal properties (Table II), for example, whether a material will bear a load at a given temperature. The decomposition temperatures, or the flash or self-ignition temperatures from ASTM D1929 for polymers are well below flame temperatures, which are 800-1200 °C. Most materials will ignite by direct flame exposure.

While cigarettes are used to test upholstered furniture, and incendiary pills (methenamine) are used to test carpets and rugs, the most common ignition tests use Bunsen burners, whether the sample is horizontal, at 45°, at 60°, or vertical.

A test series of particular importance in the plastics industry is the test series

Table II Thermal characteristics of thermoplastics

Polymer	Glass Transition Tg {°C}	Temperature resistance Short term {°C}	Temperature resistance Long term {°C}	Vicat-softening point B {°C}	Decomposition range {°C}	Flash-ignition temperature ASTM D1929 {°C}	Self-ignition temperature ASTM D1929 {°C}	Heat of Combustion ΔH {kJ/kg}
Polyethylene (Low Density)	-125,-20	100	80	75	340-440	340	350	46500
(High Density)		125	100					46000
Polypropylene	26,-35	140	100	145	330-410	350-370	390-410	42000
Polystyrene	100	90	80	88	300-400	345-360	490	36000
ABS		95	80	110	250-350	390	480	36000
PVC (rigid)	80	75	60	70-80	200-300	390	455	20000
Polyvinylidene chloride	-18	150			225-275	>530	>530	10000
Polytetrafluoroethylene	-113,127	300	260		510-540	560	580	4500
Polymethyl methacrylate	105	95	70	85-110	170-300	300	450	26000
Polyamide (Nylon 6)	75	150	80-120	200	310-350	420	450	32000
Polyethylene terephthalate	70	150	130	80	285-305	440	480	21500
Polycarbonate	149	140	100	150-155	420-600	520	no ignition	31000
Polyoxymethylene	-85	140	80-100	170	220	350-400	400	17000

After Reference 5

of Underwriters Laboratories, UL-94. In the horizontal test, a specimen is mounted horizontally and ignited at one end by a flame. Specimens which burn slower than prescribed or don't burn to a line during the test achieve an HB rating. In the vertical test specimens are mounted vertically and ignited at the bottom end by a flame from a burner. Time to extinguish is measured upon flame withdrawal and the presence or absence of flaming dripping from the specimen is noted. The resultant data allows one to assign V-2, V-1, or V-0 ratings. In a third test, a larger flame is used on a vertical specimen with the burner inclined 20° from the vertical. This latter test allows assignment of a 5-V test rating. Such tests are ignitability tests. They don't measure flame spread, or heat release, or ease of extinction, but whether ignition is achieved by a sample subjected to a small burner flame under prescribed conditions.

Such tests are not relevant for carpets, or wall coverings, or roof tiles, but are very relevant for polymers used in or as enclosures for appliances, where one wants to insure that the appliance is not a source of fire. Televisions are a good example. In the early 1970's the US Consumer Product Safety Commission found that there were 800 life threatening TV fires in the US per year (about 20,000 total fires) out of 120 million units in place. This 7 in a million rate was determined to be of unreasonable risk. At the time the antenna bracket, turner bracket, and TV enclosure were non-ignition resistant (HB) plastic, while other materials in the set were V-2. While a mandatory process was undertaken by CPSC, the voluntary standards process of UL moved more quickly with V-2 (July 1, 1975), V-1 (July 1, 1977) and V-O (July 1, 1979) rated materials being required by enclosures, antenna brackets, and turner brackets by UL Standard 1410. Having done an extensive series of TV tests at the time, the author's personal view was that with V-O materials the incidence of TV fires should fall by 1 to 2 orders of magnitude, which has been the case. Fire deaths due to TV fires have dropped from 200 per year to 10-20. There are other such success stories for Bunsen burner ignitability tests. The user needs to understand such tests are only ignitability tests and understand the applications for which they are relevant.[6]

Other Bunsen burner tests include ASTM D-635, which is similar to UL-94 (HB) and used for light transmitting plastics in US building codes. For aircraft FAR Part 25 tests include a horizontal test (transparencies), a vertical test (seat backs, wall liners), a 45° test (cargo and baggage compartment liners), and a 60° test (wire and cable). While other tests are also used, open flame ignitability tests have a broad range of applications.

Flame Spread Given sustained ignition one is concerned about spread of flame from the point of origin. This is particularly important for wall or ceiling materials in buildings where flame spread on a material could carry fire rapidly away from the items first involved.

In the US the Tunnel Test (ASTM E-84) is used to measure surface flame spread of building materials. Specimens are mounted on the ceiling of a 25 foot long tunnel and subject to a flame from a large gas burner. Flame spread is measured visually. Material performance is put into categories and it is these categories which are used to classify interior materials in US building codes. Materials or assembly surfaces are rated A (1) 0-25 flame spread, B(2) 26-75, and C(3) 76-225. Wood is nominally class C (red oak 100, plywood 150). Enclosed vertical exitways require class A, and rooms in certain public buildings class B. Such ratings have been established and justified by experience. Smoke is also measured in the Tunnel Test and a smoke developed limit of 450 required in each of the above classes.

While 20 inch wide 25ft long specimens are justified for wall linings, for many other products such samples don't exist. The ASTM E-162 radiant panel test was developed to provide an assessment of downward flame spread for 6" by 18" samples facing a 670°C radiant panel, with the sample inclined at an angle of 30° from the panel. Since irradiance decreases down the specimen, the time progress of ignition down the specimen serves to measure critical ignition energy (flame spread, F). Thermocouples in the stack above the specimen serve as a measure of heat release rate (Q). FxQ is used as an index, I_s. Mathematics have been adjusted to give numerical results comparable to the Tunnel Test for coated wood products. ASTM E162 is used for a variety of applications. It is cited in UL 1950 for large computer room computer parts (over 10 sq. ft.) for example.

Given the need for horizontal flame spread information for vertical surfaces for use in computer fire models, another apparatus, a lateral ignition and flame travel (LIFT) apparatus has been developed (ASTM E1317 and E1321).

Whether E84, E162 or E1321, all flame spread tests have problems testing the surface flame spread characteristics of thermoplastics. Thermoplastics, other than floor mounted tend to melt and flow carrying heat and flame away from the source and making results difficult to compare with other materials. One solution is to test materials on the floor of the Tunnel Test, for example.

A bench scale radiant panel test apparatus is used to test carpets in a horizontal mode. ASTM E648 is used to access the critical radiant flux (minimum heat flux at which fire propagates) for carpets used for institutional occupancies.

Heat Release Rate Much recent fire test work has focused on heat release rate apparatus. Rate of heat release determines the size of a fire. Heat release rate data are key to computer fire models. The first practical heat release rate apparatus was developed by Professor Edwin Smith at Ohio State University. ASTM E906 tests specimens in a horizontal or vertical mode to a range of heat fluxes (up to 100kw/m^2). Heat release rate curves, smoke release rate curves, and ignitability data are obtained (Figure 5).

A more sophisticated apparatus, ASTM E1354, the cone calorimeter was subsequently developed at NIST. A key feature of the NIST work is the recognition that for most materials the amount of heat released per unit of oxygen consumed is nearly a constant. Given that oxygen concentrations are far more easily determined than heat output over time, the oxygen depletion calorimeter has proved more versatile. Many E906 units have also been modified for oxygen depletion assessment. A multitude of materials have been tested and data reported using both apparatus. E906 results are now used for the selection of wall and ceiling linings for transport category commercial aircraft (FAR Part 25) (not using oxygen depletion). The cone calorimeter has been considered in Europe as a tool for the regulation of ceiling and wall linings in buildings and is under active study for furniture, as discussed elsewhere in this volume.

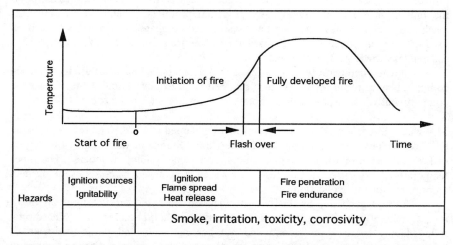

Figure 4. Sequence of events or phases in fire. Not all fires involve all phases. For example, a fire may ignite, spread to a limited area and go out. In others fire spreads, flashover is reached, and the fire is contained only by the ability of the compartment to resist penetration or the fire vents through openings. After reference 4

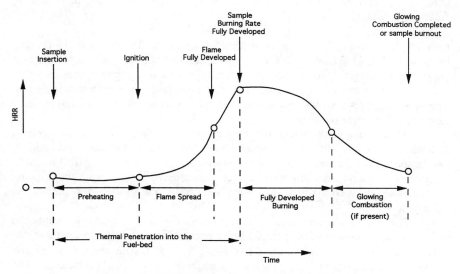

Figure 5. Sample heat release rate versus time curve.

Oxygen depletion calorimetry can be used for more than bench scale tests. If products are tested under a hood and oxygen measured in the stack exiting the hood the rate of heat release of a variety of actual products, whether TV's, business machines, chairs, cables, etc., can be assessed and compared. Such tests have been used to show the effects of flame retardants on the fire performance of real products - substantially lower peak heat release rates.

Oxygen depletion techniques have also been applied successfully to full scale compartment or room tests. What is required is an exhaust hood at the exit to the compartment which captures all the combustion products from the room.

Fire Endurance The ASTM E119 was first developed in 1918. The test is probably the most widely used and recognized in the United States for determining the degree of fire endurance of walls, columns, floors, and other building members. The time-temperature curve used in this test, Figure 6, is achieved by a series of burners appropriate for the application.

Ratings generated by this test specification are based on time intervals of 30 min, 45 min, 60 min, 90 min, 2 hr, and there after, in hourly increments. These time ratings apply to tests of bearing and non-bearing walls and partitions, tests of columns, alternative tests of protection for structural steel columns, tests of floors and roofs, tests of loaded restrained beams, alternative classification procedures for loaded beams, alternative tests for protection of solid structural beams and girders, and performance of protective membranes in walls, partitions, floor, or roof assemblies. The ratings given refer to the time - temperature curve used. If the time - temperature curve in real scale is different, then the assembly performance will be different accordingly.

Ease Of Extinction One of the most common tests used for plastics is the Oxygen Index test (ASTM D2863). In the oxygen index test the minimum percentage of oxygen in an oxygen/nitrogen mixture is determined for a top ignited specimen that will just sustain combustion. At a lower oxygen percentage the specimen will go out. The standard test is used for rigid plastics and for fabrics and films. The test has also been run on powders and liquids.

While some have called the oxygen index an ignition test, the ignition parameters are not rigidly controlled and what is measured is after sustained ignition, i.e., under what percentage of oxygen does the sample burn for at least 3 minutes. Oxygen index is a measure of ease of extinction. Some call oxygen index the "limiting oxygen index" or "LOI". The oxygen index is already a limit so the word "limiting" is redundant.

The oxygen index is sensitive to the presence of fillers, to thickness, and to the melt/flow characteristics of the polymer being tested. If the sample is heated then the oxygen index falls. At 300°C the OI is about 50% of that at room temperature. Some have used a bottom ignited sample rather than the standard top ignition. The OI is substantially lower for a bottom ignited sample, as expected, due to heat feedback to the specimen. In Europe an alternative form of the test is to determine the temperature at which the OI is 21% (the percentage of oxygen in air)

Oxygen index readily shows the effects of flame retardants, and of changes in

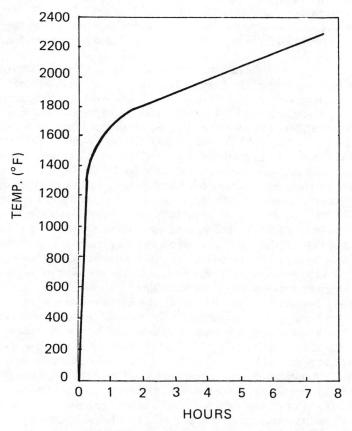

Figure 6. ASTM E119 time-temperature curve. Reference 7. Copyright ASTM. Reprinted with Permission

flame chemistry. For example the oxygen indices (Table III) for a series of polymethylenes range from 15.6 for pentane to 17.4 for polypropylene. A series of polyphenyls range from 16.3 for benzene to 32.0 for polyphenyl. The effectiveness of halogens are easily accessed. Oxygen index has been widely used as a research tool, since it uses small samples of material, and as a quality control tool. It has been used for specification purposes in the telecommunications industry. Table IV presents OI values on a large number of polymers.

A correlation has been made with UL-94 ratings, Figure 7. HB materials generally have an OI less than 23. V-2 and V-1 materials have a narrow range, OI's of 24-30, and V-O materials, as to be expected, cover a broad range from 25 and above.

Smoke Smoke is here defined as visual obscuration. Smoke is measured for construction materials in the Tunnel Test (E-84). As an alternative for light transmitting plastics, Rohm and Haas Co. was instrumental in gaining acceptance of the XP-2 smoke chamber, ASTM D2843. The loss of transmission in a horizontal light beam is measured over time as a small sample is exposed to a small flame source. NBS (now NIST) developed the NBS Smoke Chamber (ASTM E-662) to resolve some of the technical issues presented by the XP-2 smoke chamber. Samples are exposed to a radiant flaming exposure, with or without a pilot flame, resulting in either a flaming or smoldering condition depending on the ignitability of the sample. The light beam is vertical rather than horizontal to avoid effects of smoke layering.

The NBS smoke chamber is probably in the most extensively used test for smoke, both here in the USA and in some countries in Europe. It provides both a flaming condition and a smoldering condition, and some materials (e.g. wood) perform differently depending on this condition. (The E84 tunnel would be a flaming condition for wood.) Improvements to the NBS smoke chamber are underway, e.g., horizontal specimen mounting, cone radiant heater, mass loss.

From its inception the NBS smoke chamber was controversial. Cellulosic products tend to give high smoke under non-flaming conditions while high performance plastics give maximum smoke under flaming conditions. Which conditions represent real conditions? Wood products which give low smoke in the Tunnel Test now showed high smoke under the smoldering test condition.

The Arapahoe smoke apparatus (ASTM D4100) measures smoke gravimetrically. Smoke is collected on filter paper from a small sample exposed to a small burner. Results correlate with NBS smoke chamber data under flaming conditions. Its use versus the other three ASTM tests has been minimal.

Corrosivity Of special concern for polymers used in high technology applications is the generation of corrosive gases. A detailed chapter is given elsewhere in this volume. Readers are referred to that extensive discussion and analysis.

Large Scale Tests Full scale validation testing is expensive, but is the prime mechanism to be assured that bench scale results translate to real product performance. Many companies have evaluated their products through full scale: TV's, computers, aircraft interiors, rail car interiors, building products, to name a few.

Table III Contribution to Oxygen Index of major organic structural features

Polyphenyl and Polymethylene	OI
Benzene	16
Biphenyl	18
p-Terphenyl	19
Quaterphenyl	26
Polyphenyl	32
Pentane	16
Hexane	16
Cyclohexane	16
Decane	16
Hexadecane	16
Mineral oil (USP)	16
Paraffin wax	17
Polyethylene	17
Polypropylene	17

Relative effectiveness of halogen based on units of six carbons

	Alliphatic materials OI /atom	Aromatic materials OI /atom
F	<1	1.5
Cl	<1	4.3
Br	4.7	7.4
I	4.4	6.5

After Reference 8

Table IV. Oxygen Index values of polymers

Polymer		Oxygen Index		
Polyacetal		14.8		
Polyethylene Oxide		15		
Polymethylmethacrylate	FR (A)	17.3		
				25
Polyethylene	FR	17.4		
				22-29
Polypropylene	Glass (B)	17.4		
	FR			18
	Glass FR			23-29
				25-28
Polystyrene	FR	17.6	-	18.3
				22-28
Styrene/Acrylonitrile (SAN)	Glass	18		
	Glass FR			19
				27-28
Polybutadiene		18.3		
Acrylonitrile-Butadiene-Styrene (A	FR	18.3	-	18.8
	ABS/PVC			26-29
	ABS/PC			27-32
				21
Cellulose Acetate	FR	19		
				27
Polyethyleneterephthalate		20		
Chlorinated Polyethylene		21		
Phenolic	FR	21		
				26-31
Polyvinylfluoride		22.6		

Table IV. Continued.

Polymer		Oxygen Index		
Polybutyleneterephthalate	Glass	23		
	FR Glass		20	
				33
Nylon 6,6	Glass	24-29		
	Mineral Filler			
	FR		21-24	
			21	
Nylon 6,12		25	30-33	
Nylon 6	Glass	25-26		
	FR Glass			
			22-23	
Polycarbonate	Glass	25-27	28	
	FR			
			27-30	
Polyphenylene Oxide	Modified	28-29	29-32	
	FR			
			23-26	
Polydichlorostyrene		30	27-31	
Polysulfone		30	-	32
Dichloropolysulfone		41.1		
Polyvinylidine Fluoride		43.7		
Polyvinylchloride (rigid)	Plasticized	45	-	49
Tetrachloropolysulfone		50.9	23-31	
Carbon		50	-	63
Polyvinylidine Chloride		60		
Polytetrafluoroethylene		95		
Plywood		23		

Reference 9, Copyright Technomic Publishing Co. Reprinted with Permission
(A) Range for fire retardant formulations
(B) Range for glass fiber reinforced formulations.

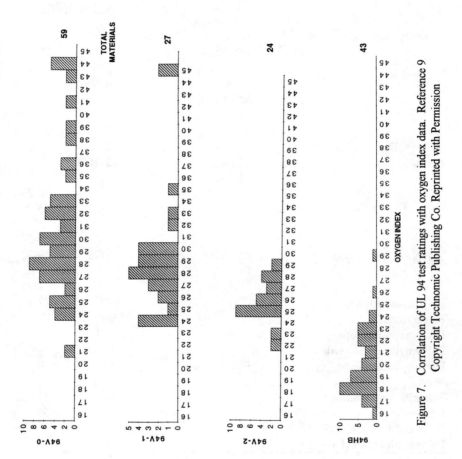

Figure 7. Correlation of UL 94 test ratings with oxygen index data. Reference 9 Copyright Technomic Publishing Co. Reprinted with Permission

Codes and Standards

Most codes and standards used in the US are developed through the voluntary consensus standards process. For building codes, three model code groups develop codes which are adopted by localities by ordinance. Voting members of the model code groups are local building code officials. Codes are based upon experience and evolve through a rational process of code change. Tests cited are tailored to the application. For example, transparencies used as light diffusers must fall out of a ceiling well below their auto ignition temperature. And depending on performance in D635 the percentage area of ceiling permitted varies. For interior finish, flame spread results are determinative. For fire resistive wall and ceiling assemblies E119 results are required. For roofing materials a burning brand test and other requirements apply. The point is that different fire performance parameters apply depending upon the application. Tests are not arbitrary but have evolved over time through a voluntary consensus process. As such they tend to not be overly stringent and to be somewhat behind the state of the art in fire test technology. They have, however, "done the job".

One important point from a commercial point of view is that being better is generally not good enough. If one does not cross a line in a code or standard, it is unlikely that a customer will pay for a better product. A customer will pay more if it does just cross the line in a code or standard and is thus "better".

European Harmonization Standards and codes are fairly static in the US and Canada. The same cannot be said for Europe. The European Union mandates harmonization. Indeed, each European country has had its own tests. The results are that materials acceptable in one country may not be acceptable in another. On the positive side the need for harmonization provides the opportunity to provide rational test protocols based upon the latest in bench and large scale test methodology and computer fire models. The negative side is the reluctance of countries and industries to accept new tests, which necessarily will mean that some materials and assemblies previously approved may lose their approval or their relative performance position.

A fire classification of building products as proposed by Eurefic would have classified wall and ceiling linings by performance in large scale tests or by cone calorimeter and room corner tests (ISO/DIS 9705) with results transformed using a calculation model. Testing of 11 representative products show the difficulty in a new test scheme, however rational (Table V).

The Eurefic approach (cone calorimeter) has not been accepted by the European Commission (Regulators Group). The approach that is being taken now is the use of six Euroclasses for products, based on several ISO tests (ISO 1182 - noncombustibility, ISO 1716 - calorific content, and ISO/DIS 11925-2 - small scale ignitability) and a new "single burning item" heat release/flame spread test (currently in development). All of this is in response to the Construction Products Directive which was issued in December, 1988, by the Commission.

In contrast to the Construction Products Directive, the European Commission has issued only a draft Furniture Directive; by which there is a prenormative research project underway (which includes the cone calorimeter) to determine whether or not the draft directive can prevail. It is not clear at this time what final decision will be

Table V. A COMPARISON OF PRESENT EUROPEAN CLASSIFICATION OF MATERIALS VERSUS RESULTS FROM ROOM/CORNER TEST ISO 9705

Product	EUREFIC With smoke	EUREFIC Without smoke	Classification in Europe England	Classification in Europe France	Classification in Europe Germany*	Italy**	Denmark
Painted gypsum paper plasterboard	A	A	1	M1	B1*	1	B
Ordinary plywood	E	E	3	M4	B2/B3	3,4	B
Textile wall covering on gypsum paper plasterboard, 500 g/m2	D	D	3	M2	B2/B3	2	B
Melamine faced high density non-combustible board	D	A	1	M1	B1*	1	B
Plastic faced steelsheet on mineral wool	C	A	1	M1	B1	2	B

Material							
FR particle board type B1	D	D	1	M1	B1	1	UC
Combustible faced mineral wool	UC	UC	4	M1	B2/B3	2	UC
FR particle board	C	B	1	M1	B1	1	A
Polyurethane foam covered with steel sheets	E	E	1	M2	B1	2	UC
PVC wall carpet on gypsum paper plasterboard, 1 250 g/m2	E	D	1	M2	B2/B3	2	UC
FR polystrene, 25mm	UC	UC		M1	B1	1	

Based on Brandschacht only, B1 means that subrequirements in Brandschacht for A2 are met.

** Spread of flame category only.

Materials tested and classified according to present European regulations by official test laboratories in each of the countries. As smoke requirements are not included in all countries, the EUREFIC classes are given both with and without smoke criteria. Classification in various countries is compared to 1991 EUREFIC proposal. Only Denmark has classification requirements on smoke. UC is unclassified Reference 10 taken by the European Commission, bearing in mind that only the UK has a current regulation for upholstered furniture. The prenormative research project for furniture is described in detail elsewhere in this volume. Results from this program will likely find easier acceptance in that fire tests for upholstered furniture are not as entrenched in the market place.

ISO and IEC Tests The International Standards Organization (ISO) and the International Electrotechnical Commission (IEC) each have a series of test methods for assessment of fire parameters. The cone calorimeter, LIFT apparatus, room corner test, oxygen index, Setchkin apparatus are all ISO tests. Two significant ignitability tests have been published by IEC, the needle flame and glow wire tests. In the former a sample is subjected to a small flame from a hypodermic needle. In the latter a sample is subjected to a glowing wire at a prescribed force. The temperature of the wire can be varied over a wide range. Europeans are pressing for the use of these tests in IEC standards for electrical and electronic equipment.

Toxicity

A Technical Report (TR) 9122 Toxicity Testing of Fire Affluents has been issued recently within ISO. The technical report discusses the conditions under which a material or product should be burned or thermally decomposed in a laboratory so as to be relevant. It discusses quantification of the toxic potency of smoke evolved and the relevance of animal models to exposures to humans. In most cases it is recognized that a small number of gases can account for the apparent toxic potency of a material (a detailed discussion of the N-gas model and its extension to NO_2 is given elsewhere in this volume).

ISO TC92/SC3 has completed its first toxicity standard - ISO 13344 Determination of the Lethal Toxic Potency of Fire Effluents. This test method provides for the selection of an appropriate combustion model.

Specific smoke toxic potency standards are now nearing completion in ISO and in ASTM. A combustion module is specified. Calculation methodology for calculation of LC_{50} values from analytical data is provided. The ASTM test requires that the LC_{50} be confirmed experimentally with a minimal set of animal (rat) exposures.

Bench scale tests determine the potency of gases under specific test conditions. Bench scale tests tend to be over ventilated for the fluxes used. The result is to underestimate the role of carbon monoxide. This is particularly important from the point of view of post-flashover fires. ASTM has a proposal to provide the methodology to provide for a CO correction of bench scale test results.

The third most common gas in fire after CO and CO_2 is hydrogen cyanide. The relative toxicity of hydrogen cyanide and CO are dependent upon the animal model. Rats may over emphasize gases other than CO.[11] A recent report indicates for human fire cases that CO appears to explain fire deaths. (See also Stemler, et al, in this volume).

One of the actions of fire retardants is to produce products of incomplete combustion, if only CO. Thus the toxic potency of fire retardant materials may be higher than the non-fire retardant offset. Yet the rate and extent of burning may be much less for the fire retardant material. The result is that bench scale toxic potency tests don't give directly the result one really desires, which is toxic hazard. That is the most difficult fact about the discussion of smoke toxicity. Despite the advent of bench scale methodology to access toxic potency, how to use these results to improve life hazard remains the subject of debate.

This Volume

Much work is underway to improve the fire performance of materials through a detailed understanding of polymer degradation chemistry. New analytical techniques are facilitating that analysis. Creative chemists are at work trying new approaches and developing new more thermally stable organic structures. What is known about the fundamental fire performance parameters of materials is being used in increasingly sophisticated mathematical fire models. That understanding is permitting the development of tests which provide data directly useable in fire models and the development of specific materials goals for such high technology applications as aircraft. And finally we are beginning to see the development of science based regulations for furnishings and building materials (particularly in Europe). Each of these topics is covered by one or more papers by leading experts in this volume. In addition a panel of experts discusses aspects of smoke and smoke toxicity.

Literature Cited

1) Hall, Jr., J.R.; Cote, A.F. In *Fire Protection Handbook,* 17th ed., Cote, A.F.; Linville, J.L., Eds.; National Fire Protection Association: Quincy, MA, 1991; pp1-1, 1-24.
2) DeHaan, J.D. *Kirk's Fire Investigation*, 3rd ed., Brady: Englewood Cliffs, N.J., 1991; p18.
3) Nelson, G.L. *Chemistry* **1978**, 51, 22-27.
4) Troitzsch, J., *International Platics* Flammability Handbook, Hansen: Munchen, Germany, 1983; p11.
5) *Ibid*, p23.
6) Hoebel, J.F.; Neily, M.L.; Ray D R.; Smith, L.E. in *Proceedings Twentieth International Conference on Fire Safety,* Hilado, C.J., ed.; Product Safety Corp.; Sunnyvale, CA, 1995.
7) E119, Standard Test Methods for Fire Tests of Building Construction and Materials, 1993 Annual Book of Standards; ASTM: Philadelphia, PA, 1993, Vol 4.07, p327.
8) Nelson, G.L. *Intern. J. Polymeric Mater.* **1979**, 7 127-145
9) Nelson, G.L.; Webb, J.L. In *Advances in Fire Retardant Textiles*, Bhatnagar, V.M., Ed.; Progress in Fire Retardancy Series; Technomic Publishing Co.: Lancaster, PA, Vol. 5; pp 271-370.
10) Sundstrom, B. In *Conference Proceedings of the First International EUREFIC Seminar 1991*; Interscience Communications, Ltd: London, U.K., 1991; pp 23-33.

11) Nelson, G.L. In *Carbon Monoxide and Human Lethality - Fire and Non Fire Studies*; Hirschler, M.M.; Debanne, S.M.; Larsen, J.B.; Nelson, G.L.; Eds.; Elsevier Science Publishers: Essex, U.K., 1993; pp 3-60.

Bibliography

a) Cullis, C.F.; and Hirschler, M.M. *The Combustion of Organic Polymers*; Oxford University Press: Oxford, UK, 1981; 420pp.
b) DeHaan, J.D. *Kirk's Fire Investigation*; Brady: Englewood Cliffs, NJ, 1991; 416pp.
c) Drysdale, Dougal. *An Introduction to Fire Dynamics*; John Wiley and Sons: Chichester, UK, 1985; 424pp.
d) *Conference Proceedings of the First International EUREFIC Seminar 1991*; Interscience Communications, Ltd: London, UK, 1991; 132pp..
e) *Fire Protection Handbook*, 17th ed.; Cote, A.E. ; Linville, J.L., Eds.; National Fire Protection Association: Quincy, MA, 1991; 2608 pp.
f) Hilado, C.J. *Flammability Handbook for Plastics*, 4th ed., Technomic Publishing Co: Lancaster, PA, 1990; 265pp.
g) *Carbon Monoxide and Human Lethality- Fire and Non-Fire Studies*; Hirschler, M.M.; Debanne, S.M.; Larsen, J.B.; Nelson, G.L.; Eds.; Elsevier Applied Science:London, UK, 1993; 425pp. Available from Chapman and Hall.
h) Lyons, J.W. *Fire*; Scientific American Library: New York, N.Y. 1985; 170pp.
i) Lyons, J.W. *The Chemistry and Uses of Fire Retardants*; Wiley Interscience: New York, N.Y., 1970; 462pp.
j) Schultz, N. *Fire and Flammability Handbook*; Van Nostrand Reinhold Co: New York, N.Y., 1985; 486pp.
k) Troitzsch, J. *International Plastics Flammability Handbook*, Hanser Publishers: Munchen, Germany, 1983; 500pp.

RECEIVED March 6, 1995

STATE-OF-THE-ART PHOSPHORUS OR HALOGEN FLAME RETARDANTS

The highest volume commercial flame retardants involve halogen or phosphorus. Such additives for large-volume plastics remain important and effective.

Simple additives have a number of negative attributes. Polymer properties are negatively affected, and the additive may leach out of the plastic. The synthesis and incorporation of suitable monomers into the polymer backbone can help to avoid this negative result. The first two chapters in this section outline such an approach. Incorporation of bis(4-carboxyphenyl)phenylphosphine oxide in Nylon 6,6 reduced the peak rate of heat release from 1200 to 500 kW/m^2 at 40 kW/m^2 and 30% additive leading (Wan, McGrath, and Kashiwagi). This approach was somewhat less effective for a high-heat polymer such as polycarbonate (Knauss, McGrath and Kashiwagi). Incorporation of triphenylphosphine oxide bisphenol in polycarbonate reduced the peak rate of heat release from 540 to 420 kW/m^2 at 40 kW/m^2 and 10% additive loading.

An approach to higher molecular weight flame retardants without direct polymer incorporation is the synthesis of an oligomer of the flame retardant. An oligomer can reduce the ability of the flame retardant to volatilize and should reduce the negative impact of the additive on polymer properties. Deanin reports results on this approach for high-density polyethylene, poly(phenylene oxide)–polystyrene, and polycarbonate, for polyaryl phosphates in these systems. Flame-retardant effectiveness increased with flame-retardant molecular weight.

For halogen one can also incorporate additives into the backbone or use polymeric or oligimeric additives. Mixtures of additives can achieve desired fire retardancy at lowered total additive levels. To improve a product's physical properties Markezich and Aschbacher's chapter provides evidence for improved action of mixed halogen flame retardants with antimony oxide. Another approach is the use of inorganic materials such as magnesium hydroxide. At 5% magnesium hydroxide loading improved oxygen-index performance is seen with a specific chlorinated flame retardant.

A final approach using more traditional flame retardants is the incorporation of char-forming agents. This is particularly important for additives like phosphorus that act at least in part in the solid phase. Two chapters discuss the use of ammonium polyphosphate with char-forming materials in polypropylene. Peak rate of heat release values are reported to fall to one-third or one-fourth with a 20–30% additive package.

The effectiveness of halogen or phosphorus flame retardants is expected to increase as researchers learn to tailor the system through increased understanding of the mechanism of decomposition of the polymer and the detailed chemistry of the flame retardant.

Chapter 2

Triarylphosphine Oxide Containing Nylon 6,6 Copolymers

I-Yuan Wan[1], J. E. McGrath[2,4], and Takashi Kashiwagi[3]

[1]IBM Almaden Laboratories, 650 Harry Road, San Jose, CA 95120
[2]Department of Chemistry and National Science Foundation Science and Technology Center, High Performance Polymeric Adhesives and Composites, Virginia Polytechnic Institute and State University, Blacksburg, VA 24061-0344
[3]Building and Fire Research Laboratory, National Institute of Standards and Technology, Gaithersburg, MD 20899-0001

> A hydrolytically stable triarylphosphine oxide containing dicarboxylic acid monomer, bis(4-carboxyphenyl) phenyl phosphine oxide $P(O)(Ph)(C_6H_4COOH)_2$, was synthesized via Friedel-Crafts reactions and chemically incorporated into the poly(hexamethylene adipamide) backbone to produce melt processable, improved flame-resistant copolymers. The content of triarylphosphine oxide comonomer in the melt synthesized copolymers was controlled from 0-30 mole%. The copolymers were crystallizable at 10 and 20 mole% incorporation of the phosphine oxide comonomer and produced tough solvent resistant films. The crystallinity was totally disrupted at 30 mole%, but the Tg values systematically increased from 58°C to 89°C. Dynamic TGA results in air at 10°C/minute showed that the char yield increased with phosphine oxide content. Cone calorimetric tests in a constant heat environment (40 kW/m^2) were employed to investigate the fundamental flame retardancy behavior of the copolymers. Significantly depressed heat release rates were observed for the copolymers containing phosphine oxide, although carbon monoxide values appeared to increase. ESCA studies of the char show that the phosphorus surface concentration was significantly increased relative to copolymer composition. It was concluded that the triaryl phosphine oxide containing nylon 6,6 copolymers had improved flame resistance and that tough melt processable films and fibers could be produced from these modified copolyamides.

Phosphorus compounds are known fire retardants (1-4), which are most commonly introduced to polymers as additives. Although economically attractive, phosphorus additives may detract from the polymers in terms of stability and performance. For example, the additives can potentially be extracted from the polymer matrix, especially in textile fiber applications (4). Such problems can be avoided by chemical incorporation of reactive phosphorus comonomers into the polymer

[4]Corresponding author

backbone. It has been reported that phosphorus containing aliphatic diamines and dicarboxylic acids have been successfully incorporated into polyamides to improve flame resistance (5,6).

The utilization of triaryl phosphine oxide monomer in demonstrably high molecular weight homo- or co-polymeric systems has been a relatively recent development and extensive research in the area is currently underway. For the past few years, our research group has been interested in synthesizing new monomers by practical methods which allow for the incorporation of tri- or diaryl phosphine oxide structures into both specialty and high volume polymeric materials. Some of the polymeric systems studied include polyimides (7), poly(arylene ether)s (8-10), epoxies (11), polycarbonates (12) and polyamides (13, 14). In general, the triaryl phosphine oxide moiety affords thermal and oxidative stability coupled with high glass transition temperatures. Amorphous morphological structures are normally produced from triaryl phosphine oxide containing homopolymers due to the non-coplanar nature of the phenyl phosphine oxide bond. However, semicrystalline copolymers can also be prepared by controlled incorporation of the triaryl phosphine oxide comonomer (10). This strategy has been extended to synthesize crystallizable high volume polyamide copolymers such as nylon 6,6. It appears to have some promise for affording new high performance flame-resistant materials for textile and engineering applications. The flammability of the polymers was tested by using cone calorimetry and, qualitatively, by simple Bunsen burner tests. These initial results are provided in this paper.

EXPERIMENTAL

<u>Monomer preparation:</u> Adipic acid and hexamethylene diamine (HMDA) were purchased from Aldrich. The acid was used without further purification and the HMDA was purified by vacuum distillation. Pure nylon 6,6 salts were prepared from these two monomers in ethanol followed by recrystallization from an ethanol/water mixture.

Bis(4-carboxyphenyl)phenyl phosphine oxide was synthesized by a series of three step reactions starting from dichlorophenyl phosphine sulfide. The first step was a Friedel Craft reaction, which is described as follows. The dichlorophenylphosphine sulfide (160g, 758 mmole) and toluene (300 ml) were added in a 4-neck flask and heated to 70°C under nitrogen flow. At this temperature, aluminum trichloride (210g, 1.755 mole) was added into the flask in aliquots over 1 hour. Then, the temperature was raised to 110°C and the reaction mixture was stirred for 7 hours. After the reaction was completed and cooled, the mixture was poured into ice water. Chloroform was used to extract the product, and the organic layer was washed twice with potassium carbonate solution. Magnesium sulfate was used to dry the organic layer, which was then filtered and the chloroform was stripped off. A brown-yellow product was slowly formed after the solvent was removed. The product was then washed with hexane several times to decolorize it. The yield of the product, bis(4-methylphenyl)phenylphosphine sulfide (BMPPS), was about 80%.

The second step was to oxidize the phosphine sulfide group to phosphine oxide, which was almost quantitative. For example, the phosphine sulfide monomer BMPPS (186g, 576 mmole) was dissolved in acetic acid (750 ml) and 50% hydrogen peroxide (80g) was added into the solution dropwise. The temperature was raised to 85-90°C then decreased to 70°C after the addition of hydrogen peroxide. The reaction was allowed to continue for another 2-4 hours. The solution was filtered through celite to remove precipitated particles. Chloroform was used to extract the product, followed by washing with water and drying with magnesium sulfate. The solvent was removed and a viscous fluid, bis(4-methylphenyl)phenyl phosphine oxide (BMPPO) resulted.

The next step was to oxidize the methyl groups to carboxylic acid groups by using potassium permanganate (15). Thus, BMPPO (152g, 496.2 mmole) was added in a 4-neck 2 liter flask equipped with an overhead stirrer and a condenser. Pyridine (600 ml) and water (300 ml) were added into the flask and the temperature was raised to 70°C. The potassium permanganate ($KMnO_4$) (580g, 3.7 mole) was added in 6-8 aliquots at 30 minute intervals and after the last addition, the temperature was raised to 80-90°C for 12 hours. At the end of the reaction, the pyridine was removed by steam distillation and the excess $KMnO_4$ was filtered off. The red solution was then acidified with concentrated hydrochloric acid to pH-5. The bis(4-carboxyphenyl)phenyl phosphine oxide (BCPPO), collected was a white product. A second oxidation was frequently needed and was conducted with a stoichiometric amount of aqueous sodium hydroxide solution. The sodium hydroxide was used to dissolve the diacid (BCPPO) as the dicarboxylate salt. A slight stoichiometric excess (by NMR spectra) of $KMnO_4$ was added into the solution and the reaction was allowed to proceed at 80°C for 6-8 hours. Then, the solution was filtered, acidified and the product was collected. Yields of the BCPPO were about 74% after the second oxidation.

Polymerization: The triphenylphosphine oxide dicarboxylic acid was successfully incorporated into nylon 6,6 copolymers at 0, 10, 20 and 30 mole% concentrations via conventional melt processes. In a typical copolymerization reaction, pure nylon 6,6 salts were prepared from adipic acid and hexamethylene diamine in ethanol, followed by recrystallization from a methanol/water mixture. Then, the triphenyl phosphine oxide dicarboxylic acid monomer and an equimolar amount of hexamethylene diamine were dissolved into water in a glass vessel.

The pH value of this solution was adjusted to 7.5 and a controlled amount of the nylon 6,6 salt was added into the solution at a concentration of about 60-70%. Copolymerization was conducted in a two stage reaction. In the first stage, the materials were transferred into a Parr® reactor, purged with argon, pressurized to 75 psi and then slowly heated to 250-260°C. During the course of the copolymerization, the pressure increased to 250-350 psi and it was maintained at this pressure for 2-3 hours. For the second stage, the pressure was slowly reduced to atmospheric over 1 hour, and then vacuum (0.5 torr) was applied to the system at 270-280°C. After 1-2 hours, a light yellow viscous nylon 6.6/triphenyl phosphine oxide co-polyamide was formed, which was cooled and isolated.

Characterization: The purity of the monomers were analyzed by ^{1}H and ^{31}P-NMR. Intrinsic viscosity values of the copolymers were determined in m-cresol at 25°C. Differential scanning calorimetry (DSC) and dynamic thermal gravimetric analysis (TGA) were performed by using Perkin-Elmer 7 series at a heating rate of 10°C per minute. The dynamic mechanical analysis (DMA) measurements were also conducted by the same instrument at a heating rate 5°C per minute, a frequency of 1 Hz and a 3-point bending test mode. The samples for DMA and other measurements were compression molded from dry copolymers at 270°C, followed by annealing at 155°C for 5 minutes. ESCA (electron spectroscopy for chemical analysis) measurements were performed with a Perkin-Elmer 5400 instrument. The samples for the ESCA measurements were 20% copolymers which were heated in air at different testing temperatures for 2-5 minutes prior to analysis.

Cone calorimetry test: The tests were performed at the National Institute of Standards and Technology by utilizing a cone calorimeter under a constant heat condition of 40 kW/m^2. The sample used was a compression molded film with a size of 10cm x 10cm x 0.3cm.

RESULTS AND DISCUSSION

The synthesis of the monomers and copolymers are described in Scheme 1 and 2, respectively. The incorporation of the triarylphosphine oxide comonomer was controlled from 0-30 mole%, since higher levels were not expected to produce crystallizable, fiber forming copolymers.

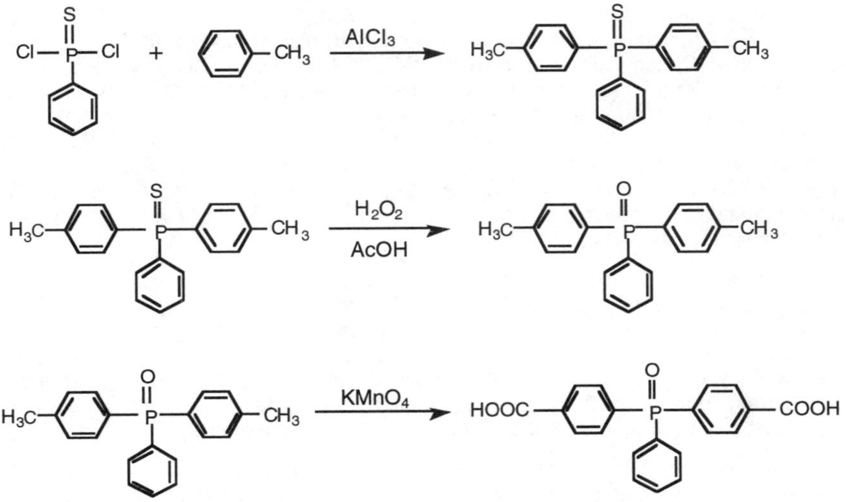

Scheme 1 Synthesis of Bis(4-carboxyphenyl) Phenyl Phosphine Oxide

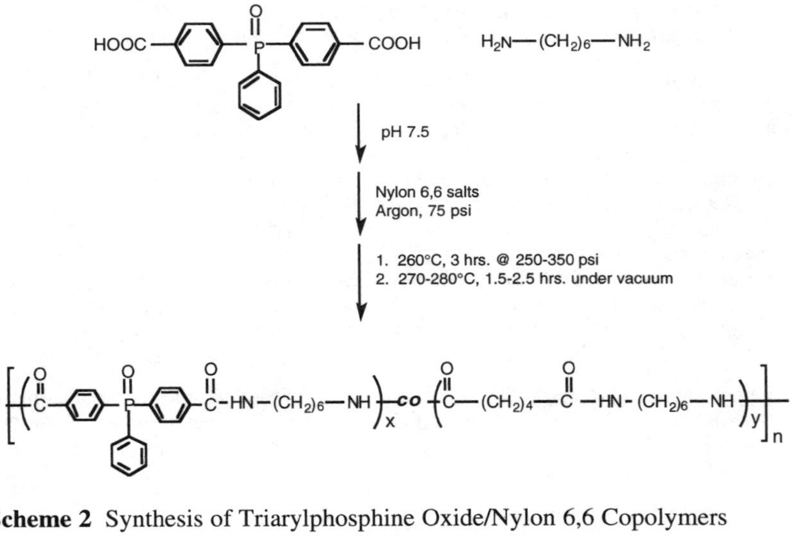

Scheme 2 Synthesis of Triarylphosphine Oxide/Nylon 6,6 Copolymers

Characterization of the triarylphosphine oxide containing copolymers is summarized in Table 1. The intrinsic viscosity of the copolymers was measured in m-cresol at 25°C and the data suggested that the copolymers have relatively high molecular weights.

Table 1 Characterization of Triarylphosphine Oxide Containing Nylon 6,6 Copolymers

P(O)% in N66	$[\eta]^1$	Tg^2 (°C)	Tm^2 (°C)	TGA^3 (°C)	Char % (750°C)	Tg^4 (°C)
0	1.32	58	253	410	0	64
10	2.40	59	214	405	3.8	80
20	2.68	75	208	403	7	86
30	1.59	89	no	402	8.5	--

Note:
1. In m-cresol at 25°C
2. DSC, 10°C/minute
3. Air, 10°C/minute
4. DMA, Loss Maxima

As shown in Table 1 and in the DSC thermograms (Figure 1), the copolymers exhibit a crystalline melting transition if the phosphine oxide comonomer is lower than 30 mole%. At 30 mole%, the copolymer is virtually amorphous. The

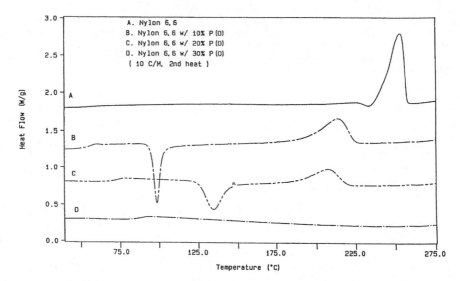

Figure 1 DSC of triarylphosphine oxide containing nylon 6,6 copolymers (heating curves)

crystallizable nature of the 0-20 mole% copolymers can also be seen from the DSC thermograms on cooling (Figure 2). In general, the nylon 6,6 homopolymer has a very fast crystallization rate. However, once the triarylphosphine oxide comonomer is incorporated into the copolymer backbone, the crystallization rate is decreased (Figure 2) and the crystal melting temperature (Tm) is depressed (Figure 1), as expected.

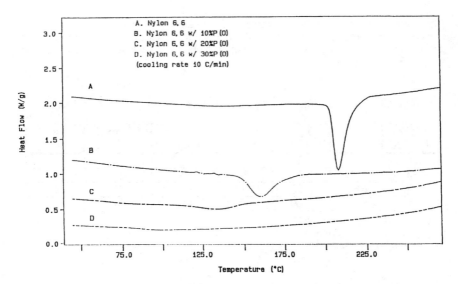

Figure 2 DSC of triarylphosphine oxide containing nylon 6,6 copolymers (cooling curves)

Nevertheless, tough solvent resistant films were obtained by compression molding and preliminary efforts at drawing fibers from the melt appear to be successful. The glass transition temperatures (Tg) and the char yields of the copolymers increased as a function of the phosphine oxide content. The increased char yield of the copolymers is shown in the dynamic TGA thermograms (Figure 3). The char yield in air is generally considered to strongly correlate with the flame retardancy of the material.

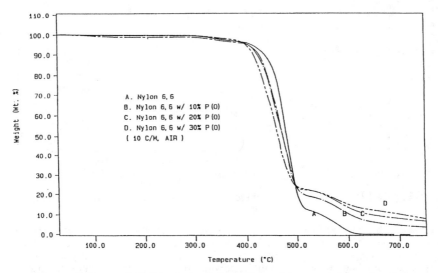

Figure 3 Dynamic TGA of triarylphosphine oxide containing nylon 6,6 copolymers in air

The DMA traces (Figure 4), indicate that the storage modulus decreases as the temperature increases. The storage modulus dramatically decreased above the Tg of the copolymers, and then reached a plateau region due to the semi-crystalline nature of the 10 or 20% copolymers. Melt flow occurs as the temperature exceeds the crystal melting transition of the copolymers. From the DMA trace of the 30% copolymer, it was concluded that the copolymer was amorphous since flow occurred at temperatures above the Tg of the copolymer.

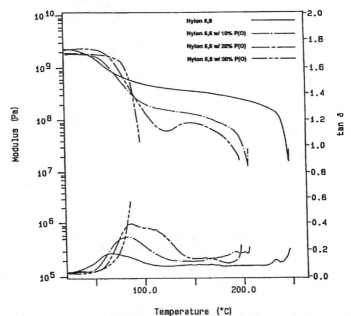

Figure 4 Dynamic mechanical behavior of compression molded triarylphosphine oxide containing nylon 6,6 copolymers (3 point bending)

Cone calorimetry results for the triarylphosphine oxide containing copolymers show that the 10% copolymer had a similar weight loss versus time profile as the nylon 6,6 and that the 20 and 30% copolymers had reduced weight loss rates (Figure 5). However, both the heat release rate and heat of combustion of the copolymers were significantly reduced compared to nylon 6,6 (Figure 6,7). These results indicated that the flame resistant properties of the copolymers have been significantly improved. However, the generation of soot and carbon monoxide increased under the test condition (Figure 8,9). The phosphine oxide moiety may act as a flame retardant in both the condensed and gas phase. ESCA results (Figure 10a & 10b), show that the phosphorus concentration on the surface of the 20% copolymer at room temperature is low but it is significantly increased after it was exposed to air at 550°C (i.e., close to the cone calorimetric test temperature) for 5 minutes. This result suggests that the phosphorus content played a role in the condensed phase. (The fluorine peaks in Figure 10a were contamination from the Teflon sheet which was used to mold the samples.) Similar results have been reported from our laboratory by Webster et al. from ESCA results (16) and Grubbs et al. by pyrolysis GC-MS data (17) of triarylphosphine oxide containing poly(arylene ethers). They found that the most of the phosphorus content stayed in

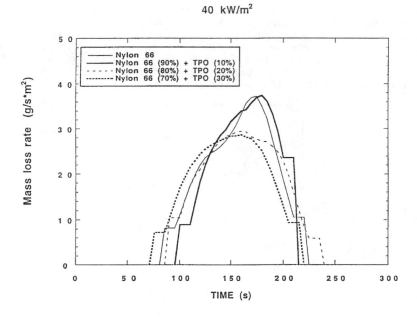

Figure 5 Weight loss versus time of triarylphosphine oxide containing nylon 6,6 copolymers at 40 kW/m^2

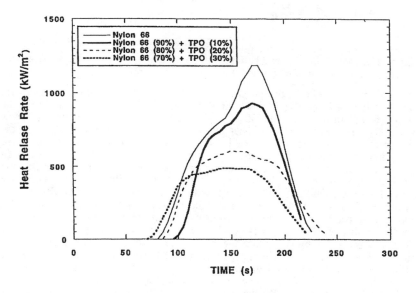

Figure 6 Heat release rate of triarylphosphine oxide containing nylon 6,6 copolymers

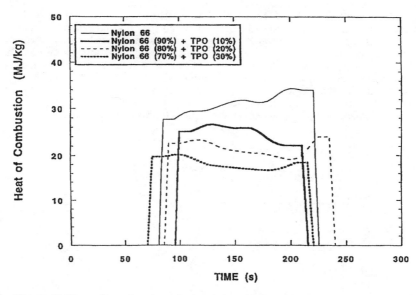

Figure 7 Heat of combustion of triarylphosphine oxide containing nylon 6,6 copolymers

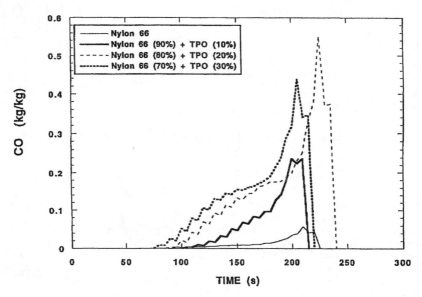

Figure 8 Carbon monoxide generation from the incomplete combustion of phosphine oxide containing nylon 6,6 copolymers

the condensed phase as char, probably in the form of phosphate, while only relatively small portions of phosphorus went to the gas phase. For a flame retardant system, large amounts of energy required for bond breakage to occur in the condensed phase (high Q_1 value) and low amounts of flammable gases evolved in the gas phase (low Q_2 value) may be desirable. Under these test conditions, the nylon 6,6 copolymers were decomposed and generated small particles. Since the phosphorus concentration on the surface was significantly increased after they were exposed to high temperature, this result strongly suggests that the chemically incorporated triaryl phosphine oxide moiety at least partially functions in the condensed phase. Detailed studies are in progress to further clarify this issue.

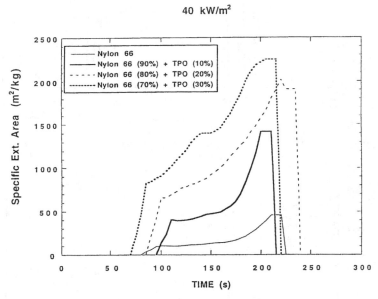

Figure 9 Specific extinction area of triarylphosphine oxide containing nylon 6,6 copolymers

CONCLUSIONS

Crystallizable triaryl phosphine oxide containing nylon 6,6 copolymers were successfully synthesized via melt polymerization in relatively high molecular weights at 10 or 20 mole%. Dynamic TGA results in air showed that the char yield increased with phosphine oxide content. Crystallinity was totally disrupted at 30 mole% incorporation, but the Tg values were systematically increased from 58°C to 89°C.

Cone calorimetric tests on the phosphorus containing polyamide copolymers exhibited significantly depressed heat release rates and heat of combustion. This property is very desirable and suggests that the copolymers are flame resistant

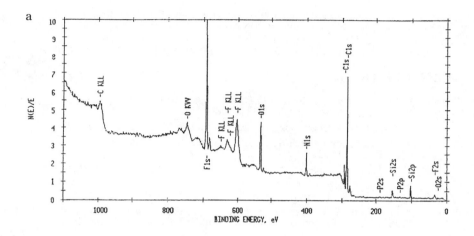

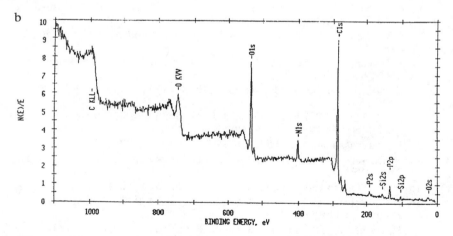

Figure 10 ESCA of 20% triarylphosphine oxide containing nylon 6,6 copolymers: a.) room temperature b.) exposed in air at 540°C for 5 minutes

materials. On the other hand, initial results showed that the "soot" and carbon monoxide generated may be increased. ESCA results show that the phosphorus on the surface of the char is significantly increased.

ACKNOWLEDGMENTS

The authors would like to gratefully acknowledge the National Institute of Standards and Technology (Contract 70NANB3H1434) for funding of this project and also to the NSF Science and Technology Center on High Performance Polymeric Adhesives and Composites (Contract # DMR 9120004).

REFERENCES

1. E. D. Weil, *Encyclopedia of Polymer Science and Technology,* Wiley-Interscience, New York, Vol. 11, (1986).
2. A. Granzow, Accounts of Chemical Research, V. 11, 5, 177, (1978).
3. A. M. Aaronson in ACS Symposium Series 486, *Phosphorus Chemistry,* Ch. 17, 218 (1992).
4. E. D. Weil in Handbook of Organophosphorus Chemistry, R. Engel, Ed., Ch. 14, (1992).
5. J. S. Ridgway, J. Appl. Polym. Sci., 35, 215, (1988).
6. J. Pellon and W.G. Carpenter, *J. Poly. Sci.*, Part A, Vol. 1, 863-876 (1963); 0. A. Pickett, Jr. and J. W. Stoddard, (Monsanto) U.S. Pat. 4,032,517, (1977).
7. Y. N. Lin, S. Joardar and J. E. McGrath, *Polym. Prepr.,* 34(1), 515, (1993).
8. C. D. Smith, H. J. Grubbs, H. F. Webster, A. Gungor, J. P. Wightman and J. E. McGrath, *High Performance Polym.,* 4, 211 (1991).
9. J. W. Connell, J. G. Smith, Jr., and P. M. Hergenrother, *Polymer.,* 36(1), 5 (1995).
10. I-Y. Wan, D. B. Priddy, G. D. Lyle and J. E. McGrath, *Polym. Prepr.* 34(1), 806, (1993).
11. S. Pak, G. D. Lyle, R. Mercier and J. E. McGrath, *Polymer,* 34(4), 885, (1993).
12. D. M. Knauss, T. Kashiwagi and J. E. McGrath, *PMSE, Polym. Prepr.,* 71, 229 (1994); see also this symposium volume.
13. R. Srinivasan and J. E. McGrath, *Polym. Prepr.,* 33(1), 503, (1992).
14. Y. Delaviz, A. Gungor, J. E. McGrath and H. W. Gibson, *Polymer,* 34(1), 210, (1993),
15. P. W. Morgan and B. C. Herr, J. *Am. Chem. Soc.,* 74, 4526, (1952).
16. H. F. Webster, C. D. Smith, J. E. McGrath and J. P. Wightman, *PMSE, Prepr.,* 65, 113, (1991).
17. H. Grubbs, C. D. Smith and J. E. McGrath, *PMSE, Prepr.,* 65, 111, (1991); H. Grubbs, PH.D. Thesis, July (1993).

RECEIVED February 28, 1995

Chapter 3

Copolycarbonates and Poly(arylates) Derived from Hydrolytically Stable Phosphine Oxide Comonomers

D. M. Knauss[1], J. E. McGrath[1,3], and Takashi Kashiwagi[2]

[1]Department of Chemistry and National Science Foundation Science and Technology Center, High Performance Polymeric Adhesives and Composites, Virginia Polytechnic Institute and State University, Blacksburg, VA 24061-0344
[2]Building and Fire Research Laboratory, National Institute of Standards and Technology, Gaithersburg, MD 20899-0001

> Hydrolytically stable bis(4-hydroxyphenyl) phenyl phosphine oxide was synthesized and utilized to produce high molecular weight polycarbonate and aromatic polyester copolymers. The glass transition temperature increased from about 150°C for the control bisphenol-A polycarbonate system to 186°C for the 50 wt. percent copolymer. The char yield via dynamic TGA in air increased from 0% for the control to 30% at 700°C for the 50% copolymer. The homopolymer had a Tg of 202°C, but only low molecular weight was achieved. In contrast, tough, transparent, high Tg polyarylates were prepared with terephthaloyl chloride that had a high char yield in air. Transparency and toughness were maintained in the copolymers, and the char yield in air increased significantly with phosphorus concentration. The materials are being characterized as improved fire resistant transparent systems and initial cone calorimetry studies do show that the heat release rate is significantly decreased. The residual carbon monoxide concentration does increase, which is consistent with the incomplete combustion.

The introduction of phosphorus compounds as flame retardants in polymers is well known (1). Most flame retardants are introduced as physically blended additives and are not chemically bonded to the polymer. Such additives can potentially be leached out of the material, which may introduce loss of protection and/or undesired hazards. Phosphorus containing comonomer flame retardants incorporated into the polymer backbone may overcome this problem. The incorporation of functionalized derivatives of triphenylphosphine oxide into a variety of polymers has recently been investigated (2-8). For example, the polycarbonate derived from bisphenol A and phosgene is an important engineering thermoplastic (9) and incorporation of various phosphorus compounds into the backbone have been examined. Some hydrolytically stable materials have been reported,[10-15] but more often incorporation of phosphorus compounds has usually created hydrolytically unstable phosphate linkages (16-23). Our objective was to

---

[3]Corresponding author

synthesize hydrolytically stable bis(4-hydroxyphenyl)phenylphosphine oxide and react this monomer with phosgene and bisphenol A to afford random or statistical copolymers. This monomer had previously been used to synthesize perfectly alternating copolycarbonates (15). Our goal was to investigate the effect of comonomer concentration on copolymer properties, including fire resistance. Some initial results of this cooperative effort are described in this paper.

Experimental

General: 4-bromophenol was obtained from Aldrich, recrystallized from carbon tetrachloride, and dried under reduced pressure. THF was vacuum distilled from a sodium / benzophenone complex. Dichlorophenylphosphine oxide, 3,4-dihydro-2H-pyran, and magnesium turnings were obtained from Aldrich and were used as received. High purity 2,2-Bis(4-hydroxyphenyl)propane (bisphenol A) was kindly provided by Dow Chemical and used as received. Methylene chloride was HPLC grade and was used as received from Baxter. Phosgene was obtained from Matheson.

Synthesis of p-bromophenyl tetrahydropyranyl ether: p-bromophenol (138.0 g; 0.79 moles) was added to a round bottom flask along with a magnetic stirbar and dihydropyran (130.0 ml; 1.42 moles) (containing 4 drops concentrated hydrochloric acid) was added dropwise from an addition funnel. After 1 hour reaction time, the product was dissolved in ether, washed three times with 10% sodium hydroxide, followed by three times with distilled water. The ether was stripped by rotary evaporation and the resulting yellow oil was recrystallized twice from 95% ethanol, to yield 151.65 g (0.59 moles; 74% yield) of white crystals with m.p. 55-56°C (lit.[23] m.p. 57-57.5°C).

Synthesis of bis(4-hydroxyphenyl)phenylphosphine oxide: Magnesium turnings (14.33 g; 0.59 moles) and a crystal of I_2 were placed in a flame dried, 3 neck, 2000 ml round bottom flask equipped with an addition funnel, condenser and overhead stirrer while purging with argon. Positive argon pressure was maintained throughout the reaction. 20 ml THF were added to the magnesium turnings and a solution of p-bromophenyl tetrahydro pyranyl ether (151.65 g; 0.59 moles) in 400 ml THF was added dropwise to the stirred mixture while cooling with a water bath. The remaining p-bromophenyl-pyranylether was washed in with 75 ml THF. The mixture stirred for three hours at room temperature and at the end of this time, only a trace of residual magnesium remained. Dichlorophenylphosphine oxide (41.1 ml; 0.29 moles) was introduced to the addition funnel along with 100 ml THF and the resulting solution was added dropwise to the Grignard reagent which had been cooled in an ice bath. After complete addition, the residual dichlorophenylphosphine oxide was rinsed in with 50 ml THF and the reaction was allowed to stir at room temperature for eight hours. The solution was then transferred to a 2000 ml round bottom flask and most of the THF was removed by rotary evaporation. Approximately 600 ml of methanol were introduced to the solution along with 20 ml of concentrated hydrochloric acid and allowed to stir for one hour to cleave the protecting group. The solvent and tetrahydropyranylmethylether were removed by rotary evaporation, producing an oil that was dissolved in 10% sodium hydroxide solution and precipitated into 1M HCl. The tan solid was filtered, washed with water and dried overnight under vacuum. The crude material was then recrystallized twice from methanol to yield white crystals, m.p. 233.5 - 234.5° C (lit.[15] m.p. 233 - 234° C).

Synthesis of phosphorus containing polycarbonate copolymers: Copolymers of controlled composition were synthesized in 50 g batches. Calculated amounts of bis(4-hydroxyphenyl)phenylphosphine oxide were added along with bisphenol A to a 500ml, 5 neck round bottom flask equipped with a mechanical stirrer, condenser, phosgene dip tube, caustic addition funnel, and pH probe. 400 ml of water, triethylamine (16 mole %) as a phase transfer catalyst, and 400 ml of methylene chloride were added along with t-butylphenol to control the molecular weight. Phosgene (1.5 equivalents) was introduced to the rapidly stirred interfacial mixture at a rate of 0.36 g/min. while maintaining the pH at 11.0 through the metered addition of a 40% caustic solution. Work-up involved washing the polymer solution with 5% hydrochloric acid followed by multiple water washes. The polymer was isolated by flashing off the methylene chloride in rapidly stirred boiling water, followed by filtration and drying at 80°C under vacuum.

Characterization: The monomers and polymers were analyzed by ^{1}H and ^{31}P NMR on a Varian 400 MHz instrument. Differential scanning calorimetry and thermogravimetric analysis were performed on Perkin-Elmer 7 series instruments at a heating rate of 10°C per minute. Intrinsic viscosity values were determined in chloroform at 25°C. Cone calorimetry was conducted under a constant external radiant flux of 40 kw/M^2, in the horizontal configuration on 10cm by 10cm by 0.3cm compression molded films.

Results and Discussion

Monomer synthesis: Bis(4-hydroxyphenyl)phenylphosphine oxide was synthesized by the reaction of the Grignard reagent derived from 2-(4-bromophenoxy)tetrahydropyran with phenylphosphonic dichloride, followed by acid catalyzed cleavage of the tetrahydropyranyl ether (Scheme 1). The bisphenol could be obtained in high purity by this method. Alternatively, a similar synthesis utilizing the Grignard reagent obtained from 4-bromoanisole and subsequent acid cleavage of the methyl ether yielded a highly colored product which could not be decolorized. More recently, it has also been possible to avoid Grignard chemistry via utilizing hydrolysis of activated aryl halide precursors (24).

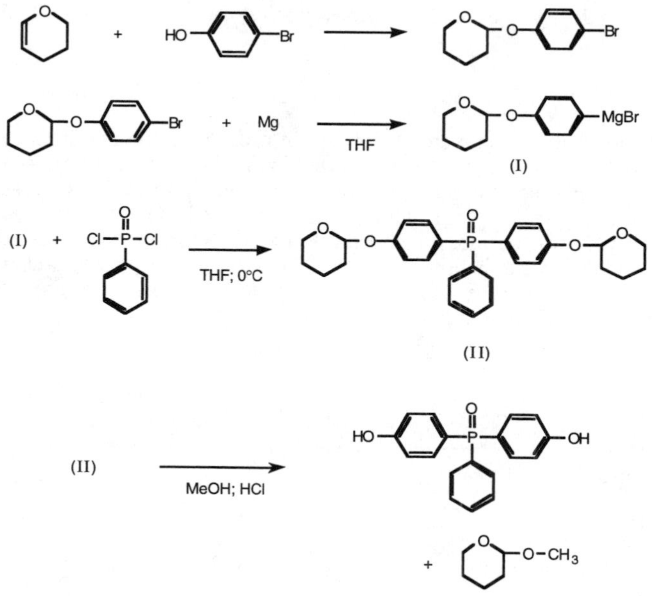

Scheme 1: Synthesis of Bis(4-hydroxyphenyl)phenylphosphine Oxide

A monofunctional phenol derived from triphenylphosphine oxide was also synthesized as a model compound and endcapper. The 4-hydroxyphenyldiphenylphosphine oxide was synthesized from the reaction of the Grignard reagent obtained from 4-bromoanisole with diphenylphosphinic chloride, followed by cleavage of the methyl ether with HBr (Scheme 2). This method yielded pure product after recrystallization.

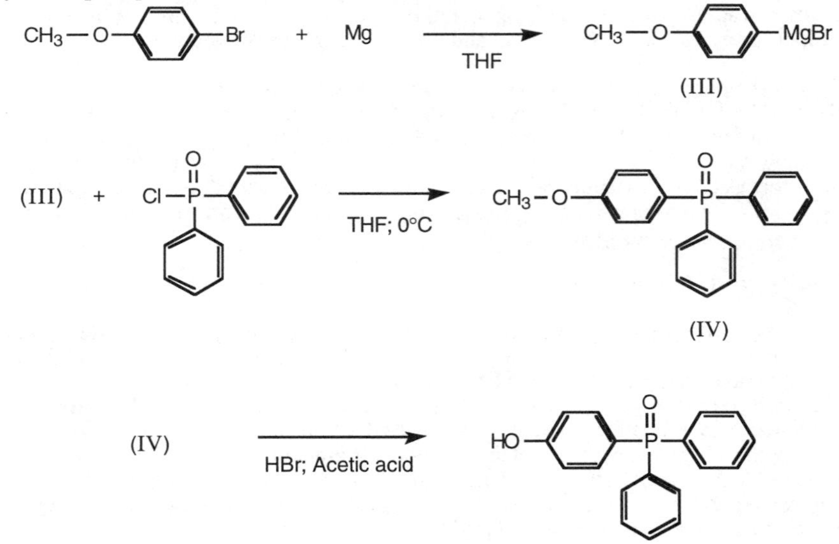

Scheme 2: Synthesis of 4-Hydroxyphenyldiphenylphosphine Oxide

Model Reactions

Some model reactions were performed in order to determine the feasibility of forming polycarbonates. It has been reported in the literature that tertiary phosphine oxide (or phosphine) compounds can react with phosgene to form tertiary phosphine dichlorides. These halogenated compounds are unstable and can be easily hydrolyzed to the tertiary phosphine oxide (Scheme 3). This reaction, if significant, could be competitive with the formation of the polycarbonate. Model reactions were performed in order to determine to what extent, if any, this might interfere with the polymerization.

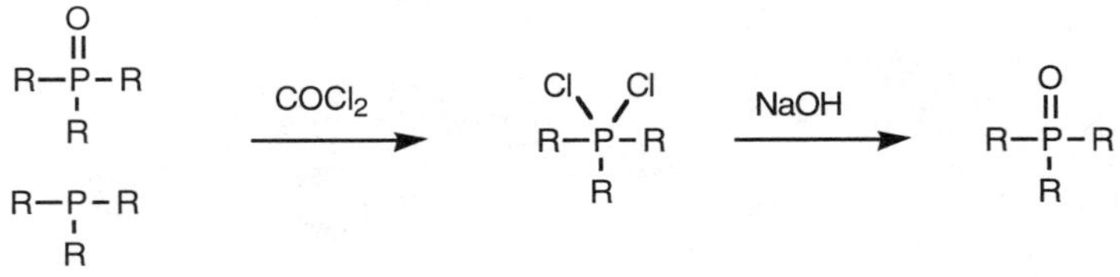

Scheme 3: Possible Formation and Hydrolysis of Tertiary Phosphine Dichloride

Under interfacial polymerization conditions, the intermediate formed from phosgenation should be quickly hydrolyzed back to the phosphine oxide. In order to model these possible reactions, triphenylphosphine oxide and triphenylphosphine were phosgenated under conditions similar to those of an interfacial polycarbonate polymerization. Under these conditions, only unreacted triphenylphosphine oxide could be isolated from the reaction mixture. Since the triphenylphosphine quantitatively yielded triphenylphosphine oxide, the hydrolysis of any triphenylphosphine dihalide generated would rapidly occur under these conditions.

It was of interest whether high molecular weight bisphenol A polycarbonate could be formed in the presence of triphenylphosphine oxide. Interfacial polymerization was conducted in the presence of a stoichiometric amount of triphenylphosphine oxide, utilizing a slight excess (1.5 equivalents) of phosgene. The polycarbonate obtained was isolated and examined by intrinsic viscosity and ^{31}P NMR. The intrinsic viscosity in chloroform was determined to be 1.45 dl/g, indicating that high molecular weight polymer was formed. ^{31}P NMR showed that no phosphorus compound was incorporated into the backbone of the polymer.

This model polymerization indicates that the reaction of phosgene with the phenolate of bisphenol A is faster than any possible reaction with triphenylphosphine oxide, but the reaction kinetics of the phenolate of the bis(4-hydroxyphenyl) phenyl phosphine oxide may not be as competitive.

The reaction of the bis(4-hydroxyphenyl) phenyl phosphine oxide with phosgene to form a polycarbonate was modeled by the synthesis of bis(triphenylphosphine oxide) carbonate from 4-hydroxyphenyldiphenylphosphine oxide. The monophenol was reacted with phosgene under the interfacial polymerization conditions, isolated from the reaction, and examined by ^1H and ^{31}P NMR. NMR analysis determined that the expected carbonate product was synthesized in approximately 98 percent purity, with the only impurity being residual starting material. This indicates that the reaction to form carbonate is indeed faster than the formation of any tertiary phosphorus dichloride side reaction product.

For comparison, bis(4-cumylphenyl) carbonate was synthesized under the same conditions. This compound was also formed in quantitative yield, with no detectable residual starting material observed by ^1H NMR. This suggests that the reaction of the triphenylphosphine oxide phenolate with phosgene may not be as fast as the reaction of phosgene with 4-cumylphenolate.

Homopolycarbonate Synthesis and Characterization

The synthesis of the homopolycarbonate of bis(4-hydroxyphenyl) phenylphosphine oxide was only partially successful. Interfacial and solution reactions were attempted, but only low molecular weight material was recovered in each case. The material was analyzed for intrinsic viscosity and by ^1H and ^{31}P NMR spectroscopy, DSC, and TGA.

The intrinsic viscosities in chloroform at 25°C ranged from 0.14 to 0.18 dl/g, while the ^1H NMR analysis (Figure 1) indicated residual phenol end groups were present. The ^{31}P NMR spectrum (Figure 2), which should depict a single peak, showed multiple unidentified phosphorus peaks to be present.

Thermal analysis of the samples by DSC indicated a T_g of 202°C with no detectable T_m. In view of the modest molecular weight of the sample, it is expected that the T_g of high molecular weight polymer would be higher than this value. Thermogravimetric analysis in air demonstrated the decreased thermooxidative stability of this low molecular weight material compared to bisphenol A polycarbonate, but did demonstrate an increase in the char yield at elevated temperature (Figure 3).

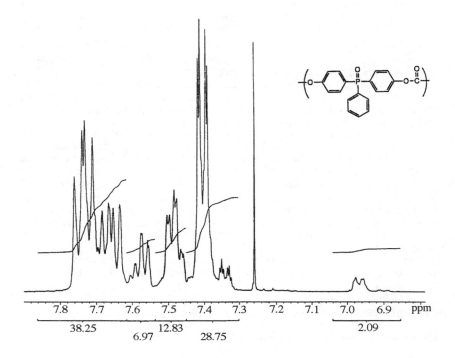

Figure 1: ^{1}H NMR Spectrum of the Polycarbonate Derived from Bis(4-hydroxyphenyl)phenylphosphine Oxide (400 MHz, CDCl$_3$)

3. KNAUSS ET AL.   *Hydrolytically Stable Phosphine Oxide Comonomers*   47

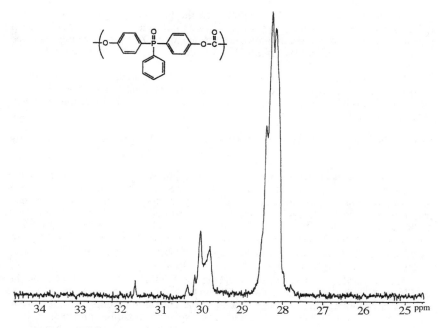

Figure 2: ^{31}P NMR Spectrum of the Polycarbonate Derived from Bis(4-hydroxyphenyl)phenylphosphine Oxide (400 MHz, CDCl$_3$)

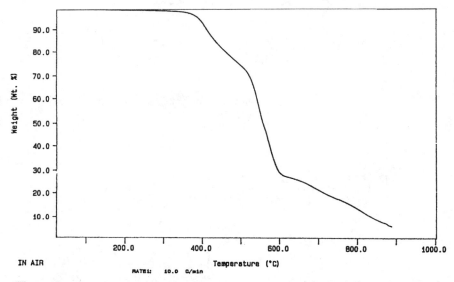

Figure 3: Thermogravimetric Analysis for the Homo-polycarbonate Derived from Bis(4-hydroxyphenyl)phenylphosphine Oxide (Air Atmosphere, 10°C/min)

An attempt was also made to synthesize the homopolycarbonate by solution polymerization in anhydrous pyridine/methylene chloride. After 1 equivalent of phosgene was added, the reaction mixture formed an insoluble gel, which persisted as the phosgene addition was continued. Addition of water to the reaction mixture produced a solution.

Although the exact nature of the gel was not determined, perhaps the tertiary phosphine dihalide was formed under the anhydrous conditions. Such linkages would be hydrolytically unstable, and could explain why the addition of water produced a solution.

Synthesis and Characterization of a Phosphorus Containing Polyarylate

The polyarylate derived from bis(4-hydroxyphenyl) phenyl phosphine oxide and terephthaloyl chloride was synthesized by interfacial methods to determine if the monomer was of sufficient purity for the synthesis of high molecular weight material (Scheme 4). A highly viscous solution resulted and analysis after isolation of the polymer yielded an IV of 0.88 dl/g. A second reaction was performed in which the molecular weight was controlled to an expected $<M_n>$ of 40.0 kg/mole, by offsetting the stoichiometry and end capping with 4-t-butylphenol. The intrinsic viscosity of this sample was determined to be 0.46 dl/g.

The polyarylate was analyzed by ^1H and ^{31}P NMR, DSC, and TGA. The NMR spectra are depicted in Figures 4, 5, and 6. Thermal analysis data for the polymer are tabulated in Table 1 and the dynamic TGA trace is depicted in Figure 6.

The homo-polyarylate obtained with this bisphenol monomer proceeds to high molecular weight and produced a tough polymer with good mechanical properties. The very high char yield at 800°C is very interesting indeed.

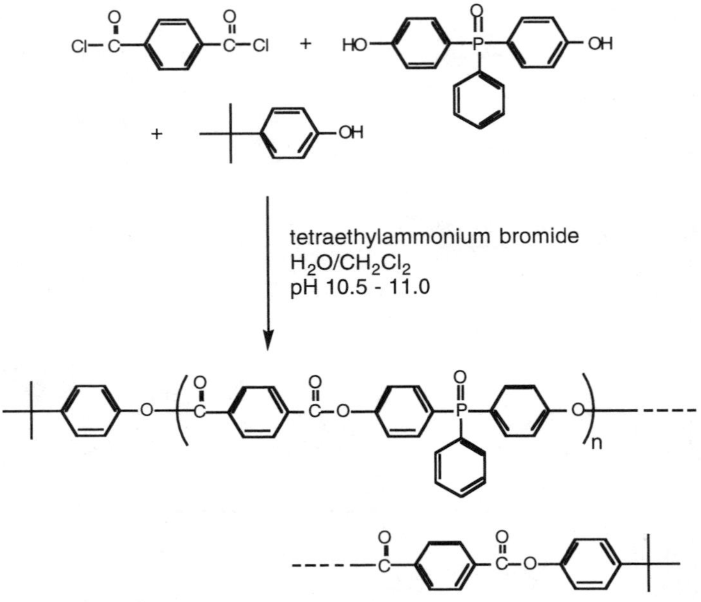

Scheme 4: Synthesis of a Phosphorus Containing Polyarylate

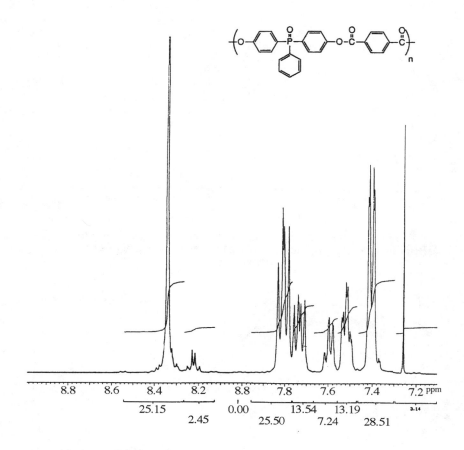

Figure 4: ^{1}H NMR Spectrum for the Polyarylate Derived from Bis(4-hydroxyphenyl)phenylphosphine Oxide and Terephthaloyl Chloride (400 MHz, CDCl₃)

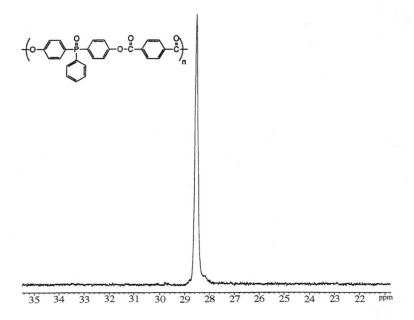

Figure 5: ^{31}P NMR Spectrum for the Polyarylate Derived from Bis(4-hydroxyphenyl)phenylphosphine Oxide and Terephthaloyl Chloride (400 MHz, CDCl$_3$)

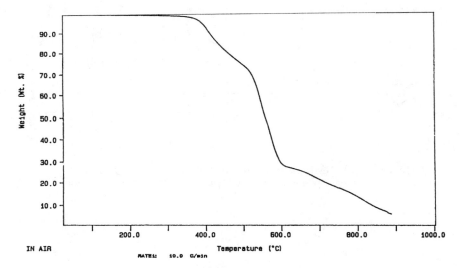

Figure 6: Thermogravimetric Analysis for the Polyarylate Derived from Bis(4-hydroxyphenyl)phenylphosphine Oxide and Terephthaloyl Chloride (Air Atmosphere, 10°C/min)

Table 1: Thermal Analysis Data for Phosphorus Containing Polyarylates

T_g (°C)	5% Weight Loss (°C)	Weight at 800°C in air (%)
246	491	48%

Synthesis of Polycarbonate Copolymers

The phosphorus containing bisphenol was reacted with phosphene and bisphenol A to afford polycarbonate statistical copolymers in varying amounts up to 50 weight percent. Copolymers were synthesized by interfacial techniques with 1, 5, 10, 25, and 50 weight percent of the repeat unit and it was possible to control the molecular weight by the addition of 4-*t*-butylphenol to the reaction mixture (Scheme 5). Bisphenol A polycarbonate homopolymers with triphenylphosphine oxide termination were also synthesized by the substitution of 4-hydroxyphenyldiphenylphosphine oxide for 4-*t*-butylphenol.

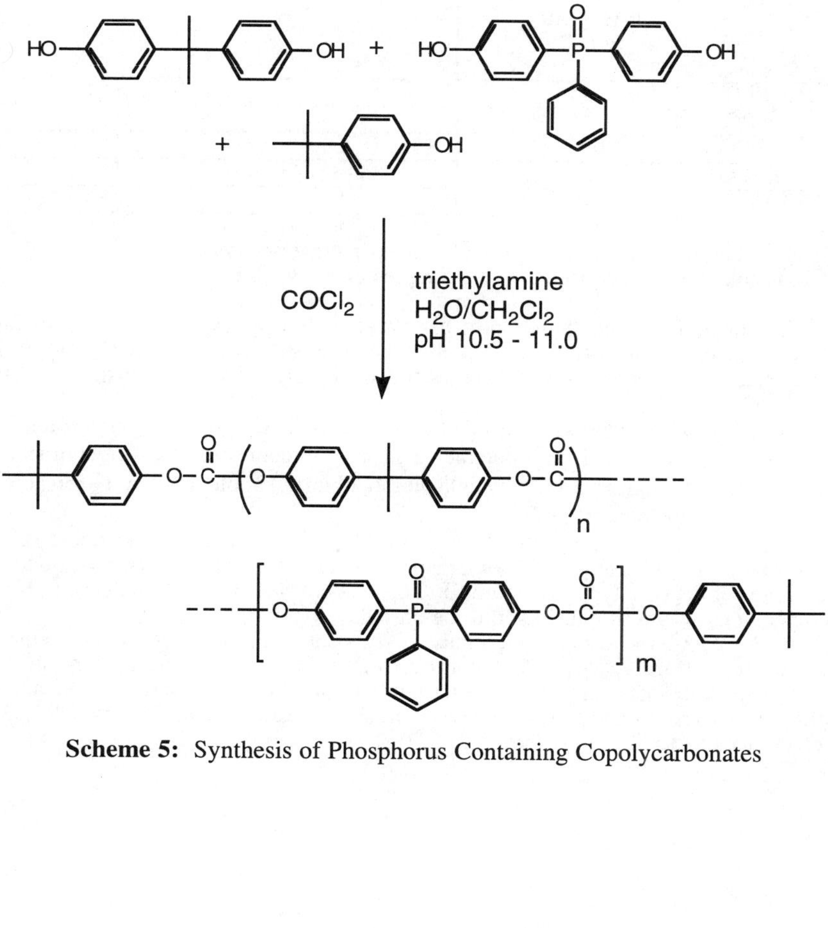

Scheme 5: Synthesis of Phosphorus Containing Copolycarbonates

Characterization of Phosphorus Containing Polycarbonates

The relative molecular weight of the polymers and copolymers were estimated by intrinsic viscosity and GPC measurements relative to polystyrene standards and the composition was determined by ^{1}H NMR. (Table 2). The molecular weight and end groups were controlled by the addition of a monofunctional phenol, either 4-t-butylphenol or 4-hydroxyphenyldiphenylphosphine oxide. In the synthesis of the copolymers, 2.4 mole percent of 4-t-butylphenol relative to the total bisphenol present was used to control the molecular weight. The intrinsic viscosity of the copolymers was found to remain constant at this level of monofunctional phenol for the copolymer containing 10 weight percent of comonomer. However, a decrease in viscosity was observed for the 25 weight percent copolymer. Accordingly, the amount of monofunctional reagent was decreased to 1.6 mole percent for the synthesis of the 50/50 copolymer, which allowed for an intrinsic viscosity closer to the other materials.

Table 2: Characterization of Phosphorus Containing Copolycarbonates

Sample	Incorporation (^{1}H NMR) (wt. %)	Intrinsic Viscosity (CHCl$_3$; 25° C) (dl/g)	T$_g$ (°C)
1% PPO*	1	0.75	154
5% PPO*	5	0.75	156
10% PPO*	9	0.72	159
25% PPO*	24	0.49	166
50% PPO**	48	0.69	186

* Molecular weight controlled with 2.6 mole percent 4-t-butylphenol
**Molecular weight controlled with 1.5 mole percent 4-t-butylphenol

The quantitative incorporation of bis(4-hydroxyphenyl)phenylphosphine oxide as a comonomer into bisphenol A polycarbonates proceeded very well and high molecular weight materials were synthesized as evidenced by the intrinsic viscosity and GPC values.

The T$_g$'s of the copolymers were determined by DSC and are listed in Table 2. An analysis of the trend in copolymer T$_g$'s as a function of the weight fraction predicts that the high molecular weight homopolymer should have a T$_g$ of about 224°C, as extrapolated from Figure 7.

Dynamic thermogravimetric analysis (TGA) in an air atmosphere was performed on the samples in order to determine the relative thermooxidative stabilities of the copolymers (Figure 8). The residual char weight remaining at high temperatures was also considered to be an indication of fire resistance. From the TGA, the copolymers were determined to be melt stable materials with nominal weight loss before 400°C, and demonstrated an observable increasing trend in the char yield with increasing phosphorus content. In fact, the char yield in air at 700°C increases from 0 percent for commercial polycarbonate to approximately 30 percent for the 50/50 copolymer.

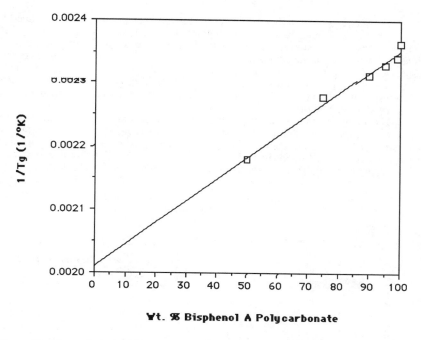

Figure 7: Linear Relationship of Copolymer Tg and Composition at 10°C/minute

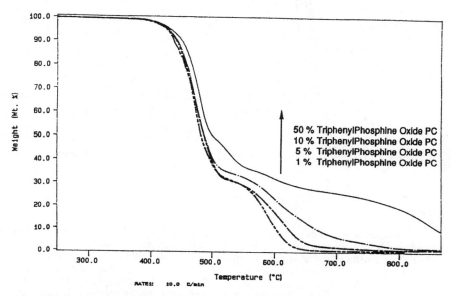

Figure 8: Thermogravimetric Analysis for Co-polycarbonates Derived from Bis(4-hydroxyphenyl) phenylphosphine Oxide and Bisphenol A (Air Atmosphere, 10°C/min)

The fire resistance of the novel copolymers has been preliminarily examined by cone calorimetry and results are graphically depicted in Figure 9. Cone calorimetry data measures the heat release rate with time and a definite improvement can be detected for the copolymers over a bisphenol A polycarbonate control sample. An improvement in the peak heat release rate is observed with as little as 1% of the phosphorus unit incorporated. It is not clear yet whether the apparent decrease in initial time for heat release to begin is important and no explanation for this behavior can be offered.

The polycarbonate with only triphenylphosphine oxide termination was also examined in comparison to the control and copolymer samples. Again, an improvement can be detected with even the small amount of triphenyl phosphine oxide introduced at the chain ends of this polymer. Further work is needed to further quantify important parameters such as smoke and total heat release.

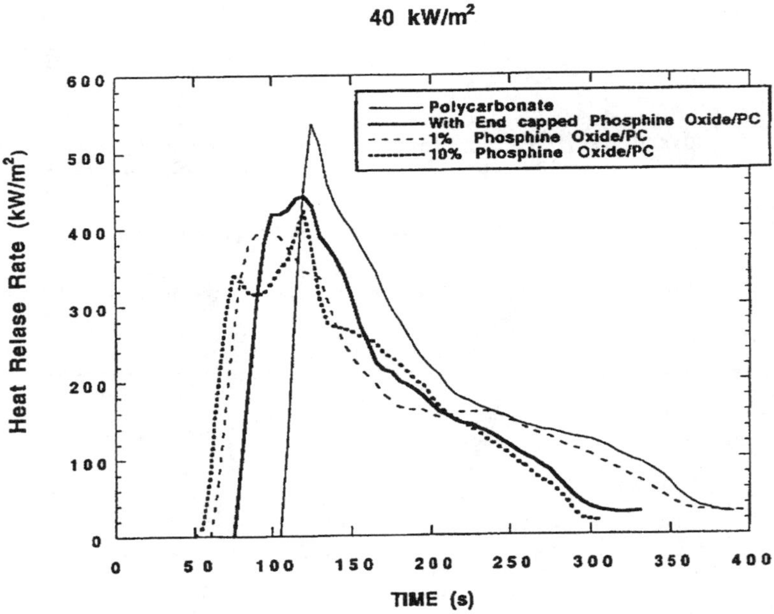

Figure 9: Cone Calorimetry Data for Phosphorus Containing Polycarbonates (10cm x 10cm x 0.3cm)

Conclusions

Phosgenation of the bis(4-hydroxyphenyl) phenyl phosphine oxide only produced low molecular weight homopolymers under the conditions investigated. However, synthesis of a series of high molecular weight copolycarbonates with bisphenol A was successfully accomplished via interfacial polymerization. Moreover, high molecular weight poly (arylates) were prepared, which showed high char yields. The copolymers were transparent, reasonably tough and were found to have high char yields and qualitatively good fire resistance. Initial results by cone calorimetry showed a decreased heat release rate and further suggest that these materials may have potential as flame retardants.

Current and Future Studies

Additional studies are in progress to further confirm and expand the results herein.

Acknowledgments

The authors would like to thank the National Institute for Standards and Technology (Contract 70NANB3H1434) and the National Science Foundation Science and Technology Center on High Performance Polymeric Adhesives and Composites (DMR-91-20004) for support of this work.

References

1. E. D. Weil in, Handbook of Organophosphorus Chemistry, R. Engel, Ed.,1992, chapter 14.
2. Y. N. Lin, S. Joardar and J. E. McGrath, *Polym. Prepr.*, **34**(1), 515, 1993.
3. C. D. Smith, H. J. Grubbs, H. F. Webster, A. Gungor, J. P. Wightman and J. E. McGrath, *High Performance Polym.*, **4**, 211, 1991.
4. J. G. Smith Jr., J. W. Connell and P. M. Hergenrother, *Polym. Prepr.*, **33**(2), 24, 1992.
5. I-Y. Wan, D. B. Priddy, G. D. Lyle and J. E. McGrath, *Polym. Prepr.* **34**(1), 806, 1993.
6. S. Pak, G. D. Lyle, R. Mercier and J. E. McGrath, *Polymer*, **34**(4), 885, 1993.
7. Y. Delaviz, A. Gungor, J. E. McGrath and H. W. Gibson, *Polymer*, **34**(1), 210, 1993.
8. I-Y Wan, T. Kashiwagi, and J. E. McGrath, *Polymeric Materials: Science and Eng.*, **71**, 233, 1994.
9. D. Freitag, U. Grigo, P.R. Muller, W. Nouvertne, *Encyclopedia of Polymer Science and Engineering*; 2nd ed.; H.F. Mark, N.M. Bikales and C.G. Overberger, Ed.; John Wiley and Sons: New York, 1988; Vol. 11, pp 648-718.
10. P. Tacke, U. Westeppe, C. Casser, U. Leyer, and H. Waldmann, CA **119**(8), 73768s, 1992.
11. J. Green, CA **108**(20), 168658t, 1987.
12. M. Shinoki, T. Matsumoto, CA **108**(18), 151187r, 1987.
13. J. C. Williams, U. S. Patent 4,680,370 (to Dow Chemical Co.), 1987.
14. D. P. Braksmayer, U. S. Patent 4,556,698 (to FMC Corp.), 1985.
15. S. Hashimoto, I. Furukawa, T. Kondo, *J. Polym. Sci., Polym. Chem. Ed.*, **12**(10), 2357, 1974.
16. S. E. Bales, U. S. Patent 4,474,937 (to Dow Chemical Co.), 1984.
17. K. S. Kim, *J. Appl. Polym. Sci.*, **28**(7), 2439, 1983.
18. CA **94**(4) 16547, 1980.
19. F. Liberti, U. S. Patent 3,711,441 (to General Electric Co.), 1973.
20. C. A. Bialous and D. B. G. Jacquiss, CA **76**(6) 86580t, 1971.
21. G. S. Kolesnikov, O. V. Smirnova, and Sh. A. Samsoniya, CA **75**(8) 49698w, 1971.
22. H. Vernaleken in, Interfacial Synthesis, F. Millich and C. E. Carraher, Eds.,1977, chapter 13.
23. W. E. Parham and E. L. Anderson, *J. Am. Chem. Soc.*, **70**, 4187, 1948.
24. S. Srinivasan and J.E. McGrath, unpublished results.

RECEIVED February 28, 1995

Chapter 4

Aromatic Organic Phosphate Oligomers as Flame Retardants in Plastics

Rudolph D. Deanin and Mohammad Ali

Department of Plastics Engineering, University of Massachusetts, Lowell, MA 01854

Polyaryl phosphate oligomers were synthesized by transesterification of triphenyl phosphate with hydroquinone and with bisphenol A. These were used as flame-retardants, in comparison with three commercial controls: triphenyl phosphate, Dechlorane^R/antimony trioxide 3/1, and decabromo diphenyl oxide/antimony trioxide 2/1, at concentrations up to 30 parts per hundred of resin. In high-density polyethylene, the polyaryl phosphate oligomers increased oxygen index from 17 up to 29, while the commercial systems reached up to 35. In modified polyphenylene oxide, the polyaryl phosphate oligomers increased oxygen index from 24 up to 36, while the commercial systems reached up to 49. In polycarbonate, the polyaryl phosphate oligomers increased oxygen index from 26 up to 39, while the commercial systems reached up to 52. Whereas the commercial systems were more effective in increasing oxygen index, the polyaryl phosphate oligomers offered promise of lower opacity, smoke, toxicity, and corrosion.

Wherever plastics are used in building and construction, electrical equipment, and transportation, flammability is a serious consideration, and it is common to add flame-retardants to control it (1,2). Of the elements which reduce flammability, organic phosphorus is generally most effective, on a weight basis, followed by organic bromine + antimony oxide, organic chlorine + antimony oxide, and inorganic hydrates (3). In a fire, bromine, chlorine, and antimony produce problems of smoke, toxicity, and corrosion. Inorganic hydrates must be used in such large amounts that they lower strength properties. Organic phosphates do not cause any of these problems; thus they should be the most desirable class of flame-retardant additives. In addition, if they are designed to be soluble in the plymer, they should not reduce its transparency.

The object of this study was to synthesize and evaluate aromatic organic phosphate oligomers as flame-retardant additives for plastics in electrical applications. Literature on preparation of polyaryl phosphate polymers generally tends toward difficult conditions and complex expensive structures (4-10). Most promising were a couple of vague references for transesterification of triaryl phosphates with aromatic diols (9,10). These were therefore adopted as the experimental starting point for the present study.

Model Reaction

When triphenyl phosphate (P) is transesterified with aromatic diols (D), such as hydroquinone or bisphenol A, and phenol (M) is removed, the reaction can be modeled as follows:

$$2\ M_3P + 1\ D \rightarrow M_2PDPM_2$$

$$3\ M_3P + 2\ D \rightarrow \underset{\underset{M}{|}}{M_2PDPDPM_2}$$

$$4\ M_3P + 3\ D \rightarrow \underset{\underset{M\ \ M}{|\ \ |}}{M_2PDPDPDPM_2} + \underset{\underset{DPM_2}{|}}{M_2PDPDPM_2}$$

and so on. Thus increasing stepwise should produce series from liquids to solids to infusible solids. This is assuming that all reactions produce a single homogeneous molecular weight. In practice, they would more likely produce a normal molecular weight distribution, so each sample would contain some smaller and some larger molecules as well.

Synthesis

In a 1-liter resin flask fitted with gas inlet, stirrer, thermometer, condenser, and electrical heating mantle, triphenyl phosphate and aromatic diol (hydroquinone or bisphenol A) were melted together in the presence of catalyst (0.18% $MgCl_2$ for hydroquinone, 0.29% sodium phenate for bisphenol A). The flask was swept with dry nitrogen, heated gradually from 50 to 300°C with constant stirring to produce transesterification, vacuum was applied and maintained, and the phenol which evolved was distilled out and condensed in an ice bath. Yields of phenol were 95-99% of theoretical; lower molecular weights were more fluid and phenol removal was more complete. At the highest molecular weights, products were so viscous as to suggest some cross-linking, at least part of the broad MWD product. Bisphenol A gave white-yellow products; whereas hydroquinone, suffering from easy quinoid formation, gave brown products. Results are summarized in Tables I and II.

Table I. Reactions of Triphenyl Phosphate with Hydroquinone

Mol Ratio TPP:HQ	Calculated Molecular Weight	Calculated Phosphorus Content, %	Viscosity of Product
1:0	326	9.5	Solid
2:1	574	10.8	Fluid
3:2	822	11.3	Fluid
4:3	1070	11.6	Fairly Viscous
5:4	1318	11.8	Viscous
6:5	1566	11.9	Viscous
7:6	1814	12.0	Highly Viscous
8:7	2062	12.0	Highly Viscous
9:8	2310	12.1	Very Highly Viscous
10:9	2558	12.1	Solid

Table II. Reactions of Triphenyl Phosphate with Bisphenol A

Mol Ratio TPP:BPA	Calculated Molecular Weight	Calculated Phosphorus Content, %	Viscosity of Product
1:0	326	9.5	Solid
2:1	692	9.0	Fluid
3:2	1058	8.8	Fluid
4:3	1424	8.7	Fairly Viscous
5:4	1790	8.7	Viscous
6:5	2156	8.6	Highly Viscous
7:6	2522	8.6	Highly Viscous
8:7	2888	8.6	Very Highly Viscous
9:8	3254	8.6	Almost Solid
10:9	3620	8.6	Solid

Compounding and Testing

Three commercial polymers typically used in electrical products were chosen for this study:

HDPE: Dow 08054-N: MI 8, D 0.954
PPO: GE Noryl 731: GP, UL 94 HB
PC: Mobay Makrolon 2658: GP, MI 12, UL 94 V-2

One hundred parts of each polymer was fused in a Brabender Banbury mixer and melt-blended with 0, 10, 20, or 30 parts of each of 6 polyaryl phosphate oligomers made from

```
 2 mols Triphenyl Phosphate + 1 mol Hydroquinone
 6    "         "            "  + 5  "      "
10    "         "            "  + 9  "      "

 2    "         "            "  + 1  "  Bisphenol A
 6    "         "            "  + 5  "      "
10    "         "            "  + 9  "      "
```

or with each of 3 commercial control flame-retardants:

3 Occidental Dechlorane 1000 + 1 M&T Thermoguard S Sb_2O_3
2 Decabromo Diphenyl Oxide + 1 M&T Thermoguard S Sb_2O_3
Triphenyl Phosphate

The lowest MW oligomers mixed poorly with the molten plastics, and caused brittleness (antiplasticization) in polycarbonate. The highest MW oligomers mixed well, and did not reduce transparency of the parent polymers. The commercial flame-retardants were solid fillers which mixed well and gave opaque blends. Banbury blends were compression molded, cut into 1/4 x 1/8 inch samples, and tested for oxygen index according to ASTM D-2863. Results are summarized in Table III.

Table III. Effects of Flame-Retardants on Oxygen Index of Commercial Plastics

Polymer	Flame-Retardant PHR:	0	10	20	30
HDPE	Dechlorane/Sb$_2$O$_3$ 3/1	17	26	28	33
	DBDPO/Sb$_2$O$_3$ 2/1	17	27	30	35
	TPP	17	20	22	23
	TPP/BPA 2/1	17	20	21	22
	6/5	17	23	25	27
	10/9	17	26	28	29
	TPP/HQ 2/1	17	20	22	23
	6/5	17	22	24	26
	10/9	17	24	26	29
PPO	Dechlorane/Sb$_2$O$_3$ 3/1	24	38	39	42
	DBDPO/Sb$_2$O$_3$ 2/1	24	40	44	49
	TPP	24	26	28	30
	TPP/BPA 6/5	24	28	29	30
	10/9	24	32	33	34
	TPP/HQ 10/9	24	33	35	36
PC	Dechlorane/Sb$_2$O$_3$ 3/1	26	37	39	45
	DBDPO/Sb$_2$O$_3$ 2/1	26	42	47	52
	TPP	26	29	33	36
	TPP/BPA 6/5	26	32	33	35
	10/9	26	37	38	39
	TPP/HQ 6/5	26	33	35	36
	10/9	26	34	36	39

Discussion

In high-density polyethylene, synthetic polyaryl phosphates raised oxygen index from 17 up to 20-29, while commercial additives gave 20-35. In polyphenylene oxide, synthetic polyaryl phosphates raised oxygen index from 24 up to 28-36, while commercial additives gave 26-49. In polycarbonate, synthetic polyaryl phosphates raised oxygen index from 26 up to 32-39, while commercial additives gave 29-52.

Increasing molecular weight of the polyaryl phosphates produced significant increase in flame-retardance. Use of hydroquinone and bisphenol A to synthesize the polyaryl phosphates produced about equivalent flame-retardance in polyethylene and polycarbonate, whereas hydroquinone phosphate was somewhat more effective than bisphenol phosphate in polyphenylene oxide.

Halogen/antimony combinations were more effective on a weight basis, because they contained a higher percentage of the flame-retarding elements (Table IV).

Table IV. Concentration of Flame-Retardant Elements in Additives

Additives	Elements	Concentration
Dechlorane/Sb$_2$O$_3$ 3/1	Cl, Sb	69.7%
DBDPO/Sb$_2$O$_3$ 2/1	Br, Sb	83.6
TPP	P	9.5
TPP/BPA 2/1	P	9.0
6/5	P	8.6
10/9	P	8.6
TPP/HQ 2/1	P	10.8
6/5	P	11.9
10/9	P	12.1

Conversely, qualitative observations indicated the polyaryl phosphates produced much less smoke than the halogen/antimony systems.

Instead of comparing flame-retardant **additives** at equal concentrations, it is interesting to compare flame-retardant **elements** at equal concentrations, by interpolating from the above experimental data (Table V, Figures l-3).
Generally phosphorus was most effective at equal concentration, followed by bromine and chlorine in that order, in agreement with many earlier studies. Surprisingly, molecular weight of the polyaryl phosphates correlated directly with effectiveness of phosphorus, even at equal phosphorus concentration; probably lower volatility kept the phosphorus in the polymer longer, permitting it to keep retarding the burning process. And bisphenol phosphate was generally somewhat more effective than hydroquinone phosphate at equal phosphorus concentration, probably for the same reason.

Conclusions

When polyaryl phosphate oligomers were added up to 30 PHR in commercial plastics, they increased oxygen index of high-density polyethylene from 17 up to 29, polyphenylene oxide from 24 up to 36, and polycarbonate from 26 up to 39. At equal loading of flame-retardant additive, commercial halogen/antimony flame-retardants gave higher oxygen index, but polyaryl phosphate oligomers offered promise of greater transparency and lower smoke, toxicity, and corrosion.

Table V. Effects of 2% of Flame-Retardant Elements on Oxygen Index of Commercial Plastics

Polymer	Flame-Retardant Additive	Flame-Retardant Elements	Oxygen Index
HDPE	Dechlorane/Sb_2O_3 3/1	Cl+Sb	20
	DBDPO/Sb_2O_3 2/1	Br+Sb	20
	TPP	P	22
	TPP/BPA 2/1	P	21
	6/5	P	26
	10/9	P	28
	TPP/HQ 2/1	P	22
	6/5	P	24
	10/9	P	25
PPO	Dechlorane/Sb_2O_3 3/1	Cl+Sb	28
	DBDPO/Sb_2O_3 2/1	Br+Sb	28
	TPP	P	28
	TPP/BPA 6/5	P	29
	10/9	P	33
	TPP/HQ 10/9	P	35
PC	Dechlorane/Sb_2O_3 3/1	Cl+Sb	29
	DBDPO/Sb_2O_3 2/1	Br+Sb	30
	TPP	P	33
	TPP/BPA 6/5	P	34
	10/9	P	39
	TPP/HQ 6/5	P	34
	10/9	P	35

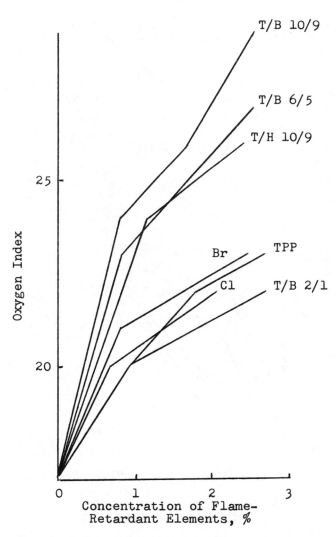

Figure 1. Efficiency of Flame-Retardant Elements in HDPE. B is BPA, H is HQ, T is TPP.

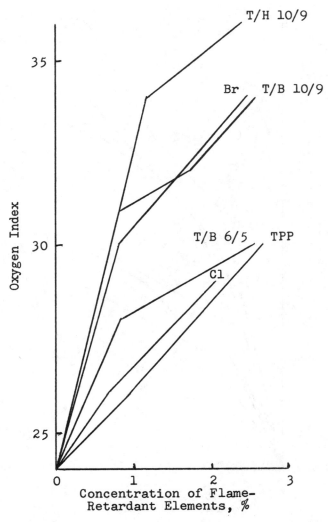

Figure 2. Efficiency of Flame-Retardant Elements in PPO. B is BPA, H is HQ, T is TPP.

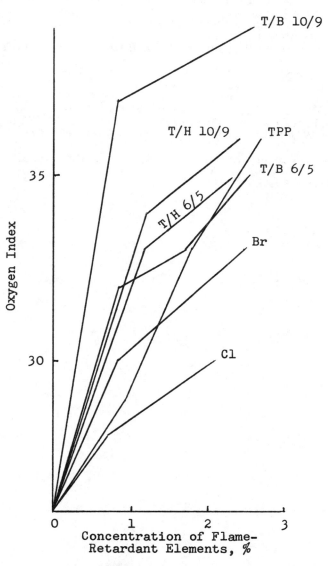

Figure 3. Efficiency of Flame-Retardant Elements in PC. B is BPA, H is HQ, T is TPP.

Literature Cited

1. Edenbaum, J. *Plastics Additives and Modifiers Handbook*; Van Nostrand Reinhold: New York, NY, 1992; Sect. X.
2. Troitzsch, H. J. In *Plastics Additives Handbook*; Gachter, R.; Muller, H., Eds.; Hanser: Munich, 1990, Ch. 12.
3. Lyons, J. W. *The Chemistry and Uses of Fire Retardants*; Wiley: New York, NY, 1970; pg. 23.
4. Weil, E. D. *Encyc. Polym. Sci. Tech..*; Vol. 11, pp. 96-126.
5. Cass, W. E. (GE). U. S. Pat. 2,616,873 (1952).
6. Zenftman, H.; McLean, A. (ICI). U. S. Pat. 2,636,876 (1953).
7. Helferich, B.; Schmidt, K. G. *Chem. Ber.* **1959**, *92*, 2051-6.
8. Stackman, R. W. *Ind. Eng. Chem., Prod. Res. Dev.* **1982**, *21*, 332-6.
9. Coover, H. W. Jr.; McConnell, R. L.; McCall, M. A. *Ind. Eng. Chem.* **1960**, *52 (5)*, 409-11.
10. Cerini, V.; Nouvertne, W.; Freitag, D. (Bayer). U. S. Pat. 4,481,338 (1984).

RECEIVED February 6, 1995

Chapter 5

Chlorinated Flame Retardant Used in Combination with Other Flame Retardants

R. L. Markezich and D. G. Aschbacher

Technology Center, Occidental Chemical Corporation, Grand Island, NY 14072

Polyolefins and ABS can be flame retarded using a mixture of a chlorinated and brominated flame retardants. There is a synergistic effect that allows flame retardant levels to be lowered, resulting in improved physical properties and lower cost formulations. The chlorinated flame retardant used is CFR, the Diels-Alder diadduct of hexachlorocyclopentadiene and 1,5-cyclooctadiene. Several different brominated flame retardants have been used which all show the same synergistic effect.

Polyolefins wire and cable formulations can also be flame retarded using a mixture of the same chlorinated flame retardant and inorganic salts, such as magnesium hydroxide. The mixture of these two flame retardants show a synergist effect in the oxygen index test. These formulations give less smoke when tested in the NBS smoke chamber. Using alumina trihydrate instead of magnesium hydroxide does not show a synergistic effect.

The synergistic action between antimony oxide and halogenated flame retardants is well known[1]. Not as well known is the synergistic action between chlorinated and brominated flame retardants to impart flame retardants properties to plastics. In 1970, R. F. Cleave[2] reported a synergistic effect between an aliphatic chlorine compound and an aromatic bromine compound in the presence of antimony oxide. Mixtures of pentabromotoluene, chlorinated paraffins, and antimony oxide appear to be more efficient than mixtures of any two alone. It does indicate that some degree of synergism exists between the two halogens when in combination with antimony oxide. In 1976, Gordon, Duffy, and Dachs[3] reported on the use of mixtures of the chlorinated flame retardant (CFR) and decabromodiphenyl oxide (DBDPO) to flame retard ABS. A more recent patent to Ilardo and Scharf[4] covers the use of mixtures of chlorinated and brominated flame retardants in polyolefins. The use of poly(tribromophenylene oxide) (BR-PPO) with the chlorinated flame retardant, CFR, gives a maximum oxygen index at 1:1 mixture of the chloro and brominated flame retardant. Figure 1 is a plot of some of the data from the patent; a FR-low density polyethylene formulation using a mixture of halogens with antimony oxide shows a maximum oxygen index at 15% of CFR and 15% BR-PPO. We have extended the work on ABS and investigated several different brominated flame retardants with the chlorinated flame retardant CFR.

Materials and Procedures

The additives used are listed in Table I. The ABS (acrylonitrile-butadiene-styrene) resins used in these evaluations were from commercial sources. The resins were dry blended with the additives, extruded using a twin-screw extruder, pelletized, dried, and then molded into test bars.

The EVA (ethylene vinyl acetate copolymer, 9% vinyl acetate) resins were also from commercial sources. The experimental samples were mixed on a two-roll mill to obtain a homogeneous sample. Compounded polymer samples were sheeted and granulated prior to molding.

Testing

Oxygen Index (O.I.) The test employed is the ASTM D2863.

UL 94. The Underwriters Laboratories vertical flame test.

Notched Izod. Izod impact resistance (ASTM D256).

HDT. Heat Deflection Temperature (ASTM D648).

Smoke Generation. Testing was performed in a NBS Smoke Chamber according to ASTM E-662. The samples, 3 by 3 inch by 0.085 inch thick were exposed to radiant heat plus propane microburners (flaming combustion). Smoke evolution was continually recorded during the test period. Results are reported as specific optimal density, Ds.

Table I. Materials and Sources

Abbrev.	Chemical Structure	Trade Name	Source
CFR	Diels-Alder adduct of hexachlorocyclopentadiene and 1,5 cyclooctadiene	Dechlorane Plus®	Occidental Chemical Corporation
CFR-2	Same chlorinated flame retardant but with a mean particle size of less than 2 microns	Dechlorane Plus®	Occidental Chemical Corporation
BFR-1	Brominated epoxy resin with 51% Br	YDB-406	Tohto Kasei
BFR-2	Bis(tribromophenoxy) ethane (70% Br)	FF680	Great Lakes
Antimony Oxide		Thermo-guard®S	AtoChem
MgOH	Magnesium hydroxide	Zerogen®35	Solem Division of JM Huber Corp.
Talc		Mistron ZSC	Cyprus Industrial Minerals
	Antioxidant	Agerite Resin D	R. T. Vanderbilt
	Peroxide	Luperox 500R	AtoChem

Results

FR-ABS

Table II gives several FR-ABS formulations using a mixture of chloro and bromo flame retardants. The oxygen indexes from the formulations in this table are plotted in Figure 2. The highest oxygen index is obtained when there is a 1:1 mixture of chlorine and bromine from the flame retardants (Formulation #3). This is also the only formulation that is UL-94 V-0 at both 3.2mm and 1.6mm.

Table III shows the results of using different amounts of antimony oxide with a 1:1 mixture of Cl and Br, total halogen 11%, in the formulations. Table IV shows the results with 10% total halogen, 1:1 mixture of Cl and Br. It appears from a plot of this data, Figure 3, that a maximum oxygen index is reached at about 8% antimony oxide with 11% halogen and 6% antimony oxide with 10% halogen.

Results

FR-W&C

Table V gives some typical talc filled FR wire and cable formulations using CFR as the flame retardant in combination with magnesium hydroxide. The total flame retardant level is constant at 25% with 5% Sb_2O_3. Figure 4 shows a graph of the % MgOH versus oxygen index. There is a synergistic effect between the chlorinated flame retardant and magnesium hydroxide which gives the highest O.I. with 20% CFR and 5% MgOH.

The use of alumina trihydrate, instead of magnesium hydroxide, in these talc filled EVA W&C formulations, which are cross linked, does not show the same synergistic effect as MgOH does (shown in Figure 5).

Table VI shows FR-EVA formulations that do not contain talc. The oxygen indexes of these formulations are shown in Figure 6. There is a maximum oxygen index when the mixture contains 5 to 10% MgOH with 5% Sb_2O_3.

The NBS smoke generation data for some of the FR-EVA formulation is shown in Table VII. As can be seen, the mixed CFR/MgOH (15:10) formulation gives lower smoke values than CFR alone (3# versus #1). The use of 0.1% Fe_2O_3 with CFR also gives a reduction in smoke, but not as much as the mixed CFR/MgOH system.

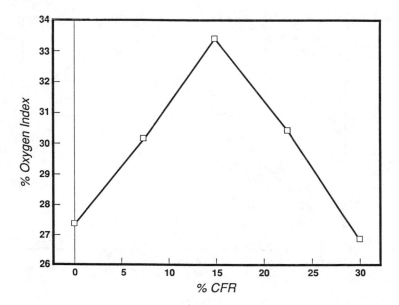

Figure 1
Oxygen Index of FR-LDPE
30% Total Halogen CFR minus % BR-PPO

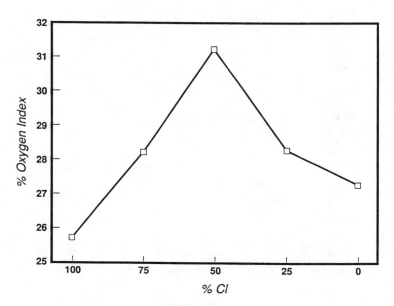

Figure 2
Oxygen Index of FR-ABS
11% Total Halogen/CFR–2 minus BFR–1

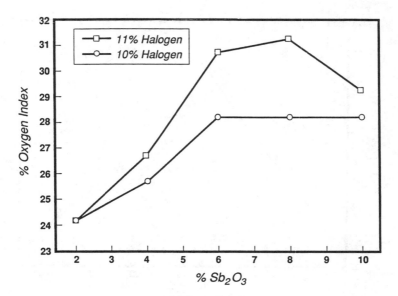

Figure 3
**Oxygen Index of FR-ABS
11% and 10% Total Halogen CFR-2/BFR-2 1:1**

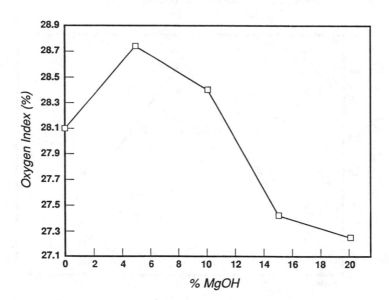

Figure 4
**Oxygen Index of FR-EVA
20% Talc/5% Sb_2O_3/25% CFR Minus % MgOH**

Table II. FR-ABS Formulations

	1	2	3	4	5
ABS	78.1	76.8	75.55	74.28	73
CFR-2	16.9	12.7	8.45	4.22	--
Sb_2O_3	5	5	5	5	5
BFR-1	--	5.5	11	16.5	22
Results					
O.I.	25.75	28.25	31.25	28.25	27.25
UL-94 3.2mm 1.6mm	V-0 NC	V-0 NC	V-0 V-0	V-0 NC	V-0 NC
Notched Izod J/M	64	82	87	107	97
% Cl	11	8.2	5.5	2.7	--
% Br	--	2.8	5.5	8.4	11
Total Halogen %	11	11	11	11	11

Table III. FR-ABS Formulations

	1	2	3	4	5
ABS	81.7	79.9	77.6	75.7	73.7
CFR-2	8.45	8.45	8.45	8.45	8.45
Sb_2O_3	2	4	6.1	8	10
BFR-2	7.85	7.85	7.85	7.85	7.85
Results					
O.I.	24.25	26.75	30.75	31.25	29.25
UL-94 3.2mm	NC	V-1	V-0	V-0	V-0
Notched Izod J/M	174	160	136	120	105
HDT deg C Annealed 24 hours/80 deg	95	96	94	96	95
% Cl	5.5	5.5	5.5	5.5	5.5
% Br	5.5	5.5	5.5	5.5	5.5
Total Halogen %	11	11	11	11	11

Table IV: FR-ABS Formulation

Weight %	1	2	3	4	5
ABS	83.2	81.2	79.2	77.2	75.2
CFR-2	7.7	7.7	7.7	7.7	7.7
Sb_2O_3	2	4	6	8	10
BFR-2	7.1	7.1	7.1	7.1	7.1
Results					
O.I.	24.25	25.75	28.25	28.25	28.25
UL-94 3.2mm	NC	NC	V-0	V-0	V-0
Notched Izod J/M	200	175	157	136	114
% Cl	5.0	5.5	5.5	5.5	5.5
% Br	5.0	5.5	5.5	5.5	5.5
Total Halogen %	10	10	10	10	10

Table V: FR-W&C Formulations

Weight %	1	2	3	4	5
EVA	47.9	47.9	47.9	47.9	47.9
CFR	25	20	15	10	5
Sb_2O_3	5	5	5	5	5
MgOH	0	5	10	15	20
Talc	20	20	20	20	20
Agerite Resin D	1.4	1.4	1.4	1.4	1.4
Luperox 500R	0.7	0.7	0.7	0.7	0.7
Results					
O.I. (%)	28.1	28.75	28.41	27.42	27.25
% Elongation	350	320	300	190	140

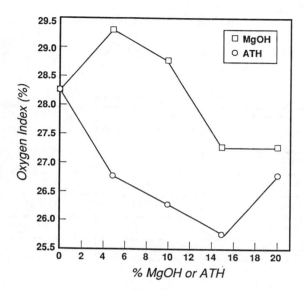

Figure 5
Oxygen Index of FR-EVA
20% Talc/5% Sb_2O_3/25% CFR Minus % ATH or MgOH

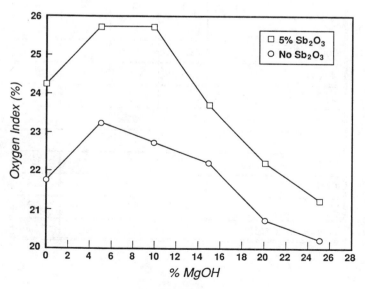

Figure 6
Oxygen Index of FR-EVA
30% CFR Minus % MgOH

Table VI: FR-EVA W&C Formulation (No Talc)

Weight %	1	2
EVA	67.9	62.9
CFR	30 to 5	30 to 5
Sb_2O_3	0	5
MgOH	0 to 25	0 to 25
Agerite Resin D	1.4	1.4
Luperox 500R	0.7	0.7

Table VII: Smoke Generation of FR-EVA
20% Talc Filled Formations (E-662) (Flaming Mode)

	1	2	3	4
CFR	25	25	15	15
Sb_2O_3	5	5	5	5
MgOH	-	-	10	10
Fe_2O_3	-	0.1	-	0.1
Results				
	Ds			
1.5 Min.	4	11	4	15
4 Min.	247	233	108	128
Max.	416	395	309	277
Time to Max. (Min.)	9.8	9.8	10.2	11.5

Conclusions

ABS resins can be flame retarded using a mixture of chlorinated and brominated flame retardants. There is a synergistic effect between chlorinated and brominated flame retardants, which can result in lower flame retardant levels needed to achieve flame retardancy.

Flame retarded wire and cable formulations using a mixture of a chlorinated flame retardant and magnesium hydroxide show a synergistic effect in the oxygen index flammability test and produces less smoke when burned.

References

1. I. Touval, Plastics Compounding, September/October, 1992.
2. R. F. Cleaves, Plast. Polym. 38 (135) 190, 1970.
3. I. Gordon, J. J. Duffy, and N. W. Dachs, US Patent 4,000,114 (1976).
4. C. S. Ilardo and D. J. Scharf, US Patent 4,388,429 (1983).

RECEIVED April 28, 1995

Chapter 6

Developments in Intumescent Fire-Retardant Systems
Ammonium Polyphosphate–Poly(ethyleneurea formaldehyde) Mixtures

G. Camino, M. P. Luda, and L. Costa

Dipartimento di Chimica Inorganica, Chimica Fisica, e di Chimica dei Materiali dell'Università, Via P. Giuria 7, 10125 Torino, Italy

Intumescent fire retardants supply a suitable, low hazard substitute for halogenated fire retardants but their effectiveness is still relatively low. Results are reported on the intumescent behavior of ammonium polyphosphate (APP)/ poly(ethylenurea formaldehyde) (PEU) system in polypropylene (PP). APP promotes the degradation of PEU at lower temperature: 315 and 400°C. Foaming of the charring material takes place in the second step of degradation as shown by dilatometric measurements and scanning electron microscopy. Preliminary investigations show that it is the intumescent char obtained from the additive which should be entrusted with the task of protecting PP from the flame.

Flammability of natural and synthetic polymers is one of the major drawbacks in their ever expanding applications. Unfortunately flammability is an intrinsic characteristic of organic substances such as the most used polymeric materials which by no means can be made completely incombustible. However, incorporation of some special structures in the polymer, either as an additive or as chemical modification of the polymer chains, improves the fire resistance of the material and allows external intervention before fire develops and propagates.

Intumescent fire retardant systems represent a relatively new class of fire retardants, developed essentially on empirical basis, whose fire retardance mechanism is still far from being well understood. Convenience in using such systems includes lack of evolution, on burning, of corrosive, toxic and sight-obscuring smokes, that can contribute to increase the overall hazard in a real fire scenario. Often they also avoid dripping of flaming material that can propagate the fire. Despite these unquestionable advantages of intumescent systems compared to traditional halogenated ones, a larger amount of an intumescent additive must be generally employed in order to meet satisfactory fire retardance standards. This generally implies a deleterious effect on the physical properties of the material and introduces processing and economic problems. Improvement of the effectiveness of intumescent systems should decrease the total amount of additive required for acceptable performance , overtaking these difficulties.

Early intumescent fire retardants have been developed for coating applications where, on burning, a charred foamed layer covers the substrate acting as a physical shield against heat transmission and mass transfer. The same concept has been extended to the protection of the polymer bulk from fire, in particular for polystyrene, polyethylene and polypropylene. Additional difficulties are encountered in this latter application of intumescence, essentially due to quantitative volatilization of the matrix to a mixture of flammable hydrocarbons, without char formation.

Relatively new systems used to impart intumescence to polystyrene and polyolefins are mixtures of ammonium polyphosphate (APP) and nitrogen containing compounds: condensation products of formadehyde with substituted ureas or melamine and products of reaction between aromatic diisocyanate and pentaerythritol or melamine(*1-3*). However, their effectiveness is generally relatively low. Thus, we have undertaken a systematic study to identify the key factors controlling fire retardance effectiveness in these intumescent systems in order to design improved fire retardant systems. Here results concerning Poly(ethylenurea formaldehyde) are reported and discussed.

Materials

APP (Exolit 422, Hoechst), prepared as a fine, white powder ($< 100\mu$), sparingly water soluble and nearly completely insoluble in organic solvents, is a linear polymer:

$$NH_4^+ \; {}^-O-\overset{\overset{O}{\|}}{\underset{\underset{NH_4^+}{O^-}}{P}}-O \left[-\overset{\overset{O}{\|}}{\underset{\underset{NH_4^+}{O^-}}{P}}-O - \right]_n -\overset{\overset{O}{\|}}{\underset{\underset{NH_4^+}{O^-}}{P}}-O^- \; NH_4^+$$

$$n \approx 700$$

X-ray diffractometry shows this sample to be 100% in the orthorhombic crystalline form II (*4*). It can be classified as a low toxicity compound, a weak skin irritant and not mutagenic. When heated to decomposition it can emit toxic fumes of NH_3, PO_x and NO_x (*5-6*).

PEU, [poly(1-methylen-2-imidazolidinone), Himont]:

$$n \approx 40$$

is a white flowing powder, moderately hygroscopic and insoluble in water and most organic solvents but soluble in chloroform and methylene chloride. It is not a skin or eye-irritant, is not mutagenic (Himont product bulletin), and is not expected to present a health hazard when used with reasonable care.

Polypropylene (PP; by Himont, Moplen FLF 20) of MFI 10-15 dg/min. (ASTM D 1238/L) was used.

All the materials were used as supplied. Binary mixtures APP/PEU were prepared by manual grinding in a mortar whereas ternary mixtures PP - APP/PEU were prepared in a Braebender mixer (AEV 330) at 190°C under nitrogen.

Experimental

Thermogravimetry (TGA) and differential scanning calorimetry (DSC) were carried out at 10°C/min. under nitrogen flow of 60 cm^3/min., unless otherwise indicated, on a DuPont 951 Thermobalance and 910 DSC Cell respectively, both driven by a TA 2000 Control System. Evolved Gas Detection Analysis (EGD) was performed by connecting a thermoconductivity probe to the thermobalance gases output.

The rate of water evolution was measured on line by connecting the thermobalance nitrogen output (30 cm^3/min.) with the water probe of a Panametrics system III hygrometer.

The rate of ammonia evolution was measured by means of an ammonia gas sensing electrode (Phoenix Electrode Company) immersed in a basic water solution (NaOH 0.1N) in which the nitrogen flow sweeping the gaseous products, evolved from the heating sample, was bubbled.

Foaming measurements were carried out in static air at a heating rate of 20°C/min. in home made equipment elsewhere described (7) using disks of 22 mm of diameter of the binary mixtures APP/PEU prepared by cold sintering.

Scanning Electron Microphotographs (SEM) were carried out with a Philiips SEM 515 instrument. Because samples were often deliquescent, in particular APP and its degradation products, attention was paid to avoid collapse of the foamed structure. Nevertheless, it was not possible to obtain the microphotographs of foamed mixtures with very high content of APP.

Infrared spectra were run on a Perkin Elmer FTIR 1710. Films on KBr were cast from viscous samples and polymer solutions whereas pellets were prepared by sinterisation of crystalline samples.

Results and Discussion

Thermal degradation. Thermogravimetry and differential scanning calorimetry were carried out on mixtures and on their separate components to elucidate possible interactions between them. Data on evolution of ammonia and water from their degradation were simultaneously collected. Parallel larger scale experiments were carried out when a higher amount of sample was required for chemical-physical characterization.

Thermogravimetry. The thermal behavior of the APP is shown in Figure 1. Three successive steps of weight loss can be recognized, taking place respectively in the range 260-420, 420-500, 500-680°C with corresponding weight loss of 13, 4 and 78% (Figure 1, A). A further 3.5% of weight is slowly lost by heating to 900°C. The amount of residue left at 680°C is 5% or 20% depending on whether a platinum or silica sample holder was respectively used. Phosphorous pentoxide was shown to attack silica and many silica-containing ceramic materials at high temperatures (8) giving thermally stable residues. For this reason high temperature treatments were performed on platinum sample holders, avoiding contact of APP with silica-containing materials. The DSC thermogram (Figure 1, B) shows an endothermic peak at 312° with a shoulder at 255°C. The peak corresponds to the first degradation phenomenon (see weight loss) whilst the shoulder can be attributed to the melting of the crystallites, as confirmed by observations at the optical microscope. Transitions amongst the several possible crystalline forms of APP weren't detected, in agreement with the thermodynamic stability of the orthorhombic form in this range of temperatures (4). The extensive degradation of APP, corresponding to the largest weight loss in TG, is strongly endothermic. An endothermic shoulder can be recognized around 560°C which could be attributed to melting of the crystalline orthorhombic O form of P_2O_5 formed on heating APP (see below), whose literature melting point is 562°C (8).

Thermal degradation of PEU occurs in two steps, as shown in Figure 2. The first step takes place between 370 and 390 °C in which 80% of the initial weight is lost at a very high rate leading to a residue that volatilises very slowly on heating at higher temperature.

Experimental and calculated thermogravimetic curves of binary mixtures APP/PEU (1/3, 1/1, 3/1 weight ratio) are reported in Figure 3. Two major weight loss steps would be expected for the mixtures if an additive behavior occurred (Figure 3, dashed curves). The steps at lower and higher temperature (380 and 630°C) should correspond mostly to PEU and APP decomposition respectively although a limited low temperature degradation of APP overlaps that of PEU and the degradation of PEU residue overlaps the APP main weight loss step. Interactions clearly occur between APP and PEU on heating because the experimental and calculated curves are not coincident. In particular the step involving the main degradation of PEU is shifted to 300-320°C, while in the high temperature range the APP decomposition seems to be delayed although a partial limited destabilization is simultaneously seen in APP-richer mixtures (Figure 3, B/E and C/F). A limited weight loss is also observed at the typical temperature of pure PEU decomposition (ca. 400°C) which is more evident in PEU-richer mixtures (Figure 3, A/D and B/E).

The comparison of calculated and experimental curves suggests that most of the PEU in the mixtures decomposes at lower temperature than expected because of the action of the APP or of its low temperature degradation product. The solid state APP-PEU interaction will not be quantitative and the unaffected PEU would then decompose at the expected temperature thus explaining the weight loss occurring in the mixtures at about 400°C. Alternatively this weight loss might be due to the decomposition of a product resulting from the APP/PEU interaction at 300-320°C. The apparent stabilization of APP observed in the mixtures might be due to thermally stable chemical bonds between APP and PEU formed on heating.

Gas evolution. Results on gas evolution from APP, PEU and their mixture 1/1 weight are shown in Figure 4. Infrared shows that gases evolved from degradation of these samples contain only ammonia and water during the first and the second step of degradation (below 500°C). Ammonia is evolved from pure APP essentially in the first step of weight loss (Figure 1A and Figure 4A, dashed line). The amount of ammonia eliminated through this step (9.5 weight %) corresponds at about 50% of the total

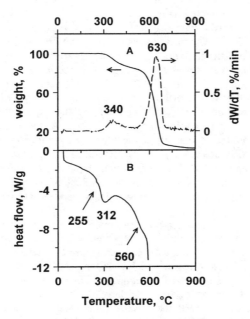

Figure 1. Thermal behavior of APP. A: Thermogravimetric curve, solid line; Derivative curve, dashed line. B: differential scanning calorimetric curve.

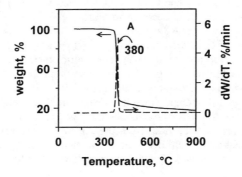

Figure 2. Thermogravimetry of PEU. Integral curve, solid line; Derivative curve, dashed line.

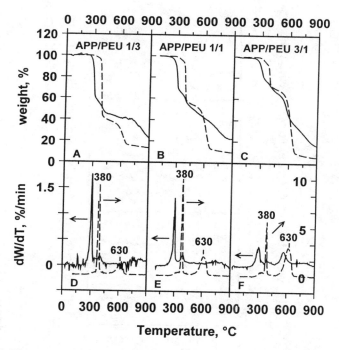

Figure 3. Thermogravimetry of mixtures APP/PEU, derivative and integral curves: Experimental curves, solid lines; calculated curves, dashed lines. APP/PEU weight ratio A and D, 1/3; B and E, 1/1; C and F 3/1.

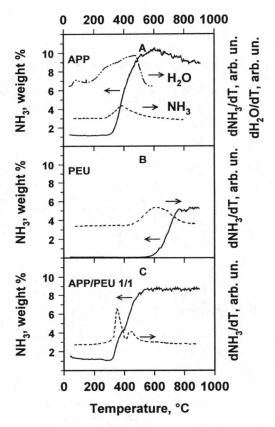

Figure 4. Ammonia evolution from: APP, A; PEU, B; Mixture APP/PEU 1/1 weight ratio, C. Solid lines refer to integral signals and dashed lines to the derivative signals. The dash-dot-dotted line in A corresponds to the derivative signal of water evolution from APP.

nitrogen contained in APP. Thus, evolution of ammonia from the ammonium ion content is limited, possibly because diffusion of ammonia through the glassy matrix created in this step of degradation may be restricted by the decreased pH. Absorbed moisture is evolved from APP at 80-100°C. In the broad peak of the rate of water evolution from APP degradation above 200°C, a shoulder (about 260°C) and a maximum (about 480°C) are recognizable corresponding to the first and the second step of APP weight loss at 250-420 and 420-500°C. The total amount of water evolved from these two steps of APP degradation is 9.7% whereas water evolved in each step cannot be directly evaluated because of the large overlap. However, the difference between weight loss and weight of ammonia evolved indicates that the water eliminated is 4.5% of APP weight corresponding to a molar ammonia to water ratio of 2, for which the following mechanism can be proposed:

$$\text{APP} \xrightarrow[-2NH_3]{\text{heat}} \text{polyphosphoric acid} \xrightarrow[-H_2O]{\text{heat}} \text{crosslinking} \quad (1)$$

Complete degradation of APP to P_2O_5 is prevented by the residual ammonia whose elimination is necessary to free the hydroxy groups for condensation. Thus it is likely that water is eliminated from the ammonium salt in the second step of APP weight loss to form phosphorimidic groups:

$$\begin{array}{c} O \\ \| \\ \sim\!P\!\sim \\ | \\ O^- \\ | \\ NH_4^+ \end{array} \xrightarrow{-H_2O} \begin{array}{c} O \\ \| \\ \sim\!P\!\sim \\ | \\ NH_2 \end{array} \quad (2)$$

A slightly more water than we have found would be required for this mechanism (about 8% instead of 5.7%). Nevertheless the solid-state nature of the reaction (2) might prevent quantitative yields.

Evolution of ammonia from PEU takes place in the second step of degradation where slow decomposition of the charred residue occurs (600°C, Figure 4B, dashed line). Ammonia represents 5% of the total weight of PEU is lost (Figure 4B, solid line) whereas no measurable water evolution was detected apart from adsorbed moisture.

In the binary mixture 1/1 9% of the sample weight is evolved in the two steps of weight loss at 300 and 400°C as ammonia (Figure 4C, solid and dashed lines), whereas the water lost throughout these steps is 3-4%. Thus slightly more ammonia and slightly less water are evolved from the mixture than predicted for an additive behavior, in which case 7.25% of ammonia and 4.85% of water evolved are calculated. While evolution of ammonia in the first step is expected to be evolved from the APP, that evolved in the second step is not predictable on the bases of PEU and APP degradation. This must indicate that an APP-PEU interaction might occur in the mixture.

Foaming behavior. The intumescent process, involving the blowing of the charring material, starts above 300°C with beginning temperature and final volume depending on the composition of the mixture as shown in Figure 5A-C. Largest foaming seems to correspond to mixtures with comparable weight content of APP and PEU. Evolved Gases Detection (EGD) from the 1/1 mixture shown in Figure 5B (dashed line) indicates that only the gases evolved in the second step of degradation (400°C, Figure 4E, solid line) are involved in the foaming process of the charring mass.

Further investigation on foaming were carried out by means of the SEM technique. Typical results are reported in Figure 6. Foaming does not occur in PEU either in the first or the second degradation step (Figure 6A, 350°C, 3 hours and Figure 6B, 450°C, 3 hours). Indeed a compact charred material is obtained from both degradation steps of PEU, with evidence of some relatively large circular craters likely to be due to the explosion of bubbles. APP/PEU mixtures give rise to a foamed structure by heating at 350°C for 3 hours (Figure 6C and 6E: outer surface of foamed char respectively from 2/1 and 1/3 mixture). The inner structure shows more relevant differences: with a lower amount of APP 1/3 mixture (Figure 6F) the walls of the cells are thinner and the volume is bigger in comparison with the cellular structure from the 2/1 mixture (Figure 6E). Thus foamed char from the 1/3 mixture should be fragile and probably unable to protect the underlying material from heat.

Charring reaction. In the first step of degradation of the mixtures a precursor of the intumescent char is created. On heating under nitrogen the APP/PEU mixtures to completion of this step (250°C, 160 minutes) other volatile products than ammonia and water are formed. This high boiling fraction can be trapped at 15°C and it has been shown by IR, NMR and GC-MS spectroscopy to contain essentially imidazolidinonic monomeric species. The residue is a physical mixture of water-soluble inorganic phosphorous moieties, such as phosphoric and polyphosphoric species, and water-insoluble charred material. In Table I the relative amount of each fraction for different mixtures is reported.

All the fractions are qualitatively similar, irrespective of the composition of the starting mixture. Thus, the chemical reactions leading to the intumescence precursor should be the same independent of the composition, even though the foaming behavior is different.

In the IR spectrum of the mixture APP/PEU 1/1 (Figure 7A, solid line) absorptions of PEU clearly dominate the IR pattern of the mixture and the presence of APP is shown by increase of absorptions, as compared to PEU (Figure 7A, dashed line), in the 1400-1500 and 1100-1300 cm^{-1} zones, where ammonium ion and phosphoryl groups respectively absorb.

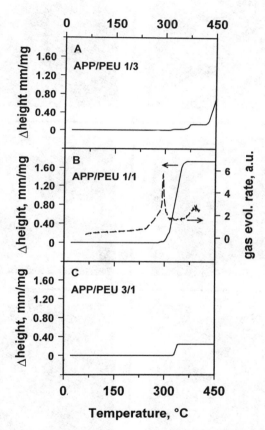

Figure 5. Foaming behavior (solid lines) as height increase (Δh) of the sample weight unit (mm/mg) of mixtures APP/PEU weight ratio: 1/3, A; 1/1, B; 3/1, C. Dashed line is the evolved gas detection, EGD signal from mixture 1/1.

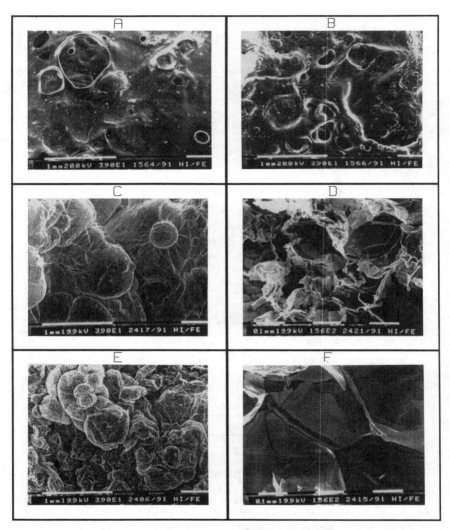

Figure 6. SEM microphotographs of the foamed residues from: A: PEU, 350°C 3 hours; B: PEU, 450°C 3 hours; C: Mixture APP/PEU 2/1, 350°C 4 hours, outer surface; D: Mixture APP/PEU 2/1, 350°C 4 hours, inner surface; E: Mixture APP/PEU 1/3, 350°C 4 hours, outer surface; F: Mixture APP/PEU 1/3, 350°C 4 hours, inner surface.

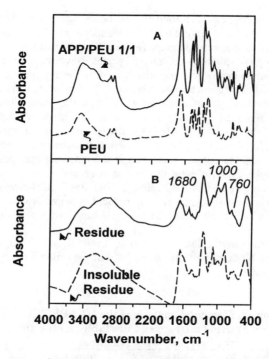

Figure 7. IR spectra of: A, PEU, dashed line; Mixture APP/PEU 1/1, solid line; B, residues of the thermal treatment to completion of the first degradation step of the 1/1 mixture: global residue, solid line; water-insoluble residue, dashed line.

Table I. First stage of degradation of APP/PEU mixtures percentage of the fractions formed

Mixture composition	Gases	High boiling products	Soluble residue	Insoluble residue
1/3	25	23	37	15
1/1	18	11	55.5	15.5
3/1	13	5.5	74.5	7

Moreover typical absorptions due to APP can be recognized in the 800-1100 cm^{-1} region and around 3400 cm^{-1}. In contrast, the IR spectrum of the global residue after completion of the first degradation step, (Figure 7B solid line), is dominated by the absoptions involving phosphorous-containing species. Nevertheless the presence of the imidazolidinonic ring in the residue is demonstrated by the sharp peak at 760 cm^{-1} and by the ureic carbonyl absorption around 1680 cm^{-1}. By hot water treatment a residue is obtained that still contains phosphorate species (Figure 7B, dashed line). Hydrolysable material has been removed by water as shown by the strong decrease of the peak attributed to P-O-C hydrolisable bonds (9) in the spectrum of the insoluble residue as compared to the global residue (Figure 7B). However the major features of the IR spectrum of the global residue are preserved in the insoluble residues. As previously stated from thermogravimetric data, the first step of weight loss involves extensive degradation of the organic part of the mixture. The attack of the phosphoric part on PEU chains, likely to be due to polyphosphoric acid moieties, lowers the temperature of degradation of 60-80°C and leads to chemical bonds of which only a fraction is hydrolisable.

From elemental analysis data, reported in Table II, which supports the above reaction path, it can be calculated that about 30% of the original phosphorous remains in the insoluble residue.

Table II. Elemental analysis of the 1/1 APP/PEU mixture: Starting mixture and insoluble residue after the first stage of degradation

Element, %	Original	Insoluble residue
C	24.50	34.15
H	5.10	4.10
N	21.32	12.05
P	15.90	8.7

Even though the chemical reaction leading to the precursor of the intumescent char should be the same irrespective of the mixture composition, the foaming behavior depends on it. Thus also physical factors should probably play an important role in the foaming process.

Effects of the polymer matrix. Investigations have been carried out in order to elucidate the role played by the polymer matrix in the intumescence process. Introduction of an intumescent additive in the polymer matrix might, in principle, affect the intumescent process either from the chemical or the physical point of view. For example mutual modification of the decomposition process or change of the melt viscosity could take place which might affect the intumescent behavior in the presence of the polymer. Alternatively the polymer could act as an inert diluent in the intumescent process.

Thermal behavior. The thermogravimetric curve (Figure 8) of the mixture containing 70% of Polypropylene (PP) and 30% of the intumescent additive APP/PEU 1/3 shows that the low temperature weight loss step due to the decomposition of the intumescent additive (327 and 292°C for the experimental and calculated curve respectively) occurs at a somewhat higher temperature than calculated assuming an additive behavior. Whereas the degradation of the polymer as well as the amount of residue either from the degradation of the additive (462°C) or at 600°C are the expected.

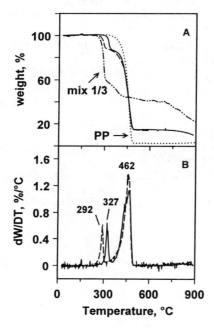

Figure 8 A: Thermogravimetric curve of : Polypropylene (PP), (dotted line); mixture APP/PEU 1/3, dash-dot-dotted line; PP (70%) + mixture APP/PEU 1/3 (30%) experimental (solid line) and calculated curve (dashed line). B: Derivatives of the thermogravimetric curves of PP (70%) + mixture APP/PEU 1/3 (30%) experimental (solid line) and calculated curve (dashed line).

Conclusions

The results show that in APP/PEU mixtures chemical reactions occur between APP and PEU on heating which are likely to be responsible for the intumescent behavior. Morphology and volume of the foamed char, which should affect the fire retardant effectiveness of the additive, depend on the mixture composition in terms of APP/PEU ratio. The thermal behavior of the additive may be somewhat modified by introduction in a polymer matrix as shown for PP

Acknowledgments

The Ministero Italiano dell'Università e della Ricerca Scientifica e Tecnologica (M.U.R.S.T., fondi 60%) and the Consiglio Nazionale delle Ricerche (CNR, Progetto Finalizzato Chimica Fine II) are gratefully acknowledged for supporting this work.

Literature cited

1. Bertelli G.; Roma P.; Locatelli R., German Patent 2,723,877 **(1977)**
2. Bertelli G.; Roma P.; Locatelli R., German Patent 2,800,891 **(1978)**
3. Marciandi F., German Patent 2,839,710 **(1979)**; Assigned to Montedison SpA CA 90, 18815u **(1979)**
4. Shen C. Y; Sthaleber N. E.; Dryoff D. R., *J. Am. Chem. Soc.* **1969**, *vol. 1*, p. 91
5. SPECTOR Hand book of toxicology
6. Sax N. I., *Dangerous Properties of Industrial Materials,* 6[a] Editions, Van Nostran Reinhold Company: US, New York, 1984
7. Bertelli G.; Camino G.; Marchetti E.; Costa L.; Casorati E., Locatelli R., *Polym. Deg. and Stab.* **1989** *vol. 25,* p.277
8. Corbridge D.E.C. *Phosphorous, an Outline of its Chemistry, Biochemistry and Technology*; Studies in Inorganic Chemistry n° 10; Elsevier Science Publishers: Amsterdam, NL, 1990, pp.95-96
9. Lin-Vien D.; Colthup N. B.; Fateley W. G.; Grasselli J. G., *The handbook of Infrared and Raman Characteristic Frequencies of Organic Molecules*; Harcourt Brace Jovanovich Publishers: US, New York 1991 pp. 270-272

RECEIVED January 4, 1995

Chapter 7

Intumescent Systems for Flame Retarding of Polypropylene

Menachem Lewin and Makoto Endo[1]

Polymer Research Institute, Polytechnic University, Brooklyn, NY 11201

Six ammonium polyphosphate (APP) based intumescent FR systems (IFR) for injection-molding grade PP were investigated and compared for OI, FR effectivity (EFF), thermal decomposition and char structure. Synergistic Effectivity (SE) is defined and found for the systems investigated to be in the range of 5.5-11.3, compared to the SE of other phosphorus-nitrogen systems (1.75), of aliphatic and aromatic bromine-antimony oxide formulations (4.3 and 2.2), and of bromine-phosphorus based systems (1.4-1.6). Correlations were obtained between the % P and the TGA residue-after-transitions (RAT). Significant linear relationships were found between OI and RAT for all cases. Cone calorimeter results for several IFR-treated PP samples are reported and compared; a correlation with OI and EFF values is noted. SEM scans of char obtained from the combustion of APP-containing intumescent PP samples were examined and differences in cellular structure discussed.

Ammonium polyphosphate (APP) with co-additives is extensively used for flame retarding polypropylene (PP) [1]. The co-additives usually consist of char-forming and blowing agents which are synergists with APP. The co-additives differ among the various commercially available formulations. For this work, it was of interest to compare the flame retardant behavior of various formulations by OI, TGA, Cone calorimetry and

[1]Current address: Showa Denko Co., Kawasaki Plastics Laboratory, Kawasaki, Japan

Table I. Phosphorus-Based Additives in Polypropylene Formulation and Results

No	FR Additives	FR Additives wt%	%P	OI	OI/%P	UL-94 1/16"	UL-94 1/8"
1	-	0	0	17.8	-	NR	NR
2	EDAP	20	3.9	26.0	2.1	NR	NR
3	EDAP	25	4.8	27.8	2.1	NR	NR
4	EDAP	30	5.9	29.8	2.0	V-2	V-0
5	EDAP	35	6.9	32.3	2.1	V-2	V-1
6	EDAP	40	7.8	34.1	2.1	V-2	V-0
7	APP	15	4.7	19.3	0.32	NR	NR
8	APP	20	6.2	19.7	0.31	NR	NR
9	APP	25	7.8	20.2	0.31	NR	NR
10	APP + Spinflam MF82	15	3.0	26.2	2.8	NR	NR
11	APP + Spinflam MF82	20	4.0	30.7	3.2	NR	V-0
12	APP + Spinflam MF82	25	5.1	33.0	3.0	V-2	V-0
13	APP + PETOL	15	3.0	21.4	1.2	NR	NR
14	APP + PETOL	20	4.0	23.6	1.5	V-2	V-2
15	APP + PETOL	25	5.1	26.4	1.7	V-2	V-2
16	APP + THEIC	15	3.0	24.6	2.3	NR	NR
17	APP + THEIC	20	4.0	27.6	2.5	V-2	V-2
18	APP + THEIC	25	5.1	32.2	2.8	V-2	V-0
19	Exolit IFR 23P	15	3.6	29.9	3.4	NR	NR
20	Exolit IFR 23P	20	4.8	34.8	3.5	V-2	V-0
21	Exolit IFR 23P	25	6.0	38.8	3.5	V-2	V-0
22	APP + PETOL t-benzoate	15	3.0	19.4	0.44	NR	NR
23	APP + PETOL t-benzoate	20	4.0	19.6	0.38	NR	NR
24	APP + PETOL t-benzoate	25	5.1	19.9	0.35	NR	NR

char morphology as well as to attempt to define amd compare the synergisms involved with other known synergisms, including bromine, nitrogen, antimony and phosphorus.

Experimental.
The PP used was an injection-grade powder (MFR = 15g/10 min., 230°C). The six intumescent formulations used were: 1. AMGARD EDAP, believed to contain APP reacted with ethylene diamine [2,3]; 2. APP without co-additives; 3. APP + Spinflam MF82 (assumed to contain poly-triazine-piperazine) [3]; 4. APP + pentaerythiritol (petol); 5. APP + trishydroxyethyl isocyanurate (THEIC); 6. Exolit IFR 23P [1]. The ratio of APP to co-additives was 2:1 in all cases. The components were mixed in the molten state in a Brabender blender of 240 ml at 200°C and 40 rpm for 12 minutes. The samples were compression-molded at 230°C, cooled to room temperature and cut to test pieces. Flame retardancy testing included OI and UL-94 tests. Formulations are given in Table I.

Cone calorimeter measurements were carried out at the National Institute of Standards and Technology on 4 samples: pure PP, PP + 30 wt. % of EDAP, PP + 20 wt. % of APP + Spinflam and PP + 20 wt. % of APP + pentaery-thritol (2:1). The samples were compression-molded to 1/8" and cut to 10 x 10 cm test pieces. The heat flux applied was 35.0 KW/m^2 in all cases. Spark ignition and grid and frame were used. The sample orientation was horizontal.

The morphology of surfaces and cross-sections of char specimens was investigated by scanning electron microscopy at enlargements of 250-1500. The char specimens were taken from the surfaces of 5 samples of PP, compounded with 25 wt. % of 5 FR formulations as described above, after combustion in the OI apparatus at the oxygen concentration of their OI.

Results and Discussion

It appears to be generally accepted that, for intumescent systems, three basic ingredients are required: a catalyst, i.e. APP, a char former i.e. pentaerythritol (petol) and a blowing agent, i.e. melamine or another nitrogen-derivative forming incombustible gases during pyrolysis and combustion. [1]. For the formulations used OI and UL-94 data are presented in Table I.

The change of OI versus weight percent of the flame retardant additives calculated as percent phosphorus is shown in Figure 1. The OI values increase linearly with the amount of additives. The slopes of the straight lines differ markedly for the various additives. The lowest slope is for APP without co-

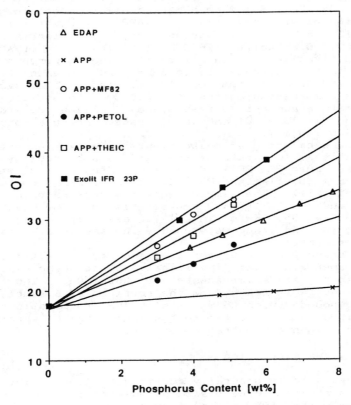

Figure 1. OI vs phosphorus content.

additives, showing the low activity of APP in spite of the high (7.8%) P content (see Table I). Substitution of 1/3 of the APP by petol increased the FR performance although, at the level of 25% additive, only an OI of 25 and UL-94 V-2 test performance achieved.

A higher FR activity is shown by AMGARD EDAP [3]. At 25% additive, an OI of 27.8 and UL-94 V-2 test performance was not obtained, whereas at a 30% EDAP level, UL-94 V-2 for test performance 1/16" and V-0 for 1/8" samples were achieved.

The results are considerably improved when THEIC, Spinflam MF82 and Exolit IFR are used. 20% of the additive in the last two cases resulted in OI values of 30 and 35, respectively, and a UL-94 V-2 test performance for 1/16" and V-0 for 1/8" samples, similar to the samples with 30% EDAP.

In Figure 1, the influence of the co-additives on the APP, which are considered synergists, is evident. The system appears to be very sensitive to the nature of the co-additive. In Figure 2, the FR effectivity, e.g. OI/% P [4] is compared for the various samples; it varies from 0.31 for APP to 2.1 for EDAP and 3.5 for Exolit IFR. These values are much higher than the general value of 1.3 given by Van Krevelen [4] for P in PP, which does not refer to intumescent systems.

Synergistic Effectivity. In Table II, the FR effectivity of APP + petol and APP + melamine are given as 1.7 and 0.92, respectively. When dividing these effectivities by the FR effectivity of APP, a "synergistic effectivity" (SE) is obtained, yielding values of 5.5 for petol and 3.0 for melamine. The SE for the combined synergists is 7.7. The values of FR effectivity and SE are considerably higher for the other intumescent systems in this study and are, in the case of Exolit IFR, 3.5 and 11.3, respectively.

Table II. FR Effectivity (EFF) and Synergistic Effectivity (SE) of APP-based Systems on PP

FR	SYNERGIST	EFF	SE
APP		0.31	
	Petol	1.7	5.5
	Melamine	0.92	3.0
	Petol + Melamine	2.4	7.7
EDAP		2.1	6.8
APP	Spinflam MF82	3.0	9.7
EXOLIT IFR 23P		3.5	11.3

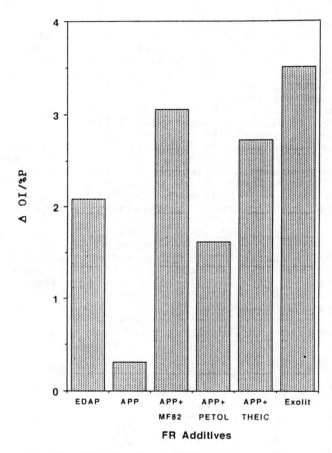

Figure 2. FR effectivity of phosphorus FR additives.

Synergistic systems are well known in the literature dealing with flame retardancy of polymers [5,6,7]. It is of interest to compare several of the known synergisms with the PP-APP intumescent system shown in Table II.

<u>Comparing Synergisms</u>. In Table III are presented data computed from the literature on a number of FR synergistic systems. The FR EFF and the SE are calculated for cotton to which Pyrovatex PC (methylol derivative of dialkyl phosphonic propionamide) together

Table III. FR Effectivity and Synergistic Effectivity of Br-Sb, Br-P amd Other Systems

POLYMER	FR	SYNERG.	EFF.	SE	REF.
Cotton	Pyrovatex		4.0		
	Pyrovatex	TMM	7.0	1.75	[8]
PP	Arom. Br		0.45		
	Arom. Br	Sb_2O_3	1.0	2.2	
	Aliph. Br		0.6		[4]
	Aliph. Br	Sb_2O_3	2.6	4.3	
PS	Aliph. Cl		0.5		[4]
	Aliph. Cl	Sb_2O_3	1.1	2.2	
PC/PET)	TPP		13.3		
2:1	BrPC		1.7		
	BrPC/TPP Blend 70:3			1.38	[10]
	BrP:70:3			1.57	
PAN	APP		1.02		
	HBCD		1.21		[9]
	APP + HBCD			1.55	

with TMM (trimethylol melamine) were applied at a 2% P and 5% N level [8]. Whereas relatively high FR effectivities are noted, the SE is only 1.75. Although the ingredients of this system - phosphorus, nitrogen and a polyhydric alcohol (cotton cellulose) - resemble an intumescent system, the SE is much lower than observed in this study.

Data on the SE of aromatic and aliphatic bromine derivatives with antimony trioxide, another conventional synergistic system, are computed from reference 4, showing SE values of 2.2 and 4.3, respectively. Similarly, for aliphatic chlorine derivatives with antimony trioxide, a SE value of 2.2 is computed for polystyrene [4].

Of particular interest are the data on bromine-phosphorus synergism. In the case of polyacrylonitrile

(PAN) treated with varying ratios of APP and hexabromocyclododecane [9], a SE value of 1.55 is obtained. This low SE value is remarkable since the bromine compound was shown to act as a blowing agent with the gaseous hydrogen bromide, released in the pyrolysis, serving to foam char.

Similar SE values are computed from data in reference 10 for a polycarbonate-polyethylene terephthalate 2:1 blend, treated with varying ratios of triphenyl phos- phate (TPP) and brominated polycarbonate. An SE value of 1.38 is found. When a brominated phosphate with the ratio of bromine to phosphorus of 70:3 is added to the same blend, it yields a value of 1.58. There are some indications, though no clear evidence, that the bromine compound may, in these cases, also serve, at least partly, as blowing agents and not as gas-phase flame retardants. The bromine-phosphorus SE values are considerably lower than the bromine-antimony ones, indicating the possibility of different flame retardant mechanisms.

Relative Importance of Co-additives. The synergistic effectivity of the PP-APP system as described in this paper is very much higher than the other known synergisms, which emphasizes the sensitivity of the intumescent system to the nature of the co-additives on one hand, and the vast possibilities for further improvement, on the other hand. It is, however, still unclear what the relative importance is of the two last co-additives. Results of a series of experiments in which the wt. % of PP and APP were kept constant at 75 and 16.6 wt. % respectively, while the proportion of melamine in the 8.4 wt. % of petol + melamine was varied, is shown in Figure 3. Both petol and melamine are clearly seen to be synergists for APP, the effect being much more pronounced for petol. When used together, they are, in fact, co-synergists. However, it is of interest to note that compositions of petol/melamine 20:80 to 80:20 % show virtually the same OI values of ca 30. This indicates that, at least in this range, melamine and petol are equivalent and interchangable. Since the petol shows a higher effectivity, the role of the blowing agent seems to be minor and necessary only in small amounts.

Mechanistic Considerations. The strong synergistic effect in intumescent systems is not suprising, considering the highly complex nature of the chemical and physical interactions between materials involved [11]. The "catalytic" action of APP is believed to consist of a series of processes occurring during the combustion: decompostion and release of ammonia and water, phosphorylation of petol and possibly also of the oxidized PP, dephosphorylation and double-bond

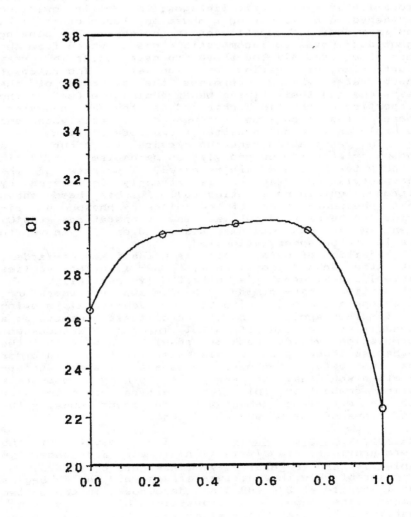

Figure 3. OI vs melamine/melamine + petol for FR polypropylene.

formation [12], participation as one of the building blocks of char. The petol is dehydrated and cross-linked via etheric linkages; the PP is oxidized and dehydrogenated during combustion, forms cross-links and is a dominant constituent in the char. The blowing agent decomposes to incombustible gases, which foam the char structure. All the above processes occur in a very short time, with various reaction rates. The ratio of these rates which determines the sequence of the reactions and their timing has a dominant effect on the properties of the final char and on the FR behavior. These rates might be influenced by catalysts and possibly engineered to better flame retardancy.

The dehydration reaction appears to occur by a phosphorylation-dephosphorylation mechanism [5,6,7], in which water molecules are released. The rate of the phosphorylation reaction is strongly influenced by nitrogen-containing moieties [6]. It has been shown that phosphorylation of cellulose by phosphorus tri-amide can be carried out at low temperatures (80°C) even in aqueous mediums [12]. Urea is known to accelerate phosphorylation [5].

Addition of petol, which is not a flame retardant but rather an additional fuel, to APP is effective (see Table I and Figures 1-3), albeit only partially. The dehydration char-forming mechanism appears, therefore, to be operative even without the additional blowing agent. The application of petol-tetra-benzoate as a co-additive instead of petol in the intumescent formulation yields negative results, showing that the mechanism indeed proceeds via esterification. In order for the petol-t-benzoate to be active, a transester-fication would have to take place, forming a phosphate during combustion. This transesterfication is too slow for an efficient dehydration to occur under the conditions of the OI and UL-94 tests.

Thermogravimetric Analysis. The results of the thermogravimetric analysis (TGA) in air are shown in Table IV and Figures 4-7. In Figure 4, several TGA diagrams of PP with additives are shown. The PP begins to decompose at 270-280°C and decomposes in one or two stages. The slope of the transitions of all additive-containing samples is smaller than that of PP. In the case of APP with Spinflam, the slope of the second transition increased with the amount of flame retardant agent present. The slopes of other samples are nearly unaffected by the amount of the flame retardant additive. The final residue at 650°C does not seem to correlate with the amount of additive or with % P. The highest amounts of residue are obtained for Exolit, APP with MF82 and EDAP.

A significant linear correlation, however, was found between the residue-after-transitions (RAT) and the amount of the flame rletardant additive (Fig. 5).

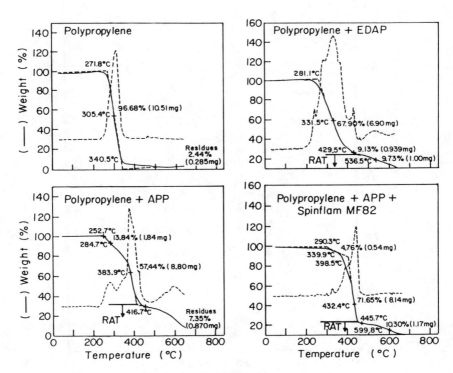

Figure 4. TGA diagrams.

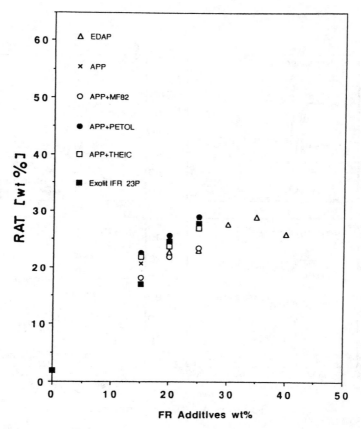

Figure 5. TGA residue-after-transitions (RAT) vs wt. % of FR additives.

Table IV. TGA Results of Phosphorus-Based Additives in Polypropylene

No	FR	FRwt%	1st Transition				2nd Transition			RAT [wt%]	Residue at 650°C [wt%]
			Ts[°C]	Tp[°C]	wt%	Slope [wt%/deg]	Tp[°C]	wt%	Slope [wt%/deg]		
1	PP	0	272	305	96.7	-1.52				1.9	2.4
2	PP + EDAP	20	281	363	74.2	-0.60	434	3.1		22.7	9.2
3		25	261	332	67.9	-0.60	430	9.1		23.0	10.7
4		30	257	336	59.1	-0.49	434	13.2		27.7	9.0
5		35	273	351	59.3	-0.55	440	11.7		29.0	8.6
6		40	263	338	62.2	-0.58	431	11.8		26.0	14.7
7	PP + APP	15	260	293	33.4	-0.36	387	45.9	-1.15	20.7	7.5
8		20	251	282	27.1	-0.42	380	47.8	-0.54	25.1	7.7
9		25	253	285	13.8	-0.25	384	57.4	-1.39	28.8	7.4
10	PP + APP +Spinflam MF82	15	307	322	9.2		407	72.7	-0.88	18.1	4.5
11		20	297	324	6.7		425	71.4	-1.09	21.9	9.6
12		25	290	340	4.8		432	71.7	-1.40	23.5	11.3
13	PP + APP + PETOL	15	269	307	77.4	-0.80				22.6	5.9
14		20	261	300	74.3	-0.81				25.7	6.5
15		25	259	301	71.0	-0.70				29.0	5.9
16	PP + APP + THEIC	15	269	310	78.1	-0.90				21.9	8.2
17		20	266	306	76.3	-0.88				23.7	6.7
18		25	262	307	73.1	-0.78				26.9	6.6
19	PP + Exolit IFR 23P	15	264	360	83.1	-0.54				16.9	5.8
20		20	260	362	75.4	-0.49				24.6	14.7
21		25	258	298	72.1	-0.58				27.9	13.5

The RAT increases with the % additive in all cases. This behavior appears compatible with an intumescent mechanism.

It is reasonable to assume that the actual combustion and flaming reaction occurred during the transitions. The char remaining directly after the transitions has just been formed into a foaming structure and did not yet have time to be further oxidized at the higher TGA temperatures and deteriorate and possibly collapse. The amount of this "primary" char clearly depends on the amount and nature of the flame retardant additive in the sample. It is to be expected that this primary char is linked to the flammability of the samples. Therefore, it is of particular interest and not surprising to note the highly significant linear relationship between the OI and % RAT (r=0.99; see Figure 6). This relationship appears to be typical for the intumescent mechanism and may serve as a clue for recognizing and characterizing the nature of the flame retardant activity. It was obtained for all six formulations investigated. The slopes, OI/RAT, of the lines of Figure 6 are given in Figure 7 for the various formulations. When comparing this data to the data on the flame retardant effectivity, some similar trends can be discerned.

<u>Cone Calorimetry</u>. The results of the cone calorimeter testing of four samples are shown in Table V and Figures 8, 9 and 10. Uncompounded polypropylene yielded the lowest residue and the highest total heat release. The peak rate of heat release was the highest and it was reached in the shortest time; combustion was rapid. The amounts of heat released decreased with the degree of flame retardancy, possibly due to a slower and lower polymer-fuel supply to the flame and to the fact that the char did not combust. It is typical for the flame retardant samples that the peak rate of heat release is delayed by the formation of char.

The fact that the amounts of carbon monoxide were higher for the flame retardant samples while the CO_2 amounts were lower indicates incomplete combustion in the flame retardant samples, which correlates with the high char residue.

Two interesting observations can be made from Figures 8 and 10: (1) the time of peak rate of heat release is delayed proportionately to the effectivity of the flame retardant formulation; (2) the heat release occurs in 2 stages. The first stage occurs at about the same time for all 3 formulations. Its peak is smaller than the main second peak. It decreases with the increase in flame retardant effectivity. The mass loss rate and the specific smoke extinction area behave in a similar manner.

The reason for the two peaks resides in the dynamics of the combustion and the char formation. It is believed that a rapid layer of cellular foamed char is formed at first. The pressure of the combustion gases produces a char structure which is impermeable to the gases as well as to the molten polymer. With an increase in pressure, the char surface barrier is pushed away from the main bulk of the polymer and a gas bubble is formed which separates and insulates the barrier from the rest of the polymer. The continued heating breaks the barrier after some time; burning is resumed and another layer of char is produced [8].

Figure 9 illustrates a remarkable correlation between most of the cone calorimeter values and the OI.

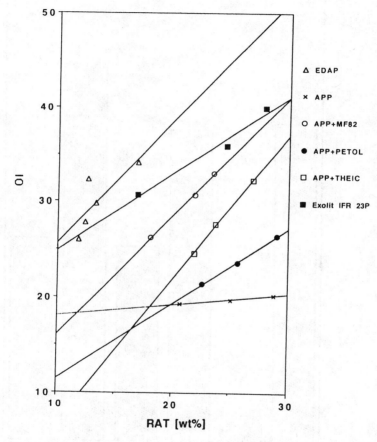

Figure 6. OI vs TGA residue-after-transitions (RAT).

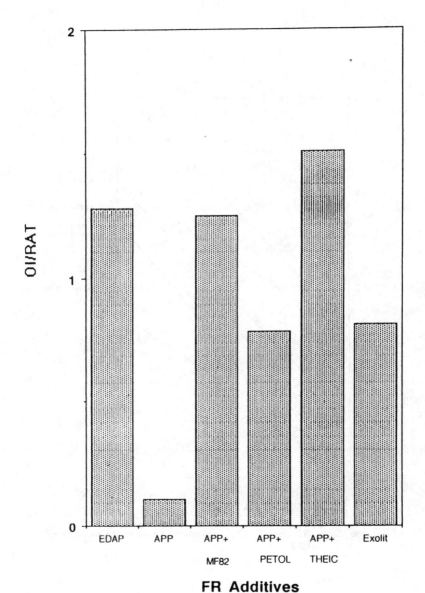

Figure 7. OI/RAT for FR additives.

Table V. Results of Cone Calorimeter

DATA		PP	PP + EDAP 30%	PP + APP+MF82 20%	PP + APP+PETOL 20%
Initial Mass	g	26.7	28.1	28.3	27.4
Final Mass	g	2.5	9.4	12.2	5.6
Residue	wt.%	9.4	33.5	43.1	20.4
Ignition Time	sec.	62	47	47	41
Time of Peak RHR*	sec.	165	330	460	245
Peak RHR	kw/m^2	575.3	143.8	136.9	205.4
Total Heat Release	MJ/m^2	91.32	58.85	51.21	69.09
Heat Release at 300 sec.	kw/m^2	246.76	99.17	40.38	151.15
Mass Loss Rate at 300 sec.	g/s*m^2	6.34	3.27	1.46	4.63
Heat of Combustion at 300 sec.	MJ/kg	38.94	30.35	27.71	32.62
Specific Ext. Area at 300 sec.	m^2/kg	615.52	568.26	215.34	735.82
Carbon Dioxide at 300 sec.	kg/kg	3.17844	2.11012	1.95889	2.39999
Carbon Monoxide at 300 sec.	kg/kg	0.03008	0.03184	0.04183	0.06174
OI	%	17.8	29.8	30.7	23.6
UL-94 1/16"	----	NR	V-2	NR	V-2

*Rate of Heat Release

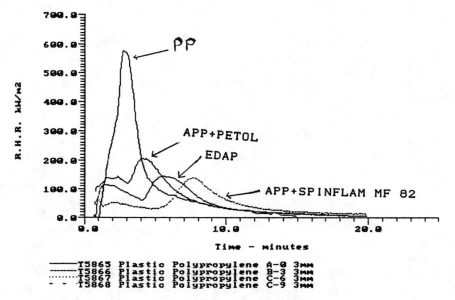

Figure 8. Rate of heat release (RHR) vs time for PP and PP treated with intumescent FR formulations.

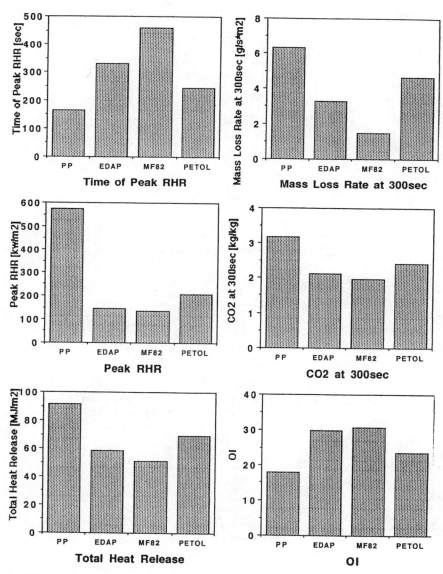

Figure 9. Cone calorimeter and LOI values for PP and 3 PP samples treated with IFR formulations.

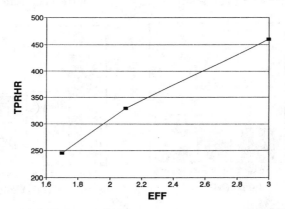

Figure 10. Time of peak of rate of heat release (TPRHR) vs FR effectivity (EFF).

The sequence of the OI values in the tested samples is rather close to the inverse sequence of the total heat released, of mass loss rate, of the peak RHR and to the sequence of the time of peak RHR. It appears, therefore, that the OI measurements allow a qualitative prediction as to the behavior of an intumescent flame retardant-treated polymer in the Cone calorimeter.

Char Morphology. SEM scans of char from the surfaces of five PP samples, treated by the intumescent systems applied in this study, are presented in Figures 11-13. The cell structure of the char is clearly visible, both on the surface as well as on the cross-section scans. The cells appear to be closed and their diameters range from 7-40 microns in the surface scans and 15-45 microns in the cross-sectin scans. Differences can be seen in the general appearance and in the ranges of cell diameters of the various scans. An attempt to define these difference is made in Table VI. The widest range for cell diameters is found in the PP + APP + petol formulation, in which no blowing agent was present. Although a cell structure is developed, only a part of the surface is foamed. In the case of PP + EDAP and PP + Exolit, the whole surface of the char scans is foamed and the cell dimensions are in a relatively narrow range.

Table VI. SEM Observations of the Chars

Sample	Diameter of cell microns		Observation	
	Surface	Cross section	Surface	Cross Section
EDAP	20-27	20-40	Nearly uniform cell diameter	Nearly complete foaming
MF82	7-20	30-45	Foaming forms membrane	Partial foaming
PETOL	27-40	15-45	Wide diameter. Clear foaming.	Partial foaming Disturbed structure
EXOLIT	10-20	15-30	Rigid foam structure. Clear foaming	Nearly complete foaming
THEIC	23-37	27-40	Unclear foaming	Unclear foaming Expanded as whole

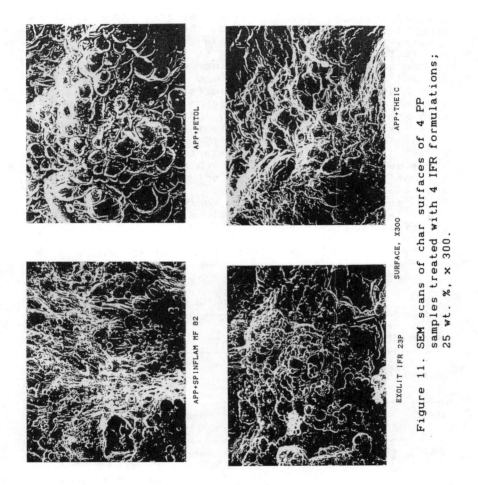

Figure 11. SEM scans of char surfaces of 4 PP samples treated with 4 IFR formulations; 25 wt. %, × 300.

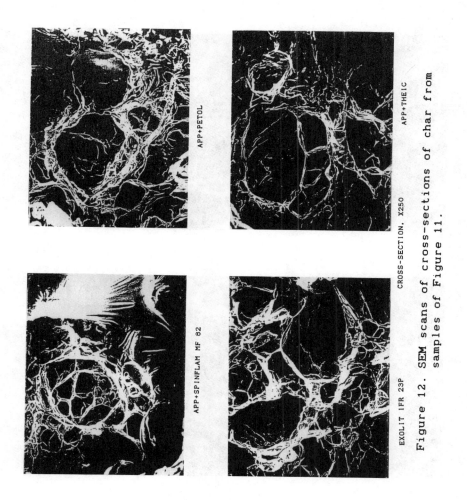

Figure 12. SEM scans of cross-sections of char from samples of Figure 11.

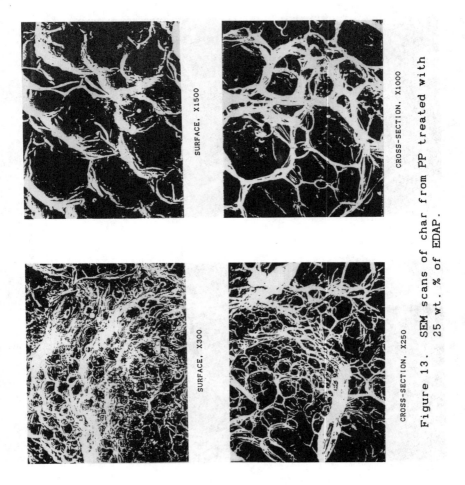

Figure 13. SEM scans of char from PP treated with 25 wt. % of EDAP.

In several cases, such as in the sample treated with Spinflam MF82, a ballooning or void was observed below the crust of the foamed char. This appears to correspond to the first peak in RHR observed in the Cone calorimeter testing, and seems to corroborate the above consideration on the two-stage intumescent process in APP-treated samples [13].

Conclusions

The flammability behavior of injection-grade polypropylene treated with six phosphorus-based flame retarding additive formulations was investigated by OI, TGA, Cone calorimetry, SEM and UL-94 methods. Linear correlations between the weight percent of the additives and of percent phosphorus and OI were established in all cases. The flame retardant effectivity (EFF), OI/% P increased from 0.31 to 3.5 in the following order: APP, APP + petol, EDAP, APP + THEIC, APP + Spinflam MF82, Exolit IFR.

The Synergistic Effectivity (SE) was defined as the ratio of EFF of the flame retardant additive with the synergist co-additive to the EFF of the flame retardant additive alone. The SE was found for the formulations investigated to be in the range of 5.5-11.3 and compared to the SE of other phosphorus-nitrogen systems (1.75), of aliphatic and aromatic bromine-antimony oxide formulations (4.3 and 2.2), of chlorine-antimony oxide (2.2) and of bromine-phosphorus-based systems (1.4-1.6).

Significant linear correlations were found for all formulations between the OI values and the values of the residue-after-transitions (RAT) obtained in TGA measurements carried out in the presence of air.

Cone calorimetry measurements showed that the peak rate of heat release is delayed proportionately to the EFF of the flame retardant formulations. Heat release occurs in two stages, indicating a stepwise formation of char layers during the combustion in the Cone calorimeter.

An inverse correlation was found between the sequence of the OI values of the tested samples and the sequences of the total heat release, of mass loss rate and of the peak rate of heat release, indicating that the OI measurements allow a prediction of the combustion behavior of an intumescent flame retardant treated-polymer in the Cone calorimeter.

The SEM scans of the chars obtained in the combustion of polypropylene samples treated with the intumescent flame retardants reveal a well-defined foamed cell structure in all cases, both on the surface as well as on the cross-section of the chars. The cell dimensions were in the range of 7-40 microns in the surface scans and 15-45 microns in the cross- sections, depending on the formulations.

Literature Cited

1. Sharf, D., Nalepa, R., Heflin R. and Wus, T. Fire Safety J. **19**, 103 (1992).
2. Weil, E. in Rec. Adv. in FR of Polym. Mat., Vol. 3, M. Lewin, Ed., BCC, p. 1 (1992).
3. Goin, C.L and Huggard, M.T., ibid, Vol. 2, p. 94 (1991).
4. Van Krevelen, D.W., in App. Poly. Symp. J. App. Polym. Sci. **31**, M. Lewin, Ed. p. 269 (1977).
5. Lewin, M., in Chemical Processing of Fibers and Fabrics, Part 2 - Functional Finishing, M. Lewin and S.B. Sello, Eds., Marcel Dekker, 1983, pp. 1-141.
6. Khanna, Y.P. and Pearce, E.M., in Flame Retardant Polymeric Materials, Vol. 2, M. Lewin, S.M. Atlas and E.M. Pearce, Eds., Plenum, 1978, pp. 43-61.
7. Lewin, M. and Sello, S.B., ibid, Vol. 1, (1975), pp. 19-136.
8. Willard, J. and Wondra, A.E., Tex. Res. J. **40**, 203 (1970).
9. Ballisteri, A., Montaudo, G., Puglisi, C., Scamporrino, E. and Vitallini, D., J. Appl. Polym. Sci., **28**, 1743 (1983).
10. Green, J., Rec. Adv. in FR of Polym. Mat., M. Lewin, Ed., Vol. 4, BCC, p. 8 (1993).
11. Camino, G., Costa, L. and Trossarelli, L., Polym. Degr. Stab. **12**, 213 (1985).
12. Basch, A. and Lewin, M., Tex. Res. J., **45**, 245 (1975).
13. Pagliari, A., Cichetti, O., Bevilacqua, A. and Van Hees, P., Proc. of Flame Retardants '92 Conf., London, 22-23 Jan., 1992, Elsevier Applied Science, pp. 41-52.

RECEIVED January 4, 1995

METALS AND COMPOUNDS AS FLAME RETARDANTS

Using metals or metal compounds with higher volume polymers is one of the more interesting areas in flame retardancy. The chemistry is specific to a given polymer and generally involves new routes to cross-linking or changes in degradation mechanism.

Starnes and co-workers use activated copper metal to reduce smoke in poly(vinyl chloride). Mechanistic work shows rapid reductive coupling promoted by copper. The result is PVC cross-linking and lower smoke.

Chandrasiri and Wilkie looked at the reaction of tin (or sulfur) compounds with poly(methyl methacrylate). Free radicals formed from the additives interact with the degradation pathways which can facilitate the formation of nonvolatile, cross-linked species.

Pearce and co-workers studied the effects of zinc chloride on styrene-acrylonitrile. Zinc chloride has been shown to complex with the nitrile group, modifying the degradation mechanism, and yielding a thermally stable triazine ring that cross-links the main chain.

These three examples of cross-linking provided by chemistry tailored to specific polymers opens the way for developing new approaches to smoke reduction and flame retardancy.

Chapter 8

Reductive Coupling Promoted by Zerovalent Copper
A Potential New Method of Smoke Suppression for Vinyl Chloride Polymers

J. P. Jeng, S. A. Terranova, E. Bonaplata, K. Goldsmith, D. M. Williams, B. J. Wojciechowski, and W. H. Starnes, Jr.

Applied Science Ph.D. Program and Department of Chemistry, College of William and Mary, Williamsburg, VA 23187–8795

> When they are exposed to activated forms of copper metal at moderate temperatures, allylic chloride models for structural segments in PVC experience rapid reductive coupling and thereby are converted into mixtures of diene hydrocarbons. The activated copper that promotes this process can be either (a) a slurry formed by the reduction of $CuI·P(Bu)_3$ with lithium naphthalenide or (b) a film produced by the thermal decomposition of copper(II) formate. Both the slurry and, apparently, the $Cu°$ formed by pyrolysis also cause the crosslinking of PVC itself. Moreover, at 200 °C, the crosslinking of solid polymer samples is promoted by copper powder of very high purity (99.999%). Since PVC crosslinking causes smoke suppression, the results of this study suggest that $Cu°$-promoted reductive coupling will tend to inhibit smoke formation when the polymer burns.

Many compounds of copper are well-known to be effective smoke suppressants for poly(vinyl chloride) (PVC) (*1*). Even so, their commercial use for that purpose apparently has not been extensive. Available evidence shows that, in general, copper additives cause "early crosslinking" of the polymer during its thermolysis (*2*). This process tends to inhibit smoke by retarding the formation of volatile aromatics such as benzene that can burn in the vapor phase (*3*). However, the crosslinking chemistry of copper compounds is not understood completely and is controversial for that reason.

Metal-containing additives for PVC frequently retard both smoke and flame by acting as Lewis-acid catalysts for reactions that lead to crosslinking (*2-4*). Unfortunately, when the Lewis acidities of such additives are high, these compounds also tend to cause char breakdown when high temperatures are reached (*5*). This cationic cracking process (*4*) may enhance flame spread enormously by generating volatile aliphatic fragments that burn with great facility (*4,5*).

0097–6156/95/0599–0118$12.00/0
© 1995 American Chemical Society

Figure 1 summarizes results obtained in a study of the cracking reaction (*4*) that was carried out with the Lewis-acid smoke suppressant (*2—6*), MoO_3, and with the much stronger Lewis acid, MoO_2Cl_2, that can be formed *in situ* from MoO_3 and the HCl that is generated by PVC thermolysis. Note that both of the organic substrates gave major amounts of low-molecular-weight alkanes, which are well-known to be excellent fuels.

Copper compounds are, in general, not strong Lewis acids. Thus is it not surprising that previous work has revealed significant differences between their behavior and that of MoO_3 in reactions with organic substances that are models for PVC (*2,6*). Other studies have suggested that the high activity of copper-containing smoke suppressants may derive, in part, from unusually high crosslinking-to-cracking ratios that result from the presence of Lewis acids that are weak (*7,8*). A further mechanistic possibility is that copper additives promote the crosslinking of PVC by reductive coupling (*9*). Such a process can be represented by the redox cycle shown as equations 1 and 2 (*4,9*), in which all of the metal ligands are not

$$2RCl + 2Cu^n \rightarrow R\text{-}R + 2Cu^{n+1}Cl \qquad (1)$$

$$2Cu^{n+1}Cl + \text{-}CH=CH\text{-} \rightarrow \text{-}CH=CCl\text{-} + HCl + 2Cu^n \qquad (2)$$

depicted. Operation of this mechanism would be especially attractive on technological grounds, because it does not require Lewis acidity and thus avoids the problem of cationic char cracking.

No conclusive evidence has been reported for the reductive coupling of PVC by copper additives. However, it has been known for some time that, in burning PVC, higher-valent copper is reduced to $Cu^°$ (*9,10*), which should be produced in a highly active state and thus should be especially effective in reaction 1.

The present chapter reports the use of activated copper metal to effect the reductive coupling of model organic chlorides and the crosslinking of the polymer. Our results suggest that the reductive coupling of PVC is indeed a viable process that is worthy of further study within the context of smoke suppression and fire retardance.

Experimental Section

General. Copper(II) formate (Pfaltz & Bauer) and the other starting materials were either used as received or purified by conventional methods. The PVC (Aldrich; inherent viscosity, 1.02) contained no stabilizers or other additives. Activated copper slurry was prepared by reducing the $CuI \cdot P(Bu)_3$ complex in THF or ether with a 10 mol % excess of lithium naphthalenide, according to a published method (*11,12*). Compounds not obtained from commercial sources were synthesized by standard procedures. All reactions were carried out under argon, and reaction products were identified from their GC retention times and mass spectral cracking patterns, using pure reference substances for comparisons as required. The GC/MS analyses were performed with a Hewlett-Packard System (Model 5988A) equipped with a fused-silica capillary column containing dimethyl- and diphenylpolysiloxane in a ratio of 95:5.

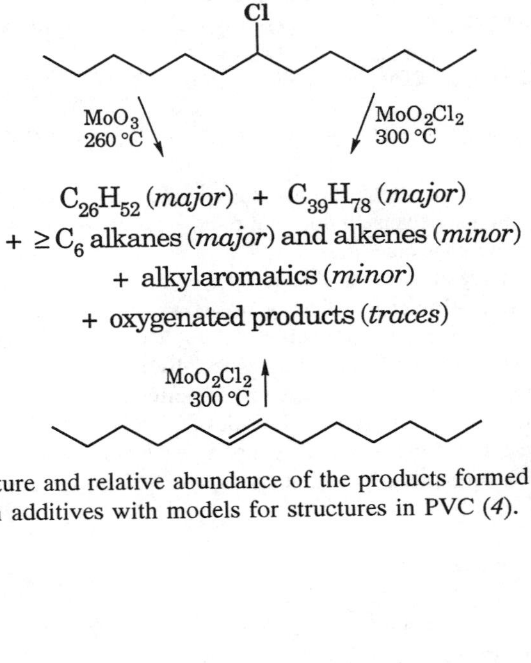
Cl
MoO3
260 °C
MoO2Cl2
300 °C
C26H52 (major) + C39H78 (major)
+ ≥ C6 alkanes (major) and alkenes (minor)
+ alkylaromatics (minor)
+ oxygenated products (traces)
MoO2Cl2
300 °C

Figure 1. Nature and relative abundance of the products formed in reactions of molydenum additives with models for structures in PVC (*4*).

Reactions of Activated Copper Slurry with Model Organic Chlorides. A model organic chloride (10 mmol) or a mixture of two chlorides (5 mmol of each) was added via syringe to a copper slurry (8 mL) prepared by reducing 10 mmol of CuI·P(Bu)$_3$. After 5 min of reaction at constant temperature, the mixture was subjected to GC/MS analysis.

Reactions of Activated Copper Film with Model Organic Chlorides. Activated copper film was obtained by decomposing copper(II) formate (1.50 g, 9.8 mmol) at 200 °C for 2—3 min. When the flask containing the film had cooled to room temperature, a model organic chloride (9.8 mmol) or a mixture of two chlorides (4.9 mmol of each) was injected via syringe. The reaction mixture was subjected to GC/MS analysis after a reaction time of 1 min.

Reactions of Activated Copper Slurry with PVC in Solution. Activated copper slurry in THF (15 mL), prepared by reducing 20 mmol of CuI·P(Bu)$_3$, was added by syringe to a refluxing solution of PVC (2.00 g, 32 mmol of monomer units) in the solvent selected (35 mL for THF, 50 mL for the other solvents). In all of the experiments except those where THF was the only solvent used, the THF was allowed to boil off rapidly at the start of the reaction period. After the chosen reflux interval, the insoluble material was isolated by filtration or decantation, washed thoroughly and repeatedly with several portions of concentrated ammonium hydroxide (in order to remove copper salts), and subjected to Soxhlet extraction with THF for 24 h. The resultant insoluble polymer was dried under vacuum for 24 h at 60 °C. Its weight was then determined and used to calculate the degree of gelation that had occurred.

Dehydrochlorination Rates and Gel Contents of Thermally Degraded Samples of Solid PVC. By using an agate mortar and pestle, a 1.00-g sample of PVC was intimately mixed with 0.10 g of a copper additive. The mixture was subjected to dehydrochlorination at 200±2 °C under flowing argon, and the rate of acid evolution was monitored by acid-base titrimetry performed with a Brinkmann Metrohm 702 SM Titrino apparatus. According to the procedure described above, degraded samples were extracted with hot THF, dried, and weighed in order to determine their gel percentages. The weights were corrected for the presence of insoluble copper species.

Results and Discussion

Reactions of Activated Copper with Model Organic Chlorides. Activated copper slurry was obtained from the $Li^+C_{10}H_8^{-\cdot}$ reduction of CuI·P(Bu)$_3$ (*11,12*). In our hands, it was found to effect the rapid reductive coupling of a wide variety of simple organic halides (chlorides, bromides, iodides) in THF or ether at temperatures ranging from 0 to 67 °C. In general, the results of these experiments were consistent with those obtained in a similar study that was reported (*12*) while our work was in progress.

Allylic chloride structures occur in undegraded PVC and also are formed when the polymer experiences degradation (Starnes, W. H., Jr.; Girois, S. *Polymer*

Yearbook, in press, and references cited therein). Thus we were especially pleased to find that the copper-promoted coupling of allylic chlorides was quite facile under the mild conditions that we used. Some representative results obtained with allylic chlorides appear in the third column of Table I.

Table I. Reductive Coupling Products Obtained from Allylic Chlorides and Activated Copper

chloride(s)	product(s)[b]	yields,[a] % Cu° slurry[c]	yields,[a] % Cu° film
3-chloro-1-butene	C_8H_{14}	19±2	20±2
trans-4-chloro-2-pentene	$C_{10}H_{18}$	36±3	24±2
1:1 (mol/mol) mixture of	C_8H_{14}	24±2	26±1
3-chloro-1-butene and	C_9H_{16}	30±1	12±3
trans-4-chloro-2-pentene	$C_{10}H_{18}$	23±2	24±2

SOURCE: Reprinted with permission from ref. 13. Copyright 1994.
[a]GC area percentages based on amount(s) of starting chloride(s); values shown are averages derived from duplicate runs. [b]Mixtures of several isomers. [c]Values obtained from experiments performed in THF at ca. 65 °C.

Control experiments showed that our organic halides could be reductively coupled by lithium naphthalenide when no copper slurry was present. However, reductive coupling also was effected (albeit in lower yields) by slurries that had been freed of any unchanged lithium naphthalenide by repetitive washing with fresh solvent. The yield reductions observed in these cases may have resulted primarily from decreases in slurry activity (11,12) caused by the washing process.

Encouraged by these findings, we then performed similar experiments with copper films that had been generated by the pyrolysis of copper(II) formate (14). These films also were effective coupling promoters, as is shown by the data in the fourth column of Table I.

Reaction of Activated Copper Slurry with PVC in Solution. The ability of the copper slurry to cause gelation of PVC was studied by using solutions of the polymer in the solvents that are listed in Table II. Gel contents were obtained by

Table II. Gel Contents of PVC Treated with Cu° Slurry in Solution

solvent	temp,°C	time, h	gel,[a]%
THF	66±2	2.0	79±2
anisole	155±2	2.0	92±2
o-dichlorobenzene	174±2	2.0	88±2
phenyl ether	257±2	0.5	90±2

SOURCE: Reprinted with permission from ref. 13. Copyright 1994.
[a]Mean values obtained from duplicate runs.

determining the weight of the polymer that was not dissolved by hot THF, and control runs were used to show that *no* gelation occurred when the copper slurry was absent. The high gel percentages reported in the table are striking in view of the rather low molar concentrations of polymer and copper that were used. Also noteworthy are the relatively minor effects of temperature upon the extent of gelation at the three lowest temperatures studied.

We have no direct information about the chemical nature of the crosslinks that were formed in the experiments of Table II. However, in view of the results of our model-compound studies, it certainly seems reasonable to believe that a major role in the crosslinking process was played by reductive coupling.

Reaction of PVC with Copper Additives in the Solid State. Preliminary work in our laboratory had shown that a highly purified form of copper powder (nominal purity, 99.999%; supplied by Aldrich Chemical Co.) was able to promote the reductive coupling of model allylic chlorides (Jeng, J. P., unpublished observations). We therefore wished to compare the gelation effectiveness of this form of copper with that of the copper resulting from copper(II) formate pyrolysis. The requisite experiments were carried out with solid samples, and the results are exemplified in Figure 2. They show that both of the additives caused rapid gel formation which, in the case of the formate, evidently was brought about by the Cu° formed *in situ*.

Rates of acid evolution were determined with solid PVC samples under the conditions used to acquire the data that are plotted in Figure 2. Figure 3 shows some rate curves that typify those obtained. The initial rapid rate observed with copper(II) formate may have resulted from the evolution of formic acid formed by decomposition of the additive, but that point was not established conclusively. Nevertheless, it is apparent that the formate caused no significant rate enhancement after ca. 20—40 min and that the commercial copper powder caused a modest rate reduction throughout the reaction period. Those observations are not in accord with Lewis-acid catalysis (which would have increased the rate (*4*)), but the effect of the powder is consistent with the mechanism of equations 1 and 2, which predicts a rate diminution caused by the coupling of allyl groups (*4*). However, we regard this evidence for that mechanism as being suggestive instead of decisive.

It is interesting to note that copper(II) formate has been suggested to have potential utility as a thermal stabilizer for vinylidene chloride copolymers (*15*). The effects of the thermal decomposition of this additive on its stabilizing properties are, at present, unclear.

Conclusions

Activated forms of Cu^0 cause the reductive coupling of allylic chlorides that are models for structures in PVC. The coupling occurs quite rapidly under very mild conditions and has been observed with (a) a copper slurry formed by the reduction of $CuI \cdot P(Bu)_3$ with lithium naphthalenide and (b) a copper film resulting from the pyrolysis of copper(II) formate. Both the slurry and, apparently, the film cause the crosslinking of PVC itself. In the case of the slurry, this process has been observed with solutions of the polymer in various solvents at temperatures ranging from ca. 66 to 257 °C. In reactions of copper(II) formate with solid PVC, accelerated crosslinking occurs at the decomposition temperature of the additive (200 °C) and

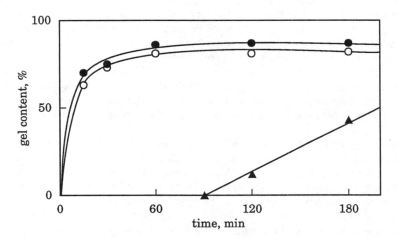

Figure 2. Gel contents of solid PVC samples degraded at 200±2 °C under argon: ▲, no additive; ○, with Cu° powder (purity, 99.999%); ●, with Cu(II) formate. See text for details. (Reproduced with permission from ref. 13. Copyright 1994 Business Communications Co.)

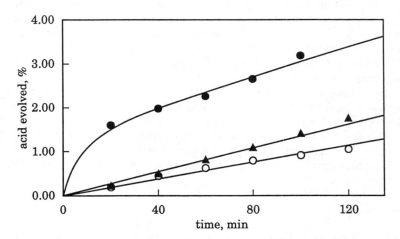

Figure 3. Acid evolution rate curves for the degradation of solid PVC samples at 200±2 °C under argon: ▲, no additive; ○, with Cu° powder (purity, 99.999%); ●, with Cu(II) formate. See text for details. (Reproduced with permission from ref. 13. Copyright 1994 Business Communications Co.)

thus is believed to result from the presence of copper metal formed *in situ*. Moreover, at this temperature, the crosslinking of solid polymer samples is promoted by a commercial form of copper powder that has a very high purity (99.999%). The crosslinking of PVC is well-known to cause smoke suppression. Thus the results of the present study suggest that copper-promoted reductive coupling will tend to prevent the formation of smoke during PVC combustion.

Acknowledgment

We thank the International Copper Association for partial support of this work.

Literature Cited

1. Kroenke, W. J. *J. Appl. Polym. Sci.* **1981**, *26*, 1167.
2. Lattimer, R. P.; Kroenke, W. J.; Getts, R. G. *J. Appl. Polym. Sci.* **1984**, *29*, 3783.
3. Starnes, W. H., Jr.; Edelson, D. *Macromolecules* **1979**, *12*, 797.
4. Starnes, W. H., Jr.; Wescott, L. D., Jr.; Reents, W. D., Jr.; Cais, R. E.; Villacorta, G. M.; Plitz, I. M.; Anthony, L. J. In *Polymer Additives*; Kresta, J. E., Ed.; Plenum: New York, NY, 1984; p 237.
5. Edelson, D.; Lum, R. M.; Reents, W. D., Jr.; Starnes, W. H., Jr.; Wescott, L. D., Jr. In *Proceedings, 19th International Symposium on Combustion*; The Combustion Institute: Pittsburgh, PA, 1982; p 807.
6. Wescott, L. D., Jr.; Starnes, W. H., Jr.; Mujsce, A. M.; Linxwiler, P. A. *J. Anal. Appl. Pyrolysis* **1985**, *8*, 163.
7. Starnes, W. H., Jr.; Huang, C.-H. O. *Polym. Prepr. (Am. Chem. Soc., Div. Polym. Chem.)* **1989**, *30* (1), 527.
8. Huang, C.-H. O.; Starnes, W. H., Jr. In *Proceedings, 2nd Beijing International Symposium on Flame Retardants*; Geological Publishing House: Beijing, China, 1993; p 168.
9. Lattimer, R. P.; Kroenke, W. J. *J. Appl. Polym. Sci.* **1981**, *26*, 1191.
10. Lattimer, R. P.; Kroenke, W. J. In *Analytical Pyrolysis: Techniques and Applications*; Voorhees, K. J., Ed.; Butterworths: Woburn, MA, 1984; p 453.
11. Ebert, G. W.; Rieke, R. D. *J. Org. Chem.* **1988**, *53*, 4482.
12. Ginah, F. O.; Donovan, T. A., Jr.; Suchan, S. D.; Pfennig, D. R.; Ebert, G. W. *J. Org. Chem.* **1990**, *55*, 584.
13. Starnes, W. H., Jr.; Jeng, J. P.; Terranova, S. A.; Bonaplata, E.; Goldsmith, K.; Williams, D. M.; Wojciechowski, B. J. In *Proceedings, 5th Annual BCC Conference on Flame Retardancy;* Business Communications Co.: Norwalk, CT, 1994; in press.
14. Kőrösy, F. *Nature* **1947**, *160*, 21.
15. Howell, B. A.; Rajaram, C. V. *J. Vinyl Technol.* **1993**, *15*, 202.

RECEIVED November 28, 1994

Chapter 9

Effect of Some Tin and Sulfur Additives on the Thermal Degradation of Poly(methyl methacrylate)

Jayakody A. Chandrasiri and Charles A. Wilkie[1]

Department of Chemistry, Marquette University, Milwaukee, WI 53233

A detailed understanding of the course of the reaction between an additive and a polymer will lead to useful information that may enable the effective design of a flame retardant for that polymer. In this study we present an interpretation of the reaction between poly(methyl methacrylate) and tetrachlorotin, phenyltin trichloride, diphenyltin dichloride, triphenyltin chloride, tetraphenyltin, and diphenyl disulfide.

The thermal degradation of poly(methyl methacrylate), PMMA, has been studied for many years. It has been well understood that essentially the only product of this degradation is monomer but it has never been clear exactly how this monomer is produced. The problem is that if the initial step is a random scission, then both a primary macroradical and a tertiary macroradical are produced and it is not reasonable to assume that these will both depolymerize at the same rate to give the same products. Recently Kashiwagi (*1*) and Manring (*2*) have proposed solutions to this problem. Kashiwagi has suggested that random main chain scission occurs first and that the tertiary macroradical depolymerizes as expected. The primary macroradical, on the other hand, loses the side chain to give a macromolecule with an unsaturated chain end which degrades to monomer. Manring suggests that the initial step is cleavage of the side chain, producing only a tertiary macroradical which then degrades to give monomer.

Over the past several years this research group has examined the effects of a wide variety of additives upon the thermal degradation of poly(methyl methacrylate). Additives that have been investigated include red phosphorus (*3-4*), Wilkinson's catalyst, ClRh(PPh$_3$)$_3$, (*5-6*), Nafions (*7*), copolymers of 2-sulfoethyl methacrylate and methyl methacrylate (*8*), and a variety of transition metal halides, MnCl$_2$, CrCl$_3$, FeCl$_2$, FeCl$_3$, NiCl$_2$, CuCl$_2$, and CuCl (*9-11*). The goal of this

[1]Corresponding author

work has been to understand, in detail, how a wide variety of additives effect the thermal degradation of PMMA so that this information may be used to design an additive that will prevent the thermal degradation of the polymer. In this paper we report on some of the more recent investigations using some tin additives (*12-13*) and diphenyl disulfide (Chandrasiri, J. A.; and Wilkie, C. A., *Polym. Degrad. Stab.*, in press).

EXPERIMENTAL

Thermal degradation is usually studied in this laboratory in the absence of oxygen by pyrolysis in sealed tubes at the appropriate temperature. Typically the additive alone, or a 1:1 by mass mixture of the additive with polymer, is placed in a sealed tube and thoroughly evacuated. After the tube has been sealed off from the vacuum line, it is placed in the oven at 375°C for some time period. The tube is then carefully opened and the contents are separated into four fractions and analyzed by conventional techniques. The four fractions are non-condensable gases (methane and CO are typically present in small amounts in these studies), condensable gases, chloroform solubles, and chloroform insolubles. The gases are separated by conventional vacuum line techniques (*14*). The tube is then removed from the vacuum line and treated with chloroform to dissolve materials that are chloroform soluble. The remaining material is denoted as chloroform insoluble but may be further separated by treatment with other solvents. The gases are identified by infrared spectroscopy while the chloroform solubles may be identified by nuclear magnetic resonance (GE 300 Omega instrument), GC-MS (Hewlett Packard 5890 gas chromatograph with a Hewlett-Packard 5970 mass selective detector), infrared spectroscopy (Mattson 4020 Galaxy Series Spectrometer), etc. Chloroform insolubles are identified by similar techniques.

Degradation of the Additives Alone. Both tetraphenyltin and tin tetrachloride are thermally stable and undergo no degradation when they are heated to 375°C in a sealed tube. The major products that are obtained when phenyltin trichloride is pyrolyzed for 25 hours are: 34% $SnCl_4$, 38% $SnCl_2$, 15% bi- and poly-phenyls, 3% chlorobenzene, 10% benzene, and 1% elemental chlorine. From diphenyltin dichloride one obtains 55% $SnCl_2$, 29% bi- and poly-phenyls, and 16% benzene. From triphenyltin chloride the products are 23% $SnCl_2$, 17% elemental tin, 40% bi- and ter-phenyls, and 20% benzene (*15*). The degradation of diphenyl disulfide produces thiophenol, and diphenyl sulfide as major products with smaller amounts of elemental sulfur, diphenyl trisulfide, diphenyl tetrasulfide, and thianthrene (Chandrasiri, J. A.; and Wilkie, C. A., *Polym. Degrad. Stab.*, in press).

Degradation of PMMA in the Presence of Phenyltin Trichloride. The pyrolysis of a 1:1 by mass mixture of additive and polymer produces 32% condensable gases and 49% chloroform insoluble. The major condensable products are carbon dioxide, water, benzene, methyl chloride, isobutyric acid, and methacrylic acid. Smaller amounts of monomeric methyl methacrylate, an anhydride, and other products were also found. The chloroform insoluble fraction is separated to obtain

17% tin (II) chloride and 32% completely insoluble, non-volatile residue, identified as char.

Degradation of PMMA in the Presence of Diphenyltin Dichloride. The same products were observed as for the trichloride; 47% condensables including CO_2, benzene, methyl chloride, water, methacrylic acid, and isobutyric acid and minor amounts of methyl methacrylate, methyl 2-methylbutyrate, an anhydride, and other products; 12% $SnCl_2$, and 32% non-volatile char.

Degradation of PMMA in the Presence of Triphenyltin Chloride. The pyrolysis of this mixture produced 49% of volatiles containing CO_2, benzene, water, methacrylic acid, and isobutyric acid; the minor products consisted of methyl methacrylate, methyl 2-methylbutyrate, methyl isobutyrate, methanol, methyltin trichloride, and an anhydride; a trace of methyl chloride was also obtained. The chloroform insoluble fraction contains 6% $SnCl_2$ and 35% non-volatile char.

Degradation of PMMA in the Presence of Tin Tetrachloride. The condensable products included methyl chloride, carbon dioxide, HCl, water, methacrylic acid, isobutyric acid, methyltin trichloride with minor amounts of methyl methacrylate, methyl isobutyrate and anhydrides. When the amount of $SnCl_4$ is increased the production of the two methyl esters is reduced. In addition to $SnCl_2$, a six-membered cyclic anhydride, and char were obtained.

Degradation of PMMA in the Presence of Tetraphenyltin. The condensable products observed in this reaction consist of 17% benzene, 29% methyl methacrylate, 2% methyl isobutyrate, 3% toluene, 3% methyl 2-methylbutyrate, CO_2, and trace amounts of other materials. The chloroform soluble fraction was 5% of the total and consisted of methyl esters of phenyl substituted propionic acids, methylphenyltin compounds, and other products. The chloroform insoluble fraction contained 11% elemental tin and 20% char.

Degradation of PMMA in the Presence of Diphenyl Disulfide. The condensable gases consist of 6% of a mixture of CO_2, CS_2, and methyl formate, 12% methyl isobutyrate, 13% thiophenol, 4% methyl methacrylate, and 4% thioanisole and traces of methylated benzenes. The chloroform soluble fraction accounted for 47% of the total and contained diphenyl sulfide, methyl phenyl disulfide, and methyl 2-methyl-3-(phenylthio)propionate). The insoluble fraction accounted for 8% of the total and contained sulfur.

Oxygen Index. The oxygen index for these blends of additives and polymers were measured on a home built apparatus using bottom ignition as previously described (6).

RESULTS AND DISCUSSION

Pyrolysis of Additives Alone. Six additives have been examined in this study, $SnCl_4$, $PhSnCl_3$, Ph_2SnCl_2, Ph_3SnCl, Ph_4Sn and Ph_2S_2. Both tetraphenyltin and tin tetrachloride are quite thermally stable and undergo essentially no degradation when they are heated at 375°C for 2 hours. The other materials are all degraded by this treatment. For the phenyltin chlorides the initial step is the cleavage of a Sn-Ph bond with the formation of a phenyl radical and the corresponding tin-based radical. A detailed description of these degradations has been published (15) but these are the radicals that are initially produced and present in largest concentration and are the species responsible for reactions with the polymer.

The degradation of diphenyl disulfide has been studied for several years (16-22). The initial step is the cleavage of the S-S bond with the formation of two arenethiyl radicals. Subsequent reactions lead to the formation of thianthrene, diphenyl sulfide, thiophenol, and other products. Again, the arenethiyl radical is initially formed and is present in greatest concentration and is responsible for the reactions with PMMA.

Degradation of PMMA in the Presence of Phenyltin Chlorides, Ph_xSnCl_{3-x}.
The degradation of PMMA is significantly effected by the presence of phenyltin trichloride. As noted above, essentially the only product in the degradation of PMMA is monomeric methyl methacrylate. It is surprising to note that methacrylic and isobutyric acids are the dominant products and that only a small amount of methyl methacrylate is observed when the polymer is degraded in the presence of any of the phenyltin chlorides. It is clear that the degradation scheme has been completely changed by the presence of the additive. The presence of phenyl and tin-based radicals has a significant effect on the degradation. A pathway to account for the products observed in this reaction is presented below as Scheme 1. The phenyl radical may interact with the polymer to form toluene and a carboxyl radical on the polymer (eq. 1). The tin-based radical can interact with an ester group of the polymer to give a methyltin chloride and a carboxyl radical (eq. 2). This carboxyl radical may hydrogen abstract to give a methacrylic acid unit (eq. 3), or it can interact with the tin-based radical to form a tin ester (eq. 4). The tin ester can lose a chlorine atom (eq. 5) and the resulting tin based radical can combine with another carboxyl radical (eq. 6).

According to equation 3, methacrylic acid units are produced. The normal degradation pathway for polymethacrylic acid is the loss of water with the formation of anhydrides. Water, anhydrides, and methacrylic acid are identified amongst the products of this reaction but monomeric methacrylic acid is not obtained for the degradation of polymethacrylic acid. When methacrylic acid units are adjacent they may lose water with the formation of anhydrides, however when they are separated it is believed that they will simply depolymerize as is observed for the ester.

The ^{13}C NMR spectrum of the residue shows resonances that may be attributable to carbonyl, aromatic carbons, backbone carbons, and pendant methyl group. The chemical shift of the carbonyl is quite downfield (187.25 ppm) from the normal position and this may be explained as arising from the tin ester. The ^{13}C NMR spectrum of tin (IV) acetate shows the ester carbonyl at 184.7 ppm. The positions of the backbone and methyl resonances are those expected for a methacrylate.

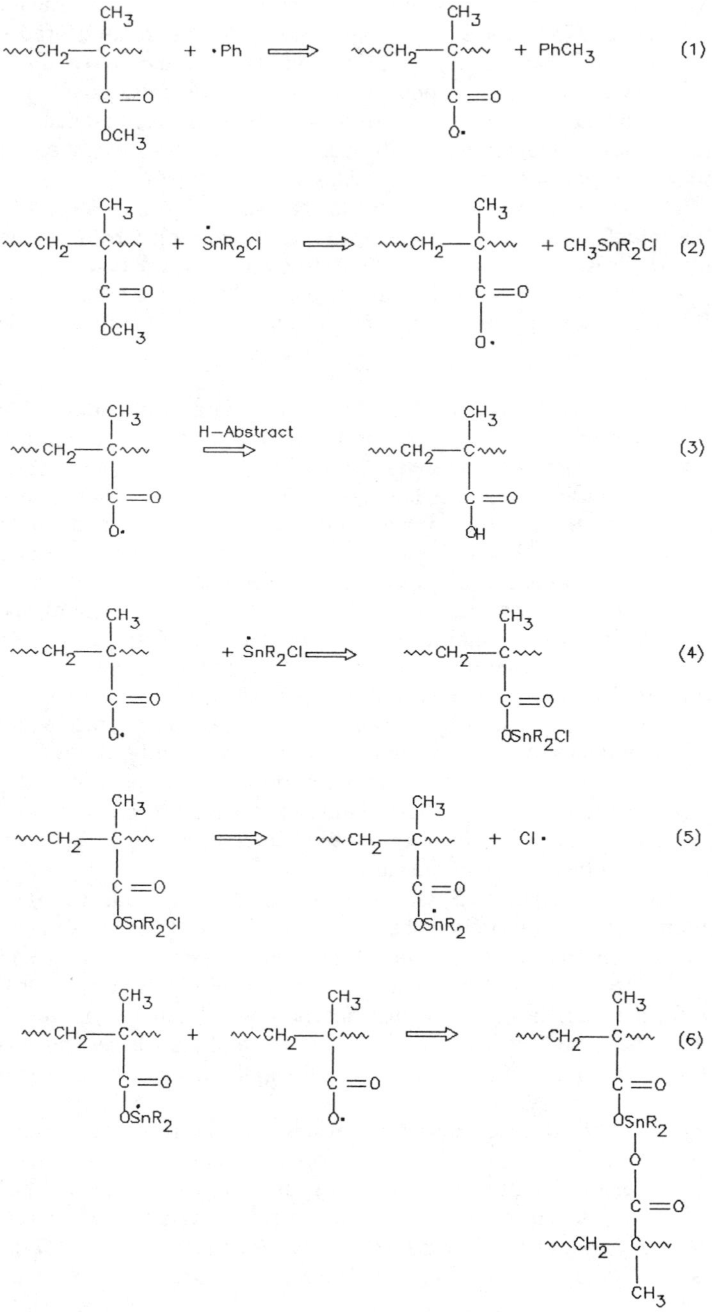

Scheme 1

Degradation of PMMA in the Presence of Tin Tetrachloride. The products of the degradation in the presence of tin tetrachloride are quite similar to those that are obtained in the presence of the phenyltin chlorides. There is, however, an important difference. The phenyltin chlorides undergo thermal degradation when they are heated alone but tin(IV) chloride does not degrade when it is heated alone but is completely consumed when heated with PMMA. A description of the interaction between these materials is shown in Scheme 2. The initial step is coordination of $SnCl_4$ to the carbonyl oxygen of the polymer (eq. 1). Similar interactions occur between thermally stable transition metal halides (9-11, 23-24) and lead to the loss of methyl chloride with the formation of the salt of the transition metal. A similar reaction occurs in this case (eq. 2). The tin ester is not thermally stable and the $SnCl_3$ radical that is produced (eq. 3) can interact with the polymer to form methyltin trichloride and a carboxyl radical (eq. 4). Subsequent reactions are reminiscent of those with the phenyltin chlorides.

Another possible pathway for the interaction of $SnCl_4$ and PMMA is the reaction of the tin compound with the monomer produced from PMMA. When a mixture of MMA and $SnCl_4$ (1.0 g each) is thermolyzed under the same conditions used for the reaction of polymer and tin(IV) chloride, all of the same products are obtained except that the anhydride is not produced. The lack of the anhydride makes it likely that the tin compound interacts directly with the polymer but does not preclude reaction with the monomer. The measured oxygen index is near 40; it is very unlikely that any amount of volatile monomer could be liberated and still have an oxygen index of this value.

Degradation of PMMA in the Presence of Tetraphenyltin. When tetraphenyltin is heated alone, a small amount of degradation occurs with the formation of benzene and elemental tin. It is likely, considering the Sn-Ph bond energy (25), that this bond does undergo some cleavage but the radical must recombine to starting material. In the presence of PMMA the phenyl radical and triphenyltin radical have other opportunities for reaction. This may be described as a mutually-assisted degradation. The acids that are produced with the phenyltin chlorides and tetrachlorotin are not observed and methyl methacrylate and benzene are the major products of this reaction; in addition methyl 2-methylbutyrate, methyl 2-methyl-3-phenylpropionate, and methyl 2-methylene-phenylpropionate are obtained. The details of the interaction are unknown at this time and no scheme can be written to describe this reaction. One would imagine that if phenyl radicals are available, they would interact as seen for the phenyltin chlorides. The absence of methacrylic acid as a product indicates that this reaction does not occur. Some of the products are attributable to reactions of a primary macroradical and this supports the Kashiwagi mechanism for PMMA degradation.

Degradation of PMMA in the Presence of Diphenyl Disulfide. The compounds that are of interest are those that contain both phenyl and methacrylate fragments and only these shall be addressed herein. The initial degradation reaction of diphenyl disulfide is the formation of arenethiyl radicals and these account for the reaction with PMMA. Of these, the most interesting is methyl 2-methyl-3-(phenylthio)propionate and its oligomers. These products arise from the interaction

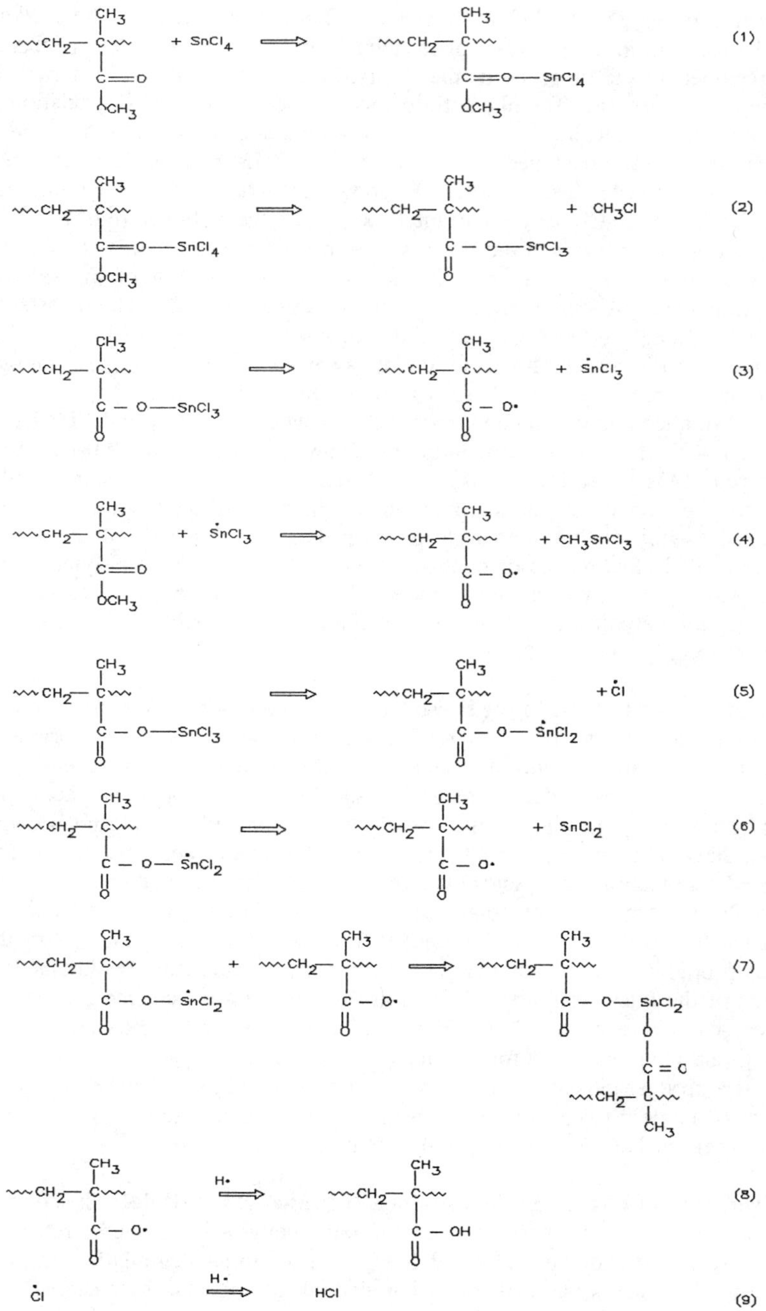

Scheme 2

between the primary macroradical and either the arenethiyl radical or diphenyl disulfide. The formation of thioanisole may arise from an S_H2 reaction of diphenyl disulfide with a methyl radical.

The relatively small amounts of thioanisole, CO, CO_2 together with the formation of ester containing oligomers indicates that diphenyl disulfide inhibits the cleavage of PMMA. Diphenyl disulfide terminates degradation of macroradicals and initiates the formation of arenethiyl radicals. This leads to the high yield of sulfur-containing oligomers and char formation that are observed in this reaction.

Oxygen Index. The oxygen index values that have been obtained for these blends of polymer and additive are shown in Table I. Except for the case of tetrachlorotin, the measured OI values indicate little ability for these additives to function as flame retardants. The improvement that is seen for $SnCl_4$ is remarkable and an additive of this type, but without the disadvantageous properties of $SnCl_4$ such as water instability, may prove useful as a flame retardant.

Table I

Oxygen Index for 1:1 blends of additive and PMMA

Additive	Oxygen Index
$SnCl_4$	40
$PhSnCl_3$	22
Ph_2SnCl_2	26
Ph_3SnCl	24
Ph_4Sn	21
Ph_2S_2	17
PMMA	17

Poly(Methyl Methacrylate) Degradation. Two mechanisms have been suggested to account for the thermal degradation of PMMA. Kashiwagi (*1*) has suggested that the main chain is initially cleaved to produce a primary and a tertiary macroradical. They have proposed that the tertiary macroradical degrades to monomer while the primary macroradical undergoes side chain cleavage which then will degrade to monomer. Manring (*2*) has proposed that side chain cleavage occurs first and that the macroradical that is formed then degrades to monomer. It is significant that primary macroradical are never obtained in the Manring scheme. At least one product in each of the degradations studies must arise from the combination of Kashiwagi's primary macroradical and a radical produced during the degradation; these products are shown in Table II. The observation of these products seems to confirm that main chain cleavage occurs first and is then followed by side chain cleavage; this supports the Kashiwagi mechanism for the thermal degradation of PMMA.

Table II

Products from Primary Macroradical

R-CH$_2$-CH(CH$_3$)COOCH$_3$

R	Additive
CH$_3$	Ph$_x$SnCl$_{4-x}$
Ph	Ph$_4$Sn
PhS	Ph$_2$S$_2$

It is difficult to imagine routes whereby these products could be formed that do not involve the primary macroradical and this offers significant support for the Kashiwagi main chain scission pathway for PMMA degradation.

CONCLUSION

The radical species that are first produced when a thermally unstable additive is used appear to be responsible for all degradation reactions and it is probably unnecessary to know the complete degradation pathway of the additive in order to understand its reaction with the polymer. None of the additives that have been studied appear to be directly useful as flame retardants for PMMA but the knowledge about reaction pathways may prove useful in the design of suitable additives. Specifically good Lewis acids, such as tetrachlorotin, may facilitate the formation of non-volatile, cross-linked ionomers that will provide the necessary stabilization under thermal conditions. The Kashiwagi main chain cleavage mechanism for the degradation is supported by these results.

LITERATURE CITED

1. Kashiwagi, T; Inabi, A.; Hamins, A.; *Polym. Degrad. Stab.*, **1989** *26*, 161.
2. Manring, L.E.; *Macromol.*, **1991**, *24*, 3304.
3. Wilkie, C.A.; Pettegrew, J. W.; Brown, C. E.; *J. Polym. Sci., Polym. Lett. Ed.*, **1991**, *19*, 409.
4. Brown, C. E.; Wilkie, C. A.; Smukalla, J.; Cody Jr.,R. B.; Kinsinger, J. A.; *J. Polym. Sci., Polym. Chem. Ed.*, **1986**, *24*, 1297.
5. Sirdesai, S. J.; Wilkie, C. A.; *J. Appl. Polym. Sci.*, **1989**, *37*, 863.
6. Sirdesai, S. J.; Wilkie, C. A.; *J. Appl. Polym. Sci.*, **1989**, *37*, 1595.
7. Wilkie, C. A.; Thomsen, J. R.; Mittleman, M. L.; *J. Appl. Polym. Sci.*,**1991**, *42*, 901.
8. Hurley, S. L.; Mittleman, M. L.; Wilkie, C. A.; *Polym. Degrad. Stab.*, **1993**, *39*, 345.

9. Wilkie, C. A.; Leone, J. T.; Mittleman, M. L.; *J. Appl. Polym. Sci.*, **1991**, *42*, 1133.
10. Beer, R. S.; Wilkie, C. A.; Mittleman, M. L.; *J. Appl. Polym. Sci.*, **1992**, *46*, 1095.
11. Chandrasiri, J. A.; Roberts, D. E.; Wilkie, C. A.; *Polym. Degrad. Stab.*, **1994**, *45*, 97.
12. Chandrasiri, J. A.; Wilkie, C. A., *Polym. Degrad. Stab.*, **1994**, *45*, 83.
13. Chandrasiri, J. A. Wilkie, C. A., *Polym. Degrad. Stab.*, **1994**, *45*, 91.
14. Shriver, D. F.; Drezdzon, M. A. *The Manipulation of Air-Sensitive Compounds*, Wiley-Interscience, New York, New York, 1986, p. 104.
15. Chandrasiri, J. A.; Wilkie, C. A.; *Appl. Organometal. Chem.*, **1993**, *7*, 599.
16. Graebe, C.; *J. Liebigs Ann. Chem.* **1874**, 174, 177.
17. Schonberg, A.; Mustafa, A. *J. Chem. Soc.*, **1949**, 889.
18. Schonberg, A.; Mustafa, A.; Askar, W. *Science* **1949**, *109*, 522.
19. Mayer, R.; Frey, H. -J. *Angew. Chem. Int. Ed. Engl.* **1964**, *3*, 705.
20. Harpp, D.N.; Kader, H.A.; Smith, R.A. *Sulfur Letters* **1982**, *1*, 59.
21. Stepanov, B.I.; Rodionov, V. Ya.; Chibisova, T.A. *J. Org. Chem. USSR*, **1974**, *10*, 78.
22. Zandstra, P.J.; Michaelsen, J.D. *J. Chem. Phys.* **1963**, *39*, 933.
23. McNeill, I.C.; McGuiness, R.C.; *Polym. Degrad. Stab.*, **1984**, *9*, 167.
24. McNeill, I.C.; McGuiness, R.C.; *Polym. Degrad. Stab.*, **1984**, *9*, 209.
25. Chambers, D. B.; Glocking, F.; Weston, M. J., *J. Chem. Soc., A*, **1967**, 1759.

RECEIVED December 28, 1994

Chapter 10

Effect of Zinc Chloride on the Thermal Stability of Styrene-Acrylonitrile Copolymers

Sang Yeol Oh, Eli M. Pearce[1], and T. K. Kwei

Department of Chemistry and Polymer Research Institute, Polytechnic University, Brooklyn, NY 11201

The effect of zinc chloride on the thermal stability of styrene acrylonitrile copolymers was studied. Our present work showed that upon the addition of zinc chloride to styrene acrylonitrile copolymers, the initial thermal stability was decreased significantly. The high temperature FT-IR study for investigating the chemistry of char formation showed that zinc chloride complexed with the nitrile group, and this, in turn, induced a modified degradation mechanism leading to thermally stable triazine ring formation which crosslinked the main chains.

Styrene-acrylonitrile copolymers are strong, rigid, and transparent, they have excellent dimensional stability and high craze resistance, and good solvent resistance (1). But their flammability requires retardation for designed end uses (2). For a better understanding of the thermal degradation of styrene acrylonitrile copolymers, several studies have appeared in the literature (3-6), but limited information has been available on the degradation processes associated with a high char forming reaction. On the other hand, it was also interesting to note studies of catalyst effects on the thermal degradation process for polymers containing nitrile groups. Several studies (7-10) showed that polymers end capped with nitriles or having nitriles pendant from the main chain could be crosslinked by a catalytic trimerization of the nitrile to thermally stable triazine rings.

Recent studies by us (12) on an unsubstituted aromatic polyamide and nitrile substituted aromatic polyamide with zinc chloride as an additive showed a significant char yield increase. The reason for the increased char formation is unknown, but we suggested that this could be due to crosslinking reactions, and/or a Houben-Hoesch type of reaction, and/or the trimerization reaction of the nitrile groups at elevated temperature (11,12).

[1]Corresponding author

In this work, the effect of zinc chloride on the thermal stability and char yield of the styrene acrylonitrile copolymers was studied. The chemistry of char formation was also investigated by employing a high temperature FT-IR technique.

EXPERIMENTAL

Purification of Chemicals. Styrene monomer (Aldrich Chemical Co.) was washed with 10% aqueous NaOH solution and purified by vacuum distillation in the presence of CuCl (40°C 18-20 *mm*-Hg). Acrylonitrile monomer (Aldrich Chemical Co.) was washed with 5% aqueous H_2SO_4 and purified by distillation under nitrogen just before use. 2,2'-Azobisisobutyronitrile(AIBN) (Polysciences Inc.) was purified by recrystallization from methanol. Zinc chloride (A.C.S. reagent grade, Aldrich Chemical Co.) was used after vacuum drying at 80°C overnight.

Random copolymers. Styrene acrylonitrile random copolymers of varying compositions were obtained by radical copolymerization at 60°C using AIBN as an initiator. The reaction time was controlled so that the conversion did not exceed 10%. After reaction, the mixture of copolymer and monomer was diluted with acetone, precipitated into methanol and dried. The copolymers were further purified by dissolution and reprecipitation in acetone and methanol, respectively.

Alternating copolymer. An alternating copolymer of acrylonitrile and styrene in the presence of dry zinc chloride was prepared as follows. Zinc chloride (37.5 *g*) was placed in a 250 *ml* single neck round bottom flask equipped with a connecting tube having a stopcock and heated at 150°C for one hour under vacuum. Acrylonitrile monomer (AN) (175 *ml*) was added to this flask and the zinc chloride was dissolved by shaking at 50-60°C. An excess of acrylonitrile was then distilled off under reduced pressure for 14 hours to yield the AN-$ZnCl_2$ complex. The composition of the complex was determined from the weights of zinc chloride initially used and of the resulting complex. Styrene monomer (35.6 *g*) was added to this complex under a N_2 blanket. The mixture spontaneously copolymerized after 1.5 hour at 40°C providing a conversion of 3.1%. The copolymer was purified by repeated precipitation from acetone solution into methanol. Complete elimination of $ZnCl_2$ was confirmed by an acetone solution of dithizone (*13*).

Preparation of SAN Containing Additive. Addition of zinc chloride: To a solution of the copolymer in acetone was added a solution of zinc chloride in acetone with vigorous stirring. The resulting copolymer solution was cast as a thin film on a glass plate. Copolymers containing various amount of zinc chloride were dried at 100°C under vacuum overnight.

Addition of hydrogen chloride: Hydrogen chloride was bubbled for 1 hour into a solution of the copolymer in acetone. The bubbled solution was stored at -5°C for two weeks. After storage, the solution was cast as a thin film on the glass plate and dried at 100°C under vacuum overnight.

Houben-Hoesch Synthesis. (1) Experiment 1A: Dry hydrogen chloride (1.0 *mole*) was passed for two hours into a stirred mixture of 0.1 *mole* of benzonitrile and 40 *ml* of anhydrous ethyl ether cooled to 0°C for 4 *hr*., after which 0.1 *mole* of resorcinol was added and the mixture was held for 14 days at 5°C. (2) Experiment 1B: 0.1 *mole* of benzonitrile and 0.03 *mole* of zinc chloride was dissolved in 40 *ml* of ether. After standing 4 *hr*. at 0°C, 0.1 *mole* of resorcinol was added and stored for 14 days at -5°C. (3) Experiment 2A and 3A: 3.9 *g* of alternation SAN and 1.0 *g* of zinc chloride were dissolved in 40 *ml* of dimethylformamide(DMF) and chloroform, respectively. After cooling to 0°C, dry hydrogen chloride was passed into the solution for 1 *hr*. The mixture was held for 14 days at -5°C. (4) Experiment 2B and 3B: A similar procedure as Exp. 2A and Exp. 2B was employed without hydrogen chloride. (5) Experiment 4A and 5A: Dry hydrogen chloride was passed for 1 hour into a stirred mixture of 3.1 *g* of benzonitrile and 1.4 *g* of zinc chloride in 20 *ml* of DMF and chloroform, respectively, and cooled to 0°C. After standing 3 hour at 0°C, 3.65 *g* of polystyrene dissolved in 20 *ml* of DMF and chloroform, respectively, was added. The mixture was stored for 14 days at -5°C. (6) Experiment 4B; A similar procedure as Exp 4A was employed without hydrogen chloride.

CHARACTERIZATION OF MATERIALS

Infrared spectroscopy. FT-IR for the samples was performed on a Digilab FTS-60 spectrometer by a thin film technique. High temperature FT-IR studies were performed using an Accuspec Model 2000 temperature, high/low pressure multimode FT-IR cell which was set in the Digilab FTS-60. For the samples for the high temperature studies, polymer solutions in acetone were coated on an aluminum plate and installed at the end face of the heating probe after the solvent was evaporated slowly at room temperature and finally dried overnight at 100°C under vacuum.

Nuclear Magnetic Resonance Spectroscopy: Proton NMR spectra were determined on a Varian EM 390 spectrometer, and the results are reported in ppm downfield from a tetramethylsilane internal reference using deuterated acetone as a solvent. Carbon-13 NMR spectra were measured at 22.5 MHz on a JEOL FX90Q spectrometer. Scans were accumulated using complete proton decoupling with a pulse angle of 90° and a pulse interval of 1 *sec*. The samples were dissolved in deuterated acetone at a concentration of 10 *mg/ml*. Tetramethylsilane was used as an internal reference. Typically, 15,000 scans were accumulated.

Thermal analysis. A Dupont 910 differential scanning calorimeter connected to a Dupont 1090 thermal analyzer was used to study the transition data and a Dupont 951 thermogravimetric analyzer connected with a Dupont 1090 thermal analyzer was employed to measure the initial decomposition temperatures, weight loss, char yield, and the temperature where maximum rate of weight loss occurred.

Viscosity measurement. Inherent viscosities of all polymers were determined in dimethyl formamide at 25°C using a Ubbelohde viscometer. The concentration of solutions was 0.5 *g/dl*.

RESULT AND DISCUSSION

Characterization of Polymer Structure, IR Spectroscopy. Abbreviated names of various copolymers under study are listed in Table I. A typical IR spectrum of the polymers is shown in Figure 1. The polystyrene spectrum showed benzene ring related bands at 3083, 3061, and 3026 cm^{-1}, —CH$_2$ stretching modes at 2848 and 2924 cm^{-1}, benzene ring vibration modes at 1600 and 1493 cm^{-1}, —CH$_2$ bending mode at 1452 cm^{-1}, —CH bending mode at 1365 cm^{-1}, and hydrogen bending mode in benzene ring at 1028 cm^{-1} (*14*).

The spectrum of styrene acrylonitrile copolymer showed a nitrile characteristic band at 2235 cm^{-1} and —CH$_2$ stretching modes which occurred at higher frequencies than those of polystyrene. The relative peak intensity of benzene ring vibration at 1493 cm^{-1} to that of —CH$_2$ bending mode at 1452 cm^{-1} was reversed due to the decreased amount of benzene ring moiety. All other characteristic peaks appeared at similar positions to those of polystyrene.

Table I. Abbreviated Names of Various Copolymers Under Study

Abbreviated Name	Full Name
PS	Polystyrene
PS10	PS containing 10 % of zinc chloride in weight
SAN	Styrene acrylonitrile copolymer
SANXX	SAN containing XX mole % of acrylonitrile
SANXXYY	SAN containing XX mole % of acrylonitrile and YY weight % of zinc chloride
ALT	Alternating SAN
ALTXX	ALT containing XX weight % of zinc chloride
ALT/HCL	ALT bubbled with hydrogen chloride gas

When zinc chloride was added to styrene acrylonitrile copolymer, the characteristic peak of the nitrile group at 2235 cm^{-1} was separated into two peaks since the peak for the nitrile group complexed with zinc chloride shifted to a higher frequency. Imoto et al. (*15*) reported with the aid of the results of Schrauzer (*16*) and Kettle et al. (*17*) that the nitrile group in acrylonitrile forms a coordination type complex with zinc chloride. They (*15-17*) also pointed out that if the nitrile group forms a complex, its nitrile band shifts to higher frequency.

To study the effect of the complex ratio on the relative intensity of nitrile peaks between "free" and "complexed" with zinc chloride, different amounts of zinc chloride were added to the alternating styrene acrylonitrile copolymer. As the amount of zinc chloride added was increased, the relative peak intensity of "complexed" to "free" nitrile group increased gradually (see Figure 2.) Also both of the nitrile peaks gradually shifted to higher frequency, i.e., from 2235 cm^{-1} to 2239 cm^{-1} for "free" nitrile group and from 2277 cm^{-1} to 2284 cm^{-1} for "complexed" nitrile groups.

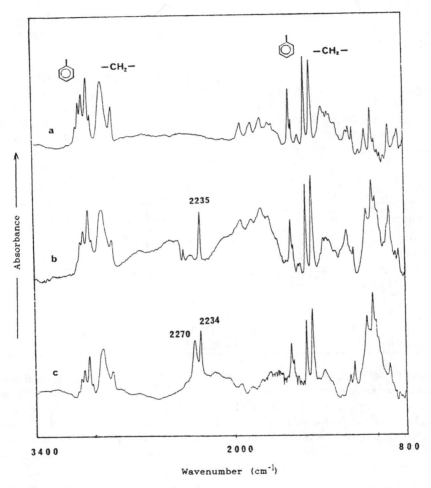

Figure 1. FT-IR spectra of (a): PS, (b): SAN50, and (c):SAN5010.

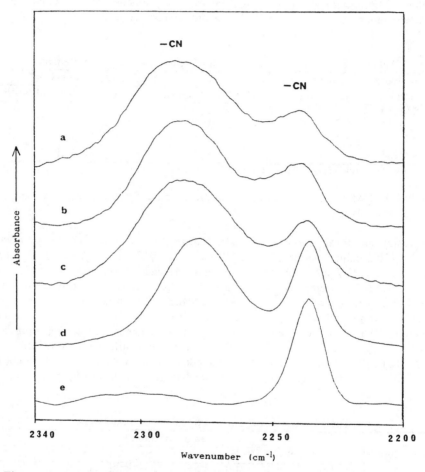

Figure 2. FT-IR spectra of (a): ALT50, (b): ALT30, (c): ALT20, (d): ALT10, and (e): ALT.

Proton NMR. The NMR spectra for various composition of styrene acrylonitrile copolymer were taken using d-acetone as a solvent. The peaks appearing at δ 7.39, δ 7.25, and δ 7.02 were assigned to the phenyl group (18). The sharp singlet at δ 2.90 was assigned to the methinic group and the peak at δ 1.85 was assigned to the methylene group (18). The compositions of the copolymers, obtained from their proton NMR spectra and calculated from the feed comonomer composition and reactivity ration (r_{AN} = 0.04, r_{ST} = 0.41) (19) are summarized in Table II.

Table II. Copolymer Composition of a Series of Random Copolymers

Copolymer samples	Monomer feed (AN *mole*%)	Polymer composition(AN mol%)	
		Calculated from r_1, r_2 [a]	Estimated from ^{1}H-NMR
SAN30	23.0	29.9	31.2
SAN40	40.0	38.9	41.4
SAN50	72.0	48.8	51.0
SAN60	93.0	59.8	63.8

[a] Reactivity ratio of r_{AN} = 0.04 and r_{ST} = 0.41 were used.

Carbon-13 NMR. The ^{13}C-NMR spectra for a series of SAN were taken using *d*-acetone as a solvent.

The assignment for the resonances of acrylonitrile (A)- and styrene (S)-central triads (AAA, 120.2 ppm; AAS or SAA, 120.8 ppm; SAS, 122.0 ppm; ASA, 142.3 ppm; SSA or ASS, 143.7 ppm; SSS, 145.9 ppm) could be made as were made by Arita et al (20). There was a clear distinction in NMR spectra between a truly alternating copolymer and a random 50 *mole*% copolymer. The triad distributions were calculated from peak area. The triad sequence distributions were also calculated according to Harwood (21) on the basis of the comonomer feed composition and reactivity r_{AN} = 0.04, r_{ST} = 0.41,) (19) and those were sufficiently coincided (see Table III). It is worth noting that the random copolymers prepared by radical copolymerization have significant amounts of adjacent acrylonitrile sequences.

Table III. The Triads Distribution of The Copolymers

Copolymer	Triad distribution [a]											
	SSS		SSA+ASS		ASA		SAS		AAS+SAA		AAA	
	Cal.	NMR	Cal.	NMR	Cal.	NMR	Cal.	NMR	Cal.	NMR	Cal.	NMR
SAN30	33.0	34	48.9	57	18.1	9	99.6	98	0.4	2	0.0	0
SAN40	14.2	17	47.0	55	38.8	28	95.8	94	4.2	6	0.0	0
SAN50	1.8	0	23.4	29	74.8	71	82.2	70	16.8	30	0.9	0
SAN60	0.1	0	5.6	0	94.3	100	42.6	28	45.3	56	12.1	16
ALT		0		0		100		100		0		0

[a] Calculated based on reactivity ratio of r_{AN} = 0.04, r_{ST} = 0.41 and monomer feed *mole* ratio; NMR observation based on ^{13}C-NMR.

Thermal Transition by DSC. The glass transition temperature data for the series of styrene acrylonitrile copolymers in this study are summarized in Table IV and in Figures 3. The glass transition temperature plotted as a function of acrylonitrile composition showed a convex curve with a maximum glass transition temperature at the 50 *mole*% composition, and also the glass transition temperature of the alternating copolymer was higher than that of the 50 *mole*% random copolymer (22) (see Figure 3).

The addition of zinc chloride to all the copolymers led to an increase in glass transition temperature, possibly due to a complex formation with the nitrile group in copolymer leading to decreased chain flexibility. The effect of zinc chloride on increasing the glass transition temperature was largest for the alternating copolymer (see Table IV). Interestingly, the glass transition temperature of the PS10 was also higher than that of neat polystyrene. Figure 4 shows the effect of the amount of zinc chloride to the glass transition temperature of the alternating copolymer. It seems that there is a saturation of glass transition temperature with increasing complex ratio of zinc chloride. ALT/HCl showed a great decrease in the glass transition temperature, possibly due to a deterioration of backbone structure by the hydrogen chloride.

Table IV. Glass Transitions of Polymers With and Without Zinc Chloride [a]

Polymers	η_{inh} [b] (dl/g)	$ZnCl_2$ (wt%)	T_g (°C) Onset	T_g (°C) Mid point	T_g (°C) End point
PS	0.13	—	101	103	106
PS10		10	101	105	110
SAN30	0.50	—	106	111	116
SAN40	0.53	—	108	113	119
SAN50	0.86	—	111	116	120
SAN60	1.13	—	111	115	117
ALT	2.81	—	114	118	120
SAN3010		10	110	116	121
SAN4010		10	112	117	136
SAN5010		10	106	115	119
SAN6010		10	112	119	123
ALT		10	120	127	132
ALT		—	114	118	120
ALT10		10	120	127	132
ALT20		20	121	128	139
ALT30		30	114	123	135
ALT/HCl		—	66	75	85

[a] Measured on a sample weight of 5±0.5 *mg* and a heating rate of 10°C/*min* under a N_2 flow rate of 80 *ml/min*.
[b] Measured on a concentration of 0.5 *g/dl* in dimethyl formamide at 25°C.

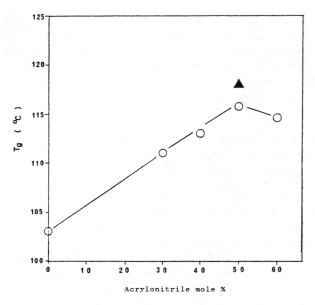

Figure 3. Glass transition temperature and composition of styrene acrylonitrile copolymers. (O): random copolymers, (▲) : alternating copolymer.

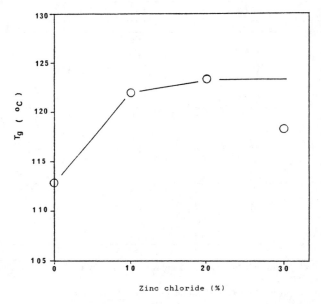

Figure 4. Glass transition temperature and alternating copolymer containing different amount of zinc chloride

Thermal Stability by TGA. The thermal stability data for the various styrene acrylonitrile copolymers in this study are summarized in Table V and their dynamic TGA thermograms are shown in Figures 5-8. The polystyrene with and without additive both decomposed without noticeable char yield above 450°C (see Figure 5.) Therefore, zinc chloride has no effect on the improvement of char yield for the polystyrene. The initial decomposition temperature (IDT) and the maximum decomposition temperature (T_{max}) were increased with the addition of additive (see Table V.) The reason for this small increase in thermal stability is not clear.

Table V. Thermal Stability and Char Yield of Polymers With and Without $ZnCl_2$ [a]

Polymers	$ZnCl_2$ (wt%)	IDT [b] (°C)	T_{max} [c] (°C)	Char yield at 900°C (%) org.	normalized [d]
PS	—	394	422	3	—
PS10	10	410	436	4	4
SAN30	—	393	419	2	—
SAN40	—	397	423	1	—
SAN50	—	391	417	2	—
SAN60	—	387	416	8	—
ALT	—	396	422	4	—
SAN3010	10	376	408	21	22
SAN4010	10	365	397	26	28
SAN5010	10	369	400	32	36
SAN6010	10	369	405	44	48
ALT	10	376	405	26	28
SAN50	—	391	417	2	—
SAN5010	10	369	400	32	35
SAN5020	20	372	403	37	45
SAN5030	30	370	400	37	51
ALT	—	396	422	4	—
ALT10	10	376	405	26	28
ALT20	20	365	395	33	40
ALT30	30	362	391	35	47
ALT/HCl	—		532	5	—

[a] Measured on a sample weight of 10±0.5 mg and a heating rate of 10°C/min under a N_2 flow of 80 ml/min
[b] Initial Decomposition Temperature
[c] Maximum Decomposition Temperature
[d] The Char Yield after normalization for the $ZnCl_2$: $\frac{CY(total) - CY(ZnCl_2)}{weight\ fraction\ of\ polymer}$

The styrene acrylonitrile copolymers without additive decomposed completely above 450°C (see Figure 6.) The SAN60 showed a noticeable amount of the char yield at 900°C. The reason for this char yield can be attributed to the intramolecular nitrile group oligomerization (6,23). The thermal stability of the alternating copolymer was higher than that of the 50 *mole*% random SAN. The similarity in the shape of all of

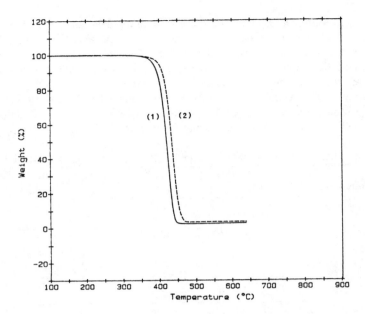

Figure 5. T_g of (1): PS and (2): PS10 at a heating rate of 10°C/*min* in N_2.

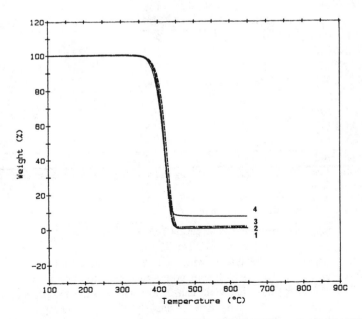

Figure 6. T_g of (1): SAN3010, (2): SAN40, (3): SAN50 and (4): SAN60 at a heating rate of 10°C/*min* in N_2.

the TGA curves and maximum decomposition temperatures suggests that there is no fundamental changes in degradation mechanism.

Upon the addition of zinc chloride to styrene acrylonitrile copolymers, the thermal stability was decreased and the char yield was increased significantly (see Figure 7 and Table V). From this observation, it appeared that zinc chloride changed the degradation mechanism of the SAN leading to high char formation. Also considering the fact that zinc chloride has no effect on the degradation of polystyrene, the modified mechanism must be mainly related to the acrylonitrile moity. This conclusion could be supported by the continuous increase of the char yield as the increase of acrylonitrile content in copolymer composition at the fixed percent of added zinc chloride. To study the effect of the complex ratio on the char yield, different amounts of zinc chloride were added to the SAN50. As the amount of zinc chloride increased, the char yield at 900°C increased and the thermal stability gradually decreased. This observation indicated that, "complexed" nitrile group played a major role in modifying the degradation mechanism. This phenomena could also be observed in a similar experiment with the ALT (see Figure 8.) ALT/HCl did not show a noticeable char yield at 900°C.

Kinetic Analysis of Thermal Degradation. Thermogravimetric analysis has been widely used as a method for evaluating kinetic parameters for polymer degradation reactions. In this study we employed Ozawa's method (24) to evaluate the kinetic parameters, since the Ozawa's method is considered the most reliable and suitable for studying the complex degradation behavior of high temperature and high char forming polymer systems. Table VI shows the result of the calculated activation energy for the initial degradation of various polymers. Typical dynamic TGA curves and the plots of the logarithms of the heating rates versus the reciprocal absolute temperatures of this polymer are shown in Figures 9 and 10.

For the neat polymers (SAN50, ALT and PS), those TGA curves could be superposed by lateral shifts. Thus, the plots of $-log$ a versus $1/T$ yielded a set of parallel straight lines. The alternating copolymer showed a higher activation energy than that of the SAN50, as expected. Polystyrene had a similar level of activation energy as the alternating copolymer. As zinc chloride was added, the TGA curves for a given copolymer deviated from superposition. Thus, a set of parallel straight lines were not obtained.

The thermal degradation of the copolymers containing zinc chloride was a complicated reaction, not governed by a single activation energy. Generally, the copolymers containing zinc chloride had higher activation energies than those of neat copolymers. The detailed activation energies for each step of weight losses are give for the ALT10 and ALT30. As the conversions of weight loss increased, the activation energies started with lower values than those of neat copolymer also increased. This observation indicated that the thermal degradation mechanism was changed by the addition of zinc chloride.

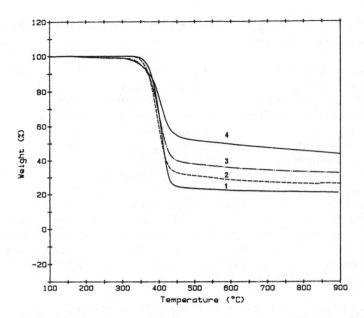

Figure 7. T_g of (1): SAN3010, (2): SAN4010, (3): SAN5010, and (4): SAN6010 at a heating rate of 10°C/*min* in N_2.

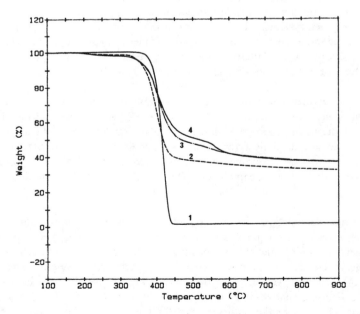

Figure 8. T_g of (1): ALT, (2): ALT10, (3): ALT20, and (4): ALT30 at a heating rate of 10°C/*min* in N_2.

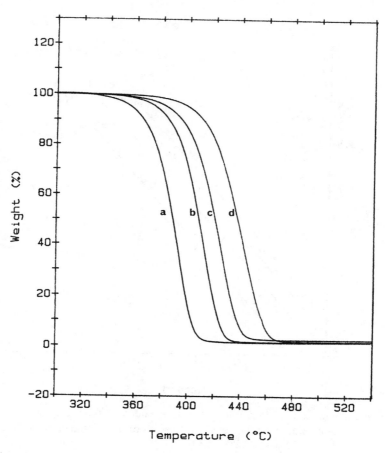

Figure 9. T_g curves for the decomposition of PS at different heating rates (a): 2°c/*min*, (b): 5°C/*min*, (c): 10°C/*min*, and (d): 20°C/*min*.

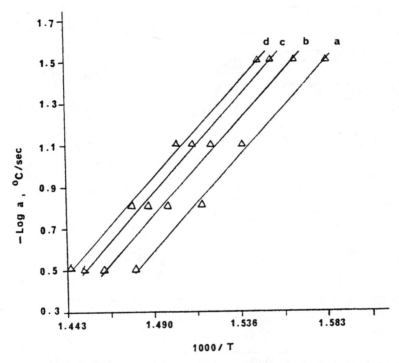

Figure 10. The plots of logarithms of heating rate versus the reciprocal absolute temperature for indicated conversions of the degradation of PS. (a): 6 %, (b): 10%, (c): 12%, (d): 16%.

Table VI. Activation Energy Calculation for The Initial Degradation of Various Polymers [a]

Polymers	Activation energy[b](Kcal/mole)		Heating rate used(°C/min)
SAN50	38 ± 1		2,5,10,20
SAN5010	46 ± 1		
ALT	42 ± 1		
ALT10	43	39 : at 8 % wt.loss	2,5,10,20
		44 : at 12% wt.loss	
		46 : at 16% wt.loss	
		46 : at 20% wt.loss	
ALT30	44	37 : at 8 % wt.loss	
		44 : at 12% wt.loss	
		47 : at 16% wt.loss	
		47 : at 20% wt.loss	
PS	43 ± 1		2,5,10,20
PS10	42 ± 2		

[a] Ozawa's method has been employed.
[b] Activation energies shown are taken as the average value for different amounts of weight loss.

Application of Houben-Hoesch Synthesis. The first stage of the Houben-Hoesch ketone synthesis employs a nitrile, hydrogen chloride (sometimes together with zinc chloride or another Lewis acid as catalyst,) and a suitably substituted aromatic hydrocarbon as substrate (25,26). The substrate, which must be activated towards electrophilic attack, is usually a polyhydroxy or polyalkoxy compound. Dry ether is the usual solvent. This stage generally results in the formation of a ketimine hydrochloride. Hydrolysis of the ketimine hydrochloride leads to ketone.

To study the possibility of a Houben-Hoesch type reaction leading to ketimine structure occurred in our studies, a series of experiments were done and the results are summarized in Table VII. Experiment 1A shows a typical example of Houben-Hoesch reaction indicated by formation of a yellow precipitate which is a ketimine hydrochloride (27). A similar experiment, with zinc chloride but without the hydrogen chloride, showed no formation of a yellow precipitate. From this observation, it could be understood that the Houben-Hoesch reaction could not occur without introduction of the dry hydrogen chloride. In other words, the role of zinc chloride was an additional catalyst for the reaction and not an indispensable component (26,27).

For the polymer solutions of the styrene acrylonitrile copolymer which have nitrile component in its molecular chain, there were no signs of reaction (see Exp. 2A-3B). If we consider higher reaction yields can be obtained when the activated substrate is added after the reaction of the nitrile with the hydrogen chloride, the styrene acrylonitrile copolymer which has nitrile component in its chain has a limitation for the Houben-Hoesch reaction (28). The solutions of the polystyrene, in which benzonitrile was added as nitrile component, also did not show any sign of the reaction (see Exp. 4A-5A).

Table VII. Application of Hoesch Reaction to PS and SAN [a]

Exp	With HCl	ppt	Exp	Without HCl	ppt
1A	Resorcinol, benzonitrile, ether	Yel.	1B	Resorcinol, benzonitrile, ether, $ZnCl_2$	No
2A	SAN, $ZnCl_2$, DMF	No	2B	SAN, $ZnCl_2$, DMF	No
3A	SAN, $ZnCl_2$, $CHCl_2$	No	3B	SAN, $ZnCl_2$, $CHCl_3$	No
4A	PS, Benzonitrile, $ZnCl_2$, $CHCl_3$	No	4B	PS, Benzonitrile, $ZnCl_2$, $CHCl_3$	No
5A	Benzonitrile $ZnCl_2$, DMF	No			

[a] Reaction condition: stored at -5°C for two weeks.

With the above knowledge, it may be inferred that the possibility of the Houben-Hoesch reaction for the SAN complexed with zinc chloride under the high temperature solid state degradation is low. The first step of the Houben-Hoesch reaction leading to a ketimine hydrochloride is a very slow reaction step which usually needs several days at low temperature (26,27). For the formation of imino chloride which is a intermediate for the reaction, excess of hydrogen chloride is needed for shifts of the equilibrium of the reaction (27). Considering the unavailability of hydrogen chloride in the SAN/zinc chloride complex at the decomposition step, it seems quite difficult for the Houben-Hoesch type of reaction to occur during the high temperature solid state degradation process.

High-temperature Infrared Studies. To study the chemistry of the char formation, a high temperature cell was attached to the Fourier transform infrared spectrometer. Polymer films coated on the surface of a aluminum plate were sufficiently thin in an absorbance range where the Beer-Lambert law was obeyed. All the spectra were taken in absorbance. The samples were preheated to 150°C and held for 3 min to remove absorbed moisture, and then heated to the initial decomposition temperature of the polymer at a heating rate of 10°C/min. During the heating, the IR spectra were taken at about 20°C intervals with 64 scans each at a resolution of 2 cm^{-1}. The scans were signal averaged in conventional manner to reduce spectral noise. When the temperature reached the initial decomposition temperate, the IR spectra were taken isothermally for 30 min to see the spectral changes at 2 *min* intervals. To support interpretation of the changes of the IR spectra at elevated temperatures, a spectral subtraction technique was also used.

Figure 11 shows infrared spectra in the 3600-850 cm^{-1} range for the SAN50 as a function of increasing temperature. The spectra were plotted on an absolute absorbance scale and consequently the changes in frequency, breadth, and relative area can be directly visualized. As the temperature increased from 150°C to 380°C all the following characteristic bands decreased in intensity continuously but at relatively slow speed: the benzene ring bands at 3060, 3026, 1599, 1492, and 1027

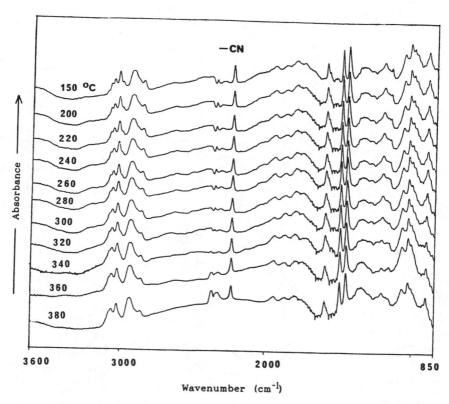

Figure 11. FT-IR spectra of SAN50 as function of temperature.

cm^{-1}, the —CH$_2$ bands at 2930, 2858, and 1452 cm^{-1}, and the nitrile band at 2235 cm^{-1}. New peaks were not observed. When the SAN50 was isothermally treated at 380°C as function of time, a faster decrease in intensity of all the characteristic peaks was observed. After 10 min, most of the peaks were greatly reduced.

The high temperature FT-IR study of the SAN5010 are shown in Figure 12. All of the characteristic peaks for benzene ring and methylene linkage gradually decreased in intensity as the temperature increased from 150°C to 380°C. The rate of decrease of intensity was much faster than that of the SAN50. Interestingly the rate of decrease in intensity of the "complexed" nitrile band was much faster than that of

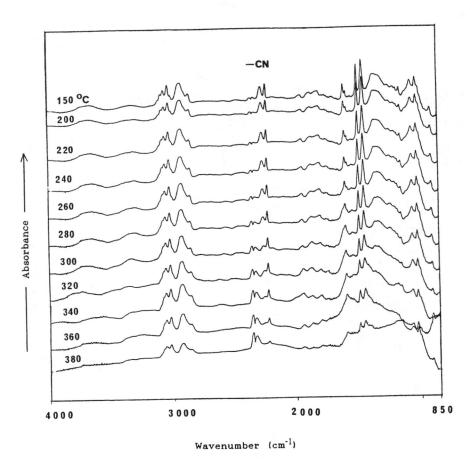

Figure 12. FT-IR spectra of SAN5010 as a function of temperature.

"free" nitrile band. This observation was confirmed by the similar experiment with the ALT10 shown in Figure 13. As the temperature increased, the relative intensity of "complexed" nitrile peak to "free" nitrile peak gradually increased. This phenomena therefore indicated that zinc chloride induced degradation reactions leading to lower initial thermal stability and modified the degradation mechanism. Another observation was the appearance of the new peaks at 1400-1560 cm^{-1}. The difference spectra shown in Figure 14 as function of temperature from 150°C to 380°C, more clearly indicated the decrease of nitrile peaks and increase of new peaks. New peaks at 1560, 1543, 1527, and 1420 cm^{-1} stand for triazine structure (Figure 15) (28-31). Therefore, the main reason for the higher char formation and lower initial thermal stability of the SAN5010 compared to those of the SAN50 could be explained as follows: zinc chloride complexed with the nitrile group, induced thermal degradation by modifying the degradation mechanism. This mechanism gave thermally stable triazine ring formation which crosslinked the main chain. The high temperature FT-IR study of the ALT10 also showed similar results as obtained from SAN5010.

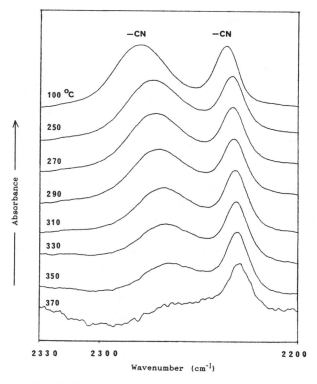

Figure 13. FT-IR spectra of ALT10 as a function of temperature.

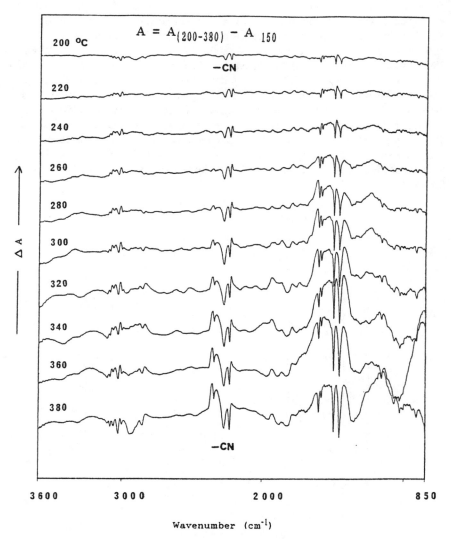

Figure 14. Difference spectra of SAN5010 as a function of temperature.

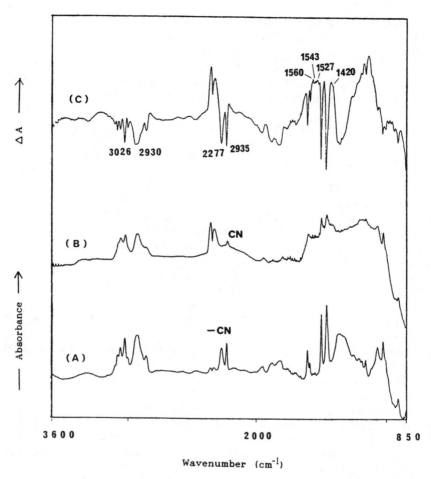

Figure 15. FT-IR spectra of SAN5010 (A): 150°C, (B): 380°C, and (C): (B)-(A).

ACKNOWLEDGMENT. We acknowledge the partial support of this research by the Petroleum Research Fund of the American Chemical Society (grant #18655-AC7).

LITERATURE CITED

1. *Encyclopedia of Polymer Sci. & Tech.*; John Wiley & Sons, Inc.: New York, NY, 1964; Vol 1, pp 425.
2. Lyons, J. W., *The Chemistry & Uses of Fire Retardants*, Wiley-Interscience: New York, NY, 1970, pp 282.

3. Grassie, N. and Bain, D. R., *J. Polym. Sci.*, **1970**, *A1*(8), 2665, 1970.
4. Andreev, A. P.; Zaitseva, V. V.; Kucher, R. V.; Zaitsev, Y. S., *Plast. Massy* **1975**, *2*, 54; *Chem. Abs.*, **1975**, *83*, 10997h.
5. Oda, R., Okuda, T., Sakai, K., and Morikage, S., *Jap. Kokai* **1975**, 75,160,392.
6. Grassie, N., and McGuchan, R., *Eur. Polym. J.*, **1972**, *8*, 865.
7. Sivaramakrishnan, K. P., and Marvel, C. S., *J. Polym. Sci. Chem. Ed.*, **1974**, *12*, 651.
8. Bruma, M. and Marvel, C. S., *J. Polym. Sci. Chem. Ed.*, **1976**, *14*, 1.
9. Hsu, L. C., *U.S. Pat.*, **1977**, 4,061,856.
10. Zeldin, M., Jo, W. H., Pearce, E. M., *J. Polym. Sci. Polym. Chem. Ed.*, **1981**, *19*, 917.
11. Whang, W. T., Ph.D. Dissertation, Polytechnic Unviersity, **1984**.
12. Kim, S. S., Ph.D. Dissertation, Polytechnic University, **1987**.
13. Yabumoto, S., Ishii, K., and Arita, K., *J. Polym. Sci. A1*, **1969**, *7*, 1577.
14. Liang, C. Y., and Krimm, S., *J. Polym. Sci.*, **1958**, *22*, 241.
15. Imoto, M., Otsu, T., and Nakabayashi, N., *Makromol. Chem.*, **1963**, *65*, 194.
16. Schauzer, G. N., *J. Amer. Chem. Soc.*, **1959**, *81*, 5310.
17. Kettle S. F. A., and Orgel, L. E., *Chem. and Ind.*, **1960**, *49*, .
18. Silverstein, Bassler and Morrial, *Spectrometric Identification of Organic Compounds*, 4th ed., John Wiley & Sons, Inc., New York, NY, **1981**, pp 220.
19. *Polymer Handbook"*, 2nd ed., Young, L. H., Eds.; Wiley-Interscience, New York, NY, 1975 Chap. II.
20. Arita, K., Ohtomo, T., and Tsurumi, Y., *J. Polym. Sci., Polym. Lett. Ed.*, **1981**, *19*, 211.
21. Harwood H. J., and Ritchey, W. W., *J. Polym. Sci.*, **1964**, *32*, 601.
22. Hirooka M., and Kata, T., *J. Polym. Sci., Poly. Lett. Ed.*, **1974**, *12*, 31.
23. Burland, W. J., and Parsons, J. L., *J. Polym. Sci.*, **1956**, *22*, 249.
24. Ozawa, T., *Chem. Sco. Jpn. Bull.*, **1984**, *38*, 1881.
25. Jeffery E. A., and Satchall, D. P. N., *J. Chem. Soc.*, **1966**, *(B)*, 579.
26. *Organic Reactions*, Adams, R., Eds.; John Wiley & Sons, Inc., New York, NY, 1949, Vol. 5, Ch. 9.
27. Zil'berman, E. N., and Rybakova, N. A., *J. Gen. Chem. USSR*, **1960**, *30*, 1972.
28. Reimschuessel, H. K., Lovelace, A. M., and Hagerman, E. M., *J. Polym Sci*, **1959**, *40*, 279.
29. Padgett, W. M., and Hamner, W. F., *J. Am. Chem. Soc.*, **1958**, *80*, 803.
30. Picklesimer, L. G., and Saunders, T. F., *J. Polym. Sci.*, **1965**, *3*, 2673.
31. Thurston, J. T., Dudley, J. R., Kaiser, D. W., Hechenbleikner, I., Schaefer, F. C., and Holm-Hansen, D., *J. Am. Chem. Soc.*, **1951**, *73*, 2981, 2984, 2992.

RECEIVED November 21, 1994

SURFACES AND CHAR

A great deal of effort in fire-retardant chemistry is directed toward the development of materials that do not contain halogenated additives. The primary approach is to increase char formation at the surface of the material, thereby protecting the underlying substrate. Understanding condensed-phase fire-retardant chemistry and applying that chemistry through the use of additives or changes in the substrate to achieve higher char yield and stability are the goals.

Char is formed from polyenes and cyclic or aromatic structures and can be enhanced by the presence of hetero atoms or inorganic materials. This section provides several examples of non-halogenated flame-retardant systems using solid phase reactions and innovative chemistry.

Gilman, VanderHart, and Kashiwagi discuss the thermal decomposition of poly(vinyl alcohol) (PVA). The fundamental condensed-phase processes and structures that lead to char formation under fire-like pyrolysis conditions are characterized. The processes involve cross-linking and aromatization. The main decomposition products of PVA are polyenes; cross-linking oligomers such as bismaleimides were used to form intermolecular cyclohexenes, substantially adding to char.

Zaikov and Lomakin report using PVA in Nylon 6,6 as a flame retardant. Indeed 20% PVA in Nylon 6,6 shows a 60% reduction in maximum rate of heat release versus Nylon 6,6 alone. Partial oxidation of PVA by $KMnO_4$ shows a further reduction in heat release rate.

A variety of potential systems exist. Weil and Zhu discuss the use of melamine in an ethylene–propylene–hexadiene terpolymer, with poly (2,6-dimethylphenylene oxide) as a char former. Considerable interest has been shown in the use of melamine as a flame retardant.

Fire is a surface phenomenon. The chemistry at the surface of an object will significantly affect ignitability and flame spread. That being the case, can the surface be changed to increase flame retardancy?

Benrashid and Nelson discuss silicone–polyurethene block copolymers. These materials show microphase segregation with substantially enhanced siliconated surfaces. Scientists can vary chemistry (siloxane, isocyanate, and chain extender), block size, and solvent or processing conditions and can achieve siliconelike fire retardancy (Ol>28) at 50% or less silicone in the polymer.

Wilkie, Dong, and Suzuki discuss grafting methacrylic acid onto acrylonitrile–butadiene–styrene (ABS) or styrene–butadiene polymers. The grafted methacrylic acid is converted to the sodium salt. The graft layer increases the time to ignition and decreases the peak heat-release rate significantly. For ABS the peak heat-release rate is reduced by > 70%.

Thermosets such as phenol–formaldehyde resins have a high degree of flame retardancy. Nyden, Brown, and Lomakin show that combustible products released during thermal degradation of phenolic resins make a significant contribution to the flammability of systems using these resins. Resins with excess phenol and after cure have much lower peak heat-release rates. Reduction of aliphatic content, removal of flammable volatiles, and enhanced cross-linking are vital.

Materials that are aromatic or have high hetero-atom content are of interest as intrinsically flame-retardant materials. Coad and Rasmussen discuss carbon–nitrogen containing materials. Keller discusses materials containing acetylenic units with particular interest in an inorganic–organic hybrid polymer. And Son and Keller report on linear siloxane–acetylene polymers as precursors to high-temperature materials.

Chapter 11

Thermal Decomposition Chemistry of Poly(vinyl alcohol)
Char Characterization and Reactions with Bismaleimides

Jeffrey W. Gilman[1], David L. VanderHart[2], and Takashi Kashiwagi[1]

[1]Building and Fire Research Laboratory and [2]Materials Science and Engineering Laboratory, National Institute of Standards and Technology, Gaithersburg, MD 20899-0001

The fundamental condensed phase processes which lead to char formation during the fire-like pyrolysis of poly(vinyl alcohol), PVA, and PVA-containing maleimides were characterized using CP/MAS ^{13}C NMR. In addition to evidence of the well known chain-stripping elimination of H_2O and the chain-scission reactions, which occur during the pyrolysis of pure PVA, evidence is presented in support of cyclization and radical reaction pathways responsible for the conversion of unsaturated carbons into aliphatic carbons. Two general mechanisms; one described as a physical encapsulation, and the other a lowering of the average volatility of certain degradation products, are proposed for the primary modes of action of maleimides on the pyrolysis of PVA.

Currently, due to concerns over the environmental effects of halogenated compounds, there is an international demand for the control of polymer flammability without the use of halogenated additives (*1,2*). An alternative to the use of halogenated fire retardants, which control flammability primarily in the gas phase, is to control polymer flammability by manipulating the condensed phase chemistry. Our approach is to increase the amount of char that forms during polymer combustion. Char formation reduces, through crosslinking reactions, the amount of small volatile polymer pyrolysis fragments, or fuel, available for burning in the gas phase; this, in turn reduces the amount of heat feedback to the polymer surface. The char also insulates the underlying virgin polymer. All of these effects combine to reduce polymer flammability.

The polymer we chose to investigate was poly(vinyl alcohol), PVA, because it is one of the few linear, non-halogenated, aliphatic polymers with a measurable (~ 4%) char yield. We have attempted to characterize the fundamental condensed phase processes and structures which lead to char formation during the fire-like pyrolysis of poly(vinyl alcohol) and then to use this information to design new strategies that do not use halogenated additives, but which increase char formation (*3,4*). Strategies for retarding the flammability of polymers through enhanced char formation require an understanding of how char forms; therefore, characterization of the polymer at several intermediate stages of decomposition is necessary. The thermal decomposition of PVA has been studied previously both in the gas and condensed phases; however, only limited characterization of the pyrolysis residues exists (5,6). Typical TGA analysis of PVA (5 °C/min, N_2), in Figure 1, shows two regions, 300-325 °C and 400-425 °C, where the weight loss rate is reduced relative to the more active portions of the thermogram. The intermediate decomposition products present in these regions have better thermal stability than PVA. We set out to characterize the structure of these pre-chars, to determine how they form, and how we might enhance their stability.

Experimental

Poly(vinyl-alcohol) Pyrolysis: PVA (99.7% hydrolyzed with aqueous NaOH, Mn = 86,000, Mw = 178,000, Scientific Polymer Products) was purified by Soxhlet extraction (MeOH, 2h) to remove NaOAC and heated (70°C, 5h) to give a white granular crystalline powder (particle size 100-1000 um) (7). This powder retained 6-7% H_2O by ^1H NMR and elemental analysis.

Previous studies have shown that the elemental composition, char yield and physical structure of polymer char, in many systems, were independent of whether they were formed via pyrolysis in nitrogen, or combustion in air (8), therefore, PVA was pyrolyzed in a flow pyrolysis apparatus using a nitrogen atmosphere (< 50 ppm O_2). This experimental setup allows study of the condensed phase decomposition processes under fire-like conditions in the absence of gas phase oxidation.

We prepared pyrolysis residues using 1-2 gram samples of PVA, at several intermediate stages of thermal decomposition. Pyrolysis was carried out for 30 minutes at each of several temperatures (250°C, 300°C, 350°C, and 400°C). The PVA was spread ~1-2 mm deep on a 5 cm x 10 cm ceramic tray for each pyrolysis. Two experiments per sample were run and the results averaged. Run-to-run differences were less than 10% of the measured value.

Poly(vinyl-alcohol)/Maleimide Pyrolysis: The maleimides N,N'-1,4-phenylenedimaleimide (1,4-PDMI, 98%, Aldrich), N,N'-1,3-phenylenedimaleimide (1,3-PDMI, 97%, Aldrich) and N-phenylmalemide (PMI, 97%, Aldrich) were used as received. The PVA/maleimide mixtures were prepared by simply grinding the maleimides with PVA, using a mortar and pestle, to give the following combinations: PVA/1,3-PDMI (10% by weight, 2 mole %), PVA/1,4-PDMI (10% by weight, 2 mole %), and PVA/PMI (12% by weight, 2 mole %). Typical particle sizes were 10-1000 um.

This approach was used because of the difficulties associated with melt blending PVA, since it's decomposition temperature is 30-40 °C below it's melting point, and because of our aversion to any solvent blending techniques. Solvent blending could adversely effect both polymer stability and flammability by altering the physical state (e.g., crystallinity) of PVA; in addition polymer-solvent reactions might occur, or other physical effects due to solvent escape might take place. All of these effects could complicate interpretation. We anticipated that some mixing would take place in the molten state as the PVA mixtures were heated. The effectiveness of the maleimides as crosslinking agents for PVA of course would depend on this mixing.

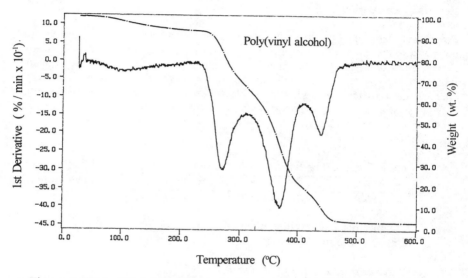

Figure 1. TGA of Poly(vinyl alcohol), 5 °C/min, broken line: weight loss vs. temperature, solid line: first derivative $\Delta\%/\Delta$ min. The plots reveal two regions, 300-325 °C and 400-425 °C, where the mass loss rate is slowed during degradation.

The mixtures were pyrolyzed in the N_2 flow pyrolysis furnace for 30 minutes at each of several temperatures (300 °C, 350 °C, and 400 °C). Two experiments per sample were run and the results averaged. Run-to-run differences for these pyrolyses were less than 4% of the measured values.

Characterization: Solid state ^{13}C NMR characterization utilized techniques of cross polarization (CP) and magic angle spinning (MAS) (25Mhz, 4Khz MAS, 1ms CP time, 3s rep. time) (9). In the decoupler blanking CP/MAS ^{13}C NMR experiments the decoupler was turned off for 40 us prior to sample observation with decoupling (10). Elemental analysis was also employed (Galbraith, two determinations per sample).

Results & Discussion

Poly(vinyl-alcohol) Pyrolysis: Upon heating PVA above the decomposition temperature the polymer begins a rapid chain-stripping elimination of H_2O shown in Figure 2 (11,12). This process coupled with melting causes the material to foam or intumesce as it decomposes. This and other decomposition reactions cause color changes and crosslinking to yield insoluble yellow to black rigid foam-like residues. The physical appearance, carbon-to-hydrogen ratio and residue yields were recorded for these pyrolyses (Table I).

Table I. PVA and Residues: Appearance, C/H Ratio and Yield

Material	Appearance	C/H Ratio	Residue Yield (%)
PVA	white granular solid	0.50	NA
Polymer residue 250 °C (30 min)	yellow-orange foam	0.52	82
Polymer residue 300 °C (30 min)	tan rigid foam	0.71	47
Polymer residue 350 °C (30 min)	dark brown foam	0.83	18
Polymer residue 400 °C (30 min)	black powder	1.00	5

Direct spectroscopic evidence of the chain-stripping elimination of H_2O was observed by comparison of the CP/MAS ^{13}C NMR spectra of PVA with the spectra of the residues. Figure 3 shows a progressive reduction in the intensity of the alcohol methine carbon (-$\underline{C}$H(OH)-) signal at ~ 70 ppm as well as the growth of signals in the olefinic and aromatic region between 120 -140 ppm as the exposure temperature increases. The spectra of the residues all show methyl signals at 15 and 20 ppm. These

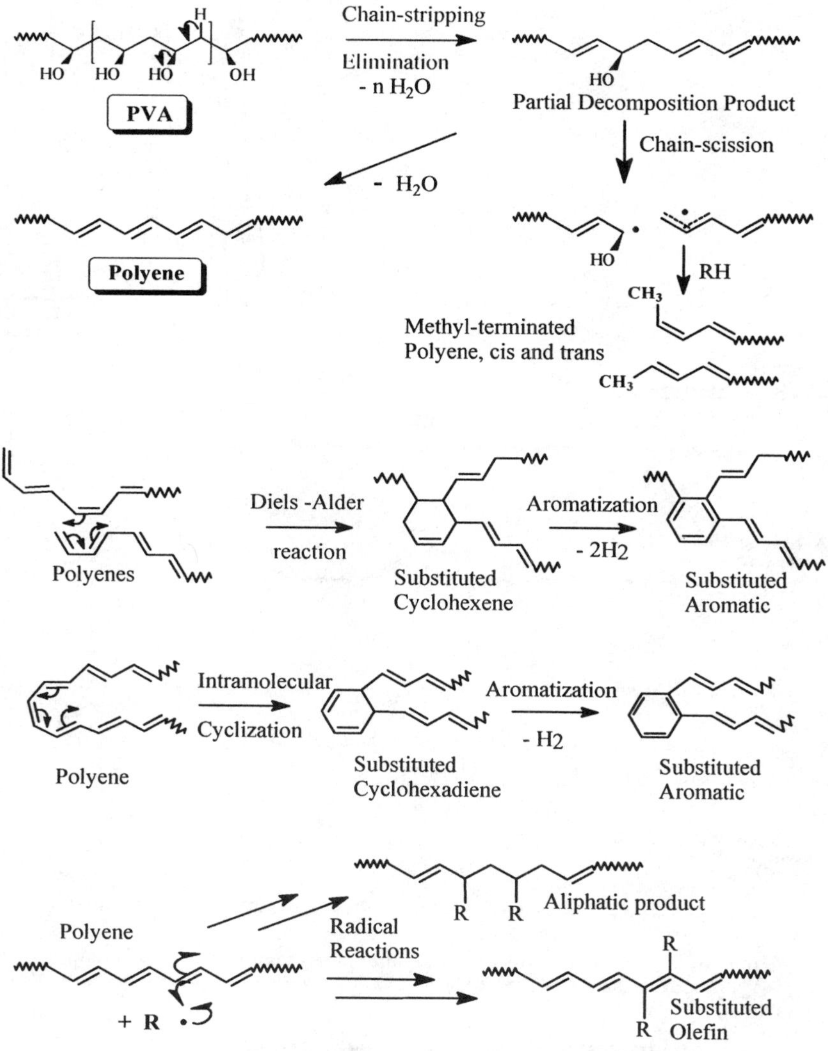

Figure 2. Proposed PVA pyrolysis reactions.

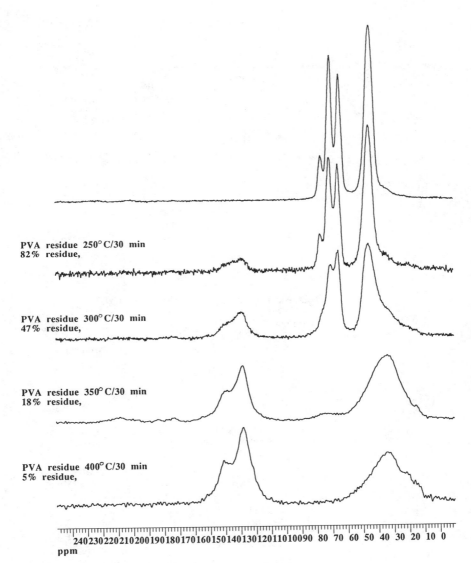

Figure 3. 25 MHz CP/MAS ^{13}C NMR spectra, normalized to the same total intensity, of PVA and PVA pyrolysis residues. Note the progressive loss of alcohol carbons (~70 ppm) and the corresponding gain in aromatic/olefinic carbons (110-150 ppm).

signals possibly arise from cis and trans allylic-methyls that may form via random chain scission reactions that can accompany elimination. This pyrolysis pathway is included in Figure 2.

At 350 °C the elimination reaction appears complete, most of the alcohol carbon signal at ~70 ppm (the fraction of alcohol-carbon intensity is only ~10% of that fraction in the original PVA) and the adjacent methylene signal at ~ 45 ppm are gone. In the 350 °C residue from integration 59% of the carbons are aliphatic (15-50 ppm). Moreover, the C/H ratio (table I) is 0.83, indicative of material that has a significant CH_2 and/or CH_3 component. It appears that at temperatures up to 350 °C the polyenes, from the elimination reaction, are converted into aliphatic groups (10-50 ppm). The Diels-Alder, intramolecular cyclization and radical reactions, shown in Figure 2, may be responsible for this conversion.

The Diels-Alder reaction has been proposed previously in a study of the acid catalyzed decomposition of PVA at 140 °C, and a similar intramolecular cyclization reaction has been proposed in the thermal decomposition of PVC (13,14). The Diels-Alder reaction, however, is the most efficient of the three processes at producing aliphatic carbons. It converts 4 of the 6 reacting sp^2 carbons into sp^3 carbons. The Diels-Alder and the intramolecular cyclization reactions produce substituted cyclohexenes and cyclohexadienes, respectively, that can aromatize to substituted aromatics.

To determine the extent to which carbon-carbon bond forming reactions, like those shown in Figure 2, occurred we employed a CP/MAS ^{13}C NMR interrupted decoupling experiment. This technique favors the observation of ~ 95% of the original intensity of the non-protonated carbons in a rigid sample; in addition about ~ 55% of the methyl carbon intensity also appears in these type of experiments. The result of applying this technique to the 350 °C residue, seen in Figure 4, shows that ~ 40% of the signal at 120-150 ppm in the normal CP/MAS spectrum is from non-protonated carbons, specifically, substituted olefinic or aromatic carbons. This indicates substantial carbon-carbon bond formation by reactions like those, shown in Figure 2, which form the substituted aromatics and or substituted olefinics. Substituted aromatics are of course the precursors to polynuclear aromatic structures found in char. Indeed, CP/MAS ^{13}C NMR analysis of PVA high temperature nitrogen-atmosphere pyrolysis char, shown in Figure 5, reveals that only these type of structures survive at high temperatures. Figure 4 also shows other non-protonated carbons, at 170 and 208 ppm, present in the 350 °C residue. The signal at 170 ppm may be assigned to the carbonyl carbon in carboxylic acid or anhydride groups and the signal at 208 ppm to ketones or the sp carbon of an allene (15). Further application of this technique revealed that the carbon-carbon bond formation reactions occur at temperatures as low as 250 °C producing substituted aromatic/olefinic carbons, see Figure 6, essentially concomitant with the chain-stripping elimination reaction. From Figure 7, we see that in the 400 °C residue ~50 % of all the carbons in the residue are aliphatic and almost all of the aliphatic carbons are still protonated. There is a strong implication of methine aliphatic carbon (CH) from these spectra. This is consistent with the overall C/H ratio of 1.00 (Table I) for the 400 °C residue, and supports the CH forming reactions shown in Figure 2, i.e., the Diels-Alder reaction, intramolecular cyclization and radical reactions.

The reactions and products shown in Figure 2 illustrate the probable mode of PVA

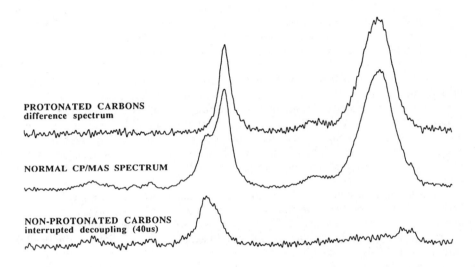

Figure 4. CP/MAS ^{13}C NMR spectra of residue of PVA pyrolyzed at 350 °C for 30 min. The spectrum of the non-protonated carbons and of a fraction (~55%) of the methyl carbons in the residue is obtained using the interrupted decoupling experiment (bottom). The difference spectrum (top) results from subtracting the non-protonated spectrum (bottom) from the normal CP/MAS spectrum (middle). The multiplication factor used in the subtraction is based on our knowledge of the average attenuation of a non-protonated carbon in the interrupted decoupling experiment.

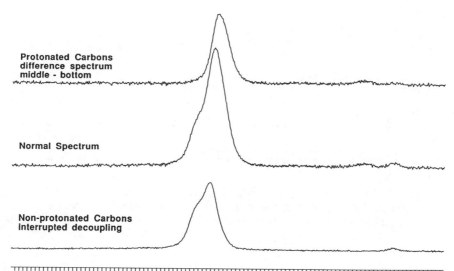

Figure 5. CP/MAS ^{13}C NMR spectra of char from pyrolysis of PVA in high temperature (700-800 °C) gasification apparatus under nitrogen. As in Figure 4 the spectrum of the protonated carbons (top) in the char was generated by subtracting the non-protonated spectrum (bottom) from the normal CP/MAS spectrum (middle). Note that the shape of the non-protonated carbon resonances differs from the corresponding shapes of the lower temperature residues (see Figures 4, 6 and 7); yet the fraction of protonated aromatic carbons is only modestly reduced at 800 °C, relative to the 350 °C and 400 °C residues.

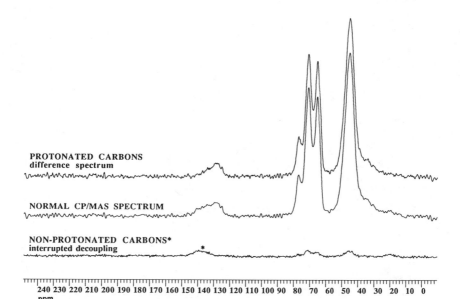

Figure 6. CP/MAS ^{13}C NMR spectra of residue of PVA pyrolyzed at 250 °C for 30 min. Interrupted decoupling spectrum (bottom) shows that even at 250 °C non-protonated carbons are present.

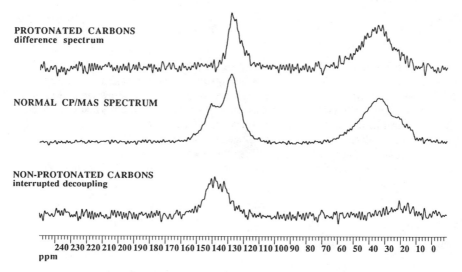

Figure 7. CP/MAS ^{13}C NMR spectra of residue of PVA pyrolyzed at 400 °C for 30 min (after Figure 4). Even at 400 °C a significant portion of the residue consist of protonated sp^3 carbons.

decomposition initially, and how these intermediate pyrolysis products subsequently react to form char.

As is often done, we would like to be able to propose overall structures for the various residues. Although we have characterized the amount of sp^2 and sp^3 carbon and the amount of non-protonated carbon, an accurate representation is not possible without a more detailed knowledge of the CH to CH_2 ratio in these residues.

Poly(vinyl alcohol)/Bismaleimide Pyrolysis: On the basis of the above information, we attempted to enhance char formation in PVA, and reduce flammability, by increasing the opportunity for crosslinking and aromatization. We focused our efforts on reactions with the main condensed-phase decomposition products, the polyenes, generated from the chain-stripping elimination reaction. No attempt was made to prevent the chain-stripping elimination reaction itself, since it produces the nonflammable volatile, H_2O.

Since there is good evidence for reaction of the double bonds of the polyenes, we introduced thermally stable crosslinking additives that we hoped would react with the polyenes to form cyclohexenes faster than the polyenes either react with themselves or undergo chain scission. Ultimately for this to be successful such additives must also facilitate aromatization of the cyclohexenes to prevent the retro Diels-Alder reaction which is otherwise favored at high temperatures (16).

Bismaleimides are a class of reactive oligomers that undergo rapid reaction (Diels-Alder, Ene, etc.) above 200 °C with materials containing carbon-carbon double bonds. Bismaleimides have been employed as chain extenders, crosslinking and aromatization agents with many polymer and composite systems and as ingredients in the preparation of low flammability materials (17, 18). The proposed reactions of these additives with the polyenes are shown in Figure 9. We examined PVA combined with each of the two bismaleimides shown in Figure 8. The bismaleimides, N,N'-1,4-phenylenedimaleimide (1,4-PDMI) and N,N'-1,3-phenylenedimaleimide (1,3-PDMI), were used to allow evaluation of the effect of structure on the bismaleimide's effectiveness. A monomaleimide, N-phenylmaleimide (PMI), was also included to compare chain branching or grafting to crosslinking.

Three mixtures, PVA/1,3-PDMI (10% by weight), PVA/1,4-PDMI (10%), and PVA/PMI (12%), were pyrolyzed in the N_2 flow pyrolysis furnace for 30 minutes at each of several temperatures. As was the case with PVA, the PVA/maleimide mixtures underwent the chain-stripping elimination of H_2O reaction and intumesced as they were heated. Figure 10 shows a plot of percent residue versus pyrolysis temperature for PVA and the PVA/maleimide mixtures. The presence of the maleimide has been factored out of these residue yields using nitrogen elemental analysis data. We assumed that the overall maleimide structure survived the pyrolysis and that the nitrogen remained intact in the imide. The PVA/maleimide mixtures all had higher residue yield relative to pure PVA at each pyrolysis temperature examined.

Pyrolysis at 300 °C. In general, if an additive can influence the decomposition process in the condensed phase at an early stage, the chances are improved for maintaining high char yields and for improving flammability properties. We focused our attention on characterizing the PVA/maleimide residues at 300 °C to determine how the maleimides improve the residue yields of PVA at this early stage in the decomposition, where a majority of the residue remains in the condensed phase.

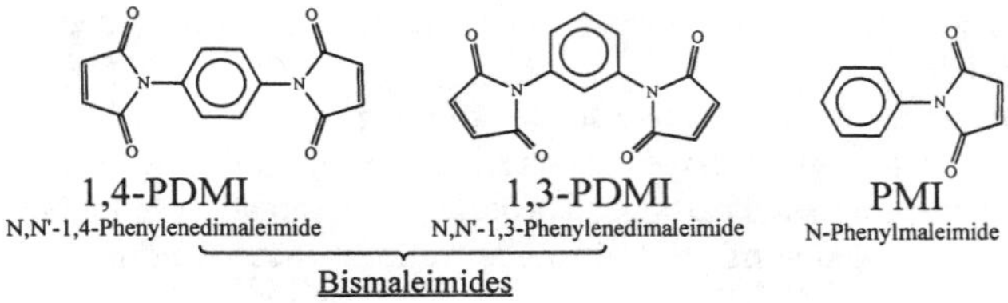

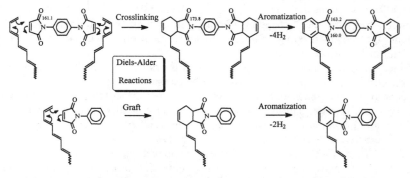

Figure 8. Structure of maleimides.

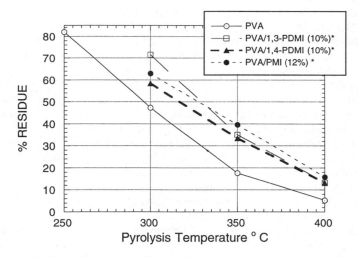

Figure 9. Diels-Alder reaction between polyene and maleimides and subsequent aromatization reactions.

Figure 10. Plot of percent residue yield vs. pyrolysis temperature for PVA and PVA with each maleimide. The * indicates that the presence of the maleimide has been factored out of these residue yields using nitrogen elemental analysis data.

Figure 10 shows how the additives vary in their relative effectiveness at stabilizing PVA. The most effective maleimide is 1,3-PDMI whose mixture with PVA gives a 71.5% residue yield at 300 °C. The mixture of PVA and PMI gave a 67% residue yield. The mixture of PVA and 1,4-PDMI shows the smallest effect with a 58.4% residue yield. The residue yield for pure PVA was 47%.

As mentioned above the effectiveness of the maleimides as crosslinking agents for PVA depends on the degree of mixing. Differences in residue yields may be due to differences in solubility and/or melting point and their effect on additive/PVA mixing. 1,3-PDMI is very soluble in organic solvents and has a m.p. of 200 °C whereas 1,4-PDMI is nearly insoluble and has a m.p. > 350 °C due to its symmetrical structure. The monomaleimide, PMI, like 1,3-PDMI is also a very soluble compound it has a m.p. of 86 °C, and a boiling point of 300 °C. These differences indicate that 1,4-PDMI may not mix as readily with PVA as 1,3 PDMI or PMI, even at elevated temperatures.

Other effects, however, must also be examined. When considering any additive/polymer system the issue of additive loss through vaporization is an important and common concern, especially when the additive is designed to act in the condensed phase. If the additive has a significant vapor pressure at the temperature of exposure its' effectiveness will be compromised. The TGA, shown in Figure 11, shows a very different thermal response for the three maleimides; between 250 °C and 350 °C 1,4-PDMI experiences a rapid, 85% weight loss whereas 1,3-PDMI shows only a 20% weight loss. Observation of the heating of 1,4-PDMI using a hot plate melting point apparatus showed that the rapid weight loss for 1,4-PDMI may be due to sublimation since the only change seen was material loss. This difference in thermal response may explain the reduced effectiveness of 1,4-PDMI. Indeed, nitrogen elemental analysis shows that there was less imide (6.9%) in the PVA/1,4-PDMI residue than in the PVA/1,3-PDMI residue (9.2%). The TGA in Figure 11 shows, in contrast to 1,4-PDMI, 95% of PMI is lost before the temperature reaches 250 °C. However, PMI gives a 63% residue yield and it is present at 9.0% in the PVA/PMI residue. These data suggest that one factor controlling the effectiveness of the maleimides may be the competition between vaporization of the maleimide and solubilization and/or condensed phase reaction of the maleimide.

CP/MAS ^{13}C NMR Results. To attempt to determine how the maleimides improved the residue yields of PVA the residues were characterized using CP/MAS ^{13}C NMR. The CP/MAS ^{13}C NMR spectra, normalized to the same total intensity, of PVA, PVA 300 °C residue, and PVA/maleimide 300 °C residues are shown in Figure 12. For all of the PVA/maleimides pyrolyzed we see evidence of some sort of maleimide reaction. The spectrum of 1,4-PDMI is inset into Figure 12 and shows that the imide carbonyl signals typical of maleimide, at 164 and 175 ppm (which are split due to ^{13}C dipolar coupling to a ^{14}N quadrupolar nucleus), are converted to a single broad signal centered at 176 ppm (see predicted solution state imide-carbonyl ^{13}C chemical shifts in Figure 9) (19). Furthermore, in the aromatic region, 120-140 ppm, of Figure 12 we see the partially resolved resonances of crystalline 1,4-PDMI transformed into a broader resonance indicative of reaction and or a non-crystalline structure.

To gain additional insight into the effect the maleimides had on the PVA pyrolysis we sought to remove the resonances attributable to bismaleimide from the PVA/1,3-

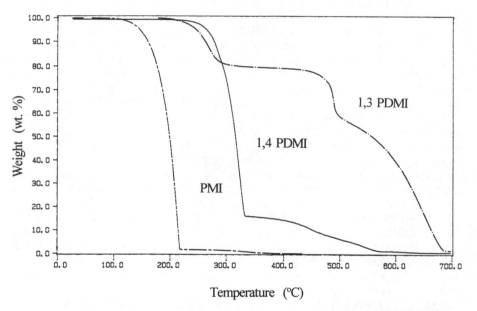

Figure 11. TGA of pure maleimides, 10 °C/min. Note that PMI experiences a rapid mass loss at low (< 200 °C) temperatures, 1,4 PDMI's mass loss is over 80% once the temperature has reached 350 °C, whereas 1,3 PDMI's mass loss is less than 50% even at 500 °C.

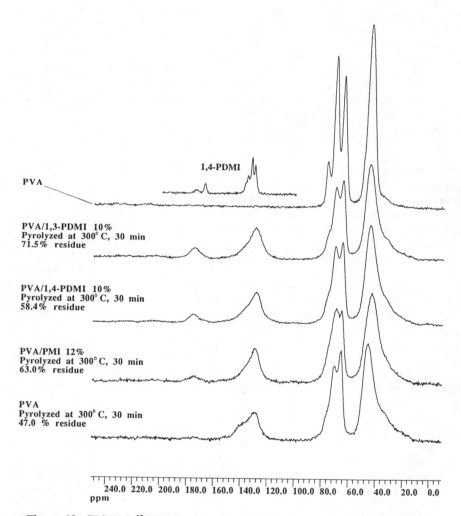

Figure 12. CP/MAS ^{13}C NMR spectra, normalized to the same total intensity, of pure 1,4 PDMI, PVA, three pyrolyzed (300 °C for 30 min) PVA/maleimide mixtures, and PVA pyrolyzed at 300 °C for 30 min. The broad peak near 172 ppm in the spectra of the mixtures arises primarily from the reacted maleimides; the aromatic/olefinic region (120-150 ppm) also contains a significant contribution from the maleimides.

PDMI residue spectrum. This allowed a more detailed comparison of the PVA/1,3-PDMI residue spectrum to the PVA residue spectrum, and was accomplished in the following manner: A sample of pure 1,3-PDMI was pyrolyzed in N_2 at 300 °C for 30 minutes, to give a residue yield of 80%. The idealized bismaleimide "homopolymer" structure, shown in Figure 13, is consistent with the spectrum of the residue from the pyrolysis of pure 1,3 PDMI shown on the bottom of Figure 14. This is based on the appearance and relative intensity of the aliphatic carbon signal centered at 45 ppm and a shift of the carbonyl band toward 175 ppm following pyrolysis. These two spectral features are consistent with polymerization at the double bond adjacent to the carbonyl. A shoulder remains at 170 ppm possibly indicative of the presence of unreacted maleimide carbonyl (~10%). The "corrected" PVA residue difference spectrum, shown at the top of Figure 14, was generated by subtracting the appropriate intensity of the "homopolymer" spectrum (bottom) from the PVA/1,3-PDMI residue spectrum (middle). The subtraction was done so that no intensity remained in the carbonyl region. The rationale which justifies doing this subtraction is based on the fact that the bissuccinimide repeat unit, shown in brackets in Figure 13, is common to both the homopolymer product and the product of the reaction of bismaleimides with polyolefins, as in Figure 9; hence, both the aliphatic and aromatic region of the homopolymer spectrum should give a good approximation of the spectral contributions resulting from reaction of 1,3-PDMI with unsaturation in PVA decomposition products. Also, no substantial carbonyl signals appeared in the pure PVA residue spectrum; therefore the main source of intensity in the carbonyl region should be from 1,3-PDMI.

Figure 15 shows a method of comparing the "corrected" PVA residue spectrum to the normal PVA residue spectrum where the intensities are scaled to their relative residue yields. This allows comparison of the residues on the basis of equal initial amounts of PVA. The resulting difference spectrum, shown on the bottom of Figure 14, appears to be that of intact PVA, indicating that less of the PVA has decomposed in the PVA/1,3-PDMI residue. Furthermore, integration of the normal PVA residue spectrum reveals that the ratio of alcohol carbon signal (-CHOH-) to the aromatic/olefinic carbon signal is 2.2 to 1. However, this ratio is 3.3 to 1 in the "corrected" PVA residue spectrum. Recalling that as pure PVA decomposes this ratio grows smaller and smaller until all the alcohol carbon signal is gone, as seen in Figure 3, these data confirm the fact that the PVA/1,3-PDMI residue is less decomposed than the normal PVA residue. These two samples, however, when compared on the basis of the same starting weight of PVA, contain the same amount of decomposition products (aromatics/olefinics at 120-140 ppm and aliphatics at 10-35 ppm), as evidenced by the nearly complete absence of these type of signals in the difference spectrum in Figure 14. An identical analysis of the 1,4-PDMI 300 °C residue spectra gave the same results.

It is not entirely obvious why, for a given initial amount of PVA, the PVA/1,3-PDMI residue shows the same amount of decomposition products as PVA does; yet, the overall residue yield is greatly increased.

An encapsulation mechanism involving bismaleimide may explain both the reduction in PVA decomposition and a greater retention of otherwise volatile decomposition products. The mechanism involves primarily a homopolymerization of the bismaleimide as a coating on the PVA, possibly accompanied by some crosslinking

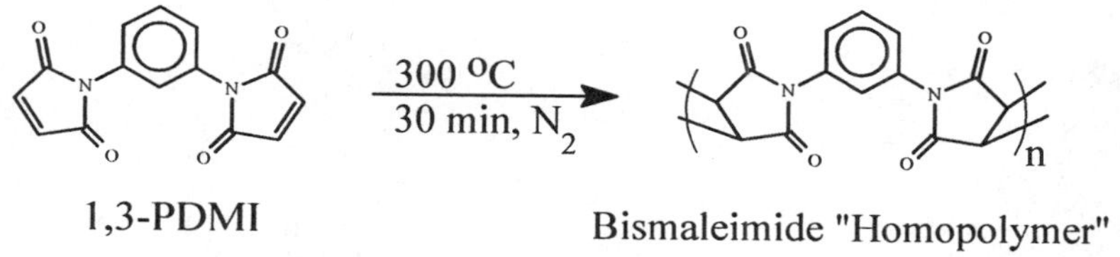

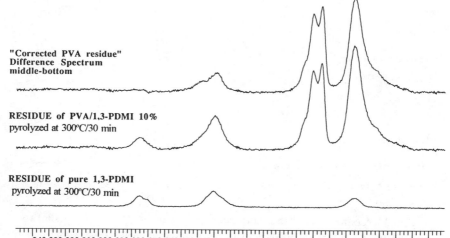

Figure 13. Proposed homopolymerization reaction of 1,3 PDMI, under the indicated conditions, to give the resulting bissuccinimide repeat unit in the polymer structure.

Figure 14. The "corrected" PVA residue difference spectrum (top) generated by subtracting the appropriate intensity of the "homopolymer" spectrum (bottom) from the PVA/1,3-PDMI 300 °C pyrolysis residue spectrum (middle).

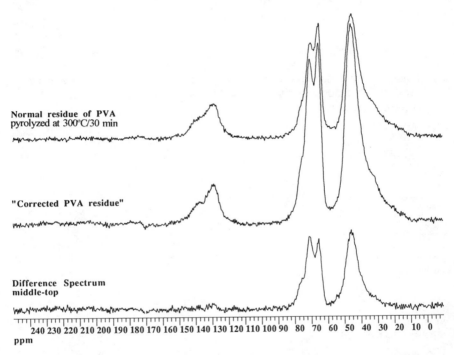

Figure 15. Comparison of the "corrected" PVA residue spectrum (middle) to the normal PVA residue spectrum (top) where the intensities are scaled to their relative residue yields. The resulting difference spectrum is shown on the bottom and appears to be that of intact PVA, indicating that less of the PVA has decomposed in the PVA/1,3-PDMI pyrolysis residue.

of PVA at the interphase between the two polymers. This highly crosslinked coating, similar in structure to the bismaleimide homopolymer shown in Figure 13, may act as a barrier preventing loss of volatiles, such as H_2O and chain-scission products, from the decomposing PVA. The existence of such a barrier may retard reactions, e.g., via Le Chatelier's principle, or it may force trapped volatiles to react with species in the condensed phase, i.e., with bismaleimide or other PVA decomposition products. In particular, it is possible that volatiles containing unsaturation reach the coating of partially polymerized 1,3-PDMI and are trapped in this layer by chemical reaction with maleimide functionality.

Relaxation Studies. We lack direct spectral evidence for reaction of 1,3-PDMI with PVA or its decomposition products. We really only know that most of the imide carbonyls in the PVA/1,3-PDMI residue spectrum are adjacent to sp^3 instead of sp^2 carbons following the pyrolysis. Since the mechanistic mode of action for the maleimides was envisioned to be chemical rather than physical and therefore depended on how well the maleimides mixed with PVA, it seemed relevant to inquire qualitatively about the level of mixing of the maleimides with PVA.

Given the structural differences between the maleimides and PVA we might anticipate differences in intrinsic relaxation times for protons in these compounds. One can use either the rotating frame proton relaxation time, $T_{1\rho}^H$, or the longitudinal proton relaxation time, T_1^H, as a time scale over which heterogeneity of composition is probed. Proton spin diffusion is a transport property (20) of spin polarization which tends to keep polarization per spin uniform in a given locale. In contrast, local differences in relaxation times promote the production of polarization gradients. If, for example, on the time scale of the shortest observed $T_{1\rho}^H$ (~5 ms) or a time scale like 0.75 s which is of the order of T_1^H, spin diffusion is not capable of keeping polarization/spin equal, then one can infer heterogeneity of composition on a distance scale related to these times. Distances of polarization transport are as follows; 5 ms in $T_{1\rho}^H$ corresponds to ~ 1.5 nm, and 0.75 s in T_1^H corresponds to ~ 27 nm. Since domains have boundaries on at least two sides, spin diffusion can usually cover a domain about twice these characteristic dimensions in such a time (21). Lineshape changes in spectra due to differences in $T_{1\rho}^H$ or T_1^H behavior suggest domains bigger than, about, two-thirds of such domain sizes, i.e., 2 nm for a $T_{1\rho}^H$ of 5 ms and 36 nm for T_1^H of 0.75 s.

We found the following results and offer the corresponding interpretations. First, in the pure PVA 300 °C residue there was an inhomogeneity on a T_1^H scale of 0.75 s; PVA decomposition products had a shorter T_1^H. Hence on a 36 nm scale or above, the PVA degradation was not homogeneous. For the PVA/1,3-PDMI residue, the carbonyl region (mainly from 1,3-PDMI) relaxed at a different rate (on a scale of 0.75 s) than did PVA-related resonances, including some intensity in the 120-140 ppm range. Figure 16 shows two spectra taken at different repetition times along with the difference spectrum. (The experimental spectra result from different number of scans; however, they are normalized to the same number of scans.) The absence of a signal in the difference spectrum in the carbonyl region, where 1,3 PDMI contributions dominate, indicates that most of the pyrolysis products of the 1,3 PDMI are in a region where T_1^H is less than 300 ms. The meaning of the difference spectrum in Figure 16 is that there are regions of the sample, which give rise to these resonances, where T_1^H 's are significantly longer than 300 ms. Included in these later regions one can find PVA

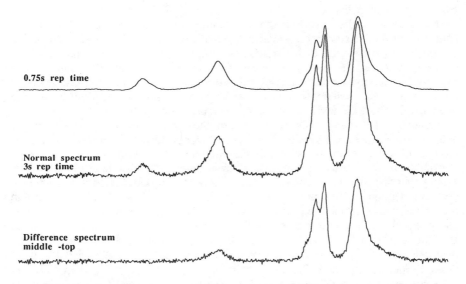

Figure 16. Comparison of CP/MAS ^{13}C NMR spectra of PVA/1,3 PDMI pyrolyzed at 300 °C for 30 min taken at different repetition times (as indicated) and the difference spectrum (bottom). The different lineshapes indicate a chemical inhomogeneity on a scale larger than 36 nm. The disappearance of the maleimide line at 172 ppm in the difference spectrum points to a significant, albeit not necessarily complete, phase separation of the maleimide phase.

olefinic/aromatic decomposition products (120-140 ppm) as well as resonances associated with undegraded or mildly degraded PVA. It should be clearly understood that the data in Figure 16 prove the inhomogeneous mixing of the 1,3 PDMI pyrolysis products; on the other hand, these data are insufficient to show whether partial mixing between PVA and 1,3 PDMI occurred. If there were only two kinds of chemically distinct regions in the sample, one might be able to address this question. Given the chemical inhomogeneities in the pyrolysis products of pure PVA cited above, one might expect three or more regions of chemical inhomogeneity in this sample. It would require a lot more data to sort out the spectra corresponding to each region. We will not attempt to determine whether some PVA pyrolysis products are spatially proximate to the pyrolysis products of 1,3 PDMI, even though this is an interesting question. Suffice it to say that the inhomogeneous mixing of the 1,3 PDMI pyrolysis products demonstrates that PVA and 1,3 PDMI melts are either not highly miscible or that the homopolymerization of PDMI or the pyrolysis of PVA prevents full mixing. Since the proposed mechanism of chemical intervention by maleimide, with the PVA decomposition products, requires thorough mixing of ingredients, we must maintain considerable skepticism that 1,3-PDMI is available to completely react with all of the sites of PVA decomposition. One should bear in mind, however, that even a partial solubility might be efficacious in reducing the decomposition rate of PVA particularly if the elimination of H_2O is autocatalytic, that is, if the presence of unsaturation from a previous elimination reaction, adjacent to an alcohol, accelerates subsequent elimination. Then reaction of even small quantities of bismaleimide with this unsaturation could significantly slow the rate of chain-stripping elimination.

For the PVA/1,4-PDMI 300 °C residue inhomogeneities are also visible on 36 nm or larger scale. For the PVA/PMI 300 °C residue inhomogeneities at the 36 nm scale were by comparison quite minor, although inhomogeneities were evident on the 2 nm scale via $T_{1\rho}^H$ experiments. An important perspective on these T_1^H experiments is that inhomogeneities can be demonstrated only when the intrinsic T_1^H's in the different regions have contrast with one another. Hence the absence of contrast in the PVA/PMI 300 °C residue spectra on a 0.75 s time scale does not automatically imply that mixing is homogeneous on the 36 nm scale. However, as noted previously, the level of retention of PMI derived products in the PVA/PMI 300 °C residue implies that PMI penetrated the PVA in significant amounts and that it mixed more thoroughly with PVA or its decomposition products than did the other maleimides.

Residue Morphology. Examination of the morphology of the residues supports the above observations. Visual inspection, using an optical stereo-microscope (30X), of the PVA 300 °C residue shows a continuous crosslinked, multi-cellular, glassy, tan-colored foam, a material which at this scale appears reasonably homogeneous except for the bubbles. PVA/1,3-PDMI residue, however, looks somewhat heterogeneous. It is made up primarily of a fused mass of 1-3 mm size foam spheres mixed with and coated by dense glassy reddish-brown material. The presence of two different morphologies supports the notion that the bismaleimide and PVA melts are not highly miscible. The formation of the foam spheres implies that the PDMI caused isolation (e.g., by melt wetting in the case of 1,3 PDMI or sublimation in the case of 1,4 PDMI) of PVA particles from one another. If the PDMI formed a melted coating on each PVA particle, then as the coating crosslinked it would behave as an elastomer allowing expansion of the molten particle as the PVA decomposes and generates volatiles. Finally, expansion might cease when the crosslink density got too high.

The morphology of the PVA/1,4-PDMI residue is similar to that of the PVA/1,3-PDMI residue except the foam spheres are not fused together at all and are only lightly "frosted" by a small fraction of the dense glassy reddish-brown material. We were actually able to separate the foam spheres and dense glassy materials by sieving (80 mesh). CP/MAS ^{13}C NMR analysis, similar to that done with the PVA/1,3-PDMI residue (i.e., Figure 14), revealed that the foam spheres (89% of the residue) were less decomposed than the dense glassy reddish-brown material (11% of the residue) even though the foam spheres contained less 1,4-PDMI residue (~4%, by elemental analysis) than the dense glassy reddish-brown material (~37%, by elemental analysis). This appeared, at first, quite puzzling and antithetical to our proposed encapsulation mechanism. It is, however, consistent with our explanation if one considers the possibility that PVA's decomposition products, especially those with lower molecular weight, like those shown in Figure 2, are more miscible with the 1,4-PDMI melt than PVA itself is, and that more extensive PVA decomposition in this material may be explained by a trapping of the decomposition products via chemical reaction with maleimide functionality. To determine, however, if the higher concentration of PVA decomposition products in the additive rich portion of the residue was due to degradation of PVA in the presence of high concentrations of additive, we pyrolyzed a PVA with 36% 1,3-PDMI. The corrected residue yield was in fact 78%, an increase of 7% relative to pyrolyses using 10 % of this bismaleimide. An analysis of the PVA/1,3-PDMI(36%) residue spectrum identical to that done for the PVA/1,3-PDMI(10%) residue, i.e., as in Figures 14 and 15, showed that overall, less PVA decomposes in the presence of the higher level of additive and that when the spectra are normalized to the same starting weight of PVA in the mixture, more decomposition products were present in the PVA/1,3-PDMI(36%) residue. The higher level of decomposition products present supports the idea that they are more miscible with and more easily trapped by the bismaleimide portion of the residue.

The PVA/PMI residue appears much more homogeneous than the PVA/bismaleimide residues and is comparable to the dense reddish-brown material present in the other PVA/bismaleimide residues with a small amount of entrapped bubbles. This morphology supports the implication that PMI mixed more thoroughly with PVA and its decomposition products than did the bismaleimides. However, more thorough mixing did not result in as strong an increase in residue yield as would be expected if molecular accessibility to the maleimide functionality was important to retarding PVA decomposition. Perhaps this relates to the monofunctional rather than difunctional structure of the PMI relative to the PDMI's. Although monofunctionality appears sufficient to retain most of the PMI in the residue. On the other hand, if the PMI is much more accessible for reaction with sites of unsaturation caused by water loss from PVA, then the extent of PVA decomposition observed tends to question the idea that maleimide reaction can arrest an autocatalyzed chain-stripping reaction, unless PMI reacts with itself in preference to reaction with unsaturation at PVA sites. A more likely scenario is that the maleimides react with unsaturation in PVA decomposition products, thereby lowering their average volatility. Whether this happens in the region where the decomposition initially takes place, as it may in the PVA/PMI mixture, or by having unsaturated volatiles get trapped by reaction in a phase-separated domain, as is likely in the PVA/PDMI mixtures, the net effect on residue yield is similar.

Gasification and Cone Calorimetry. The goal of this research was to characterize the fundamental condensed phase processes and structures which lead to char formation during polymer pyrolysis and to use this information to design new strategies which increase char formation and improve polymer flammability properties. An increase in char yield at intermediate temperatures, although promising, does not guarantee a like improvement in combustion char yield or flammability properties. To evaluate the "real" effect the maleimides have on PVA flammability we exposed PVA and two of the PVA/maleimide mixtures to fire-like non-oxidizing conditions in a gasification apparatus which exposes compression molded samples (40g, disks) to a radiant heat source similar to that experienced by materials in a real fire, 40 kW/m^2, in a N$_2$ atmosphere. The results, shown in Figure 17, reveal a significant reduction in the maximum mass loss rate for the PVA/1,3-PDMI and PVA/PMI mixtures as compared to PVA alone. The strongest effect, that of 1,3-PDMI, was a 20% increase in the time to complete pyrolysis. However, the final char yields were all very similar, ~ 5%.

Comparison of the flammability of PVA and the PVA/1,3-PDMI mixture was done in a Cone Calorimeter. This apparatus also exposes compression molded samples (40g, disks) to a radiant heat source (40 kW/m^2), but the sample is exposed to the air and is ignited by a spark source. As would be expected from the above results, there was a delay in the "time of peak heat release," 367 ± 18 seconds for PVA/1,3-PDMI versus 222 ± 33 seconds for PVA alone. However, there were no other significant differences in the flammability behavior for PVA as compared to the PVA/1,3-PDMI mixture (mean heat release rate ~ 470 kW/m^2, total heat released ~ 170 MJ/m^2, heat of combustion ~ 19 MJ/kg). The char yield, however, was slightly increased from 3% to 5% by the presence of 1,3-PDMI.

Conclusions

CP/MAS ^{13}C NMR analysis has allowed characterization of the condensed phase processes which occur during the pyrolysis of PVA and PVA combined with maleimides. The maleimides effect a stabilization of the PVA at 300-400 °C. Although this stabilization improves the pyrolysis residue yields in the 300-400 °C regime, it does not significantly increase the char yields under combustion or high temperature (700-800 °C) pyrolysis conditions. Presumably, this is due to the fact that the maleimides' primary modes of action are physical encapsulation and a lowering of the average volatility (probably by chemical trapping) of certain degradation products as opposed to the formation of thermally stable aromatic crosslinks as was initially envisioned. The moderate increase in "time of peak heat release" resulting from the addition of maleimide, might be viewed as useful if it provided enough of a delay to allow a formulation or product to pass an application specific fire test. However, the results obtained here might be significantly different when reduced to practice in bulk materials, since the effectiveness of the physical encapsulation mechanism depends upon coating the surface of particles, and such a morphology

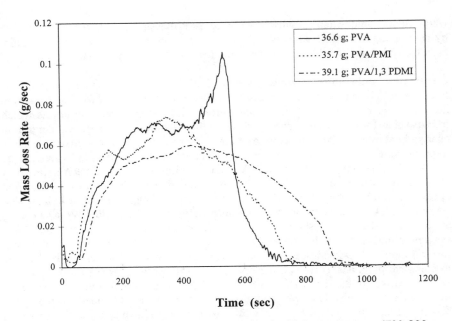

Figure 17. Plot of mass loss rate vs. time for the high temperature (700-800 °C) pyrolysis of PVA, PVA/PMI, and PVA/ 1,3 PDMI in a gasification apparatus which exposes samples (40g, disks) to a high intensity radiant heat source (40 kW/m^2), in a N$_2$ atmosphere.

would not exist in a bulk material. On the other hand a mechanism which reduces the volatility of degradation products might maintain its effectiveness and impart some protection to the bulk. To gain a greater reduction in flammability, however, will require a more through understanding of and control over the condensed phase chemistry.

Acknowledgements
The authors are grateful for the skillful assistance of Mr. Michael Smith, Mr. Jack Lee, and Mr. John Shields. We also thank Dr. Phil Austin for the gasification apparatus work and Dr. Barry Bauer for use of the TGA.

Literature Cited.
1. Dumler, R.; Lenoir, D.; Thoma, H. and Hutzinger, O. *Journal of Analytical and Applied Pyrolysis*, **1989**, 16, pp 153-158.
2. Lindsay, K. F. *Modern Plastics*, February, **1994**, p 54.
3. Van Krevelen, D.W. *Polymer*, **1975**, *vol 16*, pp 615-620.

4. Brauman, S. K. *J. of Fire Retardant Chemistry*, November, **1979**, *vol 6*, pp 248-265.
5. Tubbs, R. K.; Wu, T. K. In *Poly(vinyl alcohol), Properties and Applications;* Finch, C.A., Ed.; John Wiley & Sons: London, 1973, chapter 8.
6. Zhang, X.; Takegoshi, K. and Hikichi, K. *Polymer*, **1992**, vol 33, pp 718-724.
7. Certain commercial equipment, instruments, materials, services or companies are identified in this paper in order to specify adequately the experimental procedure. This in no way implies endorsement or recommendation by NIST.
8. Brauman, S. K.; *J. of Fire Retardant Chemistry*, November, **1979**, *vol 6*, pp 266-275.
9. O'Donnell, D. J. In *NMR and Macromolecules, Sequence, Dynamic, and Domain Structure;* Randall, J. C. Jr., ED.' ACS Symposium Series 247; American Chemical Society, Washington, D. C. 1984; pp 21-41.
10. Opella, S. J.; Frey, M. H. *J. Am.Chem.Soc.* **1979**, *vol 101*, p 5854.
11. Cullis, C.F.; Hirschler, M.M. *The Combustion of Organic Polymers,* Clarendon Press, Oxford, 1981, pp 117-119.
12. Anders, H.; Zimmerman, H. *Polymer Degradation and Stability*, **1987**, *vol 18*, pp 111-122.
13. Muruyama, K.;Takeuchi, K. and Yanizaki, Y. *Polymer,* March, **1989**, *vol 30*, pp 476-479.
14. Cullis, C.F.; Hirschler, M.M.; *The Combustion of Organic Polymers,* Clarendon Press, Oxford, 1981, pp 143-145.
15. ^{13}C NMR assignments taken from: Kalinowski, H. O.;Berger, S. and Braun, S. *Carbon-13 NMR Spectroscopy*, 1988, John Wiley & Sons Ltd., Eds.; pp 132-203., and C-13 Module for Chem Windows, Softshell International software program, version 3.02.
16. Sauer, J.; Sustmann, R. *Angew. Chem.,Int. Ed. Engl.* **1980**,*vol 19*, p 779.
17. Gruffaz, M.; Rollet, B. *U.S. Patent 4,016,114*, April 4, 1977.
18. Carduner, K. R.; Chattha, M. S. In *Crosslinked Polymers, Chemistry, Properties and Applications*, ACS Symposium Series, American Chemical Society, Washington, D. C., 1987, Vol. 367, pp 379-380.
19. VanderHart, D. L.; Tan, L. S. *Materials Research Society Symposium Proceedings*, **1989**. *vol 134*, p 560.
20. Abragam, A. *The Principles of Nuclear Magnetism;* Oxford University Press: London, 1961; Chapter V.
21. Havens, J. R.; VanderHart, D. L. *Macromolecules*, **1985**, *vol 18*, p 1663.

RECEIVED March 31, 1995

Chapter 12

New Types of Ecologically Safe Flame-Retardant Polymer Systems

G. E. Zaikov and S. M. Lomakin

Institute of Chemical Physics of Russian Academy of Sciences, 117977 Kosygin Street 4, Moscow, Russia

New types of ecologically safe flame retardant composition based on poly(vinyl)alcohol and poly(vinyl)alcohol oxidized by $KMnO_4$ were proposed for nylon 6,6. These systems can promote the formation of char by intermolecular crosslinking (" synergetic carbonization"). The Cone Calorimeter tests indicated the improvement of fire retardant properties for the compositions of nylon 6,6 with poly(vinyl)alcohol and poly(vinyl)alcohol oxidized by $KMnO_4$ in comparison with pure nylon 6,6.

The subject of ecological safeness of polymer flame retardants has become a major problem in the modern polymer industry. The different types of polymer flame retardants based on halogens (Cl, Br), heavy and transition metals (Zn, V, Pb, Sb) or phosphorus-organic compounds may reduce risk during polymer combustion and pyrolysis, yet may present ecological issues.
　　The fire retardancy of polymers can be achieved by different ways:
1. By modifying the pyrolysis scheme to produce non-volatile, or non- combustible products that dilute the flame oxygen supply.
2. By smothering the combustion through dilution of the combustible gases, or the formation of a barrier (char) which hinders the supply of oxygen.
3. By trapping the active radicals in the vapor phase (and eventually in the condensed phase).
4. By reducing the thermal conductivity of the material to limit heat transfer (char).
　　　　In our research we have focused on 2,3 and 4. An ecologically-safe flame retardant system (high temperature polymer-organic char former) based on polyvinyl(alcohol) (PVA) in NYLON 6,6 is proposed.

High temperature polymer-organic char former

Our study has been directed at finding ways to increase the tendency of plastics to char when they are burned. There is a strong correlation between char yield and fire resistance. This follows because char is formed at the expense of combustible gases and because the presence of a char inhibits further flame spread by acting as a thermal barrier around the unburned material. The tendency of a polymer to char can be increased with chemical additives and by altering its molecular structure. We have studied polymeric additives (polyvinyl alcohol systems) which promote the formation of char. These polymeric additives usually produce a highly conjugated system - aromatic structures which char during thermal degradation and/or transform into cross-linking agents at high temperatures.

Decomposition of PVA goes in two stages. The first stage, which begins at 200° C, mainly involves dehydration accompanied by formation of volatile products. The residues are predominantly polymers with conjugated unsaturated structures. In the second stage, polyene residues are further degraded at 450°C to yield carbon and hydrocarbons. The mechanism involved in thermal decomposition PVA has been deduced by Tsuchya and Sumi *(1)*. At 245°C water is split off the polymer chain, and a residue with conjugated polyene structure results:

$$(-\underset{\underset{OH}{|}}{CH}-CH_2)_n-\underset{\underset{OH}{|}}{CH}-CH_2- \longrightarrow (-CH=CH_2)_n-\underset{\underset{OH}{|}}{CH}-CH_2- + H_2O$$

Scission of several carbon-carbon bonds leads to the formation of carbonyl ends. For example, aldehyde ends arise from the reaction:

$$-\underset{\underset{OH}{|}}{CH}-CH_2-(-CH=CH_2)_n-\underset{\underset{OH}{|}}{CH}-CH_2- \longrightarrow -\underset{\underset{OH}{|}}{CH}-CH_2-(-CH=CH_2)_n-\underset{\underset{O}{\|}}{CH} + H_3C-\underset{\underset{OH}{|}}{CH}-$$

In the second-stage pyrolysis of PVA, the volatile products consist mainly of hydrocarbons, i.e. n-alkanes, n-alkenes and aromatic hydrocarbons (Table I) *(1)*.

Table I. Thermal decomposition products (240° C, four hours) *(1)*

Products	% by weight of original polymer
Water	33.400
CO	0.120
CO_2	0.180
Hydrocarbons (C_1 - C_2)	0.010
Acetaldehyde	1.170
Acetone	0.380
Ethanol	0.290
Benzene	0.060
Crotonaldehyde	0.760
3-pentene-2-one	0.190
3,5-heptadiene-2-one	0.099
2,4-hexadiene-1-al	0.550
Benzaldehyde	0.022
Acetophenone	0.021
2,4,6-octatriene-1-al	0.110
3,5,7-nonatriene-2-one	0.020
Unidentified	0.082

Thermal degradation of PVA in the presence of oxygen can be adequately described by a two-stage decomposition scheme, with one modification. Oxidation of the unsaturated polymeric residue from dehydration reaction introduces ketone groups into the polymer chain. These groups then promote the dehydration of neighboring vinyl alcohol units producing a conjugated unsaturated ketone structure *(2)*. The first-stage degradation products of PVA pyrolysed in air are fairly similar to those obtained in vacuum pyrolysis. In the range 260° - 280° C, the second-order-reaction expression satisfactorily accounts for the degradation of 80% hydrolyzed PVA up to a total weight loss of 40%. The activation energy of decomposition appears to be consistent with the value of 53.6 kcal/mol which is obtained from the thermal degradation of PVA *(2)*.

The changes in the IR spectra of PVA subjected to heat treatment have been reported *(2)*. After heating at 180° C in air bands appeared at 1630 cm^{-1} (C=C stretching in isolated double bonds), 1650 cm^{-1} (C=C stretching in conjugated diens and triens), and 1590 cm^{-1} (C=C stretching in polyenes). The intensity of carbonyl stretching frequency at 1750 - 1720 cm^{-1} increased, although the rate of increase of intensity was less than that of the polyene band at low temperatures. Above 180° C, although dehydration was the predominant reaction at first, the rate of oxidation increased after an initial induction period.

The identification of a low concentration of benzene among the volatile products of PVA *(2)* has been taken to indicate the onset of a crosslinking reaction proceeding by a Diels-Alder addition mechanism [2]. Clearly benzenoid structures are ultimately formed in the solid residue, and the IR spectrum of the residue also indicated the development of aromatic structures *(2)*.

Acid-catalyzed dehydration promotes the formation of conjugated sequences of double bonds (a) and Diels-Alder addition of conjugated and isolated double bonds in different chains may result in intermolecular crosslinking producing structures which form graphite or carbonization (b).

In contrast to PVA, it was found *(3,4)* that when nylon 6,6 was subjected to temperatures above 300°C in an inert atmosphere it completely decomposed. The wide range of degradation products, which included several simple hydrocarbons, cyclopentanone, water, CO, CO_2 and NH_3 suggested that the degradation mechanism must have been highly complex. Further research has led to a generally accepted degradation mechanism for aliphatic polyamides *(5)*:

1. Hydrolysis of the amide bond usually occurred below the decomposition temperature;
2. Homolytic cleavage of C-C, C-N, C-H bonds generally began at the decomposition temperature and occurred simultaneously with hydrolysis;
3. Cyclization and homolytic cleavage of products from both of the above reactions occurred;
4. Secondary reactions produced CO, NH_3, nitriles, hydrocarbons, and carbon chars.

The idea of introducing poly(vinyl alcohol) into nylon 6,6 composition is based on the possibility of high-temperature acid-catalyzed dehydration. This reaction can be provided by the acid products of nylon 6,6 degradation hydrolysis which would promote the formation of intermolecular crosslinking and char. Such a system we have called "synergetic carbonization" because the char yield and flame suppression parameters of the polymer blend of poly(vinyl alcohol) and nylon 6,6 are significantly better than pure poly(vinyl alcohol) and nylon 6,6 polymers.

It is well-known that nylons have poor compatibility with other polymers because of their strong hydrogen bonding characteristics. The compatibility of nylon 6 with poly(vinyl-acetate) (PVAc) and poly(vinyl alcohol) (PVA) has been studied *(6)*. Compatibility was judged from the melting temperature depression. The results indicate that nylon 6/polyvinyl alcohol blends are partially compatible. "Compatibility" in this work does not mean thermodynamic miscibility but rather easiness of mixing blends to achieve small size domains.

The next step in our plan to improve the flame resistant properties of poly(vinyl alcohol) - nylon 6,6 system was the substitution of pure poly(vinyl alcohol) by poly(vinyl alcohol) oxidized by potassium permanganate (PVA-ox). This approach was based on the fire behavior of the (PVA-ox) itself. It was shown experimentally (Cone Calorimeter) the dramatic decrease of the rate of the heat release and significant increase in ignition time for the oxidized PVA in comparison with the original PVA.

The literature on the oxidation of macromolecules by alkaline permanganate presents little information about these redox-systems. One set of workers investigated *(7,8)* the oxidation of PVA as a polymer containing secondary alcoholic groups by **$KMnO_4$** in alkaline solution. It was reported that the oxidation of PVA by/in alkaline solutions occurs through formation of two intermediate complexes (1) and/or (2) *(8)*: The reactions (a) and (b) lead to the formation of poly(vinyl ketone) (3) as a final product of oxidation of the substrate. Poly(vinyl ketone) was isolated and identified by microanalysis and spectral data *(9)*.

$$(-CH_2-\underset{\underset{OH}{|}}{\overset{\overset{H}{|}}{C}}-)_n + OH^{\ominus} \underset{-H_2O}{\overset{K_1}{\rightleftharpoons}} (-CH_2-\underset{\underset{O^{\ominus}}{|}}{\overset{\overset{H}{|}}{C}}-)_n + MnO_4^-$$

(a) ↙ ↘ (b)

Path (a):
$$(-CH_2-\underset{\underset{O-Mn^{6+}O_4^{2-}}{|}}{\overset{\overset{H}{|}}{C}}-)_n \quad (1)$$

$$(-CH_2-\underset{\underset{O-Mn^{6+}O_4^{2-}}{|}}{\overset{\ominus}{C}}-)_n \quad OH^{\ominus}$$

$$(-CH_2-\underset{\underset{O}{\|}}{C}-)_n + Mn^{5+}O_4^{3-} \quad (3)$$

Path (b):
$$H\cdots OMn^{6+}O_3^{2-}$$
$$(-CH_2-\underset{\underset{O^{\ominus}}{|}}{\overset{\overset{H}{|}}{C}}-)_n \quad (2)$$

$$(-CH_2-\underset{\underset{O}{\|}}{C}-)_n + HMn^{5+}O_4^{2-} \quad (3)$$

Experimental

Materials. The polymers used in this work were poly(vinyl alcohol), 99% hydrolyzed, M.W. 86,000, nylon 6,6 and polypropylene, isotactic, were supplied by Scientific Polymer Products, Inc., USA. The inorganic additive was potassium permanganate, R. (BA Chemicals Ltd.).

Preparation of samples, incorporation of additive. The samples for combustion measurements (blends of nylon 6,6 and PVA, PVA-ox) were prepared in a laboratory blender at room temperature (10 min), the mixed samples were compression molded at temperature 220-240°C for 10 min.

Poly(vinyl alcohol) was oxidized by **KMnO$_4$** in aqueous solution. A 10% wt. aqueous solution of poly(vinyl alcohol) was prepared at 90°C in a laboratory vessel (2 l). **KMnO$_4$** (5% by wt. of original PVA) was added into the hot aqueous solution of PVA. After a fast reaction (1.5 - 2 min.) the solution became dark-brown in color. It was allowed to cool to room temperature. Then water was removed *in vacua* at 50°C to yield of soft elastic material. This material was heated in an oven for 24 hours at 120°C to give a hard plastic material. The resulting material was milled in a laboratory ball-mill to produce a dark brown powder.

Cone Calorimeter tests on the polymer samples, as discs (radius 35 mm), were carried out at 20, 30, 35 and 50 kW/m². Each specimen was wrapped in aluminum foil and only the upper face was exposed to the radiant heater.

Results and Discussion

Preliminary Cone tests for PVA and PVA oxidized by $KMnO_4$ were carried out at heat fluxes of 20, 35 and 50 kW/m² (Table II). It is clearly seen carbon residue (wt. %) and peak of heat release rate (Peak R.H.R. kW/m²) suggest substantial improvement of fire resistance characteristics for PVA oxidized by $KMnO_4$ in comparison with PVA. PVA oxidized by $KMnO_4$ gives about half the peak of heat release rate (Peak R.H.R. kW/m²), when compared with pure PVA. Even at 50 kw/m², the yield of char residue for PVA oxidized by $KMnO_4$ was 9.1%. One reason for this phenomenon may be explained by the ability for PVA oxidized by $KMnO_4$ - (polyvinyl ketone structures) to act as a neutral (structure 1) and/or monobasic (structure 2) bidentate ligand *(9)*.

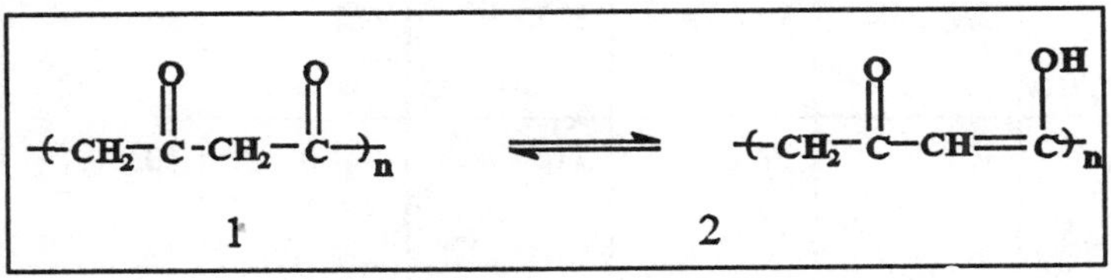

The experimental results of others (IR and electronic spectra) *(6)* provide strong evidence of coordination of the ligand (some metal ions Cu^{2+}, Ni^{2+}, Co^{2+}, Cd^{2+}, Hg^{2+}) through the monobasic bidentate mode (structure 2). Based on the above the following structure can be proposed for the polymeric complexes:

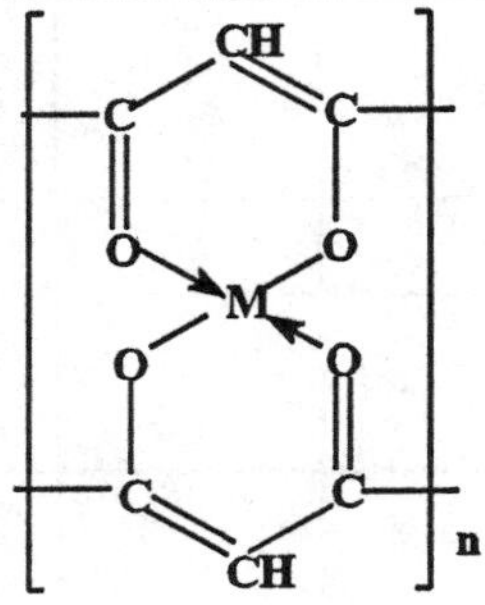

Table II. Cone Calorimeter Data of nylon 6,6 / PVA compositions

Material, Heat flux, kW/m²	Initial wt., g	Char yield, % wt.	Ignition time, sec.	Peak R.H.R., kW/m²	Total Heat Release, MJ/m²
PVA, 20	47.6	8.8	39	255.5	159.6
PVA, 35	28.3	3.9	52	540.3	111.3
PVA, 50	29.2	2.4	41	777.9	115.7
PVA-ox KMnO$_4$, 20	27.9	30.8	1127	127.6	36.9
PVA-ox KMnO$_4$, 35	30.5	12.7	774	194.0	103.4
PVA-ox KMnO$_4$, 50	29.6	9.1	18	305.3	119.8
PVA (100°C) KMnO$_4$, 20	31.1	16.3	303	211.9	124.5
PVA (200°C) KMnO$_4$, 20	35.9	25.7	357	189.0	91.1
nylon 6,6, 50	29.1	1.4	97	1124.6	216.5
nylon 6,6 + PVA(8:2),50	26.4	8.7	94	476.7	138.4
nylon 6,6 + PVA-ox(8:2) KMnO$_4$, 50	39.1	8.9	89	399.5	197.5

The result of elemental analysis of PVA oxidized by $KMnO_4$ indicates the presence of 1.5% of Mn remaining in this polymeric structure. Thus, we suggest that this catalytical amount of chelated Mn-structure incorporated in the polymer may provide the rapid high-temperature process of carbonization and formation of char.

The cone calorimeter results of two PVA samples mixed with $KMnO_4$ at 100°C and 200°C clearly showed the advantage of chemical reaction of PVA with $KMnO_4$ in the liquid phase in comparison with the solid phase mixtures (Table II).

The fire tests at 50 kW/m² for Nylon 6,6 and PVA (80:20%) compositions (typical rate of heat release curves for each sample is shown in Fig.1 confirmed the assumption of the synergistic effect of carbonization. Each of the individual polymers is less fire resistant than their composition. The scheme of "synergistic" carbonization of Nylon 6,6 and PVA is shown below. Similar trends are observed for Average Heat Release which was calculated over the total flame out period (Fig.2). The sample with PVA oxidized by $KMnO_4$ displayed even a better flame retardant properties due to the catalytical effect of Mn-chelate fragments on the formation of char (Table II). The superior Rate of Heat Release properties of PVA oxidized by $KMnO_4$ are clearly shown by comparison of the values of Maximum Rate and Average Rate of Heat Release given in Fig 2. However, a less satisfactory correlation is given in the determination of Total Heat Release date (Table II). Although, the Cone measurements indicate a decrease of Total Heat Release for nylon 6,6-PVA and nylon 6,6-PVA oxidized by $KMnO_4$ in comparison with pure PVA, the sample of nylon 6,6 with PVA oxidized by $KMnO_4$ gives a higher value of Total Heat Release than nylon 6,6 with PVA (Table II). We have qualitatively explained this fact by the influence of a catalytical amount of chelated Mn-structure incorporated in polymer on the smoldering of the polymer samples *(9)*. The flame out time for nylon 6,6 with PVA oxidized by $KMnO_4$ is larger than the flame out times of nylon 6,6-PVA and nylon 6,6 (Table III). The values of Average Heat of Combustion indicate the exothermal process of smoldering provided by chelated Mn-structures (Table III). That is why, we have found the approximately equal amount of char yield for nylon 6,6-PVA and nylon 6,6-PVA oxidized by $KMnO_4$ (Table II).

Table III. Cone Calorimeter Data of the Heat of Combustion and the Flame out time for nylon 6,6 compositions at a heat flux of 50 kW/m²

COMPOSITION	Flame out time, sec.	Avr. Heat of Com.,MJ/kg
nylon 6,6	512	31.50
nylon 6,6-PVA (80:20)	429	25.15
nylon 6,6-PVA-oxidized by $KMnO_4$ (80:20)	747	29.52

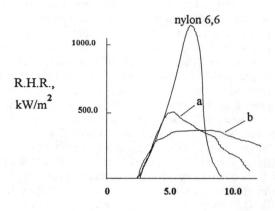

Figure 1. Rate of Heat Release vs. time for: nylon 6,6; nylon 6,6 / PVA (80%:20% wt.) - (a); nylon 6,6 / PVA oxidized by $KMnO_4$ (80%:20% wt.) - (b) at a heat flux of 50 kW/m^2.

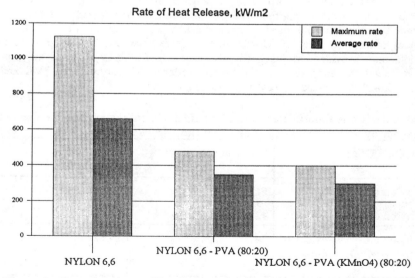

Figure 2. Cone Calorimeter Heat Release data for nylon 6,6-PVA compositions at 50 kW/m^2 of heat flux.

Scheme of "Synergistic" Carbonization of NYLON 6,6 and PVA blend

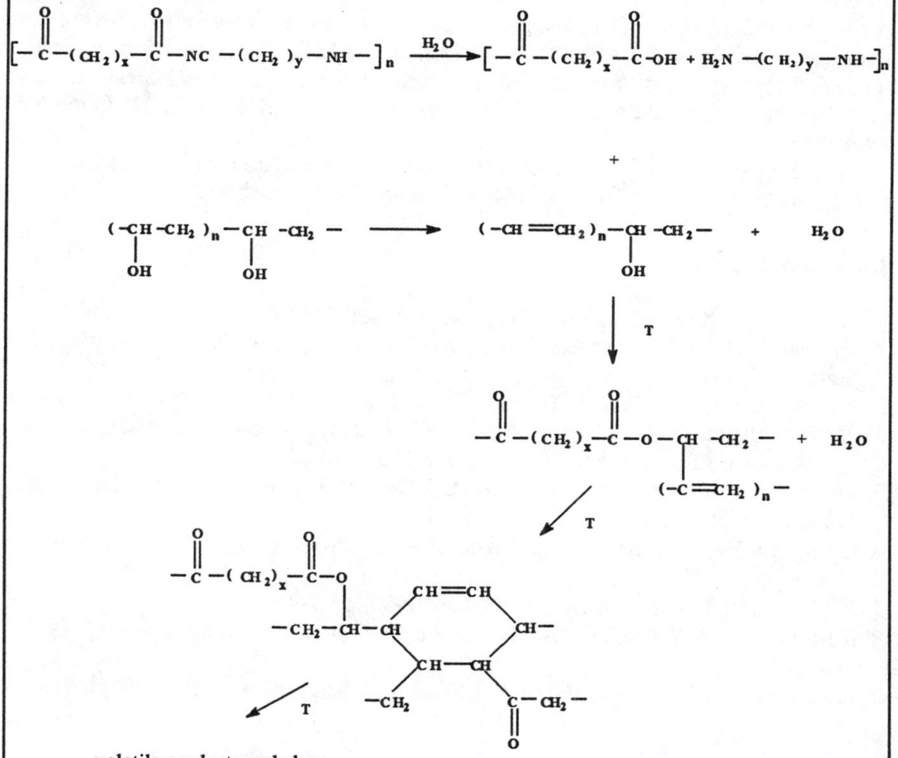
volatile products and char

Conclusions

(1) Polymer - organic char former (PVA system) incorporated in Nylon 6,6 reduced the peak rate of heat release from 1124.6 kW/m^2 (for Nylon 6,6) and 777.9 kW/m^2 (for PVA) to 476.7 kW/m^2 and increased the char yield from 1.4% (for Nylon 6,6) to 8.7% due to a "synergistic" carbonization effect. (Cone Calorimeter was operated at 50 kW/m^2 incident flux.)

(2) Cone Calorimeter data of Nylon 6,6 composition with PVA oxidized by $KMnO_4$ (Mn - chelate complexes *(9)* showed the improvement of peak rate of heat release from 476.7 kW/m^2 (composition of Nylon 6,6 with PVA) to 399.5 kW/m^2 (composition of Nylon 6,6 with PVA-oxidized by $KMnO_4$). On the other hand, the Cone Data indicated the exothermal process of smoldering for composition of nylon 6,6 with PVA-oxidized by $KMnO_4$. This reaction evidently provided by chelated Mn-structures which increases the Total Heat Release of NYLON 6,6 with PVA-oxidized by $KMnO_4$ in comparison with NYLON 6,6 with PVA.

(3) Polymeric char former such as PVA and (for example) cellulose system may present a new type of ecologically-safe flame retardant system.

Literature Cited

1. Y. Tsuchiya, K. Sumi, J.Polym.Sci., A-1, *v.7*, p.3151, **1969**
2. *Polyvinyl Alcohol.Properties and Applications*, ed. by C.A. Finch, John Wiley & Sons,
   London-New York-Sydney-Toronto, **1973**, 622 p.
3. B.G. Achhammer, F.W. Reinhard, G.M. Kline, J.Appl.Chem., *v.1*, **1951**, p.301
4. I. Goodman, J.Polymer Sci., *v.13*, **1954**, p.175
5. *Thermal Stability of Polymers*, ed. by R.T. Conley, Marcel Deccer inc., New York, **1970**, *v.1*, p.350
6. Chang-Sik Ha, Won-Ki Lee, Tae-Woo Roe, Won-Jei Cho, Polymer Bulletin 31, **1993**, pp.359-365
7. R.M. Hassan, Polymer International, *v.30*, **1993**, pp.5-9
8. R.M. Hassan, S.A. El-Gaiar, A.M. El-Summan, Polymer International, *v.32*, **1993**, pp.39-42
9. R.M. Hassan, M.A. Abd-Alla, M.A. El-Gahmi, J. Mater.Chem., *v.2*, **1992**, p.613

RECEIVED January 4, 1995

Chapter 13

Some Practical and Theoretical Aspects of Melamine as a Flame Retardant

Edward D. Weil and Weiming Zhu

Polymer Research Institute, Polytechnic University, Brooklyn, NY 11201

There is interest in finding non-halogen flame retardant additive systems for wire and cable primary insulation for use in the electric power industry. Electrical requirements for moisture-resistant primary insulation with low power loss prevent the use of relatively polar additives. Promising flame-retardant formulations have been found for EPDM or other polyolefins based on melamine, silanated amorphous kaolinite, and, in the role of char former, poly(2,6-dimethylphenylene oxide) (PPO). Tests of the hot water resistance of the formulation are encouraging. Indications regarding mode of action of melamine and PPO in this system are presented. Melamine itself appears more effective than a melamine condensation product. PPO and kaolinite together appear to form a continuous char barrier.

There is current interest in finding non-halogen flame retardant additive systems for wire and cable insulation. An unmet need in the electric power industry is for flame-retardant moisture-resistant primary insulation with low power loss. Polar additives such as ammonium polyphosphate, ATH or $Mg(OH)_2$ are probably unsuitable for primary insulation, but melamine, with zero dipole, seems plausible.

Until now, the principal uses for melamine as a flame retardant additive have been in intumescent coatings and urethane foams, usually with phosphorus compounds.[1-7] Some published evidence indicates that melamine can act as a flame retardant without need for a phosphorus compound.[9-13]

Encouraging flame retardancy results were obtained in polyolefins by a three-component formulation containing melamine as the principle flame retardant, along with powdered poly(2,6-dimethylphenylene oxide) (PPO) and a silanated calcined kaolinite.

Although melamine required a high level of addition to meet the flame retardancy goals, it is inexpensive and has little effect on processing. We believe it is the principal flame retardant in our formulation, and will discuss its mode of action shortly. PPO has the precedent of being recognized as a char forming component in HIPS-PPO blends, but it has not to our knowledge been used in polyolefins as an additive. Silanated kaolin or kaolinite is an amorphous calcined natural mineral which has been used extensively for many years in wire and cable insulation, being favorable from a power loss standpoint.

Experimental

The ingredients used in our study were the following:

-NORDEL 2722 (DuPont), an ethylene-propylene-hexadiene terpolymer. This is an elastomer commonly used in cured form as a wire and cable insulation. Several other polymers used are referenced below in the discussion.
-Dicumyl peroxide (Akzo PERKADOX BC).
-Poly(2,6-dimethyl-1,4-phenylene oxide) (PPO; GE Plastics BLENDEX HPP 820, 821 and 823, intrinsic viscosities 0.40, 0.28 and 0.49 respectively).
-Melamine (superfine grade, Melamine Chemicals Co.).
-Silanated calcined kaolinite (Engelhard's TRANSLINK 37).

Mixing was done using a Brabender Plasticorder and sheets of 1/8" thickness were pressed using a Carver press at 180°C and 10,000 psi. Curing was accomplished at 180°C for 20 minutes. Samples were cut from the sheet for oxygen index (ASTM D-2863) and UL 94 tests.

Oxygen index (OI) was determined by ASTM D-2863 using a Stanton Redcroft FTA Flammability Unit (Tarlin Scientific). The UL 94 vertical burning test was done by the standard procedure developed by Underwriter Laboratories.

We recognized that the ultimate flame retardancy goal for power wire and cable in the U. S. is most often IEEE 383, which requires cable fabrication and is impractical for a laboratory formulation project; however, many workers in this field consider oxygen index (target ≥ 30) and UL 94 vertical burning test (target V0) to be useful indicators. We considered that the use of both of downward-burning and upward-burning methods would give a more reliable indication of flame retardant behavior than either test alone. We sought formulations which would give good flame retardant performance by both oxygen index and UL 94. Since such small scale tests do not reliably predict full scale fire behavior, nor do they even reliably predict larger scale laboratory tests, the results of the present study can be considered only a precursor to larger scale evaluations. The results presented herein are certainly not to be construed as predictive of fire hazard.

We did not evaluate electrical properties in the phase of our investigation reported herein. Our strategy was to utilize only non-ionic low-polarity components in the development of flame retardant formulations and to postpone electrical measurements to a later phase.

TGA measurements were completed on a DuPont 951 thermogravimetric analyzer under 1% O_2 in N_2 with a heating rate of 20°C/min and a gas flow rate of 50 ml/min. DSC experiments were done on a DuPont DSC 910 differential scanning calorimeter with a heating rate of 20°C/min under 1% O_2 in N_2 at flow rate of 30 ml/min. IR was carried out with a Shimadzu IR-435 IR spectrophotometer. Optical microscopy was performed with a Nikon Optiphot microscope and a high intensity fiber optic illuminator (FIBER-LITE 3100, Dolan-Jenner Industries, Inc.). Cone calorimetry was performed at VTEC Laboratories Inc. Smoke density and toxic gas content of the smoke were measured at Polyplastex International Inc. using an NBS Smoke Chamber and Draeger tube measurements of the gases.

Results and Discussion

The most successful results, in terms of achieving OI value above 30 and UL 94 ratings of V1 or preferably V0 at 1/8" thickness, were with combinations of melamine, PPO, and silanated calcined kaolin. Table I shows that only the ternary system of above three components provides satisfactory flame retardancy for EPDM. Any combination of two components, melamine-PPO, melamine-kaolin, or PPO-kaolin was not effective in reducing the flammability of EPDM. On the other hand, the introduction of a third component into the EPDM formulation while keeping the total additive loading constant substantially increased OI values and reached the desired UL 94 ratings.

Table I. Flammability of EPDM Filled with Melamine, PPO, and Kaolin[a]

Total loading of ingredients (phr[b])	Melamine (phr)	PPO[c] (phr)	Kaolin (phr)	OI	UL 94[d]
160	100	60	0	25.7	Burning
160	100	0	60	25.9	Burning
160	60	0	100	25.4	Burning
160	0	60	100	22.9	Burning
160	60	40	60	32.4	V1
180	100	0	80	26.8	Burning
180	80	40	60	33.4	V0
180	100	20	60	32.2	V0
200	100	0	100	27.7	Burning
200	100	60	40	34.0	V0
220	100	0	120	29.3	Burning
220	100	60	60	35.0	V0
220	90	50	80	33.3	V0

[a] All formulations contain 100 phr of EPDM (NORDEL 2722) and 3 phr of dicumyl peroxide (DCP) as crosslinking agent. [b] phr = part per hundred resin. [c] GE BLENDEX HPP 821 is used. [d] Rating at 1/8" thickness unless otherwise noted.

Effect of Component Loading Level on the Flame Retardancy. The effect of varying each of the components was studied in an effort to achieve the optimum flame retardancy for EPDM. Increase of PPO level, holding the kaolin at 60 phr and melamine at 100 phr, gradually raised the OI values (Table II). At a PPO loading of 5 phr, the formulation showed compete burning in the UL 94 vertical burning test but extinguished before burning the entire length in the horizontal configuration. The formulation gave a UL 94 V0 rating at a PPO loading of 20 phr and above.

Table II. Effect of PPO Level on the Flammability[a]

PPO (phr)	OI	UL 94 rating
0	25.9	Burning
5	26.8	Burning
20	32.2	V0
30	33.7	V0
45	34.6	V0
60	35.0	V0

[a] All formulations contain 100 phr EPDM, 60 phr kaolin, 100 phr melamine, and 3 phr DCP. PPO used is HPP 821.

The study on the variation of the melamine loading, holding the kaolin at 60 phr and PPO also at 60 phr, showed a similar improvement of the flame retardancy (Table III). The addition of 30 phr or less melamine resulted in self-extinguishment in the horizontal burning test. The formulation was rated V0 with 50 phr or more melamine.

Table III. Effect of Melamine Level on the Flammability[a]

Melamine (phr)	OI (%)	UL 94 rating
0	25.8	Burning
10	29.2	Burning
30	30.3	Burning
50	32.7	V0
100	35.0	V0

[a] All formulations contain 100 phr of EPDM, 60 phr of PPO (HPP 821), 60 phr of kaolin, and 3 phr of DCP.

The loading level of kaolin had interesting effects on the flammability (Table IV). The OI value gradually increased with the increase of the kaolin level. At a loading of 5 to 20 phr, the formulation self-extinguished in the horizontal burning mode but burned totally in the vertical test. A V0 rating was obtained at kaolin loadings equal to or above 40 phr. However, at high loadings (equal to or above 100 phr), the OI value decreased instead.

Kaolin is a nonflammable additive, and is expected to reduce the flammability by a fuel dilution effect in the EPDM formulation. The above observation is contrary to this prediction, and suggests an interaction among the three major additives in our formulations. Another possible explanation is for the poorer flame retardancy at high kaolin loadings is that the cure is inihibited.

Varying the amount of dicumyl peroxide (DCP) (and thus the degree of crosslinking) showed that a plateau was reached at an OI of 33-33.7 with 1.5-2.5 phr of DCP (curve A in Fig. 1), or at an OI of 34.0 at 2.0 phr of DCP (curve B in Fig. 1).

The OI decline at higher peroxide levels might be caused by the presence of peroxide decomposition products such as acetophenone, the presence of which was evidenced by its strong characteristic odor. An experiment was conducted to test this hypothesis. The test specimens were placed in an oven at 100°C for 48 h and 72 h respectively before the OI and UL 94 tests were performed. Results in Table V indicated that the oven aging at 100°C removed effectively the flammable peroxide decomposition products after 72 h and thereby improved the flame retardancy of the EPDM formulation.

Evaluation of Different Polyolefins, Char Forming Additives, Clays, and Melamine Derivatives. A series of experiments were carried out to verify the usefulness of other polyolefins and additives (Table VI). The important observations from these experiments were:
(1) The ternary system of PPO-melamine-kaolin provided satisfactory flame retardancy for peroxide-cured ethylene vinyl acetate copolymer (ELVAX 460, DuPont), peroxide-cured polyethylene (PETROTHENE NA-951, Quantum Chemical Co.), and moisture-cured ethylene vinyltrimethoxysilane copolymer (DFDA-5451 NT, Union Carbide).
(2) Replacing PPO by polyphenylene sulfide (PPS, regular grade RYTON, Phillips 66), 2-imidazolidone-formaldehyde resin (SPINFLAM MF-80 without ammonium polyphosphate, Himont), epoxy-novolac resin (EPON DPS 164, Shell), and dipentaerythritol (Aqualon Co.) resulted in poor OI and failure by UL 94.
(3) Burgess clay (BURGESS KE, also a silanated calcined kaolin) gave the same flame retardancy as TRANSLINK 37 clay. Fused silica (SILTEX 22, Kaopolite Inc.) had an unfavorable result in flame retardancy when it was used to replace the kaolin.
(4) Some melamine salts, melamine cyanurate and melamine phosphate, may be used to replace part or all of the melamine.

Evaluation of Smoke Generation, Toxic Gases in Smoke and Rate of Heat Release. Table VII indicated that the formulation of EPDM, PPO, melamine, and kaolin produced toxic gases with a computed toxicity index of 0.49, measured in the flaming mode. This is not exceptionally high, despite the nitrogen content of the formulation.

The test method used was a modified version of ASTM E-662 as conducted by Polyplastex Corp. with a 5 minute heat exposure time (this modified version is more commonly used to test materials for aircraft cabin use).

The toxic gas concentration was measured by Dräger tubes during the first 5 minutes.

Table IV. Effect of Kaolin Level on Flammability[a]

Kaolin (phr)	PPO (phr)	OI	UL 94 rating
0	60	25.7	Burning
5	60	25.8	Burning
20	60	29.3	Burning
40	60	34.0	V0
60	60	35.0	V0
60	30	33.7	V0
80	30	33.9	V0
100	30	32.3	V0
120	30	32.2	V0

[a] All formulations contain 100 phr of EPDM, 100 phr of melamine, and 3 phr of DCP. PPO used is HPP 821.

Table V. Post-heating Effect on Flammability[a]

Heating time[b] (hours)	OI (%)	UL 94 rating[c]
0	33.1	Burning
48	34.2	V1
72	34.2	V0

[a] The formulation contains 100 phr of EPDM, 100 phr of melamine, 60 phr of PPO (HPP 820), 60 phr of kaolin, and 3 phr of DCP.
[b] Heating temperature: 100°C.
[c] Rating at 1/16" thickness in the vertical burning test.

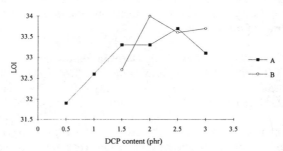

Figure 1. Crosslinking effect on the flammability. Curve A: 100 phr of EPDM, 100 phr of melamine, 60 phr of HPP 820, and 60 phr of kaolin. Curve B: 100 phr of EPDM, 100 phr of melamine, 30 phr of HPP 821, and 60 phr of kaolin.

Table VI. Flammability of Various Polyolefins and Additives[a]

Polyolefin (100 phr)	Char forming additive (phr)	Clay (phr)	Melamine or its derivatives (phr)	OI	UL 94 rating
EPDM	PPO (HPP 821) (30)	TRANSLINK 37 (60)	Melamine (100)	33.7	V0
EVA	PPO (HPP 821) (30)	TRANSLINK 37 (60)	Melamine (100)	35.0	V0
LDPE[b]	PPO (HPP 821) (30)	TRANSLINK 37 (60)	Melamine (100)	32.4	V0
EVS[c]	PPO (HPP 821) (60)	TRANSLINK 37 (60)	Melamine (100)	33.0	V0
EPDM	PPS (40)	TRANSLINK 37 (60)	Melamine (80)	25.7	Burning
EPDM	SPINFLAM MF-80 (40)	TRANSLINK 37 (60)	Melamine (80)	24.5	Burning
EPDM	Dipentaerythritol (40)	TRANSLINK 37 (60)	Melamine (80)	23.0	Burning
EPDM	EPON DSP 164 (50)	TRANSLINK 37 (80)	Melamine (90)	23.5	Burning
EPDM	PPO (HPP 821) (20)	BURGESS KE (60)	Melamine (100)	32.3	V0
EPDM	PPO (HPP 821) (20)	SILTEX 22 (60)	Melamine (100)	31.4	Burning
EPDM	PPO (HPP 821) (60)	TRANSLINK 37 (60)	Melamine (50), melamine cyanurate (50)	33.7	V0
EPDM	PPO (HPP 821) (20)	TRANSLINK 37 (60)	Melamine pyrophosphate (100)	27.0	V0
EPDM	PPO (HPP 821) (50)	TRANSLINK 37 (80)	Melamine cyanurate (90)	37.3	V0

[a] All formulations were cured with 3 phr of DCP unless otherwise noted.
[b] LDPE = low density polyethylene (PETROTHENE NA-951, Quantum Chemical Co.)
[c] EVS = ethylene vinyltrimethoxysilane copolymer. This formulation containing 93 phr of EVS (DFDA-5451NT, Union Carbide), 7 phr of dibutyl bis (1-oxododecyl) oxystannane (1-5 wt.%) in polyethylene (DFDB-5480NT, Union Carbide), and other additives listed in the table was cured in hot water at 100°C for 24 h.

Table VII. Analysis for Toxic Gases in Smoke[a]

Gas	Measured conc. ppm (flaming)	Measured conc. pm (nonflaming)	Lethal conc. (LC_{50}) ppm[b]
CO	200	10	4000
HCN	6	0	150
$NO + NO_2$	100	0	250

[a] The formulation contains 100 phr of EPDM, 60 phr of HPP 820, 100 phr of melamine, 60 phr of kaolin, and 2 phr of DCP.
[b] See reference: J. J. Pickering, *Proc. Intl. Conf. Fire Safety*, **12**, 260-270 (1987).

The smoke generation of the formulation was evaluated by both the NBS smoke chamber (as smoke density, ASTM E-662) and rate-of-heat-release calorimetry (as average specific extinction area). Results in Table VIII showed the low smoke generation of our formulation, which was comparable to commercial hydrate formulations in this respect.[14] Table IX shows rate-of-heat-release data.

Table VIII. Evaluation of Smoke Generation[a]

Method of test	$Ds_{1.5}$[c]	$Ds_{4.0}$[d]	Ave. SEA[e]
NBS smoke chamber[b] (flaming mode)	18	121	-
NBS smoke chamber[b] (nonflaming)	2	62	-
Cone calorimetry (25 KW/m² heat flux)	-	-	280.7
Cone calorimetry (50 KW/m² heat flux)	-	-	270.0

[a] For information on the formulation, see Table VII. [b] ASTM E-662. [c] $Ds_{1.5}$ = smoke density at 1.5 min. [d] $Ds_{4.0}$ = smoke density at 4.0 min. [e] Ave. SEA = average specific extinction area.

The Max. RHR values at 50 kW/m2 are in the range (328-497) reported by Malin[14] for polyolefins with halogenated flame retardant additives; the TTI values appear somewhat longer than the range (61-116 sec.) reported for these halogen systems and towards the high end of the TTI range reported by the same author for non-halogen (metal hydroxide) flame-retarded polyolefins.

In conclusion, the formulation of EPDM, PPO, melamine, and kaolin showed unexceptional smoke toxicity (based on gas analysis), low visible smoke generation, and delayed ignition.

Table IX. Data from Cone Calorimetry Tests[a]

Heat flux (KW/m^2)	Max RHR[b] (KW/m^2)	THR[c] (MJ/m^2)	TTI[d] (s)	Ave HOC[e] (MJ/kg)	Ave MLR[f] (g/m$^2 \cdot$s)	Ave SEA[g] (m^2/kg)
25	250	87.7	480	20.2	2.99	280.7
50[h]	380	126.7	132	21.0	5.64	262.8
50[h]	449	150.4	128	24.8	5.83	277.2

[a] The formulation tested contains 100 phr of Nordel 2722, 60 phr of HPP820, 100 phr of melamine, 60 phr of Translink 37, and 3 phr of DCP.
[b] Max RHR = maximum rate of heat release. [c] THR = total heat release.
[d] TTI =- time to ignition. [e] Ave HOC = average heat of combustion.
[f] Ave MLR = average mass loss rate. [g] Ave SEA = average specific extinction area.
[h] Replicates.

Evaluation of the Water Absorption and Processability. The formulation of EPDM, PPO, melamine, and kaolin showed favorable hot water resistance. In hot water at 82°C (180 F) for 7 days under conditions of the UL 83 test method, only 0.9% weight uptake occurred.

The processability of the formulation was evaluated by its apparent viscosity measured by means of the Brabender Plasticorder. The apparent viscosity of pure EPDM, flame retardant EPDM, and flame retardant EPDM containing a processing aid are listed in Table X.

As the data in the following table show, the flame retardant EPDM formulation had about 50% higher viscosity than pure EPDM. Through measuring the apparent viscosities of several formulations containing PPO, melamine, or kaolin at a constant total loading of additives, we found that the viscosity increase of the flame retardant formulation was probably mostly due to the presence of kaolin. Melamine slightly increased the apparent viscosity of the formulation, while PPO seemed to have little effect on the apparent viscosity. Addition of low density polyethylene as a processing aid reduced the viscosity. The presence of additional silane coupling agent (in addition to that already on the TRANSLINK 37) did not seem to affect the viscosity.

Mode of Action of Melamine. Melamine is capable of acting as a flame retardant in several ways which we have recently reviewed.[6] It can sublime endothermically with consumption of 0.23 Kcal/g (0.96 KJ/g)[7] and this physical transformation has been recognized as one of its modes of action. Under TGA conditions (on 1% oxygen in nitrogen), melamine is practically all volatilized by itself (curve A in Fig. 2) or mostly vaporized in the cured EPDM formulation (curve B in Fig. 2) when heated to a temperature well below that of the thermooxidative degradation of EPDM (curve C in Fig. 2). Melamine vapor can be seen to condense to a copious white smoke during

Table X. Viscosity of EPDM Formulations

composition of the EPDM formulation (phr)						Viscosity (kpoise)[a]
EPDM	PPO	Melamine	Kaolin	Additive	DCP	
100	-	-	-	-	3.0	9.51
100	HPP 821 (220)	-	-	-	3.0	8.25
100	-	220	-	-		10.90
100	-	-	220	-		17.10
100	HPP 821 (50)	90	80	-	2.5	14.20
90	HPP 821 (50)	90	80	LDPE[b] (10)	2.5	13.07
100	HPP 820 (60)	100	60	-	1.5	13.01
90	HPP 820 (60)	100	60	LDPE (10)	1.5	12.81
80	HPP 820 (60)	100	60	LDPE (20)	1.5	12.61
100	HPP 820 (60)	100	60	UCARSIL RC-1[c] (1)	2.0	13.14
100	HPP 820 (60)	100	60	UCARSIL RC-1 (1.5)	2.0	13.21

[a] Standard deviation: 0.16 kpoise. [b] LDPE = low density polyethylene (PETROTHENE NA-951, Quantum Chemical Co.).
[c] UCARSIL RC-1 is a silane coupling agent (OSi Specialties, Inc.), widely used in cable and wire formulation to improve the wet electric properties.

sublimation. Such particulates may provide further flame retardant action by several physical modes. The vapors comprise an energy-poor fuel. Melamine can dissociate endothermically to three moles of cyanamid at about 610°C[8] and this chemical dissociation may make a further contribution to its flame retardant action.

Melamine can also undergo self-condensation, almost certainly endothermically, with release of ammonia and formation of a series of condensation products, first melam, then melem, finally a crosslinked polymer, melon.[15] The released ammonia, while combustible, also is an energy- poor fuel. The release of ammonia has been associated with flame retardant effects. Ammonia has been shown to have flame inhibitory action by several modes.[16,17] Finally, the condensation products, such as melon, can provide a low combustibility barrier, a sort of nitrogenous char. The TGA experiment of a crude melon showed the very high thermooxidative stability of this material (curve E in Fig. 2), which clearly survives the degradation of EPDM under 1% oxygen in nitrogen (curve D in Fig. 2).

The following experiments point to the sublimation/vapor phase change/ammonia release as likely to be the most important effect: comparable formulations were prepared with either 100 phr of either melamine itself or a crude melamine-free melon. The formulation containing the melamine was found to give an OI of 22.7 whereas the formulation with the melon gave OI of 20.7. Likewise, the burning rate of the melamine formulation, ignited as a horizontal UL 94 test bar was 8.4 mm/min, whereas with the melon formulation, it was 14.2 mm/min.

Burning 808 mg of the formulation of EPDM(100 phr)-melamine(100 phr) at an oxygen concentration of 3 units above the OI yielded 55 mg (6.8%) of residue. The infrared spectrum of this residue showed bands characteristic of melon at 760, 800, 1230, 1400 and 3120 cm^{-1} (Fig. 3). The elemental analysis showed that the residue contained 56.70% N, 36.94% C, and 2.19% H, corresponding to 91% melon. The result indicated that the melamine underwent about 16.6% conversion to residual non-volatile solid, i.e., melon, and about 83.4% vaporization (sublimed melamine plus volatile thermolysis or combustion products).

Mode of Action of PPO. In the formulation containing EPDM (100 phr), PPO(HPP 820, 60 phr), kaolin (60 phr), and melamine (100 phr), we observed that the residue after burning at the oxygen concentration of 3 units above OI amounted to 25.1% of the original weight to which melamine contributed 1.9%, the calcined kaolin contributed 17.1% and the PPO contributed 6.1%. These percentages were calculated from elemental analysis of the residue which showed 20.59% C, 0.71% H and 4.68% N. This nitrogen content corresponds to 7.5% yield of melon based on the original melamine. Therefore, it appears that 92.5% of the melamine was vaporized or converted to volatile products, in reasonably good agreement with the fate of the melamine in the kaolin-free PPO-free experiment.

We can also get some information on the fate of the PPO (i.e. the char yield from the PPO) from this experiment. We can deduct from the weight of the 25.1% residue the amount (18.58%) corresponding to the kaolin (assuming it retains its original weight during the combustion) and we can further deduct the weight of melon (0.0747 × 25.1 = 1.87%) in the residue calculated from the 4.68% N content of the residue (equivalent to 7.47% melon), we arrive at 4.65% (25.1 - 18.58 - 1.87) of the original

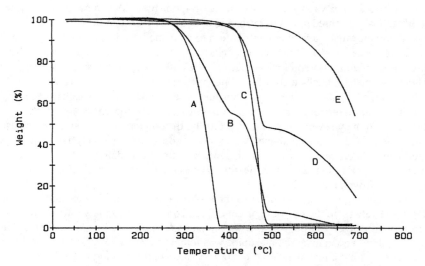

Figure 2. Thermogravimetric curves, obtained at a heating rate of 20°C/min under 1% oxygen in nitrogen atmosphere. A: melamine; B: 1:1 formulation of EPDM and melamine; C: EPDM (NORDEL 2722); D: 1:1 formulation of EPDM and melon; E: melon.

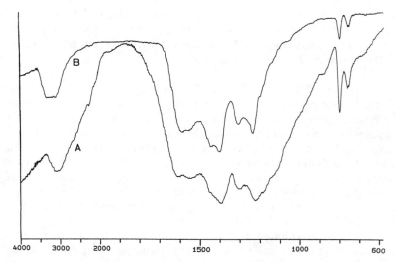

Figure 3. Infrared spectra (in KBr). A: residue from burning 1:1 formulation of EPDM and melamine. B: melon, sample obtained from Chemie Linz.

formulation weight representing char from the PPO. Since the original formulation had 18.58% PPO, this amount of char represents a 25% weight yield from PPO, assuming the EPDM does not leave char. A 25% char yield from PPO is in reasonable agreement with the 29% reported char yield from PPO under TGA conditions in nitrogen.[18] The char-forming ability of PPO was shown by GE investigators to be part of the flame retardant mode of action of organophosphate-flame-retarded PPO-HIPS blends.[7]

Figure 4 represents DSC thermograms (under 1% O_2 in N_2 and at 20°C/min) of EPDM formulations. Upon heating, the cured EPDM showed a large exothermic peak starting at about 210°C and extending to 480°C, caused by thermal degradation of EPDM (curve A). The introduction of melamine into the EPDM formulation suppressed some part of the exothermic peak, giving two endothermic peaks at 362°C and 380°C respectively (curve B). The flame retardant formulation of EPDM-PPO-melamine-kaolin showed three endothermic peaks and no exotherm up to 550°C (curve C).

The suppression of the exothermic peak by melamine is believed to be related to the flame retardant effect of melamine, because melamine may act as an energy-poor, vapor phase flame retardant through the sublimation and disassociation as we discussed above. The ternary system of PPO-melamine-kaolin effectively suppresses the exotherm of EPDM degradation in the temperature range of 210°C to 550°C, and thus provides satisfactory flame retardancy for EPDM.

Morphology of Burning Residues from Various Formulations. The morphological study on burning residues (or chars) was performed with an optical microscope and an optical fiber top illuminator. We observed that the residue from burning of the formulation of EPDM-PPO-melamine-kaolin appeared to be a dark cement-like solid with good coherence and continuity (Fig. 5). None of the formulations of either EPDM-PPO-melamine, EPDM-melamine-kaolin, or EPDM-PPO-kaolin produced as coherent residues as the EPDM-PPO-melamine-kaolin did (Fig. 6-8). In cases of EPDM-PPO-melamine and EPDM-melamine-kaolin, the residues had a cracked appearance, while the formulation of EPDM-PPO-kaolin showed almost no cracks (Fig. 8). All residues were obtained from burning the corresponding formulation at an oxygen concentration of 3 units above the OI.

We consider that an important function of the PPO in this formulation is to provide a "binder" for the kaolin during burning, thus creating a barrier. The melamine condensation product (melon) does not appear effective for this purpose, but PPO does appear to have the requisite properties. Separate experiments in which PPO was burned by exposure to UL 94 type flame showed that PPO melted before it charred. The molten PPO might flux with kaolin clay and help form a coherent residue. We hypothesize that the property of melting before charring may be an important requisite of a char former for effectiveness in forming a barrier to stop flame propagation.

Conclusions

A halogen-free phosphorus-free polyolefin formulation showing substantial flame retardancy by two small scale tests was developed based on melamine, a commercial polyphenylene oxide and a commercial calcined amorphous kaolin. Limited cone

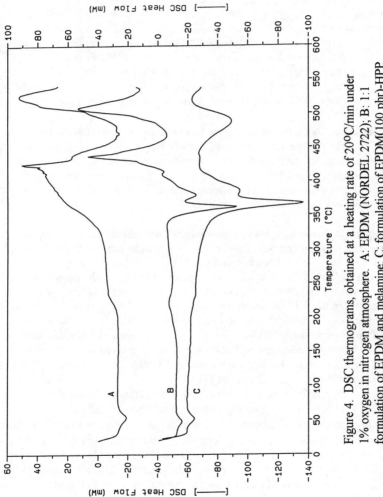

Figure 4. DSC thermograms, obtained at a heating rate of 20°C/min under 1% oxygen in nitrogen atmosphere. A: EPDM (NORDEL 2722); B: 1:1 formulation of EPDM and melamine; C: formulation of EPDM(100 phr)-HPP 821(30 phr)-melamine(100 phr)-kaolin(60 phr).

Figure 5. Optical micrograph of the residue from burning formulation of EPDM-PPO-melamine-kaolin, at 300 X.

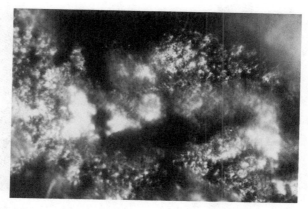

Figure 6. Optical micrograph of the residue from burning formulation of EPDM-PPO-melamine, at 300 X.

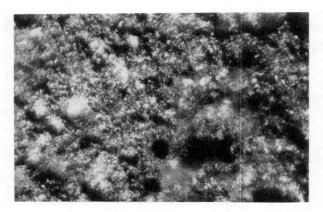

Figure 7. Optical micrograph of the residue from burning formulation of EPDM-melamine-kaolin, at 300 X.

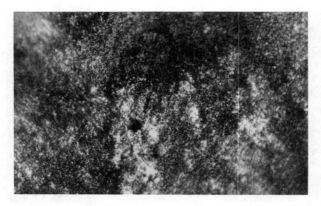

Figure 8. Optical micrograph of the residue from burning formulation of EPDM-PPO-kaolin, at 300 X.

calorimetry tests showed delayed ignition. The formulation showed encouraging hot water resistance, and it appeared to be readily processable.

Some mode of action proposals are made. Since melamine itself was more effective than the melamine condensation product, endothermic effects of melamine (sublimation, vapor dissociation, and ammonia release) appear more important than the possible barrier action of non-volatile condensation products. Some experimental observations indicate possible importance of the char former melting before charring; the role of the PPO appears to be to melt and char thus forming, together with the kaolin, a barrier layer.

Acknowledgment

The authors wish to thank the Electric Power Research Institute, and in particular Mr. Bruce S. Bernstein, Technical Advisor, for support and for permission to publish these results. We also thank Prof. Eli M. Pearce, Director of the Herman F. Mark Polymer Research Institute, for his assistance during the course of this project.

Literature Cited

1. H. L. Vandersall, *J. Fire Flamm.* **2**, 97-140 (1971).
2. M. Kay, A. F. Price and I. Lavery, *J. Fire Ret. Chem.* **6**, 69-91 (1979).
3. A.M. Batt and P. Appleyard, *J. Fire Sci.* **7**, 338-363 (1989).
4. J. Hume, K. Pettett and J. Jenc, in Flame Retard.'90, [Proc.], Elsevier, London, 1990, pp. 234-241.
5. O. M. Grace, R. E. Mericle and J. D. Taylor, "Melamine Modified Polyurethane foam," Proc. SPI Annu. Tech./Mark. Conf., 29th ("Magic of Polyurethanes"), SPI, 1985, pp. 27-33.
6. E. D. Weil, C.-H. Huang, N. Patel, W. Zhu and V. Choudhary, paper presented at BCC Conf. on Recent Advances on Flame Retardancy, Stamford, CT, May 1994.
7. G. M. Crews, *Plastics Compounding*, Sep.-Oct. 1992, 41-3.
8. J. S. MacKay (to American Cyanamid Co.), U. S. Pat. 2,566,231 (1951).
9. Hitachi Chemical Co., Ltd.), Jpn. Kokai Tokkyo Koho 58138739 (Aug. 17, 1982); *Chem. Abstr.* **99**(26), 213802u.
10. K. Kobayashi and H. Matsuo (to Shinko Chemical Co.), Jpn. Kokai Tokkyo Koho 52152949 (Dec. 19, 1977); *Chem. Abstr.* **88**(24), 171110g.
11. Oesterreichische Stickstoffwerke A.-G., French Pat. 2,096,230 (Mar. 17, 1972); *Chem. Abstr.* **77**(20), 127486h.
12. G. Bertelli and R. Locatelli (to Montedison), Eur. Pat. Appl. 65160 (Nov. 24, 1982); *Chem. Abstr.* **98**(10), 73419b.
13. J. S. MacKay (to American Cyanamid Co.), U. S. Pat. 2,656,253 (Oct. 20, 1953).
14. D. S. Malin, "Cone Corrosimeter Testing of Fire Retardant and Other Polymeric Materials for Wire and Cable Applications", paper presented in Fire Safety Conference, San Francisco, CA, January 1994.

15. D. R. Hall, M. M. Hirschler and C. M. Yavornitzky, Fire Safety Science - Proc. 1st Intl. Symp., C. E. Grant and P. J. Pagni, eds., Hemisphere Publishing, Washington, DC, 1986, pp. 421-430.
16. D. R. Miller, R. L. Evans and G. B. Skinner, *Comb. Flame* 7, 137-142 (1963).
17. B. S. Haynes, H. Jander, H. Mätzing and H. G. Wagner, 19th Symp. (Intl.) on Combustion, The Combustion Institute, 1982, pp. 1379-1385.
18. A. Factor, "Char Formation in Aromatic Engineering Polymers", chapter 19 in "Fire and Polymers", G. Nelson, Ed., ACS Symposium Series 425, Washington, D. C., 1990, pp. 274-285.

RECEIVED January 19, 1995

Chapter 14

Flammability Improvement of Polyurethanes by Incorporation of a Silicone Moiety into the Structure of Block Copolymers

Ramazan Benrashid and Gordon L. Nelson

College of Science and Liberal Arts, Florida Institute of Technology, Melbourne, FL 32901-6988

In recent years there is significant demand for flame retardant polymers which do not contain halogen or other additives. New polyurethane block copolymers containing silicone as the soft segment have been synthesized. This class of block copolymers microphase separates allowing formation of a siliconated surface. ESCA and X-ray analyses confirm enhanced siliconated surfaces for these block copolymers. Thermal analysis (TGA)shows these materials are thermally more stable than polytetrahydrofuran polyurethane and polyethyleneglycol polyurethane block copolymers (reference materials). Thermal stability of these siliconated block copolymers depends upon the content of the silicone soft segment. Oxygen index, a convenient technique for evaluation of the flame retardancy of polymers, shows siliconated polyurethanes have higher oxygen index values compared to reference materials. Siliconated block copolymers with higher polydimethylsiloxane content have higher oxygen index values. The oxygen index values also depend upon the diisocyanate used. For example, block copolymers made of hydroxy-terminated polydimethylsiloxane, $H_{12}MDI$ and 1,6-hexanediol show higher oxygen index values compared to block copolymers made of hydroxy terminated poly- dimethylsiloxane, TDI and 1,6-hexanediol. This difference is related to the extent of soft block segregation.

Polymers used in engineering applications should withstand a variety of external stresses, e.g., heat, fire, moisture, ozone, corona, etc. Indeed materials which can be specially tailored by chemistry and by processing are required for many applications. Many materials to be used successfully require significant fire retardant properties. It is increasing recognized that such materials should be halogen free, given the potential for severe damage by even a small fire in electrical and other systems when HX or other corrosive gases are released[1]. In one approach this can be done either by introducing silicone or phosphorus, which have inherent flame retardancy in the backbone of a polymer (block copolymer) or by blending a silicone polymer with other polymers. This method has an advantage over heavily loading a base resin with fillers and additives to reach a desired level of flame retardancy. In the additive approach physical and mechanical properties of the base polymers are affected, and generally not for the better. Furthermore, systems heavily loaded with halogen containing materials or metal compounds are coming under scrutiny in many industries for other safety and environmental reasons.

Silicone polymers have a different backbone versus more common polymers, a backbone consisting of alternating silicon and oxygen atoms rather than carbon atoms. The side groups are similar to those found in natural rubber and many other organic polymers. The resistance of silicone rubber to high temperature, ozone, corona, weathering and other environmental factors that tend to deteriorate insulation, is attributed to the silicone-oxygen linkage. Previous work has shown that incorporation of a siloxane moiety into the polymer back-bone provides enhanced thermal stability, hydrolytic properties and low energy surface properties, gas permeability, chain flexibility, oxygen plasma resistance and blood compatibility[2-11].

Silicones contribute to flame retardancy of other polymers in two ways: 1)as a silicone flame retardant additive for thermoplastics with major application in polyolefins[12-15], or 2) by incorporation as a part of the backbone, e.g., a silicone polyimide copolymer which is a non-halogen inherently flame retardant thermoplastic. Fire resistant materials have been synthesized by insertion of siliconated materials into the structure of a variety of polymers[16-19].

The formation of intumescent char is a highly effective flame retardant mechanism. Ideally, the substrate under burn conditions is protected from catastrophic destruction by a cellular char that is formed from at least partial involvement of the polymer substrate itself. The greater the substrate contribution to the char matrix the greater the effectiveness of the char. Benefits include lower additive loading and better overall mechanical performance. Kambour and co-workers[20,21] studied the effect of the siloxane moiety on the flame retardancy of polymers. They reported that the silicone moiety has a positive effect on the oxygen index values of polymers, causes a rise in pyrolytic char, and improvement in char oxidation resistance. The improvement may stem principally from enhanced oxidation resistance arising from the silicon retained in the char and converted to a continuous protective silica layer.

In this paper we discuss the synthesis of flame retardant, thermally stable silicone urethanes block copolymers made as shown in Figure (1) and the evaluation of their thermal stability, flame retardancy and the effect of segregation on the flammability behaviour of the block copolymers.

Experimental

Materials. Dihydroxy-terminated polydimethylsiloxanes (OHPDMS), poly-dimethyl siloxane-aminopropyl terminated (NH_2PDMS) (different molecular weight), 1,3-bis(hydroxypropyl)tetramethyldisiloxane (OHTMS), 1.3-bis(3-amino-propyl) 1,1,3,3-

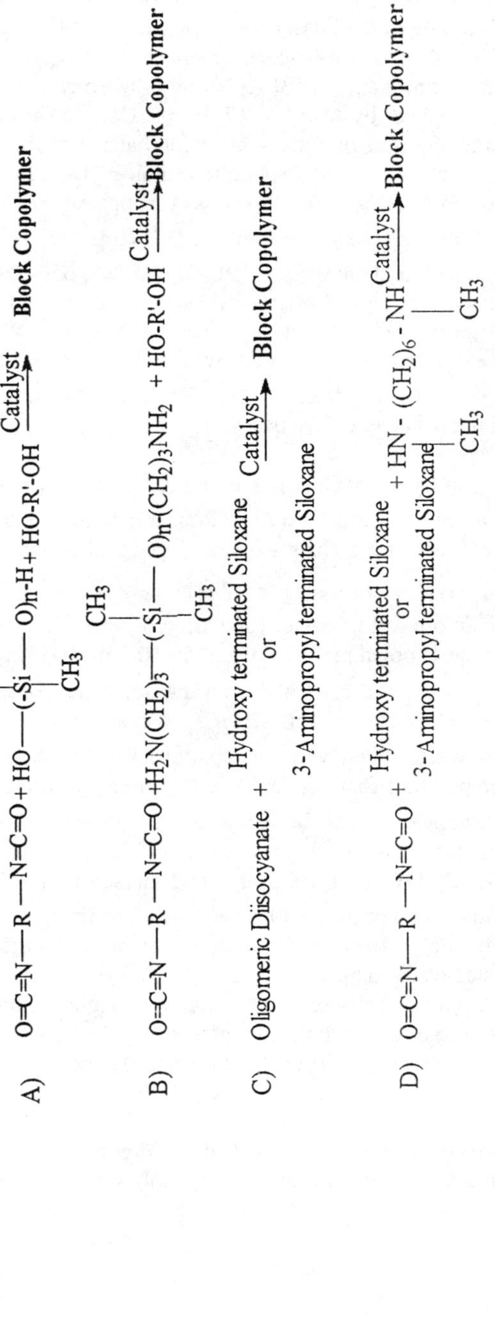

A) $O=C=N-R-N=C=O + HO-(-Si(CH_3)_2-O)_n-H + HO-R'-OH \xrightarrow{\text{Catalyst}}$ **Block Copolymer**

B) $O=C=N-R-N=C=O + H_2N(CH_2)_3-(-Si(CH_3)_2-O)_n-(CH_2)_3NH_2 + HO-R'-OH \xrightarrow{\text{Catalyst}}$ **Block Copolymer**

C) Oligomeric Diisocyanate + Hydroxy terminated Siloxane or 3-Aminopropyl terminated Siloxane $\xrightarrow{\text{Catalyst}}$ **Block Copolymer**

D) $O=C=N-R-N=C=O$ + Hydroxy terminated Siloxane or 3-Aminopropyl terminated Siloxane + $HN(CH_3)-(CH_2)_6-N(CH_3)H \xrightarrow{\text{Catalyst}}$ **Block Copolymer**

Figure 1. Synthesis Scheme for Silicone Urethane Block Copolyurethanes.

tetramethyldisiloxane (NH$_2$TMS), and 1,4-bis(dihydroxy-dimethylsilyl)-benzene (OHDMSB) were purchased from Hüls America. The oligomeric materials were degassed in a vacuum oven at 30 °C for 48 hr. Dicyclohexylmethane-4,4' diisocyanate (H$_{12}$MDI), diphenyl-methane 4,4'di- isocyanate (MDI), toluene diisocyanate (TDI), and isophorone diisocyanate (IPDI) were supplied by Miles Corporation. 1,6-Hexanediol (HDO), 1,4-benzenedimethanol (BDM), 2,2-bis'(4-hydroxyphenyl)hexafluoropropane (FBPA), and hydroxy terminated poly(1-4-butoxy)ether, were supplied by Aldrich Chemical Co. Dibromoneopentyl glycol (Saytex FR-1138) and the diol of tetra - bromophthalic diol, Saytex (RB-79), were supplied by Ethyl Corporation. Phosphorated diol FRD was supplied by FMC. Phosphorated polyol (Vircol) (M.Wt. 545) was supplied by Albright & Wilson. Polyamine 1000 (Eq. Wt. 555-625) , Polyamine 650 (Eq. 355-475) and Ultracast^{TM} PE 35 (Eq. Wt 1150-1250) and Ultracast^{TM}PE 60 (Eq. Wt. 650-750) were supplied by Air Products Company. Dimethylacetamide (DMAC) was stirred over MgO for one week, then distilled under vacuum and kept over molecular sieves 4Å, and under a nitrogen atmosphere. Methylene chloride was refluxed over CaH$_2$ and distilled immediately before use. Tetrahydrofuran (THF) and 1,4-dioxane were distilled from benzophenone ketyl immediately before use.

Synthesis of Block Copolymers (Group A or B). Block copolymers were prepared by a technique called "a one shot technique"[22], from a diol terminated polysiloxane and corresponding diisocyanate and a chain extender, mixed at room temperature under a dry atmosphere (N$_2$). Synthesis details are published elsewhere[23]

To a specific amount of oligomer (dihydroxy terminated poly-dimethylsiloxane, or aminopropyl terminated polydimethylsiloxane) in 100 mL CH$_2$Cl$_2$ were added HDO dissolved in 15 mL DMAC, and bis(4,4'-diisocyanatocyclohexyl)methane (H$_{12}$MDI) or toluene diisocyanate (TDI) in 100 mL CH$_2$Cl$_2$. Several drops of catalyst solution (dibutyltin dilaurate) were added and the solution was mechanically stirred under nitrogen at room temperature for specific period of time. Completion of the reaction was monitored by disappearance of the isocyanate IR absorption at 2270 cm^{-1}. The solvent was evaporated in *vacuo* leaving a viscous oil. The polymer was dissolved in 50 mL 1:1 CH$_2$Cl$_2$/DMAC, THF or 1,4-dioxane, and the solution was cast into films on glass plates. The films were removed from the glass after drying and stored for at least 4 weeks before test. For reference, the side of the film which faces the glass was considered the backside of the film.

All Group A and B polymers and films were made using the same general procedure. The variations in reaction parameters are reported in Tables (1 and 2). GPC on one set of copolymers (Group B) showed Mw of 118000-150000 and Mn 37000-103000.

Synthesis of Diisocyanate Terminated Oligomers (C-51). 2,2-Bis(4-hydroxyphenyl)hexafluoropropane 20 g, (0.059 mol) was dissolved in 80 mL dried

Table 1. Synthesis of Block Copolyurethanes: Group A

Polymer	Oligomer weight	Wt g	CH2Cl2 mL	Wt (g) Diisocyanate	Wt (g) Diol	Cosolvent DMAC mL	Soft block %	Film Flexibility
A-11	36000 OH PDMS[1]	20	250	13.8 H_{12}MDI[2]	6.3 HDIOL[3]	15	50	-
A-12	2000 OHPE[4]	20	250	13.8 H_{12}MDI	6.3 HDIOL	15	50	-
A-17	36000 HOPDMS	10	180	12.8 H_{12}MDI	20.7 HDIOL	20	35	-
A-18	36000 OHDMSO	7.5	150	20.5 H_{12}MDI	9.3 HDIOL	20	20	-
A-20	36000 OHPDMS	4	200	24.8 H_{12}MDI	11.2 HDIOL	20	10	+
A-21	18000 OHPDMS	10	180	19.6 H_{12}MDI	10.4 HDIOL	20	25	-
A-22	18000 OHPDMS	4	180	24.8 H_{12}MDI	11.2 HDIOL	20	10	+
A-25	18000 OHPDMS	3	200	25.5 H_{12}MDI	11.5 HDIOL	15	7.5	+
A-26	18000 OHPDMS	1.5	200	19.6 H_{12}MDI	8.9 HDIOL	20	5	+
A-27	18000 OHPDMS	1	200	26.9 H_{12}MDI	12.1 HDIOL	15	2.5	+
A-30	36000 OHPDMS	1	180	13.1 H_{12}MDI	15.9 HDIOL	15	5	+
A-34	36000 OHPDMS	1	180	26.9 H_{12}MDI	12.1 HDIOL	15	2.5	+
A-35	36000 OHPDMS	2	200	17 H_{12}MDI	7.6 HDIOL	15	7.5	+
A-39	36000 OHPDMS	4	200	20.2	15.9 FRD	20	10	-
A-42	36000 OHPDMS	3	180	16.1 TDI[5]	10.9 HDOL	20	10	+
A-43	18000 OHPDMS	3	200	16.1 TDI	10.9 HDOL	20	10	+
A-44	18000 OHPDMS	7.5	200	18.8 TDI	12.3 HDOL	20	20	+
A-45	4200 OHPDMS	6	200	14.3 TDI	9.7 HDOL	20	20	+
A-46	18000 OHPDMS	8	200	11.1 TDI	7.6 HDOL	20	30	+
A-47	4200 OHPDMS	8	200	11.1 TDI	7.7 HDOL	20	30	+
A-48	18000 OHPDMS	12	200	10.7 TDI	7.3 HDOL	15	40	+
A-49	4200 OHPDMS	12	200	10.7 TDI	7.3 HDOL	15	40	+
A-50	18000 OHPDMS	15	200	8.9 TDI	6.1 HDIOL	15	50	+
A-51	4200 OHPDMS	15	200	8.10 TDI	6.1 HDIOL	15	50	+
A-52	18000 OHPDMS	10	200	24.8 H_{12}MDI	2 OHTMS[6]	10	10	-
A-55	18000 OHPDMS	3	200	13.9 H_{12}MDI	13.1 APTMS[7]	15	10	+
A-59	36000 OHPDMS	8	200	20.2 TDI	12.9 HDIOL	20	120	-
A-60	36000 OHPDMS	12	200	16.7 TDI	11.3 HDIOL	20	30	+
A-61	36000 OHPDMS	15	200	13.4 TDI	9.1 HDIOL	20	40	+
A-62	36000 OHPDMS	15	200	11.2 TDI	8.8 HDIOL	20	50	+
A-64	18000 OHPDMS	3.6	200	16.2 H_{12}MDI	15.8 DBNPDO[8]	20	20	+
A-65	1200-2000 OHPDMS	4	200	21.5 TDI	14.6 HDIOL	20	10	+
A-66	1200-2000 OHPDMS	8	200	19.1 TDI	12.9 HDIOL	20	20	+
A-67	1200-2000 OHPDMS	12	200	16.7 TDI	11.3 HDIOL	20	30	+
A-68	1200-2000 OHPDMS	16	200	14.3 TDI	9.7 HDIOL	20	40	+
A-69	1200-2000 OHPDMS	20	200	11.9 TDI	8.1 HDIOL	20	50	+
A-70	36000 OHPDMS	20	200	11.9 TDI	8.1 HDIOL	20		+
A-71	36000 OHPDMS	5	200	12.5 TDI	32.5 TBPDO[9]	20	10	-
A-72	990[10]	8	200	19.1 TDI	11.3 HDIOL	20	20	+
A-73	990	12	200	16.7 TDI	11.3 HDIOL	20	30	+
A-74	990	16	200	14.3 TDI	9.7 HDIOL	20	40	+
A-79	36000 OHPDMS	4	200	16.5 IPDI[11]	19.5 DBNPDO	15	10	-
A-80	36000 OHPDMS	4	200	11.6 H_{12}MDI	21.7 OHTMDS	15	20	-

1) Hydoxy terminated polydimethylsiloxane, 2.) Dicyclohwxylmethane-4,4'-diisocyanate, 3.) 1,6-Hexanediol, 4) Hydroxy terminated polyethylene ether, 5) Toluene Diisocyanate, 6)Hydroxy terminated tetramethydisiloxane, 7.) 3-aminopropylterminated tetramethyl-disiloxane, 8) Dibromoneopentyldiol, 9)Tetrabromophthalic diol, 10) Poly(methyl-phenylsiloxane), 11)Isophrone diisocyanate

THF. To this solution was added 17.60 g (0.10 mol) IPDI under nitrogen over a period of 45 minutes. The solution was heated for 4 hrs at 120 °C (oil bath) under a nitrogen atmosphere. The reaction mixture was cooled to room temperature. The solvent was evaporated in vacuo leaving 37.8 g of product. The molecular weight by end group analysis[24] was 850. The above procedure was used to make all the diisocyanate terminated oligomers. The experimental variations are presented in Table (3).

Synthesis of Block Copolymers Group C, (C-53). To 11.9 g of oligomer (C-51) in 80 mL CH_2Cl_2 were added 17.24 g of hydroxy terminated poly-dimethylsiloxane (M.Wt 1500) in 80 mL THF at room temperature under a nitrogen atmosphere over a period of 45 minutes. Stirring was continued at room temperature for 2 hrs. The reaction mixture was stirred at 55-60 °C for 93 hrs. Completeness of the reaction mixture was monitored by disappearance of the isocyanate IR absorption at 2267 cm^{-1}.

The viscous solution was cast as a film on a glass plate using a 10 mil film applicator. After standing 24 hrs, the films were removed from the glass. The experimental variations are presented in Table (4).

All samples intended for surface analysis were exposed to an additional 4 days in a vacuum oven at 25-30 °C. The samples used for TGA and DSC were dried in a vacuum oven at 65-70 °C for 14 days. The polydimethylsiloxane is the soft block and the polyurethane is the hard block.

Synthesis of Block Copolymers Group D. The procedure was as described for Group A and B, except N,N'dimethylhexamethylenediamine was used as the chain extender. Table (5).

Measurements

Thermogravimetric analysis was performed on a DuPont model 951 TGA attached to a DuPont model 9900 analyzer. Version 2.2 analysis software was utilized to calculate the percent residue. The samples were analyzed in a tared aluminum pan placed in a platinum basket. The purge rate was set at 50-60 mL/min for N_2 or air and the heating rate was set for a 20°C/min increase from ambient temperature to 630°C.

Differential scanning calorimetry (DSC) experiments were performed on a DuPont model 910 DSC attached to a DuPont model 9900 analyzer using version 2.2 DSC software to analyze some of the transitions. Samples were analyzed in a crimped aluminum pan with lid. An empty aluminum pan with lid served as a reference. The purge rate was 34 mL/min N_2 and the heating rate was 10 °C/min from -75 to 150°C.

SEM and EDS analyses were performed on a model S-2700 Hitachi scanning electron microscope with an attached Kevex light element detector. Electron beam energies were 20 Kev. Data were collected from a scanned region of approximately 100 X 100 square micrometers. The X-ray detector was operated in the thin window mode at less than 20 percent dead time. A Denton Desk II Sputter coater with a Pd/Au target was employed for coating SEM samples to reduce surface charging effects.

Table 2. Synthesis of Block Copolyurethanes: Group B

Polymer	Oligomer M.Wt*	Wt g	CH$_2$Cl$_2$ mL	H$_{12}$MDI Diisocyanate g	Chain Extender g	Cosolvent mL	Soft Block %	Film Flexibility
B-63	18000	4	120	24.8	11.2 HDIOL[1]	DMAC 15	10	+
B-64	18000	9	150	24.8	11.2 HDIOL	15:50 DMAC:THF	20	+
B-65	36000	9	150	24.8	11.2 HDIOL	80 THF	20	+
B-66	18000	13.5	180	21.7	9.8 HDIOL	THF 100	30	+
B-67	18000	18	180	18.6	8.4 HDIOL	THF 100	40	+
B-68	18000	22.5	180	15.5	7.0 HDIOL	15:10 DMAC:THF	50	-
B-69	18000	27	200	12.4	5.6 HDIOL	15:90 DMAC:THF	60	+
B-72	18000	23	100	29.5	13.3 HDIOL	120 THF	5	+

1) 1,6-Hexanediol

Table 3. Synthesis of Oligomers for Group C Block Copolyurethanes

Oligomer	Diisocyanate* g	Diamine g	Diol g	M.wt
C-26	Des-W 25,3		1,4-Bis(dihydoxydimethylsilyl)benzene 12.5	1216
C-27	Dew-W 31.3	Bis(Ethylamino)-dimethyl Silane 10		1700
C-28	Isophrone 26.6	Bis(Ethylamino)-dimethyl Silane 10		1700
C-29	Isophrone 13.2		1,4-Bis(dihydoxydimethylsilyl)benzene 7.7	2700
C-30	Des w 26.5		Phosphorated diol 30	1140
C-51	Isophrone 17.6		2,2-Bis(4-hydroxyphenyl)hexafluoro-propane 20	850
C-55	Isophrone 7.8	4,4-diaminophenyl-hexa-fluoropropane 10		2067

* Mole ratio of diisocyanate/diol or diisocyanate/diamine=1.6-1.7.

Table 4. Synthesis of Block Copolyurethanes: Group C

Polymer	diisocyanate terminated Oligomers	Wt g	THF[1] mL	Oligomer	Oligomer M.Wt	Wt g	Stirring period (hrs)	Film Flexibility
C-40	2500[2]	20	50	OH PBE	2900	23.2	96	-
C-42	1140 (C-30)	17.3	100	NH_2 PDMS	2500	38	14	+
C-43	2700 (C-29)	6.9	80	OHPBE	2900	7.25	24	+
C-44	1140 (C-30)	17.3	80	OHPBE	2900	16.9	24	-
C-45	1700 (C-27)	11.5	60	NH_2 PDMS	2500	16.9	24	+
C-48	1700 (C-28)	11.5	90	Polyamine[3]	1238	8.4	40	-
C-49	1700 (C-27)	11.5	60	Polyamine[4]	950	6.4	149	-
C-50	1700 (C-27)	11.5	90	Polyamine[3]	1238	8.4	155	-
C-53	850 (C-51)	11.9	80	OH PDMS	1500	22.4	93	-
C-54	850 (C-51)	11.9	80	OH PBE	2000	28	93	+
C-57	2067 (C-55)	8.5	80	Phosphorated diol[5]	545	2.3	95	-

1) Co-solvent was CH_2Cl_2 (80-150 mL).
2) Hydroxy terminated polydimethyl siloxane.
3) 3-Aminopropylterminated polydimethylsiloxane.
4) Ultracast PE 60.
5) Vircol (Albricht & Wilson).

Table 5. Synthesis of Block Copolyurethanes: Group D

Polymer	Oligomer weight	Wt g	THF mL	Wt (g) Diisocyanate	Wt (g) N,N'-Dimethyl-hexamethylene diamine	Soft block %
D-77	OHPDMS (M.Wt. 2500)	5	180	TDI, 24.6	20.4	10
D-78	=	10	180	TDI, 21.9	18.1	20
D-79	=	15	180	TDI, 19.2	15.8	30
D-80	=	20	180	TDI, 16.4	13.6	40
D-81	=	25	180	TDI, 13.7	11.3	50

Electron Spectroscopy for Chemical Analysis (ESCA) or X-ray Photoelectron Spectroscopy (XPS) data were obtained with a Phi model 15-255G Cylindrical Mirror Analyzer (CMA) attached to a Phi model 590 Scanning Auger Microprobe. The spectra were generated with Mg K-alpha X-rays at a power of 400 watts. The analysis area covered about 1 mm^2.

Oxygen indices were performed on an original GE oxygen index tester. Oxygen index measures the ease of extinction of materials, the minimum percent of oxygen in a oxygen / nitrogen atmosphere that will just sustain combustion of a top ignited vertical test specimen. Oxygen index is one measure of flame retardancy and can be conveniently made on small quantities of sample. Oxygen index was performed on samples in powder or flake form[25]. A small porcelain cup was used which was installed on the clamp, the cup was loaded with sample, and the sample first was melted and then ignited for test.

Results and Discussion

Thermogravimetric Analysis. Introducing a silicone moiety into the structure of block copolymers increases the thermal stability of block copolymers in nitrogen and air compared a non-siliconated polyurethanes, Table (6). Char formation in air is characteristic of silicone polymers, because reaction of silicone with oxygen at high temperature leads to formation of inorganic silicon dioxide. The amount of weight residue increases with increasing the amount of silicone in the structure of block copolymer, Figure (2). EDS spectra of char resulting from air pyrolysis of polymer samples show only Si and oxygen peaks. Block copolymers containing higher silicone content show higher thermal stability, Figure (3).

DSC. The results are shown in Table (7). The DSC measurements show two thermal transitions; the one at low temperature (-45 °C) is the m.p. for the siloxane moiety of the block copolymer and the one above room temperature is related to the glass transition of the urethane hard segment. Two Tg's at -120 and -150 °C and a low m.p ~-45 °C were reported by Inoue, et.al.,[26] for siloxane PPMA block copolymers. A melting point of -50 °C was reported for a polyurethane-polysiloxane graft copolymer by Kazama , et.al.[27]

Oxygen Index. The oxygen index test of the flame retardancy of polymers is based on measuring the ease of extinction for materials in an oxygen / nitrogen mixture. The oxygen index can be measured for bar samples as well as flake, powder or liquid samples,[28,29]. The latter is useful for small amounts of sample. While oxygen index is only one fire parameter, materials with high oxygen index values are generally more flame retardant. The oxygen index value for non-siliconated ether-urethane block copolymers is in the range of 18 as shown in Table (8). We found that as the amount of siloxane in the structure of the block copolymers increases oxygen index values increase. For example oxygen indices for polymers with 2.5 to 50 percent siloxane

Table 6. Thermal Analysis of Block Copolyurethanes in Nitrogen and (Air)

Polymer	T (°C at 10% Weight Loss)	T (°C at 50% Weight Loss)	Residue (%)
A-12	316 (325)	389 (370)	1.0(0.9)
A-11	319 (310)	374(367)	0.4(22)
A-17	300 (311)	383 (402)	1.2 (4.1)
A-18	325 (323)	381 (379)	1.1(8.0)
A-27	261 (252)	353 (359)	0.4 (0.4)
A-26	267 (273)	353 (365)	0.2 (2.0)
A-25	297 (298)	356 (370)	1.6 (3.4)
A-22	314(316)	374 (370)	1.0 (3.2)
A-21	335 (335)	389 (386)	0.0 (4.2)
B-63	308 (313)	357 (357)	0.8 (2.1)
B-64	309 (318)	357 (357)	1.5 (4.1)
B-68	313 (327)	443 (435)	1.3 (5.0)
B-72	308 (313)	356 (357)	1.6 (5.8)
C-45	290 (310)	446 (454)	1.8 (3.5)
D-79	322 (319)	389 (403)	5.7 (6.1)

Table 7. DSC Data of Block Copolyurethanes.

Polymer	Soft block m.p °C	Tg °C	Tm °C
A-27	-	53	93
A-26	-45	31	-
A-25	-47	51	120
A-22	-47	47	93
A-21	-47	54	119
B-63	-41.6	68.6	126.3
B-64	-42.5	82.0	114.6
B-68	-43.7	83.1	-
C-56	-44.1	60.2	-
C-45	-	93.7	-

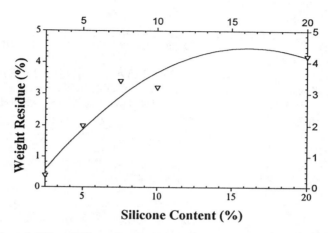

Figure 2. Effect of Silicone Content of Block Copolyurethanes on Weight Residue from Thermal Analysis in Air (A-27, A-26, A-25, A-22 and A-21).

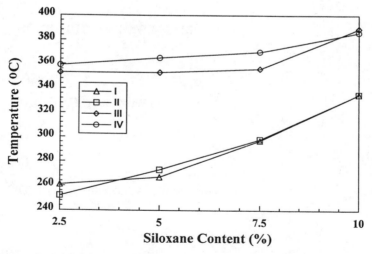

Figure 3. Effect of Silicone Content of Block Copolyurethanes on Thermal Stability of Polymers (A-27, A-26, A-25, and A-22): Temperatures Related to 10 % Weight Loss in N_2 (I), and in Air (II), and Temperatures Related to 50 % Weight Loss in N_2 (III), and in Air (IV).

Table 8. Oxygen Index Values of Example Block Copolyurethanes

Polymer	Silicone Content (%)		Oxygen Index
1-(Reference)*	0	Polyether(M.Wt 2900)+H_{12}MDI+HDIOL	18.2
2-(Reference)**	0	= =	18.6
3-(Reference)***	0	PolyTHF(M.Wt 2500)+H_{12}MDI+HDIOL	17.6
Polydimethylsiloxane		Oligomeric polydimethylsiloxane	29.8
A-34	2.5	OHPDMS(M.Wt. 36000)+H_{12}MDI+HDIOL	18.6
A-30	5	=	19.1
A-35	7.5	=	19.6
A-20	10	=	21.0
A-18	20	=	21.8
A-17	35	=	26.0
A-11	50	=	29.8
		*******	
A-27	2.5	OHPDMS(M.Wt. 18000)+H_{12}MDI+HDIOL	18.9
A-26	5	=	19.6
A-25	7.5	=	20.8
A-22	10	=	21.5
A-21	20	=	22.2
		********	
A-45	20	OHPDMS(M.Wt. 4200)+TDI+HDIOL	22.0
A-47	30	=	22.5
A-49	40	=	22.5
A-51	50	=	24.2
		*********	
A-43	10	OHPDMS(M.Wt. 18000)+TDI+HDIOL	18.6
A-44	20	=	19.6
A-46	30	=	21.8
A-48	40	=	22.5
A-50	50	=	24.3
		*********	
A-42	10	OHPDMS(M.Wt. 36000)+TDI+HDIOL	18.6
A-59	20	=	19.6
A-60	30	=	23.4
A-61	40	=	22.3
A-62	50	=	24.0
		*********	
A-65	10	OHPDMS(M.Wt.1200-2000)H_{12}MDI+1,6-HDIOL	19.5
A-66	20	=	22.8
A-67	30	=	21.3
A-68	40	=	22.8
A-69	50	=	24.3

Table 8. continued

Polymer	Silicone Content (%)		Oxygen Index
A-72	20	OHDMDPS(M.Wt. 990)+TDI+1.6-HDIOL	22.2
A-73	30	=	22.2
A-74	40	=	23.2
A-64		OHPDMS(M.Wt.18000)+H_{12}MDI +DBNPDO	25.7
A-79		OHPDMS+IPDI+DBNPDO	25.8
A-80		OHPDMS+H_{12}MDI +OHTMDS	29.8
A-39`		OHPDMS+H_{12}MDI+FRD	21.0
		**********	
B-63	10	$NH_2(CH_2)_3$PDMS(M.Wt.18000)+H_{12}MDI+1,6HDIOL	19.8
B-64	20	$NH_2(CH_2)_3$PDMS(M.Wt.36000)+H_{12}MDI+1,6HDIOL	20.7
B-65	10	$NH_2(CH_2)_3$PDMS(M.Wt.36000)+H_{12}MDI+1,6HDIOL	19.8
B-66	30	=	24.3
B-67	40	=	25.9
B-68	50	=	27.4
B-69	60	=	27.9
B-72	5	=	18.9
		**********	
C-40	-	Ultracast PE 60[a] + PBE	20.1
C-42	-	C-30 + NH_2PDMS[b]	20.7
C-43	-	C-29 + HOPBE[c]	19.8
C-44	-	C-30 + HOPBE	21.2
C-45	-	C-27 + NH_2PDMS	20.5
C-48	-	C-28 + OHPDMS	24.0
C-49	-	C-27 + Polyamine[d]	23.9
C-50	-	C-27 + Polyamine[e]	22.1
C-53	-	C-51 + OHPDMS	27.9
C-54	-	C-51 + PBE	20.5
C-57	-	C-55 + Phosphorated diol[f]	20.1
		*********	
D-77	10	OHPDMS(M.Wt2500)+TDI + N.N'Dimethylhexanediamine	25.2
D-78	20	=	27.4
D-79	30	=	28.7
D-80	40	=	30.2
D-81	50	=	30.2

*10% Soft Segment, ** 20% Soft Segment, *** 10% Soft Segment
a) Air Product Ultracast PE 60, b) 3-Aminopropyl terminated polydimethylsiloxane, c) Hydroxy terminated polydimethylsiloxane, d) Air Product Polyamine 650, e) Air Product polyamine 1000, f) Vircol (Albricht & Wilson)

content, (A34 through A11) oxygen index values rise from 18.6 to 29.8, which is close to the oxygen index value for neat polydimethyl-siloxane oligomers. The effects of siloxane content on O.I. values for four sets of block copolymers containing 10 to 60 percent siloxane soft segment are presented in Figure (4). Different systems have different O.I. values. Phase segregation in these block copolymers leads to domination of siloxane on the polymer surface. Siloxanes have solid phase activity rather than vapor phase activity and reduce flammability through increased formation of pyrolytic char and also increased resistance to char oxidation[20-21]. Char acts as an insulator between fire and bulk, and prevents fire spread. Figure (5) shows a photograph of two samples before and after oxygen index testing. The surface of the tested sample was covered with silicon dioxide. Oxygen index data for a variety of block copolymers are presented in Table (8). There is a clear difference in oxygen index resulting from $H_{12}MDI$ as the isocyanate versus TDI. Very short OHPDMS (A11 versus A69) segments result in lower O.I. values due to lower expected segregation. Block polyurethanes with N,N-dimethylhexamethylenediamine as chain extender have higher oxygen index values compared to block polyurethanes with 1,6-hexanediol as chain extender, these results correlate with the thermogravimetric results.

SEM and EDS. While detailed surface studies of these materials have been published elsewhere[30], not only is the microphase segregation phenomenon evident in block copolymers containing siloxane moiety, but the solvent from which a film is cast can have an effect on the extent of segregation and the amount of siloxane on the surface.

For materials cast as films on glass plates, surfaces facing the glass showed lower silicone compared to the surface facing air, which is probably due to a hydrogen bonding interaction between the glass surface and the hard segment of the block copolymers.

Figure (6) illustrates EDS spectra for polymer and its char. Comparison of these two spectra show that the carbon peak in the char spectrum has vanished, and that only peaks related to O and Si remain on fire exposure. Organic silicones convert to inorganic silicon dioxide.

ESCA. ESCA studies for films resulting from casting of block copolymers with different siloxane content reveal that films cast from block copolymers with high silicone content show higher silicon concentration on the surface, Figure(7). A-51 with 50% silicone moiety shows higher silicon on the surface compared to films with 40, 30 and 20 percent silicone content in which the silicon content on the surface was found to be 23.6, 10.7, 8.3 % compared to 29.2 %for A-51. The high degree of segregation of the soft block is due to a large difference between the solubility parameter of the two blocks[31-32] It is expected that the top surface of graft and block copolymers is significantly dominated by the siloxane segment component, even if the bulk siloxane is as small as 5 Wt percent[33]. The thickness of the siloxane layer ranges from 100 to 20 Å depending on the soft segment length and siloxane content[34]. It is also reported that the hard segment domains of 60-70 Å long are found embedded in the soft segment matrix[35].

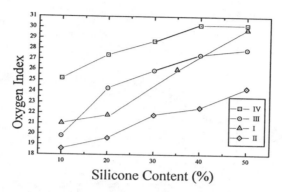

Figure 4. Effect of Silicone Content of Block Copolyurethanes on Oxygen Index: (I) A-20, A-18, A-17, A-11; (II) A-43, A-44, A-46, A-48; and A-50, (III) B-65, B-64, B-66, B-67, B-68, and (IV) D-77, D-78, D-79, D-80, D-81.

Figure 5. Photograph of Silicone Urethane Block Copolymer Samples (A) Before and (B) After Oxygen Index testing. The Surfaces of Samples After Test Were Covered with SiO_2.

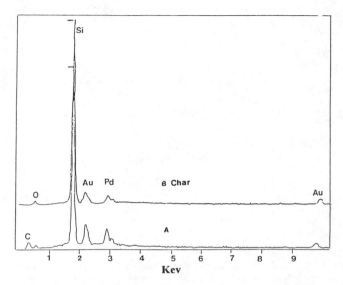

Figure 6. EDS Spectra of Block Copolyurethane A-11, (A) Neat Film, (B) Pyrolytic Char; Carbon is Absent in the Latter.

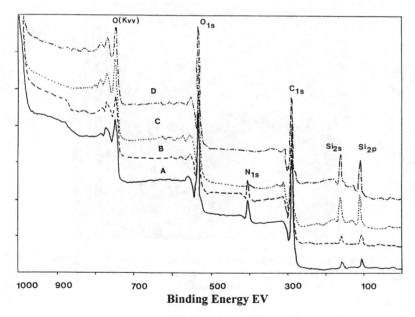

Figure 7. ESCA Spectra for Block Copolymer Films [A)A-45(20%), B)A-47 (30%), C) A-49 (40%) and D)A-51 (50%)]. Film Cast from DMAC/CH_2Cl_2. (Hard Segment Contains TDI).

It is noted that block copolymers resulting from hydroxy terminated polydimethylsiloxane, $H_{12}MDI$, and 1,6-hexandiol have better segregation compared to block copolymers obtained from hydroxyterminated polydimethylsiloxane, TDI and 1,6-hexanediol, Figure 8 versus Figure 7. The compounds shown in Figure (8) show a minimal N1s peak despite low overall silicone content. This can explain why block copolymers from $H_{12}MDI$ have higher oxygen index values compared to block copolymers obtained from TDI. In the case of first one, better segregation leads to a surface with higher silicone. Fire is a surface phenomenon, therefore having higher inherent flame retardant siloxane on the surface reduces the flammability of the block copolymers.

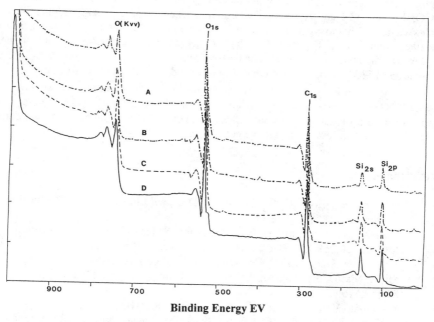

Figure 8. ESCA Spectra for Block Copolymer Films [A) A-27(2.5%), B)A-26 (5%), C)(A-25 (7.5%) and D) A-22(10%)]. Film Cast from DMAC/CH_2Cl_2. (Hard Segment Contains H12MDI).

Conclusions

Segregated block copolymers of dimethylsiloxane urethane block copolymers can be made from hydroxy or amine terminated siloxanes a diisocyanate and a chain extender, or from diisocyanate terminated oligomers and hydroxy or amine terminated siloxane oligomers. A large number of new polymers have been synthesized.

These materials are microphase segregated and surface studies show an enhanced siliconated surface. Surfaces of the films show higher silicon compared to back or bulk of the films. These novel materials show most (>95 %) of their siloxane content at the surface of films cast from solution. Soft-hard block segregation makes these materials interesting for different applications.

These materials show a low m.p. for the soft segment and a Tg above room temperature for the hard segment. By increasing the content of the siloxane moiety a more distinctive sharp peak is obtained in DSC. These materials are thermally stable, with stability increasing with increasing siloxane in the structure. Polymers with high silicone content show high oxygen index. Block copolyurethanes prepared from OHPDMS, H_{12}MDI and 1,6-hexanediol have higher oxygen index values compared to block copolymers prepared from OHPDMS, TDI, and 1,6-hexanediol, which is due to better phase separation for the former block copolyurethanes. Block copolymers made of OHPDMS, TDI and N,N'dimethylhexamethylenediamine also show higher oxygen index values. While oxygen index is only one parameter for materials to be considered as flame retardant, polymers with a higher oxygen index and silicone content have higher weight residue. The enhanced flame retardancy of these materials is due to formation of SiO_2 on the surface which insulates and protects the bulk from fire.

Silicone polyurethanes have three degrees of design freedom to achieve the desired silicone surface: chemistry (siloxane, isocyanate, chain extender), block size, and solvent or processing conditions. Clearly one can achieve significant fire retardancy (OI >28) at 50% or less silicone in the copolymer.

References:

1. Hilado, C. J.; Casey, C. J.; Chistenson, D. F. and Lipowitz, J., J. Combustion Toxicology, **1978**, vol. 5, 130.
2. Hoshino, Y. K.; Katano, H. H.; and Oskubo, S. Japan Kokai Tokkyo Hoho, Jpn. Pat. 60,258,220. Dec 20, **1985**; Chem. Abstr. 105:98559y.
3. Toga, T. Y.;and Ikeda, N. Ger. Offen, D.E., March 21, **1985**, 3.432.509.
4. Mitsui Nisso Corp. Japan Kokai, Tokkyo, Koho, Jpn. Dec. 17, **1983**, Pat. 58,217,515.
5. Kotomkin, V. Y.; Baburina V. A.; Lebedov E. P.; Bylev V. A.; Yasmikova T. E.; and Reikhsfeld V. O. Khimya i Parki Primonenie, Kremnii i Fosforogan, Soedin L, 23, **1980**, Chem. Abstr. 95:43841d.
6. Kotomkin, V. Y.; Baburina,V. A.; Lebedev V. P.; and Kercha Y. Y. Sint Poliuretanov, **1981**, 86-90.
7. Kotomkin, V. Y.; Baburina, V. A.; Lebedev, V. P. and Kercha, Y. Y.Plastic Massy, 27, **1981**, Chem. Abstr. 95:43841d.
8. Tsybul'ko, N. N.; Martinovich, F. S.; Satsura, V. M. and Mandrikova, A I.,USSR, Sept. 15, **1982**, SU 958,432.
9. Sodova, V. L.; Shepuev, E. L. ; Sergeev, L. V.; Sidorkova, T. V. and L. I. Makorova, Opt. Mekh. Prom-St, **1976**, vol.43(5) 481/24/92.
10. Ho Tai and Wynne, K. J. Preprints, American Chemical Society, Polymeric Materials Science and Engineering Division (Washington DC), **1992**, vol. 67,445.

11. Connell, J. W.; Smith Jr., J. W. and P. M. Hergenrother, P. M., J. of Fire Sciences, **1993**, 11, 137.
12. Bopp, R. C. and Miller S. US Patent, 22 May, **1979**, 4,155,898.
13. Mamoru Kondo, Jpn Kokai, Tokkyo Koho JP, 29 July, **1988**, vol. 63,183,960.
14. Frye, R. B., US Patent, August 20, **1985**, 4,536,529.
15. Winfried, P.; Jyergen, K. H.; Wolfgang S.; Christian, L.; Dieter, N.; and Werner, N., Offen DE, 14 March, **1985**, 3,347,071.
16. Factor, A.; Sannes, K. N.; and Colley, A. M., Fire and Flammability Series, part 3, edited by C. J. Hilado, Technomic Publishing Co., **1985**, vol. 20,156-162.
17. Collyer, A. A.; Clegg, D. W.; Morris, D. C.; Parker, G. W.; Wheatley G. W. and Arfield, G. C., J. Polym Sci., Part A. Polym. Chem. **1991**, vol. 29, 193-200.
18. Jadhav A. S.; Maldar, N. N.; Shide, B. M. and Vernkar S. P., J. Polym Sci., Part A. Polym Chem., **1991**, vol. 29, 193-200.
19. Lavin, K. D. And Williams D. A., International Wire and Cable Symposium Proceedings, **1986**, 286-292.
20. Kambour, R.P.; H.J. Klopfer, H. P.; and Smith, S. A., J. Appl. Polym Sci. **1981**, vol. 26(3), 847-859.
21. Kambour, R.P., J.Appl. Polym. Sci., **1981**, vol. 26(3), 861-877.
22. Zdrahala, R. J.; Gerkin, R. M.; Hager, S. L. and F. E. Critchfield, J. Appl. Polym. Sci., **1979**, vol. 24, 2041.
23. Benrashid, R.; Nelson, G. L., J. of Polym. Sci., Polym. Chem., **1994**, vol. 32, 1847-1865.
24. ASTM Procedure D-1638-74, using bromophenol blue as indicator, American Society for Testing and Materials, Philadelphia, PA.
25. Nelson, G. L.; and Webb L. L., J. Fire and Flammability, **1973**, vol. 4, 325.
26. Inoue, H.; Ueda, A.; and Nagai, S. J., Appl. Polym. Sci., **1988**, vol. 35, 2039-2051.
27. Kazama, H.; Ono, T.; Tezuka, Y. and Imai, K. Polymer, **1989**, vol. 30, 553-557.
28. Benrashid, R.; and Nelson, G. L. Proceedings of the Fire Safety and Thermal Insulation, St. Petersburg, FL, November 4-7,**1991**, 189-201.
29. Benrashid, R. and Nelson, G. L. Proceedings of Second Annual BCC Conference on Flame Retardancy, Crown Plaza Hotel, Stamford, Connecticut, May 19-21, **1992**, 47-54.
30. Benrashid, R.; Nelson, G. L.; Linn, J. H.; Hanely K. H.and Wade, W. R. J. Appl. Polym. Sci., **1993**, vol. 49, 523-537.
31. Shibayama, M. Inoue, M; Yamamoto, T. and Normura, S., Polymer, **1990**, vol. 31, 749-757.
32. Pascault, J. P. and Camberlin, J. P., Polym. Commun., **1986**, vol. 27(8) 230
33. Tezuka, Y.; Kazama, H.; and Imai, K. J. Chem. Soc., Faraday Trans, **1991**, vol. 87(1),147-152.
34. Tezuka, Y.; Ono, T.; and Imai, K. J. of Colloid Interface Sci., **1990**, vol. 16(2), 408-414.
35. Shibayama, M.; Suetsugu, M.; Sakurai, S.; Yamamoto, T.and Nomara, S., Macromolecules, **1991**, vol. 24, 6254-6262.

RECEIVED January 4, 1995

Chapter 15

Surface Modification of Polymers To Achieve Flame Retardancy

Charles A. Wilkie, Xiaoxing Dong, and Masanori Suzuki[1]

Department of Chemistry, Marquette University, Milwaukee, WI 53233

> Methacrylic acid has been grafted onto both acrylonitrile-butadiene-styrene terpolymer or styrene-butadiene block copolymer by the anthracene sensitized photoproduction of hydroperoxides. The grafted methacrylic acid has been converted to its sodium salt by treatment with aqueous sodium hydroxide. The TGA residue that is obtained at 800°C is greater than that expected based on the starting materials. Preliminary cone calorimetry results indicate that a graft layer of sodium methacrylate increases the time to ignition and also decreases the peak heat release rate. This procedure is presented as a general approach to flame retardation for a variety of polymers.

There are numerous additives that have been used as flame retardants for a wide variety of polymers. These include materials such as alumina trihydrate, which endothermically decomposes to lose water, thus removing heat and diluting the flame; halogens, which function principally by quenching the radicals that make up the flame; and phosphorus, which may function in either the vapor phase in a fashion similar to that of the halogens or in the condensed phase by either changing the mode of degradation of the polymer or by the formation of char. These three systems describe the processes that are currently known about the functioning of flame retardants. Each of these presents its unique problems. Alumina trihydrate must be used at very high loadings and this has an adverse effect on the physical

---

[1]Permanent address: Japan Synthetic Rubber Co., 100, Kawajiri-Cho, Yokkaichi, Mie, 510, Japan

properties of the polymer. The halogens are widely applicable to a variety of polymers but they may present environmental problems. Condensed phase additives change the mode of thermal degradation so that flammable gases are not produced and, instead char is obtained. These are very useful but each polymer is different and generally requires a unique additive system to achieve flame retardancy.

One approach is to investigate the effect of a wide variety of additives on the thermal degradation of a particular polymer and to use that information to design a suitable flame retardant. In this laboratory, we have followed that approach for many years and have examined the effects of many different additives on the thermal degradation of poly(methyl methacrylate). The additives that have been investigated include: red phosphorus (1-2), Wilkinson's catalyst, $ClRh(PPh_3)_3$ (3-4), Nafions (5), copolymers of 2-sulfoethyl methacrylate and methyl methacrylate (6), a variety of transition metal halides, $MnCl_2$, $CrCl_3$, $FeCl_2$, $FeCl_3$, $NiCl_2$, $CuCl_2$, and $CuCl$ (7-9), various phenyltin chlorides (10), tetraphenyl- and tetrachloro-tin (11), and diphenyl disulfide (Chandrasiri, J. A. and Wilkie, C. A.; *Polym. Degrad. Stab.*, in press).

A breakthrough in flame retardancy studies could be achieved if a general approach could be devised that 1) could be used at low loadings; 2) was compatible with the environment; and 3) had the possibility of applicability to a wide variety of polymers. In pursuing these ideas, we have concluded that a surface treatment that will form an adherent, thermally insulating char layer under thermal conditions would be advantageous. To accomplish this, one needs to 1) identify a suitable char forming material and 2) develop some process for its attachment to the surface of the polymer.

In several publication, McNeill (12-15) has described the thermal degradation of a variety of methacrylate polymers and has shown that substantial char is produced when these are thermally degraded. This data is shown in Table I.

TABLE I

Thermal Degradation of Salts of Poly(Methacrylic Acid)

Cation	Onset temperature of degradation, °C	% Residue at 500°C	Identity of Residue
H^+	200	3	"C"
Li^+	400	54	Li_2CO_3 + C
Na^+	400	64	Na_2CO_3 + C
K^+	400	66	K_2CO_3 + C
Cs^+	400	82	Cs_2CO_3 + C
Mg^{2+}	200	31	MgO + C
Ca^{2+}	280	57	$CaCO_3$ + C
Sr^{2+}	320	61	$SrCO_3$ + C
Ba^{2+}	400	70	$BaCO_3$ + C

We have decided that grafting of these polymers onto the substrate polymers offers an excellent opportunity to achieve flame retardation; methacrylic acid and its sodium salt have been selected for the initial study as the char forming monomers. Grafting offers great versatility because of the variety of ways in which it may be initiated. These include various chemical initiators, photochemical initiation, and initiation by high energy radiation. Geuskens has shown that grafting may be accomplished by the anthracene-sensitized photoproduction of hydroperoxides on the substrate and the thermal degradation of these hydroperoxides in the presence of suitable monomers (*16-17*); this grafting approach has been adopted for this study. This paper describes our recent work in this area on flame retardation of acrylonitrile-butadiene-styrene terpolymer, ABS (Suzuki, M. and Wilkie, C. A.; *Polym. Degrad. Stab.*, in press), and styrene-butadiene block copolymer, SBS (Dong, X.; Geuskens, G.; and Wilkie, C. A.; *Eur. Polym. J.*, in press).

EXPERIMENTAL

The grafting reaction was carried out as previously described (*17*, Suzuki, M. and Wilkie, C. A.; *J. Polym. Sci.: Part A: Polym. Chem.*, in press). The ABS or SBS was compression molded in a heated press to obtain films of about 200 micron thickness and anthracene was permitted to migrate into these films from a methanolic solution. Irradiation of the films with a lamp that emits between 350 and 400 nm produces hydroperoxides on the butadiene portion of the polymer. The hydroperoxidized films were heated in aqueous solutions of methacrylic acid to produce a graft layer of methacrylic acid on the film. The films were then soaked in a dilute solution of aqueous sodium hydroxide to convert the acid to the sodium salt. The percent of grafting is defined as follows:

$$\text{Weight \% Grafting} = \frac{M_g - M_{ug}}{M_{ug}} \times 100$$

where M_g = mass of grafted sample and M_{ug} = mass of sample before grafting.

Thermogravimetric analysis was carried out on a Omnitherm TGA 1000M at a scan rate of 20°C/min to a maximum temperature of 800°C. The oxygen index was measured on a home built apparatus using bottom ignition. Cone calorimetry per ASTM E 1354-92 was performed using a Stanton Redcroft/PL Thermal Sciences instrument at 25 kW/m² in the horizontal orientation. The 0.25 inch thick samples were mounted using the edge retainer frame and wire grid; the mass was approximately 75 grams. Exhaust flow was set at 24 L/sec. and the spark was continuous until the sample ignited. TGA-FTIR analysis was performed using a Cahn thermogravimetric analyzer coupled to a Mattson Instrument FTIR spectrometer. The evolved gases were sampled with a sniffer tube that extends into the sample cup and admits only some of the gases to the infrared spectrometer. Sample size used was about 40 mg with a heating rate of 20°C per minute and an inert gas purge of 30-50 cm^3/min. Evolved gases were transferred to a heated 10 cm gas IR cell by a heated quartz tube. Spectra were identified by visual identification as well as from searching spectral data bases.

RESULTS AND DISCUSSION

Three criteria have been selected to judge the efficacy of this flame retardant technique; these are: char yield in a thermogravimetric analysis experiment, cone calorimetry, and oxygen index.

Thermogravimetric Results. The TGA curve for polymethacrylic acid gives a 5% residue while a 13% residue is obtained for the polymer obtained by treatment of polymethacrylic acid with sodium hydroxide. Poly(sodium methacrylate) obtained by the homopolymerization of sodium methacrylate gives a 55% residue. This indicates that treatment of the acid with sodium hydroxide is an inefficient way to obtain the grafted sodium salt. Unfortunately it has not been possible to directly graft sodium methacrylate onto polymers at this time and the methacrylic acid followed by sodium hydroxide treatment must be followed.

For a polymer which completely volatilizes in the TGA experiment and has 10 weight % grafted methacrylic acid, one expects a residue of 5% of 10% or 0.5%. If the observed residue is greater than this, this indicates that the substrate which would normally volatilize under these conditions, has been retained within the sample.

TGA curves for unmodified SBS and a sample which has been grafted with 60 weight % methacrylic acid and converted to the sodium salt are shown in Figure 1 and the TGA data for grafted SBS is shown in Table II. Theoretical residues have been calculated using the residue obtained for the homopolymers. For the sodium salt, since the graft was formed by treatment of the acid with sodium hydroxide, the 13% residue obtained from the polymer prepared in this way was used.

TABLE II

TGA residue for MAA and NaMAA grafted SBS

weight % grafted	MAA residue	theoretical MAA residue	NaMAA residue	theoretical NaMAA residue
0	0	0	0	0
10	3.1	.5	8.7	1.3
20	4.6	1.0	13.6	2.6
30	4.3	1.5	16.0	3.9
40	3.9	2.0	25.1	5.2

One immediately notices that the residues for the grafted samples are larger than the predicted values based upon the amount of grafted methacrylic acid. The residues from sodium methacrylate grafted samples show a general increase with weight % grafted methacrylic acid, indicating that as the amount of graft layer increases, the amount of protection offered to the substrate is increased and more SBS is retained by the coating. The residues for methacrylic acid grafted samples show much more variation, it is likely that this coating is less insulating than that from sodium methacrylate and thus is less retentive of substrate.

Similar results have been obtained for grafted ABS samples. These results are presented in Table III.

TABLE III

TGA residue for MAA and NaMAA grafted ABS

weight % grafted	MAA residue	theoretical MAA residue	NaMAA residue	theoretical NaMAA residue
0	4.0	4.0	4.0	4.0
1	4.3	4.0	5.3	4.1
10	7.3	4.4	5.8	5.3
20	4.4	4.8	15.0	6.6
30	5.1	5.2	15.2	7.9
40	4.0	5.6	19.8	9.2
50	7.6	6.0	24.4	10.5

Again the TGA residues are larger than expected from the graft layer only and must indicate that the substrate participates in the char formation. For the sodium salt case, there is an increase in the char yield with an increase in the amount of grafting. Again the acid shows much more variation and the same reason is offered, the char from the sodium salt, consisting of primarily sodium carbonate with a small amount of elemental carbon, is a better insulator and more retentive of substrate. Figure 2 presents the TGA curve for an ABS sample which contains 50 weight % sodium methacrylate as well as for unmodified ABS. The initial degradation is due to the elimination of water from the methacrylic acid. The onset of degradation occurs at the same temperature in ABS, sodium methacrylate obtained from the acid by treatment with sodium hydroxide, and in the grafted sample. Grafting does not appear to effect the onset of degradation but it does change the extent of degradation.

On the whole, the TGA results indicate that there is participation of the substrate, whether SBS or ABS, in char formation. It is likely that there is some

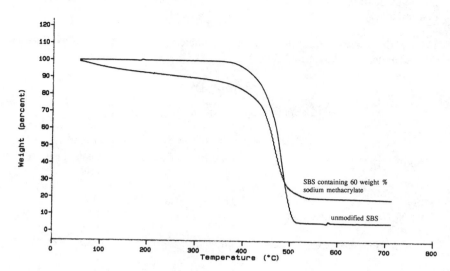

Figure 1. TGA curves for unmodified SBS and SBS to which has been grafted 60 weight % methacrylic acid which is then converted to the sodium salt. Rate is 20°C per minute under dinitrogen.

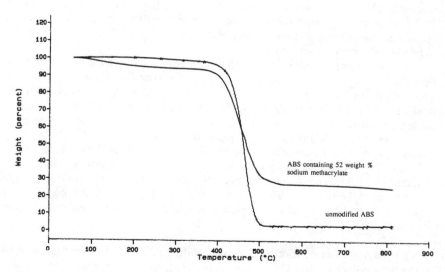

Figure 2. TGA curves for unmodified ABS and ABS to which has been grafted 52 weight % methacrylic acid which is then converted to the sodium salt. Rate is 20°C per minute under dinitrogen.

optimum coating that will achieve the maximum effect but this remains to be determined.

Cone Calorimetry. Preliminary cone calorimeter results for both grafted SBS and ABS samples are shown in Table IV.

TABLE IV

Cone Calorimeter Results for ABS and SBS

Sample	time to ignition, sec.	peak heat release rate, kW/m^2
SBS	666	960
2 wt % NaMAA grafted SBS	711	820
ABS	285	901
21 wt % NaMAA grafted ABS	460	259

The results for grafted ABS are impressive. The time to ignition is almost doubled and the peak heat release rate drops to almost a fourth of the value in unmodified ABS. The addition of only 2 weight % sodium methacrylate to SBS increases the time to ignition by 7% and decreases the peak heat release rate by 15%. It is not surprising that the result is much smaller for SBS than for ABS because of the difference in the amount of the grafted layer. Clearly the results are indicative of desirable performance.

For the grafted ABS system, additional results are available. The time to peak heat release rate is almost doubled for the modified ABS sample. The unmodified sample is essentially completely consumed by 670 seconds after the beginning of irradiation while the grafted sample had only lost 5% of its mass; during the entire 20 minute run, only 36% of the grafted sample was consumed.

The cone calorimeter results show that 1) it is more difficult to ignite samples to which sodium methacrylate has been grafted; 2) the heat release rate is reduced, for ABS dramatically reduced; 3) the rate of mass loss is substantially reduced for ABS; and 4) a substantial portion of the sample remains at the termination of the run for ABS. These offer a clear indication that enhanced char formation occurs in the grafted sample and that this provides thermal protection to the polymer.

Oxygen Index. The performance of both grafted SBS and ABS is disappointing in the oxygen index test. The oxygen index does, in fact, increase when sodium methacrylate is on the surface but the increase is marginal. The unmodified SBS has an OI of 16 and this increases to 18 for a 40 weight % grafted sample. For ABS, the results are even more discouraging because the increase is only from 18 in unmodified ABS to 19 in 40 weight % grafted NaMAA. It is clear that the OI is measuring some other property than that measured by TGA or cone calorimetery.

Chemical Interpretation of the Results. The TGA residues indicate that some of the substrate is participating in char formation. It is likely that the surface layer of sodium methacrylate is forming char and retaining the underlying polymer. TGA-FTIR results indicate that the ABS begins to degrade at higher temperatures than is observed for unmodified ABS and this means that the graft layer is offering some protection to the bulk of the polymer. During the thermal degradation of ABS, butadiene is evolved at 340°C, aromatics from the degradation of the styrene portion begin at 350°C, and monomeric acrylonitrile is observed at 400°C. In the grafted sample, butadiene and aromatics both evolve at the same temperature, 430°C, and monomeric acrylonitrile is never observed. The graft is attached to the butadiene portion of the polymer but it effects the thermal degradation of the entire polymer. This is interpreted to mean that all portions are covered by the graft layer.

CONCLUSION

The procedure of grafting onto the surface of a polymer some material that will char and protect the underlying polymer is a viable approach to flame retardancy. Work is continuing on the polymers that are discussed in this paper as well as on other polymers.

ACKNOWLEDGEMENT

The assistance of David Paul and the Monsanto Company in the cone calorimetry work is gratefully acknowledged.

LITERATURE CITED

1. Wilkie, C.A.; Pettegrew, J. W.; Brown, C. E.; *J. Polym. Sci., Polym. Lett. Ed.*, **1981**, *19*, 409.
2. Brown, C. E.; Wilkie, C. A.; Smukalla, J.; Cody Jr.,R. B.; Kinsinger, J. A.; *J. Polym. Sci., Polym. Chem. Ed.*, **1986**, *24*, 1297.
3. Sirdesai, S. J.; Wilkie, C. A.; *J. Appl. Polym. Sci.*, **1989**, *37*, 863.
4. Sirdesai, S. J.; Wilkie, C. A.; *J. Appl. Polym. Sci.*, **1989**, *37*, 1595.
5. Wilkie, C. A.; Thomsen, J. R.; Mittleman, M. L.; *J. Appl. Polym. Sci.*, **1991**, *42*, 901.
6. Hurley, S. L.; Mittleman, M. L.; Wilkie, C. A.; *Polym. Degrad. Stab.*, **1993**, *39*, 345.
7. Wilkie, C. A.; Leone, J. T.; Mittleman, M. L.; *J. Appl. Polym. Sci.*, **1991**, *42*, 1133.
8. Beer, R. S.; Wilkie, C. A.; Mittleman, M. L.; *J. Appl. Polym. Sci.*, **1992**, *46*, 1095.
9. Chandrasiri, J. A.; Roberts, D. E.; Wilkie, C. A.; *Polym. Degrad. Stab.*, **1994**, *45*, 97.

10. Chandrasiri, J. A.; Wilkie, C. A.; *Polym. Degrad. Stab.*, **1994**, *45*, 83.
11. Chandrasiri, J. A.; Wilkie, C. A.; *Polym. Degrad. Stab.*, **1994**, *45*, 91.
12. McNeill, I. C.; *Develop. Polym. Degrad.*, **1987**, *7*, 1.
13. McNeill, I. C.; Zulfiqar, M.; *J. Polym. Sci.: Polym. Chem. Ed.*, **1978**, *16*, 3201.
14. McNeill, I. C.; Zulfiqar, M.; *Polym. Degrad. Stab.*, **1979**, *1*, 89.
15. McNeill, I. C.; Zulfiqar, M.; *J. Polym. Sci.: Polym. Chem. Ed.*, **1978**, *16*, 2465.
16. Geuskens, G.; Kanda, M. N.; *Eur. Polym. J.*, **1991**, *27*, 877.
17. Geuskens, G.; Thiriaux, P.; *Eur. Polym. J.*, **1993**, *29*, 351.

RECEIVED November 21, 1994

Chapter 16

Flammability Properties of Honeycomb Composites and Phenol–Formaldehyde Resins

Marc R. Nyden, James E. Brown, and S. M. Lomakin[1]

Building and Fire Research Laboratory, National Institute of Standards and Technology, Gaithersburg, MD 20899–0001

The flammability properties of honeycomb composites, which are used in the interior cabin compartments of commercial aircraft, were examined. Analyses of the gases evolved during the thermal degradation of the components indicated that the phenol-formaldehyde resin makes a significant contribution to the flammability of these composites. The possibility that a more fire resistant formulation could be developed was examined by testing a series of resins which differed in the relative amounts of phenol and formaldehyde used in the reaction mixtures. The flammabilities of resins synthesized in excess phenol were measurably less than those synthesized in excess formaldehyde.

The burning of most polymers may be viewed from the perspective of a simple model whereby volatile hydrocarbons, which are formed during the thermal degradation of the condensed phase, are combusted in the gas phase. The basis of this model is the hypothesis that all of the available oxygen is depleted in the flames above the surface of the solid. The cycle must be initiated by heat supplied from an external source, but it is self-sustaining as long as the combustion process generates sufficient energy to further degrade the polymer into fuel. The size of the fire is measured by the net rate-of-heat released (rhr) during this process.

The distinction between the thermal degradation and gas phase combustion steps is blurred in the case of polymers, such as phenol formaldehyde (PF) resins, which contain significant amounts of covalently bound oxygen. This can effect a reduction in flammability because less heat is released during burning if the polymer has already been partially oxidized.

---

[1]Guest Researcher from the Institute of Chemical Physics, Russian Academy of Sciences, Moscow, Russia

This chapter not subject to U.S. copyright
Published 1995 American Chemical Society

Furthermore, the oxidation reactions which are responsible for the heat release, are much faster when they occur in the gas phase so that the rhr will be reduced in proportion to the fraction of hydrocarbon which is oxidized in the condensed phase. Finally, the processes involved in condensed phase oxidation also promote the formation of char (1) which provides additional fire resistance by further reducing the amount of fuel available for gas phase combustion (2).

The focus of the present investigation is on PF resins which are known to have a propensity to condensed phase oxidation and char formation. PF resins are found in plywood and other fiber based composites used in the construction of buildings, aircraft and ships. Improvements in the flammability performance of these materials is desirable for certain fire-sensitive applications, such as, in the honeycomb panels used in the sidewalls, ceilings and stowage bins of commercial aircraft.

Experimental

A. Honeycomb Composites
A composite material consisting of nomex honeycomb, fiber backing, and phenol-formaldehyde resin was obtained from the Federal Aviation Administration Technical Center. This material was identified by the code TP4/92. The original square panels were cut into circular disks with an outer diameter of 7.5 cm which was deemed most suitable for the Cone Calorimeter flammability measurements.

B. Synthetic Resins
The method used in synthesizing the PF resins is described in detail in reference 3. Reaction mixtures consisting of 0.5, 1.0 and 2.0 moles of formaldehyde (37% solution, ACS Reagent, Sigma) per mole of phenol (ACS Reagent, Sigma) were refluxed in 15 ml of 5N NaOH (ACS Reagent, Fisher) at 130 °C for approximately 2 hours. The prepolymer was washed in H_2O to remove excess salts and dehydrated by heating at 100 °C. Most of the samples were cured in air at 150 °C for 72 hours. In some cases, the resins were synthesized and cured in a N_2 atmosphere. This precaution, however, did not have an obvious effect on the distribution of products observed during thermal degradation.

C. Measurement Techniques
The NIST Cone Calorimeter was used to make all flammability measurements. This instrument measures a number of combustion-related properties including rhr and is the basis for an ASTM test method (E1354-90a) (4). Samples (~ 50g) were placed in a Pyrex dish and exposed to a heat flux of 50 kW/m^2. A high voltage arc was placed above the samples to ignite the off-gases.

Thermogravimetric analyses (TGA) were performed in a N_2 atmosphere using a Perkin-Elmer 7 Series Thermal Analysis System (Certain commercial equipment, instruments, or materials are identified in this paper in order to specify the experimental procedure. Such identification does not imply that the material or equipment identified is necessarily the best available for the purpose.). The samples, which weighed approximately 1 mg at the onset of the experiments, were heated from 30 to 800 °C at a rate of 10 °C/min.

Infrared analyses of the evolved gases (FTIR-EGA) were performed using a gas cell with a 10 cm path length and programmable paralyzer manufactured by Chemical Data Systems Instruments. Samples, ranging from about 10-20 mg in mass, were placed in a quartz tube which was surrounded by a resistance heating element. This assembly was inserted into the IR cell which was purged with N_2 and heated either isothermally at a series of temperatures between 800 and 850 °C or dynamically from 25 to 1000 °C at a rate of 50 °C/min. The progress of the pyrolysis was monitored every 5 seconds for the kinetics measurements and every 20 seconds in the dynamic heating experiments. The spectra were collected on a Mattson Galaxy 7000 Series FTIR at a resolution of 2 cm^{-1}.

The ^{13}C nuclear magnetc resonance (NMR) spectra were measured using the cross polarization (CP)/magic angle spinning (MAS) technique (5) at 25.193 MHz in a static field of 2.354 T. The spectra were acquired at a spinning rate of 4 kHz with contact and repetition times of 1 ms and 3 s, respectively.

Results and Discussion

A. Honeycomb Composites

The rhr from a sample of honeycomb composite exposed to an incident flux of 50 kW/m² is displayed as a function of burn time in Figure 1 (upper curve). Initially, these composites burn with a yellow luminous flame indicative of the presence of soot. This appears as a distinct peak in the rhr curve centered at about 45 seconds. The luminosity disappears about 15 seconds later giving way to an unstable blue flame which is associated with fluorescent emissions accompanying the oxidation of CO. The composites usually continue to smolder long after the flame disappears. This effect is indicated by the tail in the rhr curve which extends from about 70 seconds. The lower curve in Figure 1 is the rhr from a sample taken from the same panel which was heated in an oven overnight at 250 °C before it was burned. The thermograms of the resin taken before (Figure 2) and after (Figure 3) the heat treatment exhibit differences in the region between 500 and 575 °C. A comparison of the FTIR-EGA spectra of these resins is even more revealing (Figure 4). The peaks centered at about 1300 and 3000 cm^{-1}, which are most prominent in the spectrum of the untreated resin, are due to CH_4. The FTIR-EGA spectra obtained from the components of treated (Figure 5) and untreated (Figure 6) composites indicate that much higher levels of CH_4 are generated from the degradation of the resin than from the fiber backing or honeycomb.

Thus, there is a correlation between the amount of CH_4 produced in the thermal degradation of the resin and the rhr of the composite. On this basis, it seems clear that any attempt to improve the fire resistance of honeycomb composites should begin with an examination of the flammability properties of PF resins.

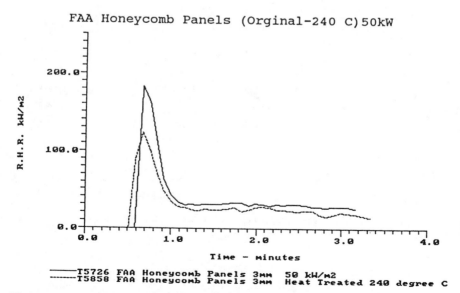

Figure 1. Comparison of rhr curves obtained by burning honeycomb composite material before (upper) and after the heat treatment.

248   FIRE AND POLYMERS II

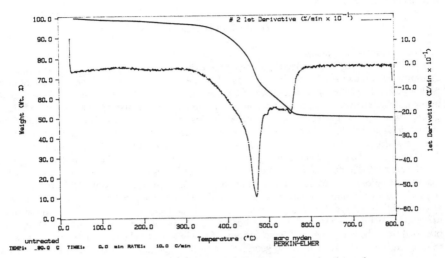

Figure 2. Thermograms of resin taken from an untreated panel.

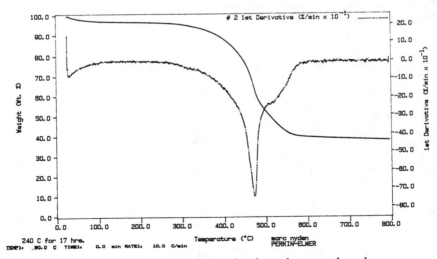

Figure 3. Thermograms of resin taken from a heat treated panel.

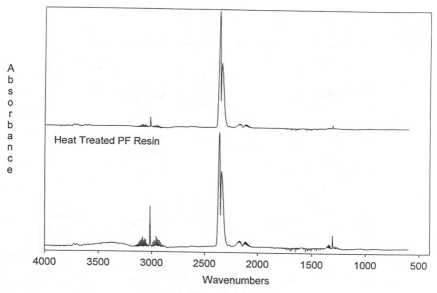

Figure 4. Comparison of FTIR-EGA spectra taken from heat treated (upper) and untreated panels.

B. Synthetic Resins

The base catalyzed reaction of phenol with formaldehyde produces stable hydroxymethylphenol intermediates which condense into branched polymers (resoles) at temperatures of between 60 and 100 °C (6). Further condensation of methylol groups occur during the curing process resulting in the formation of highly crosslinked network polymers. Methylene bridges, which are produced in head-to-tail interactions between methylphenols, are thermodynamically favored over ether linkages (7) which form when methylphenols are aligned in a head-to-head configuration or from condensation of methylphenols with phenol itself. An investigation of the degradation properties of PF resins was conducted by Conley and coworkers (7-10). They concluded that the primary degradation pathway for PF resins is oxidative in nature even in an oxygen deficient atmosphere and that thermal processes only begin to compete at higher temperatures. The spectra obtained in dynamic FTIR-EGA experiments indicate that H_2O, CO_2 and CH_3OH are evolved in the degradation of PF resins in N_2 at temperatures as low as 250 °C (Figure 7). The presence of CO is first detected at about 350 °C, while CH_4, which is the major volatile product from the thermal degradation of the resin, is evident only at temperatures above 550 °C.

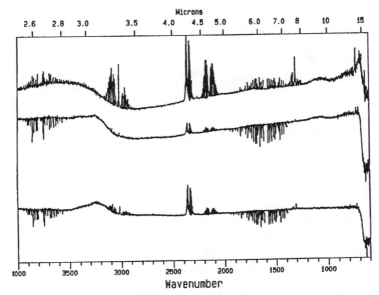

Figure 5. Comparison of FTIR-EGA spectra (1000 °C) of the honeycomb (lower), fiber backing (middle), and resin (upper) taken from an untreated panel.

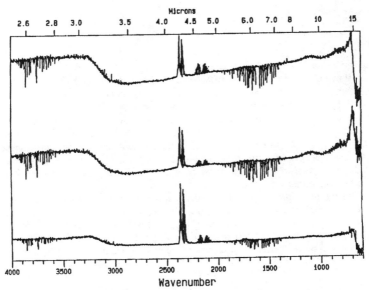

Figure 6. Comparison of FTIR-EGA spectra (1000 °C) of the honeycomb (lower), fiber backing (middle), and resin (upper) taken from a treated panel.

The rate constants (k) for the formation of CO_2 and CH_4 are reported in Table I. These values were determined by integrating characteristic peaks in the FTIR-EGA spectra measured under isothermal conditions for temperatures between 800 and 850 °C in accordance with the following model for first order thermal degradation of the polymer:

$$\ln(1-\frac{[\alpha]_t}{[\alpha]_\infty}) = -kt.$$

In this equation, $[\alpha]_t$ denotes the integrated absorbance of either product at time t and $[\alpha]_\infty$ is the asymtoptic value which is attained in the limit as $t \to \infty$. The global activation energies for CO_2 and CH_4 are 59 ±13 and 180 ± 33 kJ/mol, respectively.

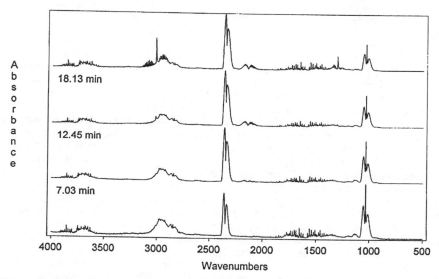

Figure 7. FTIR-EGA spectra of PF resin at 250 (bottom), 375 (7.03 min), 650 (12.45 min) and 930 °C (18.13 min)).

Table I. Rate Constants (s^{-1}) for the Formation of CO_2 and CH_4

Compound	Temperature (°C)		
	800	825	850
CO_2	0.090 ± 0.005	0.093 ± 0.030	0.095 ± 0.012
CH_4	0.058 ± 0.003	0.068 ± 0.003	0.073 ± 0.008

A mechanism has been advanced for the formation of CO and CO_2 during the oxidative degradation of PF resins. According to Conley (7), the first step is the conversion of methylene bridges into carbonyl linkages. The resulting benzophenones can either undergo thermal scission to produce CO or further oxidation resulting in the formation of benzoates which then release CO_2 in decarboxylation reactions. The evidence for this mechanism is compelling, however, the extension to oxygen deficient atmospheres, which will be the prevailing condition at the surface of burning polymers, is questionable. The postulate that the CH_4 comes from the thermal scission of methylene bridges is also suspect since this bond is known to be exceptionally strong. The bond dissociation enthalpy, based on the difference between the measured heats of formation of diphenylmethane (11) and the benzyl (12) and phenyl (13) radicals, is 389 kJ/mol. In fact, a simple mechanism whereby the methylphenol condensation process is reversed at elevated temperatures in the presence of residual H_2O can account for the evolution of all of the major volatile products. The CH_2O, which is released during the depolymerization of the resin, is known to produce CH_3OH and CO_2 via the Cannizarro reaction, while the formation of CH_4 from rearrangement of hemiacetal intermediates also seems likely; at least at the elevated temperatures considered in this study.

As is the case with many polymers, the flammability properties of PF resins will depend on how they are prepared and cured. The range of possibilities was explored by comparing flammability performance of cured and uncured resins synthesized from reaction mixtures containing different proportions of the two monomers. The results of the Cone Calorimiter measurements, which are reported in Table II, indicate that flammability increases with increasing mole fraction of formaldehyde in the reaction mixture and that the cured resins are significantly less flammable than the uncured resins. The difficulties involved in preparing large samples with uniform properties precluded the possibility of performing a rigorous uncertainty analysis based on duplicate measurements. The trends, however, are unambiguous. Indeed, the high levels of char and low heat release rates exhibited by the 2:1 resins are remarkable, considering the high flux of incident radiation used in these tests.

The relative amounts of CO_2, CH_3OH and CH_4 produced in the thermal degradation of the resins in N_2 were estimated from characteristic peaks in the steady-state FTIR-EGA spectra measured at 800 °C. The values obtained by dividing the integrated absorbance of the characteristic peak by the initial mass of the sample are listed in Table III. The CH_3OH data follow the trends observed in the flammability performance of the resins in that the highest levels were observed in the most flammable resins. This is not true in the case of CO_2. Although the concentration of CO_2 decreases with increasing phenol content, the highest levels were observed in the thermal degradation of the less flammable cured resins. There is some indication that the evolution of CH_4 may follow a similar trend. Unfortunately, the uncertainty in these measurements is too large to make a definitive statement. The variations between different samples of the same resin, which approached 50% of the measured values, were of the same order of magnitude as were the variations between the samples taken from different resins.

The ^{13}C CP/MAS spectra of the cured resins are displayed in Figure (8). The peak assignments, which were made using a computer program (14), were consistent with previously published data on phenol-formaldehyde resins (15,16). The integrated intensity of each peak, which is proportional to the number of carbon atoms corresponding to the specified assignment, is listed in Table IV. Not all of the peaks were well resolved. In some cases, the contribution of the components had to be estimated by inspection. These values are indicated by an asterisks.

These spectra indicate structural differences between the resins which are consistent with the Cone Calorimeter data. In particular, the ratio of aromatic to aliphatic carbons is highest in the least flammable (2:1) resin. The two weak resonances centered at about 57 and 75 ppm, which are discernible in the spectra of both the 1:2 and the 1:1 resins, are indicative of the presence of aliphatic OH. This is a likely source of the CH_3OH which was detected during pyrolysis of these resins. Indeed, the 2:1 resin also appears to contain significantly less aromatic OH as evinced by a reduction in the intensity of the resonance due to phenolic OH at 152 ppm.

Table II. Flammability Properties of PF Resins

Moles of P:F	Mass of Sample (g)	Time to Ignition (s)	Peak RHR (kW/m²)	Total Heat (mJ/m²)	Residue (%)
1:2 (uncured)	45.2	124	213	113	49.8
1:2 (cured)	66.3	102	174	23	76.6
1:1 (uncured)	42.2	422	164	148	44.8
2:1 (uncured)	45.0	44	116	10	93.9
2:1 (cured)	66.3	59	79	5	95.1

Table III. Integrated Peak Absorbance* at 800 °C

Moles of P:F	Uncured	Cured
CH_4 (1295-1310 cm^{-1})		
1:2	14	29
1:1	26	27
2:1	14	20
CH_3OH (970-1085 cm^{-1})		
1:2	373	149
1:1	157	44
2:1	71	12
CO_2 (2285-2390 cm^{-1})		
1:2	1278	2143
1:1	1053	1489
2:1	232	326

* Values normalized by dividing by the initial mass of the sample.

Table IV. ^{13}C NMR Integrated Peak Intensities

Assignment/ P:F Ratio	C(ar)-O-C(ar) 160-180 ppm	C(ar)-OH 152 ppm	C(ar) 130 ppm	-OCH$_2$ 55-80 ppm	CH$_3$ 15 ppm	ar:al Ratio
1:2	3.0	11.2	60.5	4.1	4.3	3.0
1:1	2.3*	9.1*	61.6	3.0	2.4	3.1
2:1	8.7*	8.7*	58.8	0.0	2.6	3.4

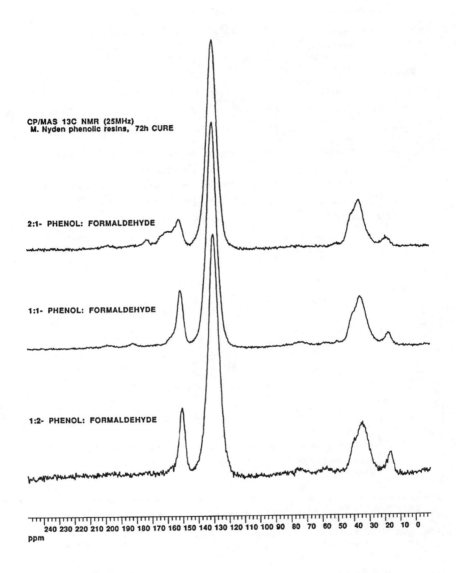

Figure 8. ^{13}C CP/MAS NMR spectra of the 1:2 (bottom), 1:1 (middle) and 2:1 PF resins(top).

This oxygen is accounted for in the spectrum of the 2:1 resin by the broad resonance extending from about 155-180 ppm which is characteristic of the presence of biphenyl ethers (15). Presumably, these ether linkages are formed as a result of self-condensation of the phenol. Thus, it is not surprising that they are more prevalent in those resins which were synthesized in an excess of this reactant. Likewise, resins synthesized in excess formaldehyde might be expected to contain significant amounts of the formaldehyde homopolymer. Pure polyoxymethylene is known to be thermally labile. Depolymerization is thought to be initiated at the chain ends at about 200 °C (17) which is consistent with the temperature at which CH_3OH and CO_2 first appear. The presence of polyoxymethylene in the 1:1 and 1:2 resins would account for the high levels of CO_2 which are evolved during curing; even when this process is carried out in N_2, as well as, providing a simple explanation for the reduced flammability of the cured resins.

Conclusions

Combustible products, particularly CH_4 and CH_3OH which are released during the thermal degradation of the PF resin, make a significant contribution to the flammability of the honeycomb composites considered in this investigation. A reduction in flammability was demonstrated in the case of resins which were synthesized from reaction mixtures containing an excess of phenol. The fire resistance of these resins is also affected by the curing process. The levels of combustible products released in thermal degradation and the peak rate-of-heat-release were significantly higher in the uncured resins than they were in the cured resins.

Acknowledgements

This work was supported by funds provided in a cooperative research agreement between the National Institute of Standards and Technology and the Federal Aviation Administration. The authors would like to specifically acknowledge the advice and encouragement extended by Constantine Sarkos and Richard Lyon of the FAA Technical Center in Atlantic City, NJ.

References

1. Nyden, M.R., Forney, G.P. and Brown, J.E., Macromolecules **1992**, 25, 1658.
2. van Krevelen, D.W., Polymer **1975**, 16, 615.
3. Martin, R.W., The Chemistry of Phenolic Resins, John Wiley, New York, 1956, 298.
4. Standard Test Method for Heat and Visible Smoke Release Rates for Materials and Products Using an Oxygen Consumption Calorimeter, **ASTM E-1354**, Philadelphia, PA, 1991, 1.
5. VanderHart, D.L., Earl, W.L. and Garroway, A.N., J. Magn. Reson. **1981**, 44, 361.
6. Knop, A. and Pilato, L.A., Phenolic Resins, Springer-Verlag, New York, NY, 1985, 25.
7. Conley, R.T., In Themal Stability of Polymers, John Wiley, New York, 1970, 457.
8. Conley, R.T. and Bieron, J.F., J. Appl. Polymer Sci. **1963**, 7, 103.
9. Conley, R.T. and Bieron, J.F., J. Appl. Polymer Sci. **1963**, 7, 171.
10. Conley, R.T.., J. Appl. Polymer Sci. **1965**, 9, 1117.
11. Pedley, J.B., Naylor, R.D. and Kirby, S.P., Thermochemical Data of Organic Compounds, Chapman and Hall, London, 1986, Second Edition.
12. Walker, J.A. and Tsang, W., J. Phys. Chem **1990**, 94, 3324.
13. Robaugh, D. and Tsang, W., J. Phys. Chem. **1986**, 90, 5363.
14. ChemWindow3, SoftShell International, Grand Junction, CO, 1994.
15. Maciel, G.E., Chuang, I. and Gollob, L., Macromolecules **1984**, 17, 1081.
16. Amram, B. and Laval, F., J. Appl. Polym. Sci. **1989**, 37, 1.
17. Madorsky, S.L., Thermal Degradation of Organic Polymers, John Wiley, New York, NY, 1964, 228.

RECEIVED November 28, 1994

Chapter 17

Synthesis and Characterization of Novel Carbon–Nitrogen Materials by Thermolysis of Monomers and Dimers of 4,5-Dicyanoimidazole

Eric C. Coad and Paul G. Rasmussen[1]

Department of Chemistry, University of Michigan, Ann Arbor, MI 48109-1055

Our preparation of materials with high nitrogen and no hydrogen content is an effort to obtain thermally stable materials with low flammability. The thermolysis of 2-chloro-4,5-dicyanoimidazole and 2-(2-chloro-4,5-dicyano-1-imidazolyl)-4,5-dicyanoimidazole functionalized with -H, and -I as leaving groups at the 1-position were examined. Thermolysis of the 2-chloro-4,5-dicyanoimidazole derivatives between 100-290°C were found to yield Tris(imidazo)[1,2-a:1,2-c:1,2-e]-1,3,5-triazine-2,3,5,6,8,9-hexacarbonitrile (**HTT**) with $(C_5N_4)_3$ composition. Thermolysis of **HTT** at 490-500°C resulted in a carbon-nitrogen material with C/N = 1.020, while the thermolysis of **HTT** at 1070°C resulted in a carbonaceous material. The examination of the thermal properties of **HTT** and its thermal decomposition products demonstrate bulk thermal stability to 350°C. The thermolysis of the 2-(2-chloro-4,5-dicyano-1-imidazolyl)-4,5-dicyanoimidazole derivatives and their products are under further investigation.

Examples of heterocyclic polymers or materials which exhibit enhanced thermal stability include semi-ladder type polymers such as polyimides, polybenzimides, polybenzimidazoles, polybenzothiazoles, polybenzoxazoles, polyoxadiazoles, and polytriazoles; and ladder type polymers such as polyimidazopyrrolones, polyquinoxalines, and polyquinizarines.[1] In these systems, the flammability is repressed by increasing the carbon to hydrogen ratio, while the thermal stability is enhanced by fusing heteroaromatic ring systems together in both the semi-ladder and ladder type polymers.

The Oxygen Index (OI) value provides a relative measure of the flammability of a substance, where the OI value is equal to 100 times the ratio of oxygen to the total amount of gas in the atmosphere required to sustain combustion. Comparison of the OI values to the weight % of nitrogen in the selected nitrogen containing materials in Table I indicate that the flammability goes down (OI value goes up) as the weight % of nitrogen increases.[2] It is also evident in Table I that some of the other nitrogen containing polymers (Nomex, Kapton, and polybenzimidazole) have high OI values. The comparison of OI values of selected polymers to their carbon/hydrogen ratio in Table II indicates to a rough approximation that as the carbon/hydrogen ratio goes up (hydrogen content decreases) the flammability goes down (OI values go up).[3]

[1]Corresponding author

Table I: Oxygen Index data of selected nitrogen containing polymeric materials

Polymeric Material	Nitrogen Content (wt. %)	OI
Polyurethane foam	4.5-5.2	16.5
Polyacrylonitrile	-	18
Nylon-6,6	12.4	24.0
Wool	16-17	25.2
Silk	18-19	>27
Aromatic polyamide(Nomex)	-	28.5
Polyimide(Kapton, Dupont)	-	36.5
Polybenzimidazole	-	41.5

Source: Data are taken from ref. 2.

This report will compare the thermal stabilities of Tris(imidazo)[1,2-a:1,2-c:1,2-e]-1,3,5-triazine-2,3,5,6,8,9-hexacarbonitrile or Hexacarbonitrile Tris(imidazo) Triazine (**HTT**) and thermolysis products from **HTT** to the thermal stabilities of thermolysis products from 2-(2-chloro-4,5-dicyano-1-imidazolyl)-4,5-dicyano-imidazole (**4**).

Table II: Comparison of Oxygen Index data to the Carbon / Hydrogen Ratios of selected polymeric materials

Polymeric Material	Carbon/Hydrogen Ratio	OI
Polyacetal	0.50	15
Polymethyl methacrylate	0.63	17
Polyethylene	0.50	17
Polystyrene	1.00	18
Polycarbonate	1.14	26
Polyarylate	1.21	34
Polyethersulphone	1.50	34-38
PEEK	1.58	35
PVC*	0.67	23-43
PPS	1.50	44-53
PVDC*	1.00	60
PTFE *	-	90

* These polymers contain halogen atoms.
Source: Data from ref. 3.

The synthesis of **HTT** (**3**) with $(C_5N_4)_3$ composition is accomplished by thermolysis of compounds **1-2** (Scheme 1).[4] In an effort to prepare polymeric materials (Scheme 2), compound (**4**) was synthesized (Schemes 3-4). TGA data for compounds **4-5** (Figure 8) demonstrate the analogous thermal transitions found for compounds **2-3** (Figure 2-3). Further work is underway to characterize the products from the thermolysis reactions in Scheme 2.

Scheme 1

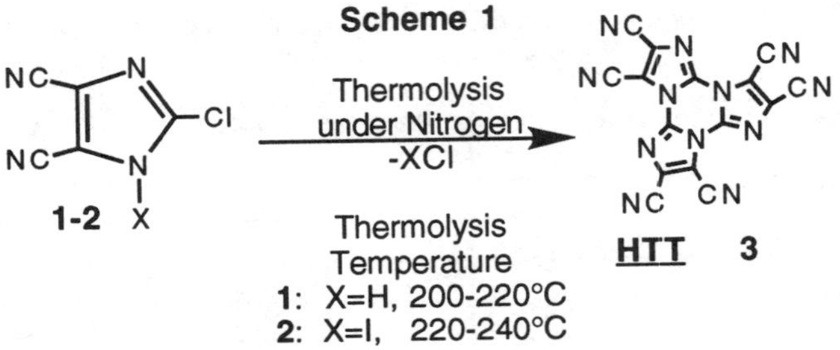

Thermolysis under Nitrogen
-XCl

Thermolysis Temperature
1: X=H, 200-220°C
2: X=I, 220-240°C

HTT 3

Scheme 2

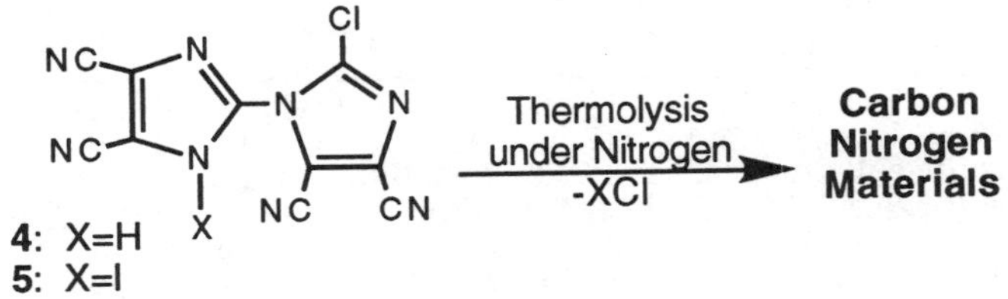

4: X=H
5: X=I

Thermolysis under Nitrogen
-XCl

→ **Carbon Nitrogen Materials**

Experimental.
General Procedures. Melting points were recorded on a Mel-Temp apparatus and are uncorrected. Infrared spectra were recorded on a Nicolet 5-DX FTIR spectrophotometer. ^{13}C NMR (75 MHz) were recorded on a Bruker AM-300 spectrophotometer. Elemental analyses were done at the University of Michigan on a Perkin-Elmer 2400 CHN analyzer or done by Oneida Research Services, Inc., Whitesboro, NY. TGA were recorded on a Perkin-Elmer Series 7 Thermal Analysis System with a heating rate of 5°C/minute, under N_2 or air. 1-methyl-2-bromo-4,5-dicyanoimidazole (**6**) was prepared as in reference 5. **HTT** (**3**) and thermolysis products from **HTT** were prepared as in reference 6.

1-methyl-2-fluoro-4,5-dicyanoimidazole (7). A reaction mixture comprised of 20.0 g (0.094 mol) of **6**, 13.68 g (0.236 mol) of spray-dried KF, a catalytic amount of 18-crown-6 ether, and 25 mL of diglyme were heated at reflux overnight. The liquid was decanted and the salt was washed with acetone. The decanted liquid and acetone washes were combined. The acetone and diglyme were removed under reduced pressure. The resulting brown oil was vacuum distilled twice with Bp 95-110°C at 0.02-0.03 mm producing 50.0 g (89%) of **7** as a clear oil which crystallized upon standing. The white solid **7** displays the following properties: TLC EtOAc R_f 0.64; Mp 46-48°C; IR(KBr): 2243, 1595, 1512, 1411, 1174 cm^{-1}; UV-vis(CH_3CN) $\lambda_{max}(\epsilon)$ 250(12120); ^{1}H NMR (300.1 MHz, $CDCl_3$, ppm) δ 3.73(s, 3H); ^{13}C NMR (75.5 MHz, $CDCl_3$, ppm) δ 150.0(d, J = 252.1 Hz), 117.3(d, J = 12.7 Hz), 110.7, 110.4(d, J = 3.9 Hz), 107.0, 31.4; MS (E/I) m/z (relative intensity) 151 (M+1, 8), 150 (M+, 100%), 149 (20), 135 (2), 122 (5), 109 (5); HRMS (EI with DCI probe) m/z calcd. for $C_6H_3N_4F$ 150.0342, obsd. 150.0338; Anal. calcd for $C_6H_3N_4F$: C, 48.02; H, 2.00; N, 37.32. Found: C, 47.96; H, 1.85; N, 37.15.

1-methyl-2-(2-amino-4,5-dicyano-1-imidazolyl)-4,5-dicyanoimidazole (9). A reaction mixture comprised of 40.0 g (.267 mol) of **7**, 68.4 g (.400 mol) of **8**, a catalytic amount of 18-crown-6 ether, and 400 mL of CH_3CN were heated for 8 h at 50°C. The CH_3CN was removed under vacuum. The resulting solid was dissolved in EtOAc and was extracted with 10% NH_4OH. The solvent was removed under

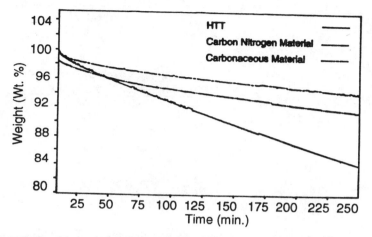

Figure 5. Isothermal TGA of HTT, Carbon Nitrogen Material from HTT, and Carbonaceous Material from HTT at 400°C under nitrogen.

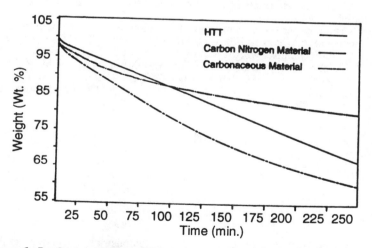

Figure 6. Isothermal TGA of HTT, Carbon Nitrogen Material from HTT, and Carbonaceous Material from HTT at 400°C under air.

Scheme 5

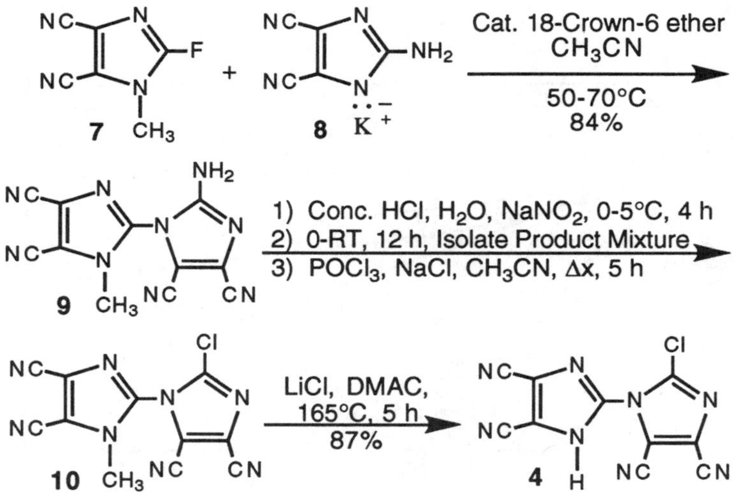

[Scheme 5 showing reaction of compound 7 (1-methyl-4,5-dicyano-2-fluoroimidazole) with compound 8 (potassium salt of 1-methyl-4,5-dicyano-2-aminoimidazole) using Cat. 18-Crown-6 ether, CH₃CN, 50-70°C, 84% to give compound 9; then 1) Conc. HCl, H₂O, NaNO₂, 0-5°C, 4 h; 2) 0-RT, 12 h, Isolate Product Mixture; 3) POCl₃, NaCl, CH₃CN, Δx, 5 h to give compound 10; then LiCl, DMAC, 165°C, 5 h, 87% to give compound 4.]

aromatic substitution reaction using **8** as a nucleophile. The amino group of compound **9** is diazotized and decomposed forming a mixture of compound **10**, and its hydrolysis products, and a small amount of the corresponding imidazolone dimer. The imidazolone dimer and hydrolysis products are converted to **10** using POCl₃. Dealkylation of **10** is accomplished using LiCl and DMAC yielding compound **4**.

The thermal properties of compound **11** were investigated to model the effect of 1,2-connectivity between 4,5-dicyanoimidazole rings. The TGA of **11** under nitrogen (Figure 7) shows weight losses of 7% by 560°C, and 70% between 560-900°C, while the TGA data under air shows weight losses of 4% by 300°C, 25% between 300-600°C, and 68% between 600-750°C. The results from the model compound demonstrate strong thermal stability to 500°C under both air and nitrogen.

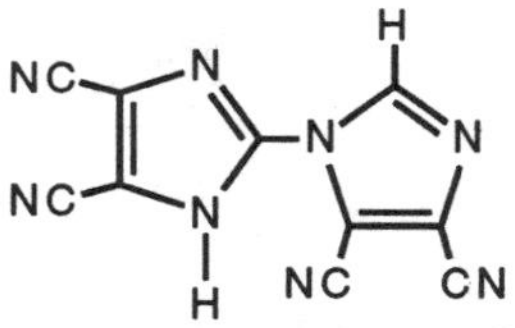

Compound 11

Preliminary TGA data for compounds **4-5** demonstrate the analogous thermal transitions found for compound **2**. The TGA of **4** (Figure 8) shows a weight loss of 14% between 95-270°C which may correspond to loss of HCl, and a weight loss of 14% between 270-480°C which may correspond to the loss of $(CN)_2$, and N_2. Heating to 900°C yields complete weight loss.

The TGA of **5** (Figure 8) shows a weight loss of 44% between 95-325°C which may correspond to loss of ICl, and a weight loss of 9% between 325-545°C which may correspond to the loss of $(CN)_2$, and N_2. Heating to 900°C yields complete

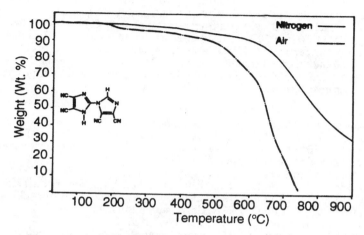

Figure 7. TGA of Compound 11 under nitrogen and air.

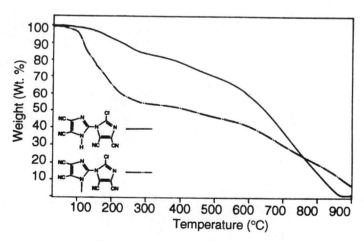

Figure 8. TGA of Compounds 4 and 5 under nitrogen.

weight loss. Further work is underway to characterize the products from these thermolysis reactions.

Conclusions.
The thermolyses the 2-chloro-4,5-dicyanoimidazole derivatives between 100-290°C were found to yield Tris(imidazo)[1,2-a:1,2-c:1,2-e]-1,3,5-triazine-2,3,5,6,8,9-hexacarbonitrile (**HTT**) with $(C_5N_4)_3$ composition. Thermolysis of **HTT** at 490-500°C resulted in a carbon-nitrogen material with C/N = 1.020, while the thermolysis of **HTT** at 1070°C resulted in a carbonaceous material. Examination of the thermal properties of **HTT** and its thermal decomposition products demonstrated bulk thermal stability to 350°C. The thermolysis of the 2-(2-chloro-4,5-dicyano-1-imidazolyl)-4,5-dicyanoimidazole derivatives and their products are under further investigation.

References.
1. Rossbach, V.; Oberlein, G., in *Handbook of Polymer Synthesis*, Ed. by Kricheldorf, H.R., pp.1197-1280, Marcel Dekker, INC, New York, **1992**.
2. Cullis, C.F; Hirschler, M.M. *The Combustion of Organic Polymers*, pp.49, 53-54, 254-255, Oxford University Press, New York, **1981**.
3. Brydson, J.A., in *Developments in Plastics Technology-4*, Ed. by Whelan, A.; Goff, J.P., pp.178-179, Elsevier Science Publishers Ltd., New York, **1989**.
4. (a) Coad, E.C.; Apen, P.G.; Rasmussen, P.G. *J. Am. Chem. Soc.*, **1994**, *116*, 391; (b) Coad, E.C. Ph.D. Thesis, University of Michigan **1994**.
5. Apen, P.G.; Rasmussen, P.G. *Heterocycles*, **1989**, *29*, 1325.
6. (a) Coad, E.C.; Rasmussen, P.G. *Proceedings of the ACS Division of Polymeric Materials: Science and Engineering*, **1993**, *69*, 321; (b) Coad, E.C. Ph.D. Thesis, University of Michigan **1994**.

RECEIVED January 11, 1995

Chapter 18

High-Temperature Copolymers from Inorganic–Organic Hybrid Polymer and Multi-ethynylbenzene

Teddy M. Keller

Materials Chemistry Branch, Code 6127, Naval Research Laboratory, Washington, DC 20375–5320

Considerable interest has been shown in the uses of polyfunctional arylacetylenes as precursors to carbon. Carbon erodes rapidly in air at temperatures as low as 400°C. Much effort is thus being devoted at developing techniques for protecting carbon/carbon composites against oxidation at elevated temperatures. Phenolic resin systems and petroleum and coal tar pitches are currently used as the carbon matrix precursor material. There are numerous problems associated with these carbon precursor materials. We are engaged in the synthesis of compounds containing three or more phenylethynyl groups substituted on the benzene ring and inorganic-organic hybrid polymers. Copolymers of a multi-phenylethynylbenzene and a hybrid polymer show outstanding flame resistance or oxidative stability. Both compounds contain acetylenic units for conversion to the copolymer. The resistance to oxidation is a function of the amount of the hybrid polymer present in the copolymer.

Carbon-carbon (C-C) composites are strong, lightweight, high temperature materials that are used as ablators in short duration rocket and reentry systems and are currently being developed for structural applications in advanced missiles, aircraft, and aerospace vehicles. Many future applications for C-C composites require operation at elevated temperatures in an oxidizing environment. Depending on the application, they are expected to be used for periods ranging from minutes to a few thousands hours at temperatures above 1000°C and approaching 2000°C. Unfortunately, there is a major problem in using such materials in an oxidizing environment. Carbon erodes rapidly in air at temperatures as low as 400°C. Great effort is thus being devoted at developing techniques for protecting C-C composites against oxidation at elevated temperatures ([1-4]). Much of this interest has arisen from recent plans of the U.S. Government to build a National Aerospace Plane

(NASP) which would require lightweight structural materials stable up to 1500°C in air. Carbon-carbon composites are known to retain good mechanical properties up to 2000°C under inert conditions.

An effective method of protecting carbon from oxidation is to establish a barrier against oxygen penetration in the form of an external coating. Most of these coatings rely on oxide films as oxygen diffusion barriers. The development of external coatings such as ceramics to protect C-C composites was initiated about 20 years ago to provide reusable thermal protection for the shuttle orbiter vehicles. Because of the large differences in thermal expansion characteristics of carbon fibers and ceramic materials, few coatings have been found to withstand thermal cycling without cracking. Thick CVD coatings of silicon carbide (SiC) are currently used to protect C-C composites at temperature up to 1300°C. Microcracks in the SiC coating can lead to catastrophic failure since penetration of oxygen through the cracks will result in rapid oxidation of the C-C composite to carbon monoxide and carbon dioxide (5). The strategy that has proven most successful dealing with cracked external coatings is to employ a boron-rich inner coating beneath the cracked outer coating that acts as the oxygen barrier. In this scheme, oxygen penetrating the crack oxidizes the boron layer to produce a compliant sealant glass (B_2O_3) that fills and seals the crack. Prominent coating combinations consist of a SiC outer coating and boron-rich inner coatings that consist of elemental boron or B_4C. The silicon-based ceramics are used for outer coatings because of their excellent oxidative stability, refractoriness, and relatively low thermal dimensional changes.

Another approach often used in combination with external barrier coating is to add elemental boron, B_4C, SiC, and phosphorous compounds to precursor carbon matrix material during processing (6-9). On exposure to air at elevated temperatures, these additives are expected to oxidize and provide in-depth oxidation protection. Experience has shown that it is difficult to achieve a uniform dispersion of the particulate additive throughout the composite and substantial amounts of the carbonaceous material is oxidized before the additive can become effective.

Considerable interest has been shown in the uses of polyfunctional arylacetylenes in the preparation of thermally stable polymers (10-14) and recently as precursors to carbon (15-17). Phenolic resin systems and petroleum and coal tar pitches are currently used as the carbon matrix precursor material. There are numerous problems associated with these carbon precursor materials such as difficulty in processability, low char yield, and lack of consistency of pitch composition. We are interested in aromatic containing acetylenic compounds as a carbon source that have low melting points, have a broad processing window which is defined as the temperature difference between the melting point and the exothermic polymerization reaction, can be easily polymerized through the acetylene units to thermosets, and lose little weight during curing and pyrolysis to carbon under atmospheric conditions. Of further importance is the fact that the processing of C-C composites from acetylene-substituted aromatics reduces the number of required impregnation cycles relative to pitch and phenolics and drastically diminishes the pressure requirements. Our strategy for the synthesis of these materials involves the preparation of multiple-substituted benzenes bearing phenylethynyl groups. Secondary acetylenes have been shown to be less reactive or exotherm at a higher temperature relative to primary acetylenes. Moreover, some primary acetylenes have been reported to react explosively (18).

We are engaged in the synthesis of compounds containing three or more phenylethynyl groups substituted on the aromatic unit and inorganic-organic hybrid polymers. Various multi-secondary acetylene substituted aromatic hydrocarbons were synthesized and evaluated as to their ease of homopolymerization and their ease of carbonization after being polymerized. Out of our studies, 1,2,4,5-tetrakis-(phenylethynyl)benzene **1** has been found to exhibit outstanding properties as a carbon precursor material (19). Moreover, a poly(carborane-siloxane-acetylene) **2** has been shown to exhibit exceptional oxidative properties to 1000°C (20). This paper is concerned with the synthesis of copolymers from **1** and **2** and characteristic studies pertaining to the oxidation stability at elevated temperatures.

Experimental

Thermal analyses were performed with a DuPont 2100 thermal analyzer equipped with a thermogravimetric analyzer (TGA, heating rate 10°C/min) and a differential scanning calorimeter (DSC, heating rate 10°C/min) at a gas flow rate of 50 cc/min. The reported glass transition temperature (T_g) was identified as the midpoint of the endothermic displacement between the linear baselines. Thermal and oxidative studies were achieved in nitrogen and air, respectively. The TGA studies were performed on melts and films of the copolymers. All pyrolysis studies were performed under atmospheric conditions. All aging studies were accomplished in a TGA chamber. 1,7-Bis(chlorotetramethyldisiloxyl)-*m*-carborane was purchased from Dexsil Corporation and was used as received.

Synthesis of 1,2,4,5-Tetrakis(phenylethynyl)benzene 1.
Phenylacetylene (4.697 g, 45.98 mmol), 1,2,4,5-tetrabromobenzene (4.113 g, 10.45 mmol), triethylamine (29.1 ml, 209 mmol), pyridine (16.9 ml, 209 mmol) and a magnetic stirring bar were added to a 250 ml round bottom flask. The flask was fitted with a septum and then chilled in an isopropanol/dry ice bath. After the flask had cooled, the mixture was degassed several times by the alternate application of partial vacuum and argon. To the flask was added palladium catalyst, which consisted of $Pd(PPh_3)Cl_2$ (0.147 g, 0.209 mmol), CuI (0.139 g, 0.731 mmol) and PPh_3 (0.294 g, 1.120 mmol). The septum was refitted and the flask was again degassed. The flask was warmed up to room temperature, then placed in an oil bath at 80°C, and stirred overnight resulting in the formation of a copious amount of a white precipitate. The product mixture was poured into 200 ml of water. The product was collected by suction filtration, washed several times with water, and dried. Recrystallization from methylene chloride and ethanol afforded 1,2,4,5-tetrakis(phenylethynyl)benzene **1** in 84% yield; mp: found 194-196°C, lit.193-194°C (18).

Polymerization and Carbonization of 1,2,4,5-Tetrakis(phenylethynyl)benzene 1 Under Inert Conditions.
The monomer **1** (15.1 mg) was weighed into a TGA pan and cured by heating under a nitrogen atmosphere at 225°C for 2 hours, at 300°C for 2 hours, and at 400°C for 2 hours resulting in the formation of a solid thermosetting

polymeric material. During the heat treatment, the sample lost 1.1% weight. Upon cooling, a TGA thermogram was taken between 30 and 1000°C resulting in a char yield of 85%.

An alternate procedure involves performing the polymerization and carbonization in one step. A sample of 1 was heated between 30 and 1000°C under inert conditions. At 1000°C, the carbon residue exhibited a char yield of 85%.

Oxidation of Carbon Formed from 1,2,4,5-Tetrakis(phenylethynyl)benzene 1.
A TGA thermogram was taken of the carbon residue between 30 and 1000°C in a flow of air at 50 cc/min. The sample started to slowly lose weight at approximately 500°C with catastrophic decomposition occurring between 600 and 800°C.

Synthesis of Poly(butadiyne-1,7-Bis(tetramethyldisiloxyl)-*m*-Carborane) 2. In a typical synthesis, a 2.5M hexane solution of *n*-BuLi (34.2 ml, 85.5 mmol) in 12.0 ml of THF was cooled to -78°C under an argon atmosphere. Hexachlorobutadiene (5.58g, 21.4 mmol) in 2.0 ml THF was added dropwise by cannula. The reaction was allowed to warm to room temperature and stirred for 2 hr. The dilithiobutadiyne in THF was then cooled to -78°C. At this time, an equimolar amount of 1,7-bis(chlorotetramethyldisiloxyl)-*m*-carborane (10.22 g, 21.4 mmol) in 4.0 ml THF was added dropwise by cannula while stirring. The temperature of the reaction mixture was allowed to slowly rise to room temperature. While stirring the mixture for 1 hour, a copious amount of white solid (LiCl) was formed. The reaction mixture was poured into 100 ml of dilute hydrochloric acid resulting in dissolution of the salt and the separation of a viscous oil. The polymer 2 was extracted into ether. The ethereal layer was washed several times with water until the washing was neutral, separated, and dried over sodium sulfate. The ether was evaporated at reduced pressure leaving a dark-brown viscous polymer 2. A 97% yield (9.50 g) was obtained after drying in vacuo. GPC analysis indicated the presence of low molecular weight species ($\approx$500) as well as higher average molecular weight polymers (Mw$\approx$4900, Mn$\approx$2400). Heating of 2 under vacuum at 150°C removed lower molecular weight volatiles giving a 92% overall yield. Major FTIR peaks (cm^{-1}): 2963 (C-H); 2600 (B-H); 2175 (C$\equiv$C); 1260 (Si-C); and 1080 (Si-O).

Pyrolysis of Poly(butadiyne-1,7-Bis(tetramethyldisiloxyl)-*m*-Carborane) 2 in Nitrogen. A sample (24 mg) of 2 was weighed into a platinum TGA pan and heated at 10°C/min to 1000°C under a nitrogen atmosphere at a gas flow rate of 50 cc/min resulting in a ceramic yield of 87%. Upon cooling back to room temperature, the ceramic material was heated at 10°C/min to 1000°C under a flow rate of air at 50 cc/min. During the oxidative heat treatment, the ceramic material gained weight (1-2 weight percent) attributed to oxidation on the surface.

Pyrolysis of Poly(butadiyne-1,7-Bis(tetramethyldisiloxyl)-*m*-Carborane) 2 in Air.
A sample (13.7 mg) of 2 was weighed into an platinum TGA pan and heated at 10°C/min to 1000°C under a flow of air at 50 cc/min resulting in a ceramic yield of 92%. The ceramic was aged at 1000°C for 4 hours resulting in a slight weight gain attributed to the formation of a protective layer enriched in silicon oxide. Moreover, the sample retained its structural integrity.

General Polymerization Procedure for Blending of 1 and 2. Various concentrations of 1 and 2 were weighed into aluminum planchets and heated at 200°C to melt 1. The resulting melt was mixed thoroughly by stirring and a sample was removed for evaluation by DSC and TGA thermal analyses. All TGA studies were performed to 1000°C.

Pyrolysis of Various Mixtures of 1 and 2 Under Inert Conditions. Mixtures of 1 and 2 were weighed into a TGA pan, cured to a thermoset, and converted into a char by heating at 10°C/min to 1000°C under a flow of nitrogen. All samples showed char yield of 80-85%.

Oxidative Stability of Chars from Blending of 1 and 2. The chars were cooled back to 50°C and rescanned to 1000°C under an air atmosphere at 10°C/min at a flow rate of 50cc/min. Steady improvements in the stability were observed as the concentration of 2 was increased.

Results and Discussion

Synthesis of Copolymer. While investigating the thermal and oxidative properties of poly(butadiyne-1,7-bis(tetramethyldisiloxyl)-*m*-carborane) 2, we became interested in using this material to protect carbon derived from acetylenic aromatic hydrocarbons against oxidation at elevated temperatures. Our scheme involves the blending of various concentration of 1,2,4,5-tetrakis(phenylethynyl)benzene 1 and 2, heating to 1000°C under inert conditions, and determining the oxidative stability of the char. Uniform dispersion of 1 and 2 in the melt could be readily achieved with polymerization through the acetylenic units affording a homogeneous polymer.

The synthesis of 1,2,4,5-tetrakis(phenylethynyl)benzene 1 and poly(butadiyne-1,7-bis(tetramethyldisiloxyl)-*m*-carborane) 2 have been reported previously (*19,20*). Compound 1 was prepared from the reaction of 1,2,4,5-tetrabromobenzene and phenylacetylene in the presence of a catalytic amount of a palladium salt. The poly(carborane-siloxane-acetylene) 2 was synthesized from the reaction of 1,7-bis(chlorotetramethyldisiloxyl)-*m*-carborane and dilithiobutadiyne.

Ph-C≡C   C≡C-Ph
       \\   /
        [benzene ring]
       /   \\
Ph-C≡C   C≡C-Ph

1

$$\left[= = -\underset{\underset{CH_3}{|}}{\overset{\overset{CH_3}{|}}{Si}}-O-\underset{\underset{CH_3}{|}}{\overset{\overset{CH_3}{|}}{Si}}-CB_{10}H_{10}C-\underset{\underset{CH_3}{|}}{\overset{\overset{CH_3}{|}}{Si}}-O-\underset{\underset{CH_3}{|}}{\overset{\overset{CH_3}{|}}{Si}}- \right]_n$$

2

The acetylenic functionality in both 1 and 2 provides many attractive advantages relative to other cross-linking centers. An acetylene moiety remains inactive during processing at lower temperatures and reacts thermally to form conjugated polymeric cross-links without the evolution of volatiles.

$$1 + 2 \longrightarrow \text{High Temperature Copolymer} \downarrow \text{Carbon/Ceramic Mass}$$

DSC Studies. Cure studies of 1 and 2 were performed by DSC analysis to 400°C (see Figure 1). A thermogram of 1 shows an endothermic transition (m.p.) at 195°C and an exothermic transition at 290°C. Upon cooling another thermogram was obtained showing a T_g at 164°C and a strong exotherm commencing at approximately 300°C. A sample of 1 that had been cured by heating under inert conditions at 225°C for 2 hours, at 300°C for 2 hours, and at 400°C for 4 hours did not exhibit a T_g. A DSC thermogram of 2 shows a small broad exotherm from about 150 to 225°C which was attributed to the presence of a small amount of primary terminated acetylenic units. This peak was absent when 2 was heated at 150°C for 30 minutes under reduced pressure. These low molecular weight components must be removed to ensure the formation of a void-free thermoset. A larger broad exotherm commencing at 250°C and peaking at 350°C was attributed to the reaction of the acetylene functions to form the cross-links. This exotherm was absent after heat treatment of 2 at 320°C and 375°C, respectively, for 30 minutes. The polymer 2 could be degassed at temperatures below 150°C without any apparent reaction of the acetylenic units. Compound 2 displays only an exothermic transition at 346°C. The exothermic transitions are attributed to polymerization through the acetylenic units. Fully cured samples of 1 and 2 did not exhibit a T_g, which enhances their importance for structural applications.

DSC analyses of blends of 1 and 2 show a homogeneous reaction initially to a thermoset. The DSC scans of the blends show only one cure exotherm for each of the compositions studied. For example, weight percent mixtures (90/10 and 50/50) of 1 and 2 display endotherms (m.p. 1) and exotherms (polymerization reaction) peaking at 195°C, 293°C and 193°C, 300°C, respectively (see Figure 2). It is apparent from the observed cure temperature for the blends that 1 being more reactive initially forms radicals that are not selective in the chain propagation reaction with the acetylenic units of both 1 and 2 (21). Charred samples that have been heat treated to 1000°C do not exhibit characteristic endothermic and exothermic transitions.

Pyrolysis Studies. The thermal stability of 1 was determined under inert conditions. During the heat treatment to 1000°C, the acetylenic compound 1 is initially converted into a dark brown thermoset polymer, which behaves as a precursor polymer for further conversion into carbon. Pyrolysis of 1 to 1000°C under inert conditions afforded a char yield of 85% and a density of 1.45 g/cc. Very little

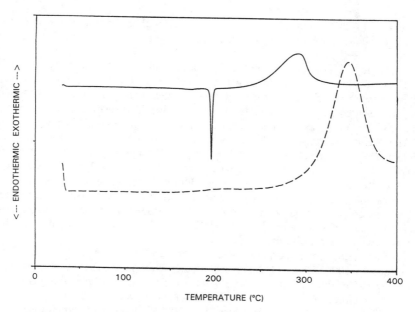

Figure 1. DSC thermogram of: 1 (solid line) and 2 (dash line).

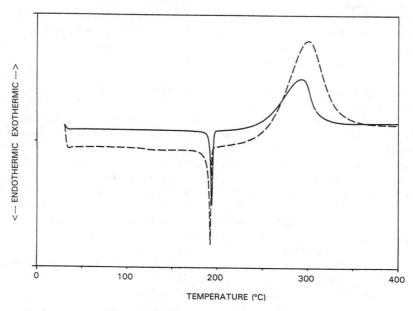

Figure 2. DSC thermogram from weight percent mixtures of 1 and 2: 90/10 (solid line) and 50/50 (dash line).

weight loss (1-2%) occurred below 500°C. A TGA thermogram of the pyrolyzed product (carbon) in air resulted in catastrophic degradation occurring between 600 and 800°C (see Figure 3).

Polymer 2 possesses exceptional thermal and oxidative stability to 1000°C (see Figure 4). It is a viscous liquid that is soluble in most common organic solvents and is easy to process into structural components. Pyrolysis of 2 to 1000°C in nitrogen, resulted in a ceramic yield of 85%. Further heat treatment of the ceramic at 1000°C for 12 hours resulted in no additional weight loss. When the ceramic material was cooled back to 50°C and rescanned to 1000°C in air, the sample gained weight ($\approx$2%) attributed to surface oxidation. A TGA thermogram of 2, which was heated to 1000°C in air, exhibited a ceramic yield of 92%. Further TGA aging studies of the ceramic in air revealed that additional weight loss did not occur and that the sample actually increased in weight as observed previously. When the aged sample was cooled and heated to 1000°C under nitrogen, no weight changes were observed. These observations show the stability of the ceramic material under both inert and oxidative conditions.

The thermal and oxidative stability of various mixtures of 1 and 2 was determined to 1000°C by TGA analysis. Studies have been performed on samples containing 0-50% by weight of 2. The scans were run at 10°C/min at a gas flow of 50 cc/min in either nitrogen or air. Samples containing various amounts of 1 and 2 afforded char yields of 85% when heated to 1000°C under inert conditions. Upon cooling, the carbon/ceramic mass was reheated to 1000°C in air. The oxidative stability of the charred mass was found to be a function of the amount of 2 present. Charred samples obtained from 5, 10, 20, 35, and 50% by weight of 2 showed chars of 12, 27, 58, 92, and 99.5%, respectively, when heated to 1000°C in air (see Figure 5). TGA scans of the oxidized chars were completely stable in air to 1000°C. These results indicate that the oxidative stability of the copolymer and carbon/ceramic mass is enhanced as the concentration of 2 is increased.

Aging Studies In Air. The carbonaceous mass produced from the pyrolysis of 1 to 1000°C under inert conditions was aged in a flow of air at 400°C and 500°C (see Figure 6). Heat treatment at 400°C resulted in an initial weight gain attributed to the absorption and interaction of oxygen with the carbon prior to oxidative breakdown. After $\approx$45 minutes, a cessation of the weight gain was observed. Shortly thereafter, the sample started to gradually lose weight. After 6 hours the sample had lost $\approx$2.5% weight. Upon exposure to air at 500°C, a carbon char commenced to lose weight immediately. Moreover, the rate of breakdown increased as a function of time. After 1 hour, the sample had lost about 9% weight.

Isothermal aging studies were performed on the ceramic formed from 2 under oxidative conditions. A sample of 2 was heated under a nitrogen atmosphere to 1000°C to afford a ceramic yield of 85%. Upon cooling, the ceramic sample was aged in air at 500, 600, and 700°C. After each aging study, the sample was cooled to room temperature. When heat treated at 500°C for 20 hours, the sample gained 0.11% weight. The sample was then aged at 600°C for 6 hours. While heating up to 50 minutes, the sample lost weight (0.25%) and then gained 0.05% weight upon heating for an additional 5 hours. After 3 hours at 600°C, no further weight loss occurred. For heat treatment at 700°C, the sample initially lost weight (0.16%)

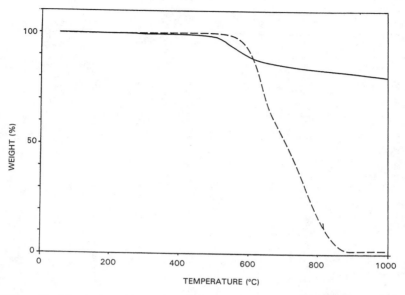

Figure 3. Thermal stability to 1000°C of: 1 (solid line) under nitrogen atmosphere and carbon char from 1 (dash line) in flow of air.

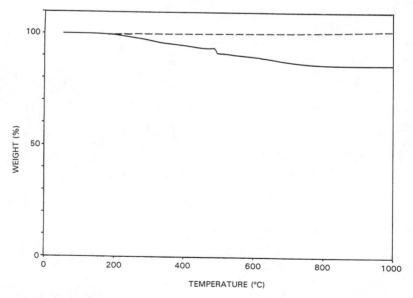

Figure 4. Thermal stability to 1000°C of: 2 (solid line) under nitrogen and ceramic char from 2 (dash line) in flow of air.

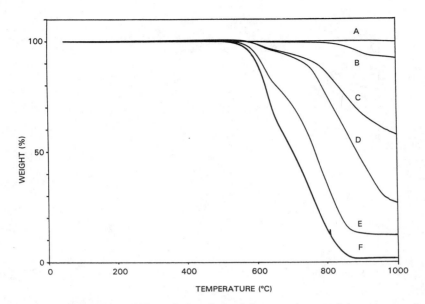

Figure 5. Oxidative stability of char formed from various weight percent of 1 and 2: **A**, 50/50; **B**, 65/35; **C**, 80/20; **D**, 90/10; **E**, 95/5; and **F**, 100/0.

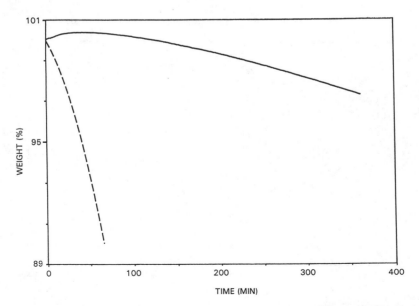

Figure 6. Oxidative aging studies on carbon char from 1 at: 400°C (solid line) and 500°C (dash line).

during the first 25 minutes. Between 25 and 300 minutes, the sample gradually increased in weight (0.29%). Another sample of 2 heated under a flow of air at 50 cc/min to 1000°C and held at this temperature for 10 hours afforded a ceramic yield of 87%. When the ceramic was further heated in air at 500°C for 12 hours, no apparent weight changes occurred. Regardless of the heat treatment, the samples retained their structural integrity except for some shrinkage during pyrolysis.

Extreme aging conditions show the importance of silicon and boron in the protection of carbon-based systems against oxidation. Two ceramic compositions prepared from 50/50 and 65/35 weight percent blends of 1 and 2 were initially processed to 1000°C under a nitrogen atmosphere to form a char yield of 83% in each case. Upon cooling, the chars were heat treated to 1000°C under a flow of air and aged for 2 hours resulting in weight losses of 4.5 and 12.5% for the 50/50 and 65/35 weight percent compositions, respectively (see Figure 7).

Since carbon commences to degrade oxidatively at approximately 400°C, similar aging studies were performed on the ceramic materials formed from various blends of 1 and 2. Two ceramic compositions prepared by heat treatment under a nitrogen atmosphere to 1000°C from 80/20 and 50/50 weight percent blends of 1 and 2 were aged in air at 400°C and 500°C. On exposure at 400°C, the char from the 80/20 mixture immediately commenced to gain weight to a maximum of 1.48%. The sample then gradually lost weight and was at 100% weight retention after 5.2 hours of heat treatment. Heat treatment of another charred sample at 500°C resulted in an immediate weight loss with a weight retention of 97.6% after 5 hours. The char from the 50/50 mixture showed outstanding oxidative stability. The copolymer quickly gained about 0.42% weight with very little further weight change during the 20 hour heat treatment. Upon increasing the temperature to 500°C, the polymer still displayed excellent stability with a weight retention of about 98.8% after isothermal aging for 20 hours. The oxidized film that developed during the heat treatment at 400°C formed a protective barrier against oxidation.

On exposure of the copolymers formed from 1 and 2 to an oxidizing environment, a protective film initially develops that deters or alleviates further oxidation at a given temperature. The formation of the oxidized film and any weight loss associated with the exposure was accelerated by heat treatment of the carbonaceous/ceramic mass to 1000°C in air. Such samples were prepared from 50/50 and 65/35 weight percent blends of 1 and 2 heated to 1000°C, consecutively, in nitrogen and air and then isothermally aged in air in sequence for 10 hours each at 600 and 700°C (see Figure 8). The chars from the 50/50 and 65/35 weight percent blends gained and lost about 0.1% and 18% weight, respectively, at 600°C. During the heat treatment at 700°C, the samples lost about 0.4% and 4% weight. The superior performance of the 50/50 blend shows that critical amounts of boron and silicon are necessary to protect a carbon-based material against oxidation.

Conclusion

Copolymers of 1 and 2 show outstanding oxidative stability. Both compounds contain acetylenic units for conversion to the copolymer. The resistance to oxidation was a function of the amount of 2 present in the copolymer. The studies show that carbon can be protected from oxidation at various temperatures by proper

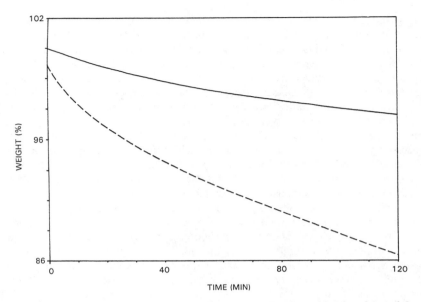

Figure 7. Oxidative aging studies at 1000°C on char from blends of 1 and 2: 50/50 weight percent (solid line) and 65/35 weight percent (dash line).

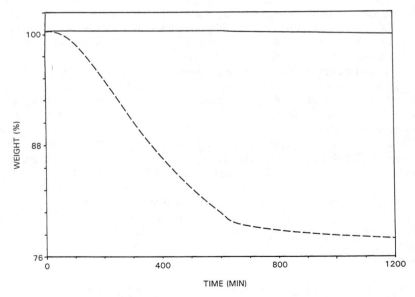

Figure 8. Oxidative studies on chars from blends of 1 and 2 preheated to 1000°C in flow of air before aging at 600 and 700°C: 50/50 weight percent (solid line) and 65/35 weight percent (dash line).

incorporation of silicon and boron units into a carbon precursor material. Further studies are underway to evaluate and exploit the copolymers as matrix materials for high temperature composites and carbon/ceramic composites.

Acknowledgments

Acknowledgment is made to the Office of Naval Research for financial support of this work. The author wishes to thank Dr. Ken M. Jones and Dr. David Y. Son for the synthesis of 1,2,4,5-tetrakis(phenylethynyl)benzene and poly(butadiyne-1,7-bis(tetramethyldisiloxyl)-*m*-carborane), respectively.

Literature Cited

1. Strife, J. R.; Sheehan, J. E. *Bull. Am. Ceram. Soc.* **1988**, *67*, 369.
2. Sheehan, J. E. *Carbon* **1989**, *27*, 709.
3. McKee, D. W. *Carbon* **1987**, *25*, 551.
4. McKee, D. W.; Spiro, C. L.; Lamby, E. J. *Carbon* **1984**, *22*, 507.
5. Kim, D. P.; Economy, J. *Chem. Mater.* **1993**, *5*, 1216.
6. Luthra, K. L. *Carbon* **1988**, *26*, 217.
7. McKee, D. W. *Carbon* **1986**, *24*, 736.
8. Jawed, I.; Nagle, D. C. *Mat. Res. Bull.* **1986**, *21*, 1391.
9. Rakszawski, J. F.; Parker, W. E. *Carbon* **1964**, *2*, 53.
10. Stille, J. K.; Harris, F. W.; Rukutis, R. O.; Mukamal, H. *J. Polym. Sci., Part B* **1966**, *4*, 791.
11. Samyn, C.; Marvel, C. *J. Polym. Sci., Polym. Chem.* **1975**, *13*, 1095.
12. Frank, H.; Marvel, C. *J. Polym. Sci., Polym. Chem.* **1976**, *14*, 2785.
13. Banihashemi, A.; Marvel, C. *J. Polym. Sci., Polym. Chem.* **1977**, *15*, 2667.
14. Hergenrother, P. M. *J. Polym. Sci., Polym. Chem.* **1982**, *20*, 2131.
15. Economy, J.; Jung, H.; Gogeva, T. *Carbon* **1992**, *30*, 81.
16. Zaldivar, R. J.; Rellick, G. S. *SAMPE Journal* **1991**, *27*, 29.
17. Stephens, E. B.; Tour, J. M. *Macromolecules* **1993**, *26*, 2420.
18. Neenan, T. X.; Whitesides, G. M. *J. Org. Chem.* **1988**, *53*, 2489.
19. Jones, K. M.; Keller, T. M. *Polym. Mat. Sci. & Eng.* **1993**, *68*, 97.
20. Henderson, L. J.; Keller, T. M. *Polym. Prep.* **1993**, *34(1)*, 345.
21. Sastri, S. B.; Keller, T. M.; Jones, K. M.; Armistead, J. P. *Macromolecules* **1993**, *26(23)*, 6171.

RECEIVED November 18, 1994

Chapter 19

Linear Siloxane—Acetylene Polymers as Precursors to High-Temperature Materials

David Y. Son and Teddy M. Keller

Materials Chemistry Branch, Code 6127, Naval Research Laboratory, Washington, DC 20375–5320

> Novel linear polymers containing alternating diacetylene and siloxane units have been prepared in high yields. The syntheses are one-pot procedures which involve the reaction of 1,4-dilithiobutadiyne with the appropriate dichlorosiloxane compound. IR and NMR analyses of the products support the proposed polymer structures. The polymers are processable and are thermally cross-linked, forming hard, void-free thermosets. The thermosets have excellent thermo-oxidative stabilities. They are stable to ~400 °C in air, and retain most of their mass when heated to 1000 °C in air or nitrogen.

Modern technology has created a demand for processable thermosetting resins that demonstrate high thermal and oxidative stability. These materials find use in many electrical and aerospace applications. Many conventional resins are derived from organic materials such as epoxy compounds, formaldehyde-phenolic resins, or propargyl compounds. Typically, these thermosets are stable to 300-400 °C in air, but decompose rapidly and completely at temperatures higher than 400 °C.

Our research in this area has focused on linear inorganic-organic hybrid polymers containing diacetylene units in the backbone (*1*). Diacetylene groups react thermally or photochemically to form a cross-linked polymer consisting of an extended conjugated network (*2,3*). Besides creating a tough networked material, these cross-links can introduce novel optical properties into the polymer such as thermochromism (*4,5*), mechanochromism (*6*), and optical non-linearity (*7,8*). To be of practical use as high temperature materials, the resins should possess certain properties such as thermo-oxidative stability and good moisture resistance. Most conventional organic polymers lack the ability to withstand temperatures in excess of 200-250 °C for extended periods in an oxidizing environment. In addition, these materials tend to have poor hot-wet mechanical properties. Combining inorganic elements with the organic elements of a polymer should enhance the thermal and oxidative stability of the polymers. Siloxyl groups are a logical choice for inclusion in these polymers as they possess good thermal and oxidative stability and high hydrophobicity (9). Furthermore, their flexibility should contribute favorably to the processability of the resulting polymers. The polymer (-C≡C-C≡C-SiMe$_2$OSiMe$_2$-)$_n$ (1), obtained via the oxidative coupling of 1,3-diethynyltetramethyldisiloxane, was previously reported by Parnell and Macaione (10). The authors reported obtaining only low molecular weight products in addition to some insoluble material. Herein, we describe an alternative high-yield synthesis of 1 and the related polymer (-C≡C-C≡C-(SiMe$_2$O)$_2$SiMe$_2$-)$_n$ (2). These polymers can be converted into thermosets thermally or photochemically, and can be pyrolyzed to form ceramic materials. The syntheses and characterizations of these new materials will be described in this report.

Experimental

All reactions were carried out in an inert atmosphere unless otherwise noted. Solvents were purified by established procedures. 1,3-Dichlorotetramethyldisiloxane and 1,5-dichlorohexamethyltrisiloxane were obtained from Silar Laboratories and used as received. n-Butyllithium (2.5 M in hexane) was obtained from Aldrich and titrated before use. Hexachlorobutadiene was obtained from Aldrich and distilled before use. Thermogravimetric analyses (TGA) were performed on a DuPont 951 thermogravimetric analyzer. Differential scanning calorimetry analyses (DSC) were performed on a DuPont 910 instrument. Unless otherwise noted, all thermal experiments were carried out at a heating rate of 10°C/min and a nitrogen flow rate of 50 mL/min. Infrared spectra were obtained using a Nicolet Magna 750 FTIR spectrophotometer. Gel-permeation chromatography (GPC) data were obtained using a Hewlett-Packard Series 1050 pump and two Altex µ-spherogel columns (size 10^3 and 10^4 Å respectively) connected in series. All values were referenced to polystyrene. ^{1}H NMR and ^{13}C NMR spectroscopy were performed on a Bruker AC-300 spectrometer using CDCl$_3$ as solvent. Elemental analyses were performed by Galbraith Laboratories, Knoxville, TN.

Preparation of 1,4-dilithio-1,3-butadiyne. A hexane solution of n-BuLi (10.6 mL of a 2.5M solution, 26.5 mmol) was added to a flask containing THF (5 mL) cooled in a dry ice/acetone bath. Subsequently, hexachlorobutadiene (0.99 mL, 6.3 mmol) was added dropwise via syringe, resulting in the formation of a heavy precipitate. After completion of addition, the cold bath was removed and the reaction mixture was stirred at room temperature for two hours. The resulting dark-brown mixture was used without further treatment.

Preparation of Polymer 1. A mixture of 1,4-dilithio-1,3-butadiyne (6.3 mmol) in THF/hexane was cooled in a dry ice/acetone bath. To this mixture, 1,3-dichlorotetramethyldisiloxane (1.24 mL, 6.3 mmol) was added dropwise over 15 min. After addition, the cold bath was removed and the mixture was stirred at room temperature for two hours. The tan mixture was poured into 20 mL of ice-cooled saturated aqueous ammonium chloride solution with stirring. The mixture was filtered through a Celite pad and the layers were separated. The aqueous layer was extracted twice with Et$_2$O and the combined organic layers were washed twice with distilled water and once with saturated aqueous NaCl solution. The dark brown organic layer was dried over anhydrous magnesium sulfate and filtered. Most of the volatiles were removed at reduced pressure and the residue was heated at 75 °C for three hours at 0.1 torr to give **1** as a thick, dark brown material (1.04 g, 92%). Polymer **1** slowly solidifies on standing at room temperature and liquefies at approximately 70 °C. ^{1}H NMR (ppm) 0.30 (s, 12H, -Si(CH_3)); ^{13}C NMR (ppm) 1.7, 1.9 (-Si(CH$_3$)), 84.9 (-Si-C≡C-), 86.9 (-Si-C≡C-). Anal. Calcd. for (C$_8$H$_{12}$OSi$_2$)$_n$: C, 53.31; H, 6.66; Si, 31.16. Found: C, 55.81; H, 7.61; Si, 27.19.

Preparation of Polymer 2. The same procedure that was used in the preparation of **1** was used in the reaction of 6.3 mmol of 1,4-dilithio-1,3-butadiyne with 1,5-dichlorohexamethyltrisiloxane (1.72 mL, 6.3 mmol). The same workup procedure yielded **2** as a slightly viscous dark brown oil (1.44 g, 90%). ^{1}H NMR (ppm) 0.11 (s, 6H, -C≡C-SiOSi(CH_3)$_2$-), 0.26 (s, 12H, -C≡C-Si(CH_3)$_2$-); ^{13}C NMR (ppm) 85.2 (-Si-C≡C-), 86.8 (-Si-C≡C-). Anal. Calcd. for (C$_{10}$H$_{18}$O$_2$Si$_3$)$_n$: C, 47.22; H, 7.08; Si, 33.12. Found: C, 48.98; H, 7.59; Si, 29.45.

Thermoset from Polymer 1. A sample of **1** (1.00 g) was placed in an aluminum pan and degassed under vacuum for 15 min at 110 °C. Afterwards, the sample was placed in a tube furnace under argon and cured at 150 for 2.5h, 200 for 5h, 300 for 2h, and 400 °C for 2h. This curing process produced a void-free, dark-brown hard solid, **3** (0.94 g, 94%).

Thermoset from Polymer 2. A sample of **2** (1.00 g) was placed in an aluminum pan and degassed under vacuum at room temperature. Afterwards, the sample was placed in a tube furnace under argon and cured at 150 for 2.5h, 225 for 5h, 325 for 2h, and 400 °C for 2h. This curing process produced a void-free, dark-brown hard solid, **4** (0.90 g, 90%).

Pyrolysis of Thermoset 3. A small disk-shaped sample of **3** (0.742 g) was placed in a tube furnace and heated to 1000 °C at a rate of 10 °C/min under argon. The temperature was held at 1000 °C for one hour before the sample was cooled to room temperature. A hard, dark disk-shaped solid, **5**, was recovered (0.551 g, 74%).

Results and Discussion

Synthesis. Polymers **1** and **2** consist of alternating diacetylene and siloxane units in the backbone. The synthesis of the polymers is a simple two-step, one-pot reaction and is adapted from a previously reported synthesis of polysilyldiacetylenes (Scheme 1) (*11*). Treatment of hexachlorobutadiene with four equivalents of n-butyllithium in THF/hexane at -78 °C generates 1,4-dilithio-1,3-butadiyne. After the mixture is stirred at room temperature for two hours and then recooled to -78 °C, the appropriate dichlorosiloxane is added dropwise to the reaction mixture. Stirring at room temperature for two hours followed by aqueous workup gives **1** and **2** in 85-95% yield (preliminary experiments performed by Dr. Leslie J. Henderson in this laboratory were inconclusive and resulted in no isolation of product). Polymer **1**

solidifies at room temperature (softening point ~70 °C) while polymer 2 is a viscous liquid. This difference in properties can be attributed to the longer flexible siloxane spacer in 2. Both polymers are easily soluble in common organic solvents such as chloroform, acetone, THF, and Et_2O.

The structures of 1 and 2 were determined using infrared (IR) and NMR spectroscopy. The IR spectra for both polymers show strong acetylenic stretching bands (2071 cm^{-1}) and strong bands in the Si-O stretching region (1045-1055 cm^{-1}). The ^1H NMR spectrum for 1 shows only one singlet corresponding to the -$SiCH_3$ protons in the polymer (δ 0.30). The ^{13}C NMR spectrum confirms the presence of the acetylenic carbons with peaks at δ 84.9 and 86.9, and also shows the -$SiCH_3$ carbons as two fine singlets at δ 1.68 and 1.86. For polymer 2, the internal -$SiCH_3$ protons are observed as a singlet at δ 0.11 and the external -$SiCH_3$ protons as a singlet at δ 0.26 in the ^1H NMR spectrum. In the ^{13}C NMR spectrum for 2, the acetylenic carbons are observed at δ 85.2 and 86.8 while the -$SiCH_3$ carbons are observed as peaks at δ 0.94 and 1.79.

The polymers were also characterized using gel permeation chromatography (GPC) and elemental analysis. GPC analysis indicated broad molecular weight distributions, with peak maxima occurring at ~10,000 (relative to polystyrene) for both polymers. The presence of lower molecular weight species can be attributed to early chain termination and formation of cyclic species during polymerization. Reaction of excess n-butyllithium with a Si-Cl bond can terminate a polymer chain. Also, cleavage of Si-O-Si bonds by alkyllithium species can result in chain breakage, resulting in lower molecular weights. Elemental analysis data were in general agreement with calculated values. The relatively large amounts of endgroups present can explain the slight discrepancies in the data. Similar elemental analysis variations were reported for poly[(silylene)diacetylenes] (12).

The thermal behaviors of 1 and 2 were examined using differential scanning calorimetric (DSC) and thermogravimetric (TGA) analysis. The DSC data (in nitrogen) for both polymers are shown in Figure 1. Large exotherms are observed at 289 °C and 315 °C for 1 and 2, respectively. These exotherms can be attributed to the diacetylenic cross-linking reactions (13). Polymer 1, a solid, reproducibly exhibits two small endotherms attributable to melting transitions between 50 and 80 °C. When 1 is heated to 100 °C, recooled to room temperature, and then reheated, the endotherms are again observed. These observations may indicate that segregated domains of hard (diacetylene) and soft (siloxane) segments of different crystallinities exist in the polymer. A similar phenomenon has been observed in polyurethane-diacetylene copolymers (4). The TGA data (in nitrogen) for both polymers are shown in Figure 2. As seen in the figure, polymer 1 has a higher char yield than 2 (1 - 74%; 2 - 58%). Polymer 2, having a higher concentration of siloxane groups in its backbone, is more susceptible to degradation at higher temperatures via formation of monomeric cyclic siloxanes (14,15). TGA data obtained in air (Figure 3) show a similar trend, but as expected, the weight losses are greater than the losses obtained during the heating in nitrogen. The initial weight losses while heating to ~300 °C in both atmospheres can be attributed to the evaporation of residual solvent and low molecular weight oligomers.

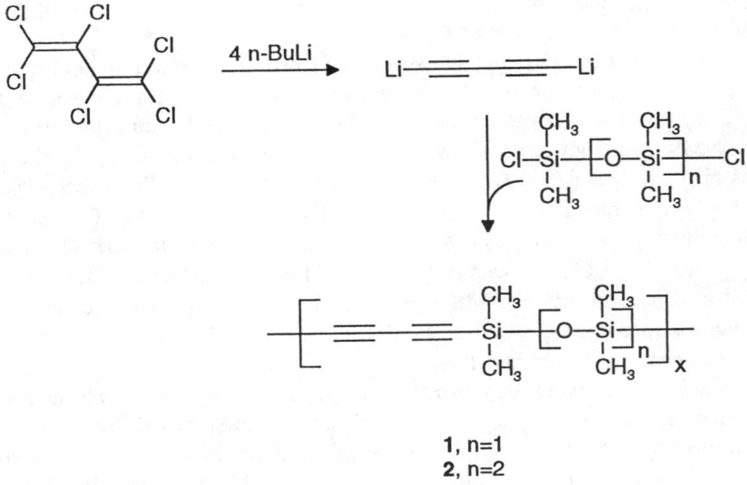

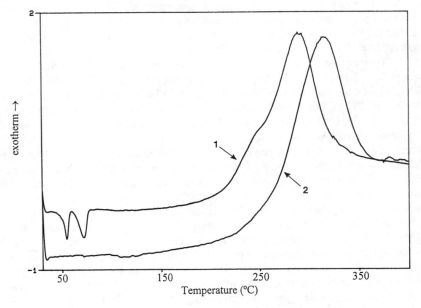

Scheme 1. Preparation of polymers **1** and **2**.

1, n=1
2, n=2

Figure 1. DSC data for **1** and **2** (10 °C/min in nitrogen).

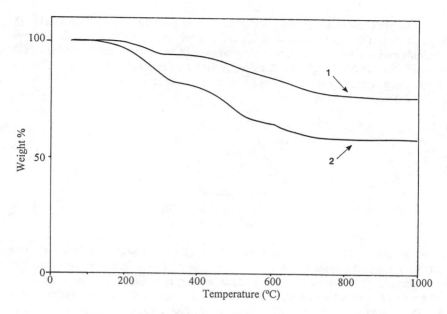

Figure 2. TGA data for **1** and **2** (10 °C/min in nitrogen).

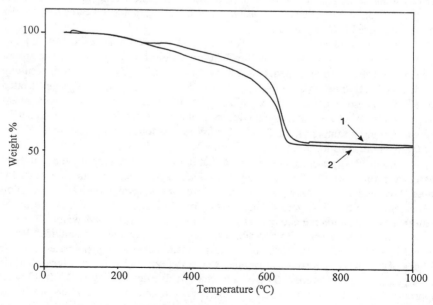

Figure 3. TGA data for **1** and **2** (10 °C/min in air).

In other synthetic studies, we attempted to prepare (-C≡C-SiMe$_2$OSiMe$_2$-)$_n$ by reaction of dilithioacetylene with 1,3-dichlorotetramethyldisiloxane. No polymer was obtained - a mixture of low molecular weight products was recovered which included the cyclic compound 2,2,5,5,7,7,10,10-octamethyl-2,5,7,10-tetrasila-1,6-dioxa-3,8-cyclodecadiyne (*16*) shown below. The structure was verified by X-ray crystal diffraction analysis.

Cure Studies. The diacetylene cross-linking reaction makes it possible to cure polymers **1** and **2** thermally. A degassed sample of **1** was heated in argon at 200, 300, and 400 °C for two hours at each temperature to give a hard, void-free thermoset (**3**) with 94% mass recovery. Some shrinkage of the sample was observed. An IR spectrum of **3** showed almost total disappearance of the diacetylenic absorption, indicating that cross-linking was complete. In a similar manner, polymer **2** was cured by heating in argon at 225, 325, and 400 °C for two hours at each temperature. A hard, void-free thermoset (**4**) was obtained with 90% mass recovery. Again, some shrinkage was observed. For both samples, curing in air resulted in lower mass retention. The lack of visible voids in both samples indicates that all loss of mass occurred before gelation.

The glass transition temperatures for **3** and **4** were found to be 144 and 170 °C, respectively. The greater chain flexibility in **2** due to the longer siloxane spacer perhaps contributes to an enhancement in the amount of cross-linking in **4**, resulting in a higher T_g. By heating a sample of **3** at 450 °C for four hours, the T_g was elevated to 195 °C. These values are comparable to those of other cross-linked silicone resins (*17*). Treatment at the higher temperatures is not necessary for obtaining higher T_g values. Curing a sample of **1** at 200 °C for five hours, followed by heating at 300 °C for twelve hours produced a thermoset with a T_g of 185 °C.

Water absorption studies indicated that both **3** and **4** absorb small amounts of water over long periods of time. Samples of **3** and **4** were placed in boiling water for three days and left in room temperature water for several months. Absorption percentages for **3** and **4** were 4.4 and 2.7%, respectively. These values are lower than those of conventional epoxy resins, which typically absorb 5-8% water.

Thermosets **3** and **4** both exhibit excellent thermo-oxidative stabilities. When heated in air to 1000 °C at 10°/min, **3** and **4** remain stable to approximately 400 °C (Figure 4). Unlike conventional organic thermosets, **3** and **4** do not lose all their mass at higher temperatures. The overall char yields after heating to 1000 °C were approximately 60% for both **3** and **4**. The residue obtained here is likely similar to the residue from the charring of the original linear polymers **1** and **2** in air. Enhanced

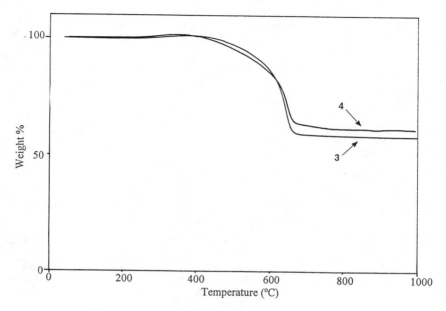

Figure 4. TGA data for **3** and **4** (10 °C/min in air).

thermo-oxidative stabilities in polysiloxanes have been attributed to conversion of silicon to silica at higher temperatures (*18*) - this is likely occurring in this case as the chars had an entirely white surface. This hypothesis is also supported by preliminary aging studies which indicated that **3** increased in mass by nearly 6% when heated at 300 °C in air. After gaining the extra mass, thermoset **3** remained fairly stable at 300 °C in air, losing mass at a rate of only 0.002%/min (Figure 5).

When a sample of **3** was heated to 1000 °C in argon, a black char, **5**, representing only 26% weight loss was recovered. The char retained the original shape of **3** without cracking. This char also possesses excellent thermo-oxidative stability. On heating to 1000 °C in air at a rate of 10 °C/min, **5** remained stable to ~600 °C, at which point gradual weight loss was observed. The overall weight loss was only 15% (Figure 6). Preliminary aging studies showed a 28% weight loss after 15 hours in air at 500 °C. Char **5** had an elemental analysis of C - 47.31%, H - 0.56%, O - 24.13%, Si - 28.00%, which corresponds to a composition of $Si_{1.00}C_{3.94}H_{0.56}O_{1.51}$. While further characterization studies have not yet been performed on **5**, it is likely that **5** is a type of silicon oxycarbide material. Silicon oxycarbide glasses have been formed by pyrolysis of carbon-containing polysiloxanes and typically exhibit very good oxidation resistance properties (*19-24*).

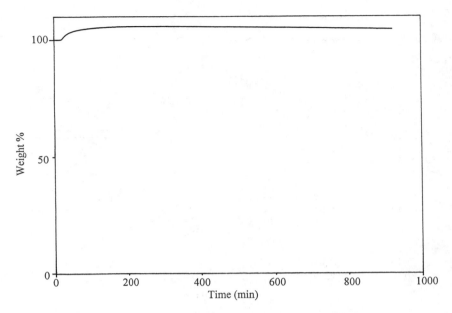

Figure 5. Isothermal aging of **3** (300 °C in air).

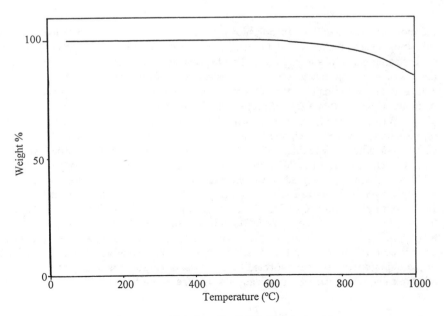

Figure 6. TGA data for **5** (10 °C/min in air).

Conclusion

New thermosetting polymers containing siloxane and acetylenic units in the backbone have been prepared. These polymers are easily processable and can be prepared simply and efficiently in high yields. Heating the polymers at elevated temperatures converts them into hard thermosets. These thermosets show stability to 400 °C in air and retain most of their mass when heated to 1000 °C in air or nitrogen. Future investigations are focusing on the mechanical properties of these promising materials.

Acknowledgments

Acknowledgment is made to the Office of Naval Research for financial support of this research. The authors are also grateful to Dr. Tai Ho for assistance with the GPC measurements and to Dr. Clifford George and Dr. Mark Erickson for the X-ray crystal analysis of 2,2,5,5,7,7,10,10-octamethyl-2,5,7,10-tetrasila-1,6-dioxa-3,8-cyclodecadiyne.

Literature Cited

1. Henderson, L. J.; Keller, T. M. *Macromolecules* **1994**, *27*, 1660.
2. Wegner, G. *Die Makromol. Chem.* **1970**, *134*, 219.
3. Kuhling, S.; Keul, H.; Hocker, H. *Macromolecules* **1990**, *23*, 4192.
4. Rubner, M. F. *Macromolecules* **1986**, *19*, 2114, 2129.
5. Beckham, H. W.; Rubner, M. F. *Macromolecules* **1989**, *22*, 2130.
6. Nallicheri, R. A.; Rubner, M. F. *Macromolecules* **1991**, *24*, 517, 526.
7. *Nonlinear Optical Properties of Organic Molecules and Crystals*; Chemla, D. C. and Zyss, J., Eds.; Academic Press: New York, 1987.
8. Nakanishi, H.; Matsuda, H.; Okada, S.; Kato, M. *Polym. Adv. Tech.* **1990**, *1*, 75.
9. Dvornic, P. R.; Lenz, R. W. *High Temperature Siloxane Elastomers*; Hüthig & Wepf: Heidelberg, 1990.
10. Parnell, D. R.; Macaione, D. P. *J. Polym. Sci., Polym. Chem. Ed.* **1973**, *11*, 1107.
11. Ijadi-Maghsoodi, S.; Barton, T. J. *Macromolecules* **1990**, *23*, 4485.
12. Bréfort, J. L.; Corriu, R. J. P.; Gerbier, Ph.; Guérin, C.; Henner, B. J. L.; Jean, A.; Kuhlmann, Th.; Garnier, F.; Yassar, A. *Organometallics* **1992**, *11*, 2500.
13. Corriu, R.; Gerbier, P.; Guérin, C.; Henner, B.; Jean, A.; Mutin, H. *Organometallics* **1992**, *11*, 2507.
14. Thomas, T. H.; Kendrick, T. C. *J. Polym. Sci., A2* **1969**, *7*, 537.
15. Hergenrother, P. M. *Angew. Chem., Int. Ed. Engl.* **1990**, *29*, 1262.
16. Kloster-Jensen, E.; Rømming, C. *Acta. Chem. Scand. B* **1986**, *40*, 604.
17. Critchley, J. P.; Knight, G. J.; Wright, W. W. *Heat-Resistant Polymers*; Plenum Press: New York, 1983; Chapter 6.
18. Kambour, R. P. *J. Appl. Poly. Sci.* **1981**, *26*, 861.
19. Babonneau, F.; Thorne, K.; Mackenzie, J. D. *Chem. Mater.* **1989**, *1*, 554.
20. Bois, L.; Maquet, J.; Babonneau, F.; Mutin, H.; Bahloul, D. *Chem. Mater.* **1994**, *6*, 796.

21. Babonneau, F.; Bois, L.; Yang, C.-Y.; Interrante, L. V. *Chem. Mater.* **1994**, *6*, 51.
22. Laine, R. M.; Rahn, J. A.; Youngdahl, K. A.; Babonneau, F.; Hoppe, M. L.; Zhang, Z.-F.; Harrod, J. F. *Chem. Mater.* **1990**, *2*, 464.
23. Burns, G. T.; Taylor, R. B.; Xu, Y.; Zangvil, A.; Zank, G. A. *Chem. Mater.* **1992**, *4*, 1313.
24. Hurwitz, F. I.; Heimann, P.; Farmer, S. C.; Hembree, Jr., D. M. *J. Mater. Sci.* **1993**, *28*, 6622.

RECEIVED November 28, 1994

Fire Toxicity

Most people who die in fires die from the effects of toxic gases, not from the heat or flames of the fire itself. That being the case much of the research over the last 15–20 years has been devoted to bench-scale toxicity tests in the hope that "less toxic" materials would be developed. Standards for bench-scale tests have now been finalized both in the International Standards Organization (ISO) and the American Society for Testing and Materials (ASTM). Results in bench-scale tests can be accounted for by a small number of gases, the "N" gas model. The chapter by Levin et al. offers a state-of-the-art discussion of the N-gas model and its extension to an additional gas, NO_2.

Bench-scale tests measure toxic potency, not toxic hazard or "toxicity". Toxic potency data must be used in conjunction with other fire parameters in a fire model to ascertain whether, for example, one material is truly better than another, even if they differ by more than the 95% confidence limits in toxic potency, or even a factor of 3 or 10.

Carbon monoxide is common to all fire atmospheres. Comprehensive studies on CO toxicity to humans in both fire and non-fire incidents show that real fire toxicity is almost solely determined by CO. There is no universal CO lethality threshold level, as frequently reported. Lethality depends upon the victim's age, disease, and physical condition. Fire and non-fire CO victims differ in these respects: fire CO victims are either much older, and much younger, or suffering from a preexisting disease, thereby increasing their sensitivity to CO exposure. CO concentrations increase in big fires and are determined more by oxygen availability and geometry and less by the chemical composition of fuels. The use of synthetic materials in our environment, therefore, has made no statistically determinable difference on fire toxicity. Bench-scale tests may be poor predictors of post-flashover toxic fire hazard because they invariably produce a high CO_2/CO ratio, grossly underpredicting the role of CO. Also bench-scale tests using rodents may overpredict the role of other agents such as HCN.

HCN, the third most common fire gas, is discussed by Stemler et al.'s chapter. Using miniature pigs as test subjects, they found blood HCN values at death 4 times that of mice and 2 times that of rats. The relative toxic potency of HCN changes with different animal species. Mayorga et al. discuss another important gas, NO_2, and its effects at both low and high blood levels.

Chapters by Caldwell et al. Ritchie et al. endeavor to highlight bench-scale tests used to more fully understand the effects of smoke. In Caldwell's research a modified cone calorimeter is used to evaluate smoke production for an advanced composite material under a variety of conditions of heat flux and air flow. In Ritchie's chapter a screening battery of neurobehavioral tests is evaluated to more fully predict human effects of nonlethal toxic exposures.

The potential exists for the formation of highly toxic species when organic materials are pyrolyzed or burned. This possibility has been a major driving force in the development of toxic potency tests in air. Lenoir et al. discuss the potential for formation of polybrominated dibenzodioxins and dibenzofurans from certain brominated flame-retardant systems. Even if formed such unique compounds are found under very limited conditions. In the final chapter, Voorhees discusses how soot analysis from real fires is used to better understand the role of specific materials present during a fire, including which materials were involved with the fire and when.

Toxic potency tests are now available, but the struggle to meaningfully use that data continues.

Chapter 20

Further Development of the N-Gas Mathematical Model
An Approach for Predicting the Toxic Potency of Complex Combustion Mixtures

Barbara C. Levin, Emil Braun, Magdalena Navarro, and Maya Paabo

National Institute of Standards and Technology,
Gaithersburg, MD 20899

A methodology has been developed for predicting smoke toxicity based on the toxicological interactions of complex fire gas mixtures. This methodology consists of burning materials using a bench-scale method that simulates realistic fire conditions, measuring the concentrations of the following primary fire gases - CO, CO_2, O_2, HCN, HCl, HBr, and NO_2 - and predicting the toxicity of the smoke using an empirical mathematical model called the N-Gas Model. The model currently in use is based on toxicological studies of the first six of the above listed primary gases both as individual gases and complex mixtures. The predicted toxic potency (based on this N-Gas Model) is checked with a small number of animal (Fischer 344 male rats) tests to assure that an unanticipated toxic gas was not generated. The results indicate whether the smoke from a material or product is extremely toxic (based on mass consumed at the predicted toxic level) or unusually toxic (based on the gases deemed responsible). The predictions based on bench-scale laboratory tests have been verified with full-scale room burns of a limited number of materials of widely differing characteristics chosen to challenge the system. The advantages of this approach are: 1. The number of test animals is minimized by predicting the toxic potency from the chemical analysis of the smoke and only using a few animals to check the prediction; 2. Smoke may be produced under conditions that simulate the fire scenario of concern; 3. Fewer tests are needed, thereby reducing the overall cost of the testing; and 4. Information is obtained on both the toxic potency of the smoke and the responsible gases. These results have been used in computations of fire hazard, and this methodology is now part of a draft international standard that is currently being voted on by the member countries of the International Standards Organization (ISO), Technical Committee 92 (TC92). In this chapter, a new 7-Gas Model including NO_2 and the data used in its development are presented.

The majority of people that die from exposure to fires are primarily affected by inhaling the toxic gases present in the smoke and not from burns. Fire death statistics examined for the years 1979 to 1985 attributed two-thirds of the victims to smoke inhalation and only one-third to burns (*1*). During these years, total fire deaths decreased 17%, but this decrease is primarily due to 34% fewer burn victims. Smoke fatalities only dropped 6%. The share of total fire deaths attributed to smoke inhalation has actually increased approximately 1% per year during this period. To further decrease deaths due to fires, the emphasis needs to be placed on reducing the number of deaths due to smoke inhalation. To accomplish this goal, we need to know the toxic products that are generated when materials thermally decompose and the toxicological interactions of the gases found in the complex chemical mixtures created by fires.

Our objectives in the development of the N-Gas Model were :

- To establish the extent to which we can explain and predict the toxicity of a material's combustion products by the interaction of the major toxic gases generated when that material is thermally decomposed in the laboratory,

- To develop a bioanalytical screening test which examines whether a material produces extremely toxic or unusually toxic combustion products and a mathematical model which predicts that toxicity,

- To predict the occupant response from the concentrations of primary toxic gases present in the environment and the time of exposure, and

- To provide data for use in computer models to predict the hazard that people will experience under various fire scenarios.

The N-Gas model using 6 gases (see Equation 1) is based on the hypothesis that a small number ("N") of gases in the smoke accounts for a large percentage of the observed toxic potency (*2-9*). This equation is based on studies at NIST on the toxicological interactions of six gases, CO, CO_2, HCN, reduced O_2, HCl and HBr. The concentrations of each of these gases necessary to cause 50% of the laboratory test animals (Fischer 344 male rats) to die either during the exposure (within exposure LC_{50}) or during the exposure plus the 14 day post-exposure observation period (within plus post-exposure LC_{50}) was determined. (The studies on HCl and HBr were conducted at Southwest Research Institute under a grant from NIST.) Similar measurements for various combinations of these gases indicated the additive, synergistic, and antagonistic toxicological effects of these gases. In this chapter, we will present the new data on NO_2 and discuss its impact on the N-Gas Model.

The LC_{50} may also refer to the mass of material loaded in the furnace or consumed by the exposure divided by the animal exposure chamber volume (g/m^3) which causes 50% of the animals to die within exposure or within exposure plus post-exposure. To reduce the number of animals necessary to determine an LC_{50} value, the material's LC_{50} is predicted by the N-Gas model. When the mass of material burned generates sufficient gaseous combustion products (those in the model) to produce an N-

Gas Value approximately equal to 1, that mass of material should be close to the material's LC_{50}. One or two animal tests can be conducted to check the predicted value and assure that an unexpected toxic gas was not produced. In most of the studies presented here, the exposure time was 30 minutes. We have also determined the LC_{50}'s of many of these gases both singly and mixed at times ranging from 1 to 60 minutes and have found that in all the cases examined that Equation 1 holds if the LC_{50}'s for the other times are substituted into the equation (5,6).

The N-Gas model prediction for 6 gases is based on the following empirical mathematical relationship:

$$N\text{-Gas Value} = \frac{m[CO]}{[CO_2]-b} + \frac{[HCN]}{LC_{50}HCN}$$

$$+ \frac{21-[O_2]}{21-LC_{50}O_2} + \frac{[HCl]}{LC_{50}HCl} + \frac{[HBr]}{LC_{50}HBr} \quad (1)$$

where the numbers in brackets indicate the time-integrated average atmospheric concentrations during a 30 minute (or other time) exposure period [(ppm x min)/min or for O_2 (% x min)/min]. Although CO_2 at concentrations generated in fires would not be lethal (the 30 minute LC_{50} of CO_2 is 470,000 ppm or 47% and the highest level of CO_2 possible in a fire is 21% and that would only happen if all the O_2 were converted to CO_2 which is highly unlikely), we found a synergistic effect between CO_2 and CO such that as the concentration of CO_2 increases (up to 5%), the toxicity of CO increases. Above 5% CO_2, the toxicity of CO starts to revert back towards the toxicity of CO by itself. The terms m and b define this synergistic interaction and in the 30 minute exposures, m and b equal -18 and 122000 if the CO_2 concentrations are 5% or less. For 30 minute studies in which the CO_2 concentrations are above 5%, m and b equal 23 and -38600, respectively. We have also shown that carbon dioxide increases the toxicity of the other gases currently included in the model as well as NO_2 (3,10). However, we found empirically that the effect of the CO_2 can only be added into this equation once. We, therefore, included the CO_2 effect into the CO factor for the following reasons: (1) we have examined the effect of many different concentrations of CO_2 on the toxicity of CO and have only examined the effect of 5% CO_2 on the other gases (5% was chosen based on data that showed that 5% CO_2 caused the greatest increase in CO toxicity) and (2) CO is the toxicant most likely to be present in all real fires.

The LC_{50} value of HCN is 200 ppm for 30 minute exposures or 150 ppm for 30 minute exposures plus the post-exposure observation period. (Exposure to CO in air only produced deaths during the actual exposures and not in the post-exposure observation period, whereas, HCN did cause numerous deaths in the first 24 hours of the post-exposure period). The 30 minute LC_{50} of O_2 is 5.4% which is subtracted from the normal concentration of O_2 in air, i.e., 21%. The LC_{50} value of HCl or HBr, respectively, for 30 minute exposures plus post-exposures times is 3700 ppm (11) and

3000 ppm (Switzer, W.G., Southwest Research Institute, personal communication).

In our pure and mixed gas studies, we found that if the value of Equation 1 is approximately 1, then some fraction of the test animals would die. Below 0.8, no deaths would be expected and above 1.3, all the animals would be expected to die. Since the concentration-response curves for animal lethalities from smoke are very steep, it is assumed that if some percentage (not 0 or 100%) of animals die, the experimental loading is close to the predicted LC_{50} value. Our results using this method show good agreement (deaths of some of the animals when the N-gas values are above 0.8) and the good predictability of this approach.

This model can be used to predict deaths that will occur only during the smoke exposure or those that will occur during and following the exposure. (The animals were not treated during this post-exposure period. It would be an interesting series of experiments to examine how the effects of various post-exposure treatments would impact on this model.) To predict only the deaths during the exposures, HCl and HBr are not included in the equation, since at concentrations normally found in fires, these two gases only have post-exposure effects. To predict deaths that would occur both during and following the exposure, one uses the mathematical model as shown.

The N-Gas Model has been developed into an N-Gas Method. This method reduces the time necessary to evaluate a material and the number of test animals needed for the toxic potency determination. It also indicates whether the toxicity is usual (i.e., the toxicity can be explained by the measured gases) or is unusual (additional gases are needed to explain the toxicity). To measure the toxic potency of a given material with this N-Gas Method, a sample is combusted under the conditions of concern and the gases in the model are measured. Based on the results of the chemical analytical tests and the knowledge of the interactions of the measured gases, an approximate LC_{50} value is predicted. In just two additional tests, six rats are exposed to the smoke from a material sample size estimated to produce an atmosphere equivalent to the approximate LC_{50} level (this can be for within exposure or both within plus post-exposure). The deaths of some percentage of the animals (not 0 and not 100%) indicates that the predicted LC_{50} would be close to the actual calculated LC_{50}. No deaths may indicate an antagonistic interaction of the combustion gases. The deaths of all of the animals may indicate the presence of unknown toxicants or other adverse factors. If more accuracy is needed, then a detailed LC_{50} can be determined.

Another important factor to consider is how well these small-scale bioanalytical approaches predict the toxicological effects observed in real-scale fire tests. Rats were exposed to the gases generated from materials thermally decomposed under various test conditions (the NIST Radiant Panel Method, the Cup Furnace Smoke Toxicity Method and large-scale room tests) (*9, 12-14*). In most cases, the N-Gas Model was able to predict the deaths correctly. In the case of PVC, the HCl factor was only included in the prediction of the total (within plus post-exposure) deaths. The model correctly predicted the results as long as the HCl was greater than 1000 ppm; therefore, it is possible that HCl concentrations under 1000 ppm may not contribute to lethality even in the post-exposure period. More experiments are necessary to show whether a true toxic threshold for HCl does exist.

Experimental Approaches

Gases. In all tests, chemical analyses were conducted to determine the concentrations of carbon monoxide (CO), carbon dioxide (CO_2), and oxygen (O_2). In tests of nitrogen-containing materials, hydrogen cyanide (HCN) was measured. In some tests, hydrogen chloride (HCl), hydrogen fluoride (HF), hydrogen bromide (HBr) and total nitrogen oxides (NO_x) were also measured to determine if sufficient quantities would be generated to warrant further monitoring. Calibration gases (CO, CO_2, HCN) were commercially supplied in various concentrations in nitrogen. The concentrations of HCN in the commercially supplied cylinders were routinely checked by silver nitrate titration (*15*), since it is known that the concentration of HCN stored under these conditions will decrease with time. Nitric oxide (NO) in nitrogen, a standard reference material, was obtained from the Gas and Particulate Science Division, NIST.

Carbon monoxide and CO_2 were measured continuously during each test by non-dispersive infrared analyzers. Oxygen concentrations were measured continuously with a paramagnetic analyzer. Syringe samples (100 µL) of the chamber atmosphere were analyzed for HCN approximately every three minutes with a gas chromatograph equipped with a thermionic detector (*16*). The concentration of NO_x was measured continuously by a chemiluminescent NO_x analyzer equipped with a molybdenum converter (set at 375°C) and a sampling rate of 25 mL/min. The change from a stainless steel converter to a molybdenum converter prevented interference from HCN. All combustion products and gases (except HCN, NO_x, and the halogen gases) that were removed for chemical analysis were returned to the chamber. The CO, CO_2, O_2 and NO_x data were recorded by an on-line computer every 15 seconds.

The halogen gases, HF, HCl, and HBr, were analyzed by ion chromatography. The combustion products were bubbled into 30 mL impingers containing 25 mL of 5 mM KOH at a rate of approximately 30 mL/min for the 30 minute tests. The flow was monitored every five minutes and averaged over the 30 minute run to determine the amount of gases collected. The resulting solution was analyzed for F^-, Cl^-, and Br^- by the modified method A-106 as described in reference (*17*). In this modified method, the eluent was changed from a 2.5 mM lithium hydroxide solution to a 5 mM KOH solution, a manual injector was used instead of an automatic injector, and a 590 programmable pump was employed instead of the 510 solvent delivery module.

For each test, the reported gas concentrations are the time-integrated average exposure values which were calculated by integrating the area under the instrument response curve and dividing by the exposure time [i.e., (ppm x min)/min or, in the case of O_2, (% x min)/min]. The calculated CO and CO_2 concentrations are accurate to within 100 ppm and 500 ppm, respectively. The calculated HCN concentrations are accurate to 10% of the HCN concentration. The calculated NO_x concentrations are accurate to 10% of the NO_x concentration.

Animals. Fischer 344 male rats, weighing 200-300 grams, were obtained from Taconic Farms (Germantown, NY).* They were allowed to acclimate to our laboratory

---

*Certain commercial equipment, instruments, materials or companies are identified in this paper to specify adequately the experimental procedure. Such identification does not imply recommendation or endorsement by the National Institute of Standards and Technology, nor does it imply that the materials or equipment identified are the best available for the purpose.

conditions for at least 7 days prior to testing. Animal care and maintenance were performed in accordance with the procedures outlined in the National Institutes of Health's "Guide for the Care and Use of Laboratory Animals." Each rat was housed individually in suspended stainless steel cages and provided with food (Ralston Purina Rat Chow 5012) and water *ad libitum*. Twelve hours of fluorescent lighting per day were provided using an automatic timer. All animals (including the controls) were weighed daily from the day of arrival until the end of the post-exposure observation period.

Determination of Single and Mixed Gas Toxicity. All animal exposures were conducted using the chemical analysis system, and the animal exposure system that were designed for the Cup Furnace and the NIST Radiant Panel Smoke Toxicity Methods (*18,19*).

The exposure chamber is a 200 liter rectangular closed chamber in which all the gases (except the small amounts removed for chemical analysis of HCN, NO_x, HCl, and HBr) are kept for the duration of the test. Six rats are exposed in each test. Each animal is placed in a restrainer and inserted into one of six portholes located along the front of the exposure chamber such that only the heads of the animals are exposed. In the tests conducted to determine various gas LC_{50} values, the desired test concentrations of the gas or gases were generated in the chamber and animal exposures started when the animals were inserted into the portholes. For material LC_{50}'s, the exposure started when the material thermal decomposition was initiated. In most of the studies, animals were exposed for 30 minutes; other exposure times were used to assure the model held for other times as well.

The toxicological endpoint was the LC_{50} values, which were calculated based on the deaths that occurred either during the exposures or the exposure plus at least a 14 day post-exposure observation period. The percentage of animals dying at each gas concentration was plotted to produce a concentration-response curve from which the LC_{50} values and their 95% confidence limits were calculated by the statistical method of Litchfield and Wilcoxon (*20*).

Nitrogen Dioxide and Mixed Gas Toxicity Results and Discussion

NO_2 Toxicity. Deaths from NO_2 in air occur only in the post-exposure period and its LC_{50} following a 30 minute exposure is 200 ppm. The seven experiments to determine this LC_{50} are shown in Figure 1 (open circles) which plots the percent lethality vs. the NO_2 concentration. The concentration which is statistically calculated to correspond to 50% lethality is the LC_{50}.

NO_2 plus CO_2. The 30 minute LC_{50} of CO_2 is 47%. Carbon dioxide plus NO_2 showed synergistic toxicological effects (*21*). Figure 1 (solid squares) shows the increase in toxicity (line shifts to the left; LC_{50} value decreases) when the animals are exposed to various concentrations of NO_2 plus 5% CO_2 (*10*). The LC_{50} for NO_2 following a 30 min exposure to NO_2 plus 5% CO_2 is 90 ppm (post-exposure deaths) (i.e., the toxicity of NO_2 doubled).

NO_2 plus CO. Carbon monoxide produces only within-exposure deaths and its 30 min LC_{50} is 6600 ppm (Fig. 2, open circles). In the presence of 200 ppm of NO_2, the within-exposure toxicity of CO doubled (i.e., its 30 minute LC_{50} became 3300 ppm, Fig. 2, solid squares). An exposure of approximately 3400 ppm CO plus various concentrations of NO_2 showed that the presence of CO would also increase the post-exposure toxicity of NO_2. The 30 minute LC_{50} value went from 200 ppm to 150 ppm (Fig. 3, solid squares). We used 3400 ppm of CO as that concentration would not be lethal during the exposure and we would be able to observe the post-exposure effects of CO on NO_2; the LC_{50} of CO (6600 ppm) would have caused deaths of the animals during the 30 minute exposure.

NO_2 plus O_2. The 30 minute LC_{50} of O_2 is 5.4% and the deaths occur primarily during the exposures (Fig. 4, open circles). In the presence of 200 ppm of NO_2, the within-exposure LC_{50} of O_2 and its toxicity increased to 6.7% (Fig. 4, solid squares). In the case of O_2, increased toxicity is indicated by an increase in the value of the LC_{50} since it is more toxic to be adversely affected by a concentration of O_2 ordinarily capable of sustaining life. Exposure of the animals to 6.7% O_2 plus various concentrations of NO_2 showed that the NO_2 toxicity doubled (i.e., its LC_{50} value decreased from 200 ppm to 90 ppm) (Fig.5, solid squares).

NO_2 plus HCN. An antagonistic toxicological effect was noted during the experiments on combinations of HCN and NO_2. As mentioned above, the LC_{50} for NO_2 alone is 200 ppm (post-exposure) (Fig. 5). The 30 minute within-exposure LC_{50} for HCN alone is 200 ppm (Fig. 6, open circles). These concentrations of either gas alone is sufficient to cause death of the animals (i.e., 200 ppm HCN or 200 ppm NO_2 would cause 50% of the animals to die either during the 30 min exposure or following the 30 min exposure, respectively). However, in the presence of 200 ppm of NO_2, the within-exposure HCN LC_{50} concentration is 480 ppm or 2.4 times the LC_{50} of HCN alone (Table 1 and Fig. 6, solid squares).

A possible mechanism for this antagonistic effect is as follows: In the presence of H_2O, NO_2 forms nitric acid (HNO_3) and nitrous acid (HNO_2). These two acids are most likely responsible for the lung damage leading to the massive pulmonary edema and subsequent deaths noted following exposure to high concentrations of NO_2. Nitrite ion (NO_2^-) formation occurs in the blood when the nitrous acid dissociates. The nitrite ion oxidizes the ferrous ion in oxyhemoglobin to ferric ion to produce methemoglobin (MetHb) (Equation 2). MetHb is a well-known antidote for CN^- poisoning. MetHb binds cyanide and forms cyanmethemoglobin which prevents cyanide from entering the cells. In the absence of MetHb, free cyanide will enter the cells, react with cytochrome oxidase, prevent the utilization of O_2, and cause cytotoxic hypoxia. If, on the other hand, cyanide is bound to MetHb in the blood, it will not be exerting its cytotoxic effect. Therefore, the mechanism of the antagonistic effect of NO_2 on the toxicity of cyanide is believed to be due to the conversion of oxyhemoglobin [$O_2Hb(Fe^{++})$] to methemoglobin [$MetHb(Fe^{+++})$] in the presence of nitrite [see Equation 2 (22) and Figure 7].

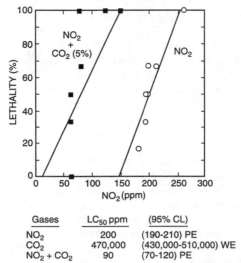

Gases	LC$_{50}$ ppm	(95% CL)
NO$_2$	200	(190-210) PE
CO$_2$	470,000	(430,000-510,000) WE
NO$_2$ + CO$_2$	90	(70-120) PE

30 Minute exposures, 5% CO$_2$, all deaths occurred post-exposure within 24 hours

Figure 1. Synergistic effect of CO$_2$ on NO$_2$ toxicity. Open circles: NO$_2$. Solid squares: NO$_2$ plus 5% CO$_2$. WE: within exposure; PE: post-exposure. CL: confidence limits.

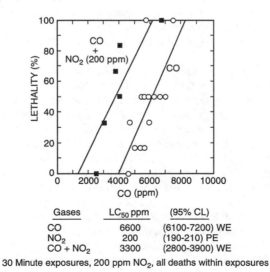

Gases	LC$_{50}$ ppm	(95% CL)
CO	6600	(6100-7200) WE
NO$_2$	200	(190-210) PE
CO + NO$_2$	3300	(2800-3900) WE

30 Minute exposures, 200 ppm NO$_2$, all deaths within exposures

Figure 2. Effect of NO$_2$ on CO toxicity. Open circles: CO. Solid squares: CO plus 200 ppm NO$_2$. WE: within exposure; PE: post-exposure. CL: confidence limits.

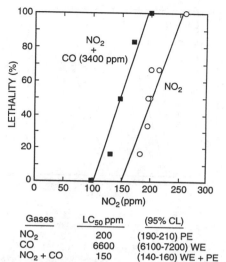

Figure 3. Effect of CO on NO_2 toxicity. Open circles: NO_2. Solid squares: NO_2 plus approximately 3400 ppm CO. WE: within exposure; PE: post-exposure. CL: confidence limits.

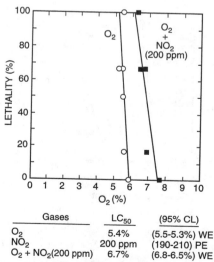

Figure 4. Effect of NO_2 on O_2 toxicity. Open circles: O_2. Solid squares: O_2 plus 200 ppm NO_2. WE: within exposure; PE: post-exposure. CL: confidence limits.

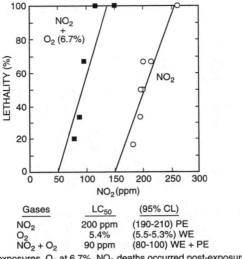

Figure 5. Effect of O_2 on NO_2 toxicity. Open circles: NO_2. Solid squares: NO_2 plus 6.7% O_2. WE: within exposure; PE: post-exposure. CL: confidence limits.

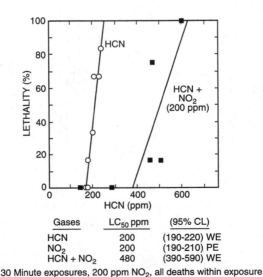

Figure 6. Antagonistic Effect of NO_2 on HCN toxicity. Open circles: HCN. Solid squares: HCN plus 200 ppm NO_2. WE: within exposure; PE: post-exposure. CL: confidence limits.

TABLE 1. ANTAGONISTIC EFFECTS OF MIXTURES OF HCN AND NO_2

Gas Concentration[a]		Gas Concentration/LC_{50}[b]				Deaths	
HCN (ppm)	NO_2 (ppm)	HCN WE[c]	HCN WE + PE[d]	NO_2 PE[d]	HCN + NO_2	WE	WE + PE
150	190	0.75	1.00	0.95	1.70[c] 1.95[d]	0/6	3/6
280	200	1.40	1.87	1.00	2.40[c] 2.87[d]	0/6	6/6
460	190	2.30	3.07	0.95	3.25[c] 4.02[d]	1/6	1/6
470	200	2.35	3.13	1.00	3.35[c] 4.13[d]	3/4	4/4
510	190	2.55	3.40	0.95	3.50[c] 4.35[d]	1/6	6/6
600	200	3.00	4.00	1.00	4.00[c] 5.00[d]	6/6	6/6

a: Time-integrated average concentration over the 30 minute exposure [(ppm x min)/min].
b: LC_{50} value (30 min exposure): HCN (WE) = 200 ppm; HCN (WE + PE) = 150 ppm; NO_2 (PE) = 200 ppm.
c: WE: Within the 30 minute exposure.
d: WE + PE: Within the 30 minute exposure plus the post-exposure observation period.

$$2H^+ + 3NO_2^- + 2O_2Hb(Fe^{++}) =$$

$$2MetHb(Fe^{+++}) + 3NO_3^- + H_2O \qquad (2)$$

Tertiary Mixtures of NO_2, CO_2, and HCN. Since the binary gas mixture studies indicated that NO_2 plus CO_2 showed synergistic toxic effects and NO_2 plus HCN showed antagonistic toxic effects, it was of interest to examine combinations of NO_2, CO_2, and HCN. In this series of experiments, the concentrations of HCN were varied from almost 2 to 2.7 times its LC_{50} value (200 ppm). The concentrations of NO_2 were approximately equal to one LC_{50} value (200 ppm) if the animals were exposed to NO_2 alone and approximately 1/2 the LC_{50} (90 ppm or twice as toxic) if the animals were exposed to NO_2 plus CO_2; the concentrations of CO_2 were maintained at approximately 5%; and the O_2 levels were kept above 18.9 %. The results of these experiments are shown in Table 2.

Earlier work (*3,10*) indicated that the presence of 5% CO_2 with either HCN or NO_2 produced a more toxic environment than would occur with either gas alone. The antagonistic effects of NO_2 on HCN shown in Table 1 indicate that the presence of one LC_{50} concentration of NO_2 (~ 200 ppm) will protect the animals from the toxic effects of HCN during the 30 minute exposures, but not from the post-exposure effects of the combined HCN and NO_2. However, results in Table 2 indicate that CO_2 does not make the situation worse, but rather provides protection even during the post-exposure period. In each of the six experiments shown in Table 2, some or all of the animals lived during the test and in 4 tests, some of the animals lived through the post-exposure period even though column 12 of Table 2 shows that the animals were exposed to combined levels of HCN, NO_2 and CO_2 that would be equivalent to 4.7 to 5.5 times the lethal concentrations of these gases. One possible reason that CO_2 seems to provide an additional degree of protection is that NO_2 in the presence of 5% CO_2 produces 4 times more MetHb than does NO_2 alone (Fig. 8) (*10*).

Mixtures of CO, CO_2, NO_2, O_2, and HCN. The initial design of these experiments was to look for additivity of the CO/CO_2, HCN, and NO_2 factors keeping each at about 1/3 of its toxic level, while keeping the O_2 concentration above 19%. When these initial experiments produced no deaths, we started to increase the concentrations of CO up to 1/3 of the LC_{50} of CO alone (6600 ppm), HCN was increased to 1.3 or 1.75 times its LC_{50} depending on whether the within-exposure LC_{50} (200 ppm) or the within- and post-exposure LC_{50} (150 ppm) is being considered, and NO_2 was increased up to a full LC_{50} value (200 ppm). The results indicated that just adding a NO_2 factor (e.g., $[NO_2]/LC_{50}$ NO_2) to Equation 1 would not predict the effect on the animals. A new mathematical model was developed and is shown as Equation 3. In this model, the differences between the within-exposure predictability and the within-exposure and post-exposure predictability is: (1) the LC_{50} value used for HCN is 200 ppm for within-exposure or 150 ppm for within-exposure and post-exposure and (2) the HCl and HBr factors are not used to predict the within-exposure lethality, only the within-exposure and post-exposure

TABLE 2. TERTIARY MIXTURES OF NO_2, CO_2, AND HCN

Gas Concentrations[a]				Gas Concentration/LC_{50}[b]								Deaths	
HCN (ppm) (1)	NO_2 (ppm) (2)	CO_2 (ppm) (3)	O_2 (%) (4)	HCN WE[c] (5)	HCN WE + PE[d] (6)	NO_2 PE (7)	$NO_2 + CO_2$ PE (8)	HCN[c] + NO_2 (9)	HCN[d] + NO_2 (10)	HCN[c] + ($NO_2 + CO_2$) (11)	HCN[d] + ($NO_2 + CO_2$) (12)	WE[c]	WE + PE[d]
380	200	51400	19	1.91	2.55	0.98	2.18	2.89	3.53	4.09	4.72	2/6	3/6
410	200	51300	19	2.07	2.75	0.98	2.18	3.05	3.73	4.24	4.93	1/6	3/6
430	200	49700	19	2.17	2.89	0.98	2.17	3.14	3.86	4.33	5.05	2/6	6/6
460	190	53300	19	2.29	3.05	0.97	2.16	3.26	4.02	4.44	5.20	2/6	4/6
500	190	54300	19	2.51	3.35	0.95	2.12	3.47	4.30	4.63	5.47	0/6	3/6
550	190	50400	19	2.74	3.65	0.97	2.14	3.70	4.61	4.88	5.79	5/6	6/6

a: Columns 1 - 4: Time-integrated average concentration over the 30 minute exposure (ppm x min/min).
b: LC_{50} value (30 min exposure): HCN (WE) = 200 ppm; HCN (WE + PE) = 150 ppm; NO_2 (PE) = 200 ppm; NO_2 in the presence of 5% CO_2 (PE) = 90 ppm.
c: WE: Within the 30 minute exposure.
d: WE + PE: Within exposure + Post-exposure observation period.
Column 5 = Column 1/200 ppm [i.e., LC_{50} HCN (WE)].
Column 6 = Column 1/150 ppm [i.e., LC_{50} HCN (WE + PE)].
Column 7 = Column 2/200 ppm [i.e., LC_{50} NO_2(PE)].
Column 8 = Column 2/90 ppm [i.e., LC_{50} NO_2 in presence of 5% CO_2 (PE)].
Column 9 = Columns 5 + 7.
Column 10 = Columns 6 + 7.
Column 11 = Columns 5 + 8.
Column 12 = Columns 6 + 8.

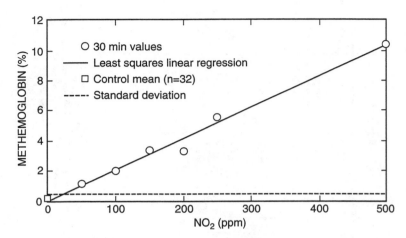

Figure 7. Methemoglobin values from 30 minute exposures to various concentrations of NO_2. Modified and reprinted from reference 10 with permission of the author.

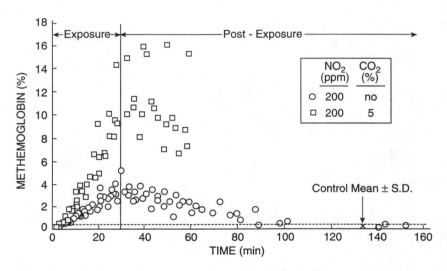

Figure 8. Methemoglobin generation during and following exposure to NO_2 alone (open circles) or NO_2 plus 5% CO_2 (open squares). Control mean ± the standard deviation (S.D.) was 0.2 ± 0.2 (n = 32 animals). Modified and reprinted from reference 10 with permission of the author.

lethality. According to Equation 3, animal deaths will start to occur when the N-Gas Value is above 0.8 and 100% of the animals will die when the value is above 1.3. The experimental results supporting this model are presented in Table 3. Results in Table 3 indicate that in those cases where the values were above 0.8 and no deaths occurred, the animals were severely incapacitated (close to death) as demonstrated by no righting reflex or eye reflex.

The N-Gas Model Including NO_2.

$$N\text{-Gas Value} = \frac{m\,[CO]}{[CO_2] - b} + \frac{21 - [O_2]}{21 - LC_{50}(O_2)} +$$

$$\left(\frac{[HCN]}{LC_{50}(HCN)} \times \frac{0.4\,[NO_2]}{LC_{50}(NO_2)} \right) + 0.4 \left(\frac{[NO_2]}{LC_{50}(NO_2)} \right) +$$

$$\frac{[HCl]}{LC_{50}(HCl)} + \frac{[HBr]}{LC_{50}(Hbr)} \qquad (3)$$

For an explanation of these terms, see the paragraph following Equation 1. Equation 3 should be used to predict the within-exposure plus post-exposure lethal toxicity of mixtures of CO, CO_2, HCN, reduced O_2, NO_2, HCl, and HBr. The LC_{50} values will be the same as those given for Equation 1 using 150 ppm for HCN and 200 ppm for NO_2. If one wishes to predict the deaths that will occur only during the exposure, the LC_{50} value used for HCN should be 200 ppm and the HCl and HBr factors should not be included. To predict the lethal toxicity of atmospheres that do not include NO_2, Equation 1 is to be used.

Conclusions

In the binary gas studies, NO_2, a toxic gas which exerts its lethal effect following an exposure, increased the toxicity of all the tested within-exposure toxic gases except HCN. The reverse was also seen - i.e., the post-exposure toxic effects of NO_2 were greater if the animals had also been exposed to CO, CO_2, or reduced O_2. The exception to these results occurred when the animals were exposed to NO_2 plus HCN, a combination which produced an antagonistic toxicological effect. The explanation for this antagonistic effect is believed to be due to the production of methemoglobin (a cyanide antidote) by nitrite ions formed from the dissociation of nitrous acid in the blood. The nitrous acid was generated by the reaction of NO_2 and H_2O in the lung.

Although the binary combinations of NO_2 and CO_2 showed that the toxicity of NO_2 doubled, the tertiary gas combination studies with HCN, NO_2, and CO_2 indicated that CO_2 did not increase the toxicity of the mixture, but may have further increased the protective effect of the NO_2. This increased protective effect of CO_2 is probably due to

TABLE 3. Mixtures of CO, CO_2, NO_2, O_2, and HCN.

Gas Concentration					N-Gas Value WE Equation 3	N-Gas Value WE + PE Equation 3	Deaths WE	Deaths WE + PE	Incapacitation (% of survivors)
CO (ppm)	CO_2 (ppm)	HCN (ppm)	O_2 (%)	NO_2 (ppm)					
1280	51800	50	19.4	70	0.6	0.62	0/6	0/6	0
1310	51800	80	19.4	100	0.71	0.73	0/5	0/5	0
2030	46800	50	19.6	90	0.8	0.81	1/6	2/6	80
2060	45800	50	19.7	100	0.82	0.84	2/6	4/6	100
2180	51000	70	19.5	70	0.83	0.85	0/6	0/6	66.7
2160	50800	50	19.4	90	0.88	0.9	0/6	1/6	16.7
2210	50900	50	19.4	110	0.93	0.95	0/6	2/6	16.7
1900	49700	160	19.4	190	1.26	1.36	4/6	6/6	100
1930	50500	190	19.3	200	1.36	1.46	4/6	6/6	100
1980	53700	210	19.1	200	1.42	1.55	6/6	6/6	N/A
2210	53200	260	19.2	200	1.58	1.75	6/6	6/6	N/A

WE: Within exposure
PE: Post-exposure
Incap: Incapacitation

the greater amount of MetHb produced by the combination of NO_2 and CO_2 than by NO_2 alone as shown by earlier work in our laboratory. Earlier work in this laboratory showed that more methemoglobin is produced by the combination of NO_2 and CO_2 than by NO_2 alone.

A new N-Gas Model was needed to move from the 6-gas model which includes CO, CO_2, HCN, reduced O_2 concentrations, HCl and HBr to a 7-gas model which also includes NO_2. Validation studies looking at a series of materials and products under conditions ranging from laboratory bench-scale to full-scale room burns indicated that, in all cases, the 6-Gas Model was able to predict the deaths correctly. In the case of PVC, the HCl factor was only included in the prediction of the post-exposure (not within-exposure) deaths and preliminary data showed that concentrations under 1000 ppm may not have any observable effect even in the post-exposure period. More experiments are necessary to show whether HCl has a toxic threshold. The 7-Gas model works when the animals are exposed to various concentrations of the tested gases; studies need to be done to ensure that the 7-Gas Model predicts the outcome when nitrogen-containing materials and products are thermally decomposed.

Caution: The values given for use in equations 1 and 3 are dependent on the test protocol, on the source of test animals, and on the rat strain. It is important to verify the above values whenever different conditions prevail and if necessary, to determine the values that would be applicable under the new conditions.

Acknowledgments

This research was supported in part by the Polyurethane Division of the Society of the Plastics Industry, Inc. and The Department of Respiratory Research, Walter Reed Army Institute of Research, M. Mayorga, Project Officer. We thank Richard H. Harris, Jr. for the chemical analysis of HCl and HBr by ion chromatography.

Literature Cited

1. Harwood, B.; Hall, J. R. Jr. What kills in fires: smoke inhalation or burns? *Fire J.* **1989,** *May/June*, pp. 29-34.
2. Levin, B. C.; Paabo, M.; Highbarger, L.; Eller, N. Toxicity of complex mixtures of fire gases. *The Toxicologist* **1990**, *vol. 10*, p. 84.
3. Levin, B.C.; Paabo, M.; Gurman, J.L.; Harris, S.E. Effects of exposure to single or multiple combinations of the predominant toxic gases and low oxygen atmospheres produced in fires. *Fundam. Appl. Toxicol.* **1987** *vol.9*, pp. 236-250.
4. Levin, B.C.; Paabo, M.; Gurman, J.L.; Harris, S.E.; Braun, E. Toxicological interactions between carbon monoxide and carbon dioxide. *Toxicology* **1987,** *vol 47*, pp. 135-164 .
5. Levin, B.C.; Gurman, J.L.; Paabo, M.; Baier, L.; Holt, T. Toxicological effects of different time exposures to the fire gases: carbon monoxide or hydrogen cyanide or to carbon monoxide combined with hydrogen cyanide or carbon dioxide. In *Proceedings of the Ninth Joint Panel Meeting of the U.S.-Japan (UJNR) Panel on Fire Research and Safety;* NBSIR 88-3753; National Bureau of Standards: Gaithersburg, MD April, 1988.

6. Levin, B.C.; Paabo, M.; Gurman, J.L.; Clark, H.M.; Yoklavich, M.F. Further Studies of the Toxicological Effects of Different Time Exposures to the Individual and Combined Fire Gases: CO, HCN, CO_2, and Reduced Oxygen; In *Polyurethane '88, Proceedings 31st Soc. of Plastics Meeting,*; Technomic Pub. Co., Inc.; Lancaster, PA, 1988, pp. 249-252.
7. Levin, B.C.; Paabo, M.; Gurman, J.L.; Harris, S.E. Toxicological effects of the interactions of fire gases. In *The Proc. of the Smoke/Obscurants Symposium X*, Adelphi, MD, 1986, Vol. II, pp. 617-629.
8. Levin, B.C.; Gann, R.G. Toxic potency of fire smoke: measurement and use. In *Fire and Polymers: Hazards Identification and Prevention;* Nelson, G.L., Ed.; ACS Symposium Series 425; American Chemical Society: Washington, DC, 1990, pp 3-11.
9. Babrauskas, V.; Harris, R.H.,Jr.; Braun, E.; Levin, B.C.; Paabo, M.; Gann, R.G. The role of bench-scale test data in assessing full-scale fire toxicity. NIST Technical Note 1284, National Institute of Standards and Technology: Gaithersburg, MD January, 1991.
10. Levin, B.C.; Paabo, M.; Highbarger, L.; Eller, N. Synergistic effects of nitrogen dioxide and carbon dioxide following acute inhalation exposures in rats. NISTIR 89-4105, National Institute of Standards and Technology: Gaithersburg, MD, June, 1989.
11. Hartzell, G.E.; Grand, A.F.; Switzer, W.G. Toxicity of smoke containing HCl. In *Fire and Polymers: Hazards Identification and Prevention;* Nelson, G.L., Ed.; ACS Symposium Series 425; American Chemical Society: Washington, DC, 1990, pp. 12-20.
12. Braun, E.; Levin, B.C.; Paabo, M.; Gurman, J.L.; Holt, T.; Steel, J.S. Fire toxicity scaling. NBSIR 87-3510, National Bureau of Standards: Gaithersburg, MD, 1987.
13. Braun, E.; Levin, B.C.; Paabo, M.; Gurman, J.L.; Clark, H.M.; Yoklavich, M.F. Large-scale compartment fire toxicity study: Comparison with small-scale toxicity test results. NBSIR 88-3764, National Bureau of Standards: Gaithersburg, MD, 1988.
14. Braun, E.; Gann, R.G.; Levin, B.C.; Paabo, M. Combustion product toxic potency measurements: comparison of a small-scale test and "real-world" fires. *J. Fire Sciences* **1990**, *vol. 8*, pp. 63-79.
15. Kolthoff, I.M.; Sandell, E.B. *Textbook of Quantitative Inorganic Analysis*, Second Ed., MacMillan Co.: New York, 1953, p. 546.
16. Paabo, M.; Birky, M.M.; Womble, S.E. Analysis of hydrogen cyanide in fire environments. *J. Comb. Tox.* **1979**, *vol. 6*, pp.99-108.
17. Heckenberg, A.L.; Alden, P.G.; Wildman, B.J.; Krol, J.; Romano, J.P.; Jackson, P.E.; Jandik, P.; Jones, W.R. *Waters Innovative Methods for Ion Analysis.* Waters Manual No. 22340, Revision 0.0, Millipore Corporation, Waters Chromatography Division, Milford, MA, 1989.
18. Levin, B.C.; Fowell, A.J.; Birky, M.M.;; Paabo, M.; Stolte, A.; Malek, D. Further development of a test method for the assessment of the acute inhalation toxicity of combustion products. NBSIR 82-2532, National Bureau of Standards: Gaithersburg, MD, 1982.

19. Levin, B.C. The development of a new small-scale smoke toxicity test method and its comparison with real-scale fire tests. In *Toxicology from Discovery and Experimentation to the Human Perspective*; Chambers, P.L.; Chambers, C.M.; Bolt, H.M.; Preziosi, P., Eds.; Elsevier Science Publishers: 1000 AE Amsterdam, The Netherlands, 1992; pp 257-264.
20. Litchfield, J.T.; Wilcoxon, F. A simplified method of evaluating dose-effect experiments. *J. Pharmacol. & Exp. Therapeut.* **1949**, *vol. 96*, pp. 99-113.
21. Levin, B.C.; Paabo, M.; Navarro, M. Toxic interactions of binary mixtures of NO_2 plus CO, NO_2 plus HCN, NO_2 plus reduced O_2, and NO_2 plus CO_2. *The Toxicologist* **1991**, *vol. 11*, pp. 222.
22. Rodkey, F.L. A mechanism for the conversion of oxyhemoglobin to methemoglobin by nitrite. *Clin. Chem.* **1976**, *vol. 22/12*, pp. 1986-1990.

RECEIVED February 6, 1995

Chapter 21

Correlation of Atmospheric and Inhaled Blood Cyanide Levels in Miniature Pigs

F. W. Stemler, A. Kaminskis, T. M. Tezak-Reid, R. R. Stotts, T. S. Moran, H. H. Hurt, Jr., and N. W. Ahle

U.S. Army Medical Research Institute of Chemical Defense, Aberdeen Proving Ground, MD 21010−5425

The LCT50 (exposure time and atmospheric concentration needed to produce 50% lethality) has been commonly used to quantify the toxicity of a gas such as hydrogen cyanide (HCN). Few studies have been performed in which blood cyanide concentrations were measured simultaneously in animals at known exposure concentrations and time. This study was an attempt to correlate which blood cyanide levels would cause lethality in miniature pigs when exposed to hydrogen cyanide (HCN) for a fixed time. An automated microdistillation assay (1) was used to continuously monitor arterial blood cyanide before, during and after the exposures to a HCN/air mixture. Seven animals were exposed to a HCN/air mixture for two minutes each, four to 1176 ±SD 70 mg/m^3, and three animals to 2125 ±SD 91 mg/m^3. Two of the three animals exposed to the high HCN/air mixture died with a peak blood cyanide concentration of about 4.1 ±SD 0.38 μg/mL. Four animals exposed to the low HCN/air mixture had a peak blood cyanide concentration of 2.94 ±SD 0.71 μg/mL. All four survived for a 24-hour post-exposure observation period before they were sacrificed. Several physiological parameters were also monitored.

No standardized *in vivo* model for cyanide inhalation exists for examining physiological parameters and for evaluating potential treatments in cyanide poisoning. Parenteral routes of cyanide administration have been preferred due to protective and precautionary requirements associated with gaseous hydrogen cyanide. Although the dog once was the primary model for studies of cyanide poisoning and treatment, public opinion precludes its routine use as an experimental animal. A summary of a number of inhalation studies of HCN in

This chapter not subject to U.S. copyright
Published 1995 American Chemical Society

several animal species with LCT50 values has been reported (2). The LCT50 values for large animals (dogs, goats, pigs, monkeys, etc.) were obtained from exposures to various concentrations of HCN and times. However, simultaneous blood cyanide levels were not measured in these animals. Therefore, this study was performed to determine the lethal concentration of cyanide in the blood of miniature pigs exposed to known atmospheric concentrations (c) of HCN for a fixed time (t) in minutes.

Methods

Animals. Seven Yucatan Miniature male pigs weighing between 14 and 19 kg were used in these experiments. Upon arrival from Charles River Laboratories, the animals were kept in quarantine for a period of 7 days and were screened for evidence of disease before use. They were maintained under an AAALAC accredited animal care and use program. Tap water was supplied **ad libitum** and Lab Porcine Chow Grower (Purina 5084, Purina Mills, Inc., Richmond, IN) was hand fed and also was available in feed bins twice a day. Groups of three or four animals were housed in the same pen on plastic coated slotted flooring. Pigs were housed individually one day prior to and three days following surgery. The room was maintained at 20-22°C with a relative humidity of 50% ± 10% using at least 10 complete air changes per hour of 100% conditioned fresh air. The room was maintained on a 12-hour light/dark lighting cycle with no twilight.

Daily training sessions were conducted to accustom the pigs to handling by attendants and restraint (initially 0.5 hr and increased daily up to 2.0 hr) in a Panepinto sling (3). While restrained in the sling, the pigs were also conditioned to an anesthesia cone which covered the nose and mouth. Since pigs have an extraordinarily keen sense of smell, a drop of almond extract on a cotton ball was introduced in the nose cone to familiarize the animal to the odor of HCN. The daily training sessions were continued for approximately 10 days pre- and also 3 days post-surgery.

Surgical Procedures. After an overnight fast, each pig received preanesthetic intramuscular injections of 0.08 mg/kg atropine sulfate, 2.2 mg/kg ketamine hydrochloride, and 2.2 mg/kg xylazine prior to aseptic surgery. An endotracheal tube (5.0 mm o.d.) was inserted and the cuff inflated, and anesthesia was maintained with a mixture of halothane and oxygen. Two silastic Dow Corning catheters (#602-26s i.d. 0.062"/o.d. 0.095" and #602-20s i.d. 0.040"/o.d. 0.085") joined together with adhesive were inserted into the right carotid artery and advanced to the aorta. One tip was slightly (1/8") in advance of the other. Precaval blood was drawn via a Dow Corning catheter (#602-175 i.d. 0.030"/o.d. 0.065") inserted into the right external jugular vein. The catheters were filled with heparin (1000 IU/mL) and tunneled subcutaneously beneath the skin to the right of dorsal midline before exiting the skin. The exposed ends of the catheters were coiled and placed between two Velcro patches, one of which was sutured to the skin. After surgery, the catheters were cleared by removal of heparinized blood and filled with fresh heparin three times daily to maintain patency. No antibiotic therapy was required to prevent postoperative sepsis.

Experimental Procedures. The pig, after an overnight fast, was positioned in a Panepinto sling in a specially designed pig restraining device which immobilized the head and nose cone. The exposures were conducted within an approved fume hood. Baseline measurements were made and blood samples were drawn during the 30 min prior to the exposure to HCN. Blood samples, timed from the beginning of exposure, were drawn at 3-, 5-, 8-, 10-, and at 15-minute intervals up to 2 hours post exposure. A 3-way sliding-Type™ valve (2870 Hans Rudolph Inc., Kansas City, MO 64114) attached to a coupler lock-ring assembly (200087) and a 2-way non-rebreathing valve (2600) with inhalation/exhalation ports allowed the animal to be instantly switched between breathing room air and the toxic gas mixture. A schematic of the exposure system in Figure 1 shows the interfacing of the valved assembly to the nose cone. Air and HCN 0.5% (4530 ppm or 5000 mg/m^3), balance nitrogen, (Matheson Gas Co., East Rutherford, N.J.) were combined in a mixing chamber to generate final concentrations which were flowed through the system at 20 liters/minute. The gas mixture exited through a CBR-M18 activated charcoal filter which adsorbed the cyanide prior to discharge into the hood. Two to three samples of the gas mixture were collected in gas tight syringes from the sampling port during the two-minute exposure period and subsequently analyzed for cyanide content. Each gas sample (1 mL) was bubbled through a solution (1 mL 0.01 M sodium hydroxide) and analyzed in the Auto Analyzer II (Technicon Auto Analyzer, Tarreytown, N.Y.). Following the exposure period the entire system was flushed with air for 4-5 minutes.

Cardiopulmonary Measurements and Blood Sampling. Aortic arterial blood pressures were recorded continuously on a Gould Recorder by attachment of a pressure transducer to one arm of the double lumen catheter in the carotid artery. Electrocardiograms were obtained from patch electrodes attached to the skin. Ventilatory rate was monitored by pneumograph. Heparinized arterial (and venous) blood samples for the measurement of pO_2, pCO_2 and pH were drawn before exposure and at intervals that would correspond to 1, 2, 3, 8, 13 minutes up to two hours post exposure. A rapid, completely automated microdistillation assay (1) was used to continuously monitor the cyanide concentration in blood. Aortic blood from the animal via the carotid catheter was drawn through the Auto Analyzer shortly before exposure to HCN. The withdrawal rate of 0.3 mL/min was continued throughout the exposure period and for 10-15 minutes post exposure. Blood specimens (1.5 mL) were also collected by syringes for cyanide analysis in the Auto Analyzer. The syringes contained 15 μL of a 15% solution of EDTA and 30 μL of a 0.5% solution of 4-dimethylaminophenol (4-DMAP). Loss of cyanide before analysis was precluded by methemoglobin formation by the 4-DMAP.

Safety Precautions During Animal Exposures. Since toxic, and possibly fatal, effects may occur if HCN is inhaled or absorbed through the skin, certain safety procedures were followed. Although the cyanide tank, the animal and all parts of the exposure system were within the confines of the hood, only two people were present in the room during exposures. Both individuals had audible alarms and wore protective suits, including coverage of head and feet, made of Saranex

23 coated TYVEX^R and long sleeved butyl gloves. They were equipped with a Scott Presur-Pak^R 2.2, 30-minute self-contained breathing apparatus, which allowed breathing under positive pressure. A third individual, similarly suited and protected with a Scott Pak, was stationed outside the room to observe the two people in the room during the exposure period. The observer was prepared to rescue in the event of an emergency.

Results

Clinical Symptoms. Once the exposure period began, no response from the animal was observed for 10-15 seconds. Struggling and vocalization occurred during the next 45-60 seconds. Thereafter, the animal's struggling efforts ceased. Neither excessive salivation nor emesis was observed in the animals. Urination and defecation were not observed during or following the exposure.

Blood Concentrations of Cyanide. Figure 2 shows the blood concentrations of cyanide in μg/mL in arterial blood during and following the inhalation exposure to two concentrations of HCN. Figure 3 shows the blood concentrations of cyanide for individual animals over time. Two animals with peak blood concentrations of about 4 μg/mL died within 10-13 minutes following exposure to high doses of HCN. The blood concentration of survivors dropped rapidly to slightly above 1 μg/mL at 30 minutes post exposure. Also of interest was the finding that the highest blood concentrations were reached 2-3 minutes after, and not during, the two-minute exposure period.

Cardiopulmonary Measurements and Electrocardiograms. Overall, there was no typical clear cut pattern of response in either arterial pressure or heart rate of pigs exposed to cyanide. Segments from a recording of physiological parameters which include blood pressure (BP) and electrocardiogram (ECG) are shown in Figure 4 for one pig prior to and following exposure to HCN. In this animal exposed to 2230 mg/m^3 for two minutes T wave enlargement occurred early in the exposure (< 2 min). Similarly, there was progressive shortening of the S-T segment with obliteration of the interval in 4 minutes along with elevation of the S-T segment. Similar ECG changes to those in Figure 4 were noted in all animals but one. Note that blood pressure was well-maintained three minutes from start of exposure. Four to five minutes after the beginning of exposure a change in breathing rate occurs which appears to separate the animals into two groups as shown in Figure 5.

Blood Gases and pH. Arterial and venous gas tensions (pO_2 and pCO_2), pH and the hematocrit in pigs exposed to HCN by inhalation were also determined. The base excess and standard bicarbonate were calculated. The changes observed in pigs appear to be similar to those reported for other species during cyanide poisoning. Both arterial and venous blood pH decreased from about 7.4 to 7.0. In most of the animals arterial and pre-caval venous blood pO_2 initially increased and in some animals the venous blood pO_2 doubled within minutes. With pCO_2 there was a less consistent typical pattern with time. Base Excess in arterial and

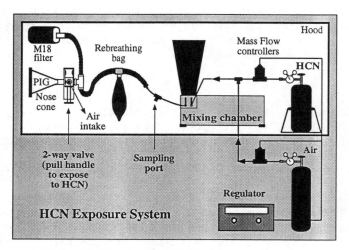

Figure 1. Schematic of HCN Exposure System.

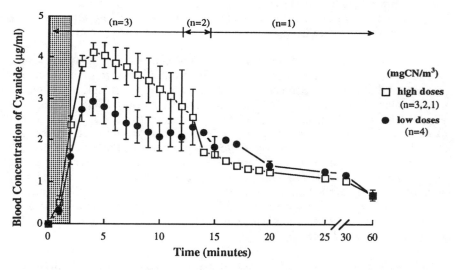

Figure 2. Mean Aortic Blood Cyanide Concentration in Miniature Pigs Exposed to HCN for 2 Minutes. The number of live animals at the higher dose with time is indicated by n. The shaded area indicates the two-minute exposure period, and the error bars are Standard Error of the Mean (±SEM).

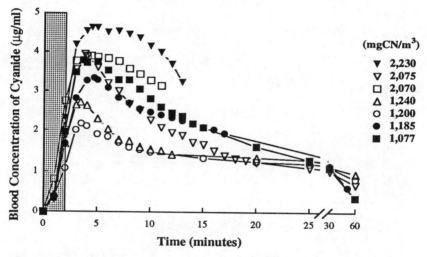

Figure 3. Aortic Blood Cyanide Concentration in Individual Miniature Pigs Exposed to HCN for 2 Minutes.

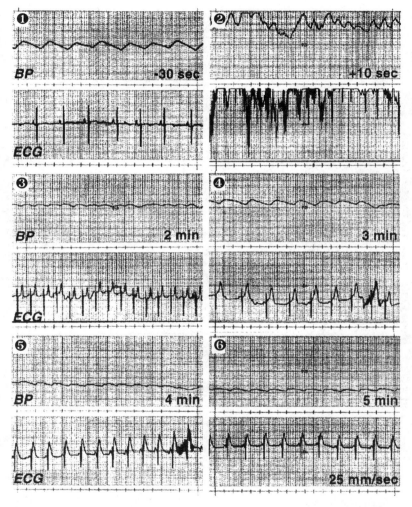

Figure 4. Blood Pressure and ECG in an Animal Exposed to 2230 mg/m^3 HCN for Two Minutes.

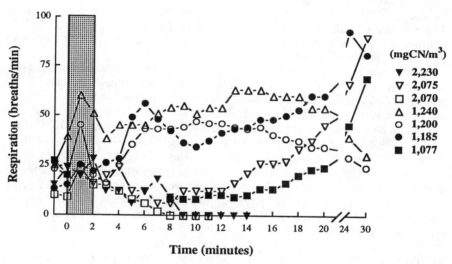

Figure 5. Breathing Rates in Individual Miniature Pigs Exposed to HCN for 2 Minutes.

venous blood decreased rapidly to approximately -30 between 15 and 30 minutes. Arterial and venous $^-HCO_3$ also decreased to approximately 4 between 5 and 30 minutes. Return of Base Excess and $^-HCO_3$ to control values was almost completed by two hours post exposure.

Discussion

Few studies have been performed which attempted to correlate exposure to known concentrations of HCN by inhalation with blood cyanide concentrations. The sensitivity of animal species to gaseous HCN can be rank-ordered by comparison of LCT50 values (2). High values suggest that goat, monkey and pig are among the relatively resistant species when compared to the values for the dog. Based on certain assumptions and data from animal studies McNamara estimated LCT50 values which places man in the category of a resistant species. The estimated LC50 for human was 3404 mg/m^3 (3083 ppm) for a LCT50 3404 mg min/m^3 for one minute and 1464 mg/m^3 (1326 ppm) for a LCT50 4400 mg min/m^3 for three minutes.

Although our study consists of limited numbers of animals the exposure of pigs to a concentration of 2125 mg/m^3 HCN for two minutes resulted in a concentration time of 4250 mg min/m^3, which is interpreted as a high resistance of the miniature pig to HCN comparable to that of other resistant species. In one study exposuring rats to lethal concentrations of atmospheric HCN, time-weighted blood cyanide concentrations above 2 μg/mL were sufficient to kill the animals (4). The present study in a large animal model suggests that higher blood cyanide concentrations are required for lethality in the miniature pig. Lethality in pigs breathing HCN (about 2125 mg/m^3 for two minutes) occurred at a blood cyanide concentration of about 4 μg/mL. An identical concentration of 4 μg/mL cyanide in whole blood of pigs (breed not given) was observed immediately after death following an intraperitoneal injection of 5 x LD50 potassium cyanide (5).

In this study maximal blood concentrations of cyanide in animals were not found during exposure but rather in 2-3 minutes following the end of exposure. This response is presumed to be due to continued uptake of cyanide contained in the lungs and from the dead space in the nose cone. This observation is in accord with life saving measures advocating rapid removal of victims from a toxic environment and initiation of resuscitation in a non-breathing unconscious victim. However, cyanide was eliminated quite rapidly from blood of pigs with a disappearance time (t1/2) of approximately 23 minutes. The disappearance time is identical to that obtained in the pharmacokinetic study in unanesthetized beagle bitches injected intravenously with potassium cyanide (6).

Rapid recovery from loss of consciousness (<10 minutes) in men exposed to HCN in industrial accidents has also been reported (7). Upon admission to the hospital blood levels of cyanide in the patients who had lost consciousness ranged from 2.7-3.38 μg/mL. In two other unconscious men, who regained consciousness rapidly, blood samples drawn 30 minutes after exposure contained cyanide concentrations of 4.67-7.68 μg/mL. These blood concentrations suggest that man is very resistant to HCN. Based on blood cyanide concentrations, the miniature pig may be considered to be a good animal model for studies of cyanide poisoning in man.

In Figure 5, a change occurs two to three minutes after the end of exposure separating the breathing rate into a pattern of either depressed or increased rate. Not only was the breathing rate lowered in the animals exposed to the three highest concentrations of HCN, but also in one animal exposed to the lowest concentration. A comparison, of individual blood concentrations of cyanide in Figures 3 and breathing rates of animals in Figure 5, suggests that the blood concentrations of cyanide were greater in animals with depressed breathing rates. Likewise, elevated breathing rates were observed when lower concentrations of cyanide (<3.5 μg/mL) were found in the blood. A blood concentration of about 4 μg/mL of cyanide appears to be an approximate separation point at which breathing rate is maintained or depressed leading ultimately to survival or death. The study suggests, at least to some degree, a correlation between inhalation concentrations, time and cyanide blood levels.

Electrocardiographic (ECG) tracings of these pigs exposed to HCN exhibited changes similar to those reported for humans executed by inhalation of hydrocyanic acid at much higher concentrations (8). The rapid changes in ECG upon exposure of an animal to HCN reflect the progressive loss of oxidative metabolism in heart muscle despite the development of high oxygen tensions in the arterial and venous bloods. These tensions which are too high to be a likely result of simple hyperventilation, with or without elevated cardiac output, suggest that oxygen consumption by tissue cells was reduced drastically. This would be the expected result of poisoning of aerobic metabolism in the tissues by cyanide binding to cytochrome oxidase. The lack of oxygen consumption by the tissues allows the venous blood to return to the lungs with an abnormally high pO_2. Here it is again exposed to ambient pO_2 of the inhaled air (approximately 150 torr) and recirculated through the tissues with little or no oxygen extraction. The sudden elimination of aerobic metabolism caused by cyanide allows the arterial and venous oxygen tensions to approach equilibrium with ambient air, if ventilation and cardiac output are maintained. By a similar mechanism the arterial and venous carbon dioxide tensions fall, since carbon dioxide is no longer produced by aerobic metabolism. However, the simple effect of diminished aerobic CO^2 production is complicated by the release of carbon dioxide from the body stores secondary to the lactic acid acidosis produced by enforced anaerobic metabolism. Variations in ventilation and cardiac output produce additional differences between animals in the recorded carbon dioxide tensions. The pronounced decreases in pH and base excess found after cyanide exposures reflect lactic acidosis of anaerobic metabolism produced by cytochrome oxidase inactivation by cyanide.

Disclaimer. The opinions or assertions contained herein are the private views of the authors and are not to be construed as official or as reflecting the views of the Army, the Department of Defense, or the US Government.

In conducting the research described in this report, the investigators adhered to the **Guide for the Care and Use of Laboratory Animals** of the Institute of

Laboratory Animal Resources, National Research Council (NIH Publ. No. 86-23, Revised 1985).

Literature Cited

1. W.A. Groff, Sr., F.W. Stemler, A. Kaminskis, H.L. Froehlich, and R.P. Johnson, "Plasma Free Cyanide and Blood Total Cyanide: A Rapid Completely Automated Microdistillation Assay," Clin. Toxicol. **1985**, 23, 133-163.
2. B.P. McNamara, "Estimate of the Toxicity of Hydrocyanic Acid Vapors in Man," Edgewood Arsenal Technical Report, EB-TR 76023, August, **1976**.
3. L.M. Panepinto, "Laboratory Methodology and Management of Swine in Biomedical Research," Swine in Biomedical Research (M.E. Tumbleson, Ed.) Plenum, NY. 1986; Vol. I, 97-109.
4. B.C. Levin, J.L. Gurman, M. Paabo, L. Baier, L. Procell and H.H. Newball, "Acute Inhalation Toxicity of Hydrogen Cyanide," The Toxicologist. **1986**, 6, 59.
5. B. Ballantyne, "Toxicology of Cyanides," Ballantyne, B. and Marrs, T.C. Eds; Clinical and Experimental Toxicology of Cyanides. Bristol;John Wright, 1987; pp 41-126.
6. J.E. Bright and T.C. Marrs, "Pharmacokinetics of Intravenous Potassium Cyanide," Human Toxicol. **1988**, 7, 183-186.
7. N.R. Peden, A. Taha, and P.D. McSorley, "Industrial Exposure to Hydrogen Cyanide: Implications for Treatment," Br. Med. J. (UK) **1986**. 293, 538.
8. J. Wexler, J.L. Whittenberger, and P.R. Dumke, "The Effect of Cyanide on the Electrocardiogram of Man," Amer. Heart J. **1947**. 34, 163-173.

RECEIVED November 21, 1994

Chapter 22

Environmental Nitrogen Dioxide Exposure Hazards of Concern to the U.S. Army

M. A. Mayorga[1], A. J. Januszkiewicz[1], and B. E. Lehnert[2]

[1]Walter Reed Army Institute of Research, Department of Respiratory Research, Washington, DC 20307–5100
[2]Pulmonary Biology–Toxicology Program, Los Alamos National Laboratory, Los Alamos, NM 87545

The U.S. Army is concerned with the health effects of acute, short-duration, high-level nitrogen dioxide (NO_2) exposure. A substantial amount of information is known about chronic, low-level NO_2, however, much is unknown about the effects and the mechanisms of injury induced by acute, high-level NO_2 exposure. The combination of experiments involving rodent and ovine models describe the concentration-dose responses, histological effects, cardiopulmonary changes and cellular effects after NO_2 exposure at rest and with exercise. The effects of high-dose NO_2 exposure and the effects of exercise in a rodent model following NO_2 exposure are described for the first time in these experiments.

High levels of nitrogen dioxide (NO_2) is of concern to the military in both training and combat scenarios. High levels of NO_2 may be encountered during the combustion of materials within aircraft, ships, submarines and armored vehicles. Peak levels of NO_2 above 2000 ppm have been measured in armored vehicles penetrated by high temperature shaped rounds, a type of ammunition. Levels of NO_2 within enclosures generated by the firing of one's own weapons may also pose a health risk, particularly with today's relatively smaller, portable and powerful weapons. Though much is known about the health effects of low-level NO_2 exposure, less is known of the effects of high-level NO_2 exposure. Furthermore, though one could surmise increased NO_2 toxicity and decreased exercise performance after NO_2 exposure in an exercising animal or human, the effects of exercise on NO_2-induced pulmonary toxicity and the effect of NO_2 exposure on exercise is also unknown. The United States Army is one of the world's leaders in the testing of procured

This chapter not subject to U.S. copyright
Published 1995 American Chemical Society

and developed weapon systems with regard to system and human performance, incapacitation, vulnerability and survivability. Testing is conducted as part of the Army's Live Fire Testing Program (LFTP). Additionally, the U.S. Army Health Hazard Assessment Program (HHAP) is responsible for establishing methodology for the identification and elimination of health hazards in weapon systems and providing analysis of potential health risks associated with a given weapon system. The HHAP is concerned with lower level-NO_2 exposure incurred in training, whereas, the LFTP is concerned with higher-level NO_2 exposure encountered in combat. Many other countries have adopted the U.S. Army's human incapacitation and the health hazard risk criteria for injury incurred through the use of various weapon systems. These criteria are based on modification of published civilian standards and ongoing toxic gas research sponsored by the U. S. Army Medical Research and Development Command. The Los Alamos National Laboratory and the Department of Respiratory Research, Walter Reed Army Institute of Research conducted a series of experiments in rats and sheep to characterize a dose-response curve, to identify threshold levels and determinants of toxicity, cellular responses, and the effects of exercise after NO_2 exposure. This paper provides a summary of these experiments which are described in further detail elsewhere ([1,2]). The impact of the results of the Army's basic research and its testing of real weapon systems, of basic research, and the establishment of toxicity criteria has led to interior structural modification of vehicles, the installation of ventilation and automatic fire extinguishing systems, and the establishment of individual soldier face masking guidelines.

EXPERIMENTAL

Materials

Adult, male, Fischer-344, specific-pathogen free rats were obtained from Sasco, Omaha Nebraska, and Harlan Sprague Dawley, specific-pathogen-free rats were obtained from Indianapolis, Inc. Rats were acclimated, housed two per cage, and provided with standard diet and water _ad libitum._ Disease-free crossbred ewes (35-50 kg) were obtained from Ovine Technologies, Inc. New Hope, PA. Sheep were housed in indoor cages, two per cage, and provided standard feed and water _ad libitum_. The sheep were fasted for 24 hr prior to experiments.

NO_2 Exposure in Rats. Groups of 8-12 animals weighing an average of 253 g were exposed to 25, 50, 75, 100, 150, 200, or 250 ppm of NO_2 for 2, 5, 15, or 30 min. In separate experiments, animals were exposed to 500-2000 ppm NO_2 for 1 min. Another group of rats, serving as controls, were exposed to air for 30 min. Nitrogen dioxide was generated from dinitrogen tetroxide (N_2O_4) and mixed with anhydrous air in a mixing chamber. The exiting NO_2 was then regulated for desired concentration and directed into a 12 port inhalation exposure chamber as shown in Figure 1. Exposure NO_2 concentrations were monitored with a dual channel IR-UV spectrophotometer (Binos Inficon, Leybold-Heraeus, Germany). The concentration of nitric oxide (NO) in the exposure mixture was less than 30 ppm. Experimental and control animals were sacrificed with sodium pentobarbital IP at 4, 8, 24 and 48 hr and the lungs excised. The lungs were weighed, including an excised right cranial lobe (RCL). The RCL was separately weighed before and after drying to provide a wet and dry RCL (RCLWW and RCLDW). The remaining lung was fixed with 10% formalin in phosphate buffered saline. The fixed lung was sectioned, stained with hematoxylin and eosin and scored for the intensity and distribution of fibrin, polymorphonuclear leukocytes (PMN), macrophages and red blood cells in alveolar spaces, and type II pneumocyte hyperplasia (Table I). Data were statistically analyzed using one-way analysis of variance and Dunnett's t-test.

NO_2 Exposure in Exercising Rats. Male Fischer-344 rats, weighing 250-300 g were trained for 19 days on a treadmill at a 15% grade with daily incremental (5 M/min) increasing velocities every 30 sec from 10 M/min to 60 M/min until the rats were able to perform a 30 min exercise (3). The duration of exercise used in all exercise experiments was 30 min. The rats were exposed at rest to 100 ppm NO_2 or air for 15 min. Exercise was performed at 1 hr pre-, immediately, 8 or 24 hr post-exposure. Another control group of rats were rested post- exposure. Animals were sacrificed at either 1, 8, or 24 hr. Groups of animals were exercised twice, immediately and at 8 hr or at 24 hr post-exposure followed by necropsy. Control animals were exposed to air for 15 min and were either rested or exercised and sacrificed at 1, 8, or 24 hr post-exposure (Table II). Lung weights were determined and histological examinations were performed as described above.

Lung Lavage Analysis of NO_2-Exposed Rats. Adult male Fischer-344 rats (250-300 g) were exposed to 100 ppm NO_2 X 15 min as described above. Lavage time points were 8, 24, 48, 72 and 96 hr post-exposure. Six sequential lung lavages with 8 ml aliquots of phosphate buffered saline (PBS) were performed on excised lungs. Recovered lavage fluid was analyzed for total cell count, cell differential, total protein by the Lowry method, and protein fractions by ion exchange high performance liquid chromatography (4).

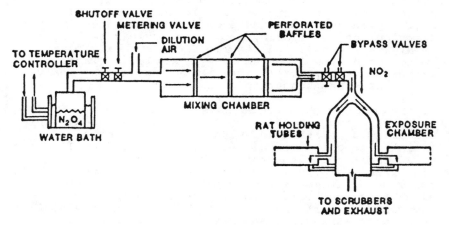

Figure 1. Diagram of rat exposure system used to administer air or nitrogen dioxide.

TABLE I
RAT HISTOPATHOLOGICAL SCORING SCHEME FOR NO_2 INDUCED LUNG INJURY

LESIONS	INTENSITY	DISTRIBUTION
Intra-alveolar fibrin	0 = not detected	0 = not present
Intra-alveolar PMNs	1 = trace to mild appearance (present in few proximal alveoli)	1 = focal (present in an occasional alveolar duct and proximal alveolar structures)
Intra-alveolar AMs	2 = moderate (present in several proximal alveoli)	2 = multifocal (present in several alveolar ducts and proximal alveolar structures)
Intra-alveolar RBCs	3 = high degree (present in many proximal and some more distal alveoli)	3 = diffuse (present in virtually all alveolar ducts and proximal alveolar structures)
Type II pneumocyte hyperplasia	4 = high intensity (present in essentially all proximal and more distal alveoli)	

PMNs: Polymorphonuclear Leukocytes
AMs: Alveolar Macrophages
RBCs: Red Blood Cells

TABLE II
RAT EXPERIMENTAL DESIGN FOR COMBINATION NO_2 EXPOSURE AND EXERCISE

Group	NO_2 Conc (ppm)	Duration of Exposure (min)	Rest (min)	Exercise in Relation to Exp (hr)	Necropsy Post Exp (hr)
1a	0 (control)	15	15		1
b	0 (control)	15	15		8
c	0 (control)	15	15		24
$2a_1$	100	15		1hr pre	8
a_2	100	15		1hr pre	24
a_3	100	15		1hr pre	1.5
b_1	100	15		immed	8
b_2	100	15		immed	24
b_3	100	15		immed	8.5
c_1	100	15		8	24
c_2	100	15		8	24.5
d	100	15		24	
3a1	100	15		immed & 8	8.5
a2	100	15		immed & 8	24
b	100	15		immed & 24	24

Nose versus Lung NO_2 Exposure in Sheep: The sheep were prepared with a chronic carotid loop one month prior to studies(2). On the day of exposure, the animals were instrumented with venous and arterial catheters, a 7-French thermodilution catheter, a nasotracheal tube (Bivona, Inc., Gary, IN), an airway pressure transducer (4-French, model PR-219, Millar Micro-Tip, Millar Instruments, Inc., Houston, TX), and an esophageal transducer (8-French, model PR-346, Millar Micro-Tip, Millar Instruments, Inc, Houston, TX). A pneumotachometer (Series 3700, Hans-Rudolf Inc., Kansas City, KS) was placed at the proximal end of the nasotracheal tube. Cardiopulmonary measurements included respiratory rate, tracheal lateral pressure, pleural pressure, tidal airway flow, core body temperature, arterial blood pressure, right-sided cardiac chamber pressures, pulmonary artery wedge pressure and cardiac output. Other measurements obtained were arterial and mixed venous blood samples, hemoglobin, methemoglobin and hematocrit. Cardiopulmonary variables derived from the direct measurements described above included tidal volume, transpulmonary pressure, lung resistance, dynamic lung compliance, systemic and pulmonary vascular resistance, and oxygen consumption (5). Sheep (n=6) were exposed to either 500 ppm NO_2 or air for 15 min through a system composed of glass, Teflon and Teflon-coated components through a nasotracheal tube or a plastic modified canine anesthesia mask. Nitrogen dioxide and nitrogen monoxide concentrations were monitored with a dual beam IR-UV spectrophotometer (Binos Inficon, Leybold-Heraeus, Germany) and a manual detector (model 8014-400A, Matheson Gas Products). Nitrogen monoxide levels were less than 1 ppm. Cardiopulmonary measurements were performed at immediately pre- and post-exposure and at 4 and 24 hr post-exposure. Necropsies were performed at 24 hr and wet to dry lung weight determined, and gross examination performed. Lungs were fixed and prepared for histological examination. Statistical analysis was performed using one-way and two-way analysis of variance for effects of gas concentration and effects over time, respectively.

Bronchoalveolar Lavage (BAL) of NO_2 Exposed Sheep.
Experimental and control sheep were prepared and exposed as described above except that exposure time was increased to 20 min and experimental time points for cardiopulmonary measurements (0, immediately post-exposure, 0.5, 6 and 24 hr post-exposure) and BAL (0.5, 6, and 24 hr) were expanded. Additionally, a baseline BAL was performed 2 wk prior to exposure. Bronchoalveolar lavage was performed in different segments of the right caudal lobe with 90-cm modified fiber-optic bronchoscope (Pentax, model FB-PP10, Precision Instrument Corporation, Orangeburg, N.Y.).

Two-hundred ml, in aliquots of, 20-30 ml of sterile normal saline were instilled and aspirated with a 30-ml syringe. The BAL fluid (BALF) volume was recorded, pooled, cooled to 4 °C, and divided into aliquots for further testing. Bronchoalveolar lavage fluid was analyzed for total cell count, cell differential, total protein and albumin (4). Necropsies were performed at 24 hr, wet/dry lung weight ratios determined, gross pathological examination performed, and tissue prepared for histological examination.

RESULTS

NO_2 Exposure in Rats. Injury, as assessed by percent change of lung wet weights (LWW), demonstrated no significant injury after exposure to 25, 50, or 75 ppm NO_2 X 2, 5, or 15 min, however, the histological changes of trace-mild, focal alveolar fibrin and moderate Type II pneumocyte hyperplasia were noted at these concentrations after ≥5 min duration. Figures 2, 3a and 3b demonstrate a general increase in injury, as measured by LWW, generalized fibrin index and generalized Type II pneumocyte hyperplasia index at NO_2 concentrations above 75 ppm and with increasing duration of exposure. NO_2 concentration appeared to have a greater effect on toxicity than duration of exposure. Lung wet weight increased with post-exposure time and plateaued at 24 hr except for the 250 ppm NO_2 exposure group, as shown in Figure 4. Percent change in RCLDW correlated with percent change in LWW. Figures 2 and 5 showed that short duration-very high concentration exposures induced more lung injury than long duration-low concentration exposures.

Lung Lavage Fluid (LLF) Analysis of NO_2-Exposed Rats. Lung lavage fluid total protein increased after exposure to 15 ppm NO_2 X 15 min at 8 hr, peaked at 24 hr and returned nearly to baseline by 96 hr, as demonstrated in Figure 6. Lung lavage fluid albumin and transferrin demonstrated a similar pattern of elevation and return toward baseline as the total protein. Figure 7 showed that significant PMN alveolar accumulation did not occur until 24 hr and was maximal at 72 hr, whereas, the maximal increase in alveolar macrophages did not occur until 48 hr post-exposure.

NO_2 Exposure in Exercising Rats. Lung wet weight, extent and severity of alveolar fibrin and RBC extravasation increased after exposure to 100 ppm NO_2 X 15 min followed by exercise immediately post-exposure or at 8 hr. (Figure 8, Table III). These changes were assessed at a 30 min post-exercise necropsy. However, when rats were exercised 8 hr following exposure and necropsied 24 hr post-exposure, LWW and histological changes, with the exception of fibrin

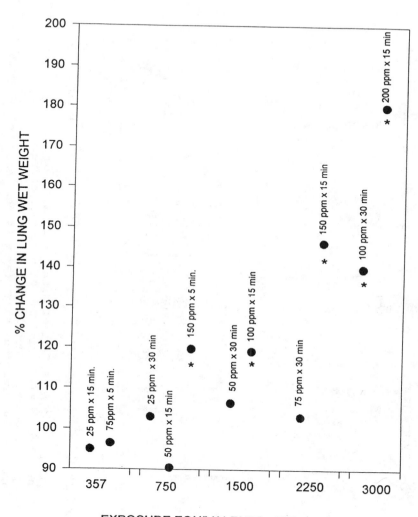

Figure 2. Effects of a 5, 15 or 30-minute nitrogen dioxide exposure on rat lung gravimetrics. Asterisks denote statistical significance ($P < 0.05$) from control (air-exposed). Redrawn and reprinted with permission of Elsevier Science Ireland, Ltd.

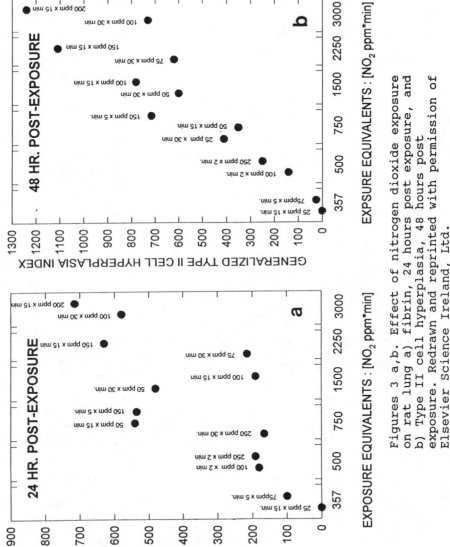

Figures 3 a,b. Effect of nitrogen dioxide exposure on rat lung a) fibrin, 24 hours post exposure, and b) Type II cell hyperplasia, 48 hours post exposure. Redrawn and reprinted with permission of Elsevier Science Ireland, Ltd.

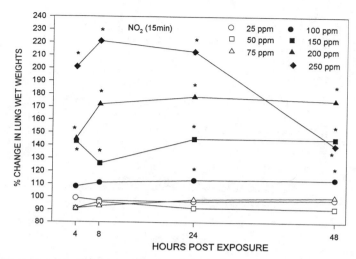

Figure 4. Time course of effect of a 15-minute exposure to nitrogen dioxide on rat lung gravimetrics. Values represent the means ± SE for 8-12 animals per group. Asterisks denote statistical significance (P < 0.05) from control (air-exposed).

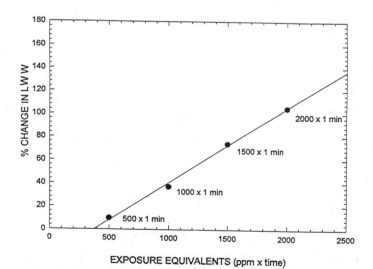

Figure 5. Effect of a 1-minute nitrogen dioxide exposure on rat lung wet weight (LWW). Each point represents the mean of 5-12 animals. Redrawn and reprinted with permission of Elsevier Science Ireland, Ltd.

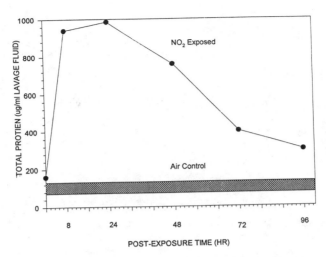

Figure 6. Effect of nitrogen dioxide exposure on rat lung lavage protein concentration over a 4-day post-exposure sampling period. Shaded area shows normal range for control (air-exposed) rats. Each point represents the mean of 4 animals. Redrawn and reprinted with permission of Elsevier Science Ireland, Ltd.

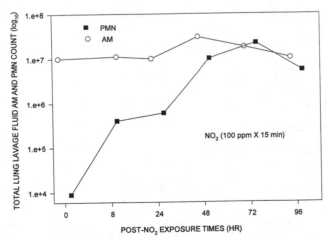

Figure 7. Effect of a 15-minute, 100 ppm nitrogen dioxide exposure on rat lung lavage alveolar macrophage (AM) and polymorphonuclear leukocyte (PMN) count over a 4-day sampling period. Each point represents the mean of 4 animals. Redrawn and reprinted with permission of Elsevier Science Ireland, Ltd.

TABLE III
OCCURRENCE OF FIBRIN AND EXTRAVASATION OF ERYTHROCYTES IN RAT LUNG POST NO_2 EXPOSURE AND EXERCISE VERSUS REST

Postexposure condition(s)	Postexposure sacrifice time (hr)	Fibrin Distribution	Fibrin Severity	Fibrin Intensity	Extravasated Erythrocytes Distribution	Extravasated Erythrocytes Severity	Extravasated Erythrocytes Intensity
Rest	1	2.5 ± 0.4	1.0 ± 0.0	1.0 ± 0.0	0	0	0
Rest	8	4.0 ± 0.0	1.2 ± 0.2	1.8 ± 0.3	2.7 ± 0.2	1.0 ± 0.0	1.3 ± 0.2
Rest	24	4.0 ± 0.0	1.0 ± 0.0	1.8 ± 0.4	3.0 ± 0.0	1.0 ± 0.7	2.0 ± 0.3
Immediate Exercise	1	4.0 ± 0.0	1.8 ± 0.3	1.5 ± 0.3	1.0 ± 0.6	0.5 ± 0.3	0.5 ± 0.3
Immediate Exercise	8	4.0 ± 0.0	1.8 ± 0.3	2.5 ± 0.3	3.0 ± 0.0	1.0 ± 0.0	2.3 ± 0.3
Immediate Exercise	24	4.0 ± 0.0	2.4 ± 0.3	3.6 ± 0.3	2.8 ± 0.2	1.8 ± 0.2	3.2 ± 0.4
Rest, exercise at 8 hr	8.5	3.8 ± 0.3	1.5 ± 0.3	2.3 ± 0.3	2.8 ± 0.3	1.0 ± 0.0	2.0 ± 0.4
Rest, exercise at 8 hr	24	4.0 ± 0.0	1.6 ± 0.3	2.4 ± 0.4	1.4 ± 0.6	0.6 ± 0.2	1.4 ± 0.6
Rest, exercise at 24 hr	24.5	4.0 ± 0.0	1.0 ± 0.0	2.3 ± 0.2	2.8 ± 0.2	1.0 ± 0.0	2.0 ± 0.3
Immediate and 8 hr exercise	8.5	4.0 ± 0.0	2.8 ± 0.3	3.2 ± 0.2	3.3 ± 0.2	1.5 ± 0.2	2.8 ± 0.2
Immediate and 8 hr exercise	24	4.0 ± 0.0	3.0 ± 0.0	3.0 ± 0.0	3.5 ± 0.3	3.0 ± 0.0	3.3 ± 0.3
Immediate and 24 hr exercise	24.5	4.0 ± 0.0	3.2 ± 0.2	3.0 ± 0.0	3.4 ± 0.3	2.8 ± 0.2	2.8 ± 0.4

Values represent the means ± SE for 4-6 animals studied per group severity, were not significantly different from NO_2-exposed, non-exercised rats necropsied at 24 hr. Lung wet weights were significantly greater in twice-exercised rats over rats exercised once and necropsied. However, the difference appeared to resolve after a 24 hr rest post-exposure prior to necropsy. Figure 9 demonstrates that exercise performance was decreased, as assessed by maximum oxygen consumption, after rats were exposed to 25, 50, and 100 ppm NO_2 X 15 min. One minute exposure to levels greater than 100 ppm NO_2 also significantly reduced exercise performance.

Nose-Only versus Lung-Only NO_2 Exposure in Sheep. The mean inspired NO_2 dose (500 ppm X 15 min) in lung-only exposed sheep was 57% greater than that of nose-only exposed sheep. Significant changes between the two groups noted included mean inspired ventilation and respiratory rate, as shown in Figures 10a and 10b. Furthermore, Figures 11a and 11b show that lung resistance elevation and dynamic compliance reduction generally occurred sooner and was more sustained in the lung-only exposed group as compared to the nose-only exposed. Histological changes of generalized patchy pneumonia and mild PMN infiltration were observed in only the lung-exposed sheep. Mild interalveolar capillary infiltration was noted in both groups.

Bronchoalveolar Lavage Fluid (BALF) Analysis of NO_2 Exposed Sheep. Bronchoalveolar fluid protein and albumin were significantly elevated in the NO_2-exposed sheep versus the air-exposed, Figures 12a and 12b. No significant changes were noted in total cell count or the percentage of PMN at the sampling times. However, a significant decrease in percentage of alveolar macrophages and increase in epithelial cells were noted at all sampling times, but, most marked at 6 and 24 hr post-exposure. These changes are reflected in Figures 13a and 13b.

CONCLUSIONS

The pulmonary effects of "high-level" (>25-30 and >50 ppm for small and large animals, respectively) NO_2 have been described as a result of U.S. Army-sponsored small and large animal experiments. In the small animal model, a complete concentration-response curve and the effect of exercise on injury have been described. Even at the lower levels of NO_2 tested in these studies (25, 50, and 75 ppm X 5 min), histological changes of NO_2 toxicity were noted. Injury as assessed by LWW and histological changes increased with increasing NO_2 concentration and exposure time. Nitrogen dioxide concentration appeared to be a more important determinant than duration of exposure. Exercise potentiated lung

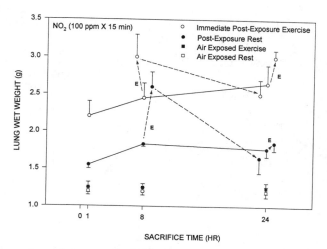

Figure 8. Effect of post-exposure rest or exercise on lung gravimetrics from air (squares)- or nitrogen dioxide (circles)- exposed rats. Values represent the means ± SE for 4-6 animals per group. Note exercise-induced lung damage enhancement in nitrogen dioxide-exposed rats.

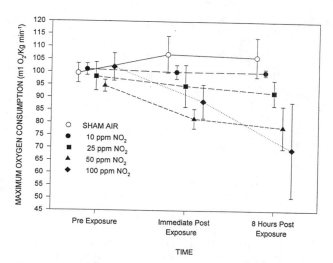

Figure 9. Effect of 15-minute nitrogen dioxide exposure on rat maximal exercise performance, indexed by maximum oxygen consumption. Values represent the means ± SE for 4-6 animals per group. Note concentration- and time-dependent performance decrement after exposure.

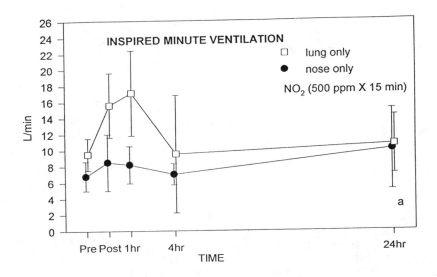

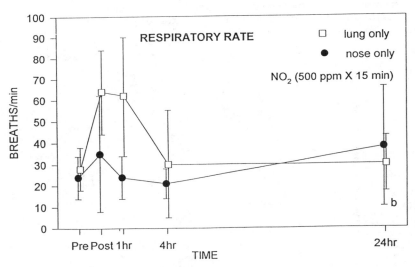

Figures 10 a,b. Effect of a 15-minute 500 ppm nitrogen dioxide exposure on a) inspired minute ventilation and b) respiratory rate from lung-only and nose-only exposed sheep over a 24-hour post-exposure sampling period. Values represent the means ± SD for 4-6 animals per group.

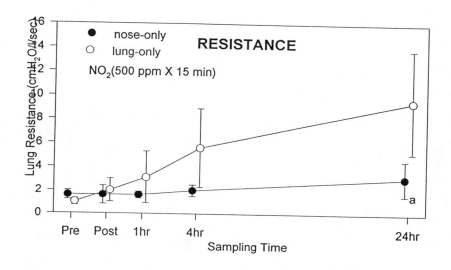

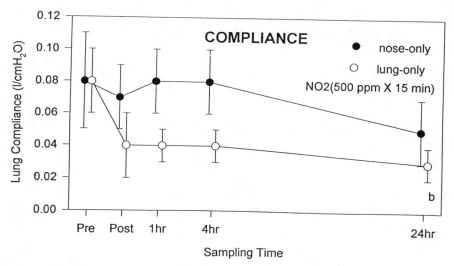

Figures 11 a,b. Effect of a 15-minute 500 ppm nitrogen dioxide exposure on a) lung resistance and b) dynamic lung compliance, from lung-only and nose-only exposed sheep over a 24-hour post-exposure sampling period. Values represent the means ± SD for 4-6 animals per group. Redrawn and reprinted with permission of Elsevier Science Ireland, Ltd.

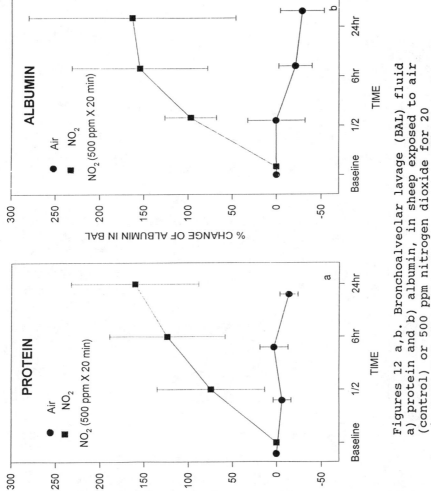

Figures 12 a,b. Bronchoalveolar lavage (BAL) fluid a) protein and b) albumin, in sheep exposed to air (control) or 500 ppm nitrogen dioxide for 20 minutes. Values represent the means ± SD for 4-6 animals per group.

22. MAYORGA ET AL. *Environmental Nitrogen Dioxide Exposure Hazards* 341

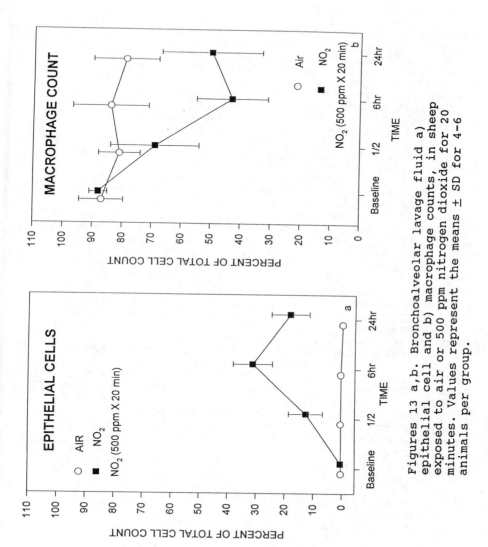

Figures 13 a,b. Bronchoalveolar lavage fluid a) epithelial cell and b) macrophage counts, in sheep exposed to air or 500 ppm nitrogen dioxide for 20 minutes. Values represent the means ± SD for 4-6 animals per group.

injury. There is a period between the exposure and 24 hr post-exposure during which exercise potentiated injury, however, this period of susceptibility disappeared after 24 hr. Threshold NO_2 exposure experiments in sheep revealed increases in airway resistance and decreases in lung compliance which was more pronounced in lung-exposed versus nose-exposed sheep. These changes were maximal at 24 hr. In both animal models, LLF and BALF analysis corroborate a pulmonary capillary endothelial leak mechanism of injury as manifested by increased total protein and albumin. Histological changes in both models include an exudative pneumonia with PMN infiltration and RBC extravasation and Type II pneumocyte hyperplasia.

The significance of this research to the Army LFTP and HHAP is evident. If extrapolations from small and large animal experiments to humans are made, several conclusions can be generated. Firstly, military personnel may experience more NO_2 toxicity as the concentration and duration of exposure increases. Secondly, military personnel exposed to high-level, short-duration NO_2 exposure in combat may be at greater risk to NO_2 lung-induced injury than personnel exposed to low-level, longer duration NO_2 exposure, which could be encountered in select training scenarios (firing from bunkers, enclosures, partially covered foxholes). Thirdly, exercise can potentiate NO_2 toxicity. The likelihood of the need of demanding physical performance is very high in both military training and combat. Fourthly, NO_2 lung toxicity can cause decrements in exercise performance which could have grave implications for the mission of an individual soldier and that of his higher command units. Lastly, mouth breathing (i.e., during exercise) of toxins which bypass the protective effect of the nasal passages may be associated with increased lung toxicity. Irrespective of the scenario, combat or training, the toxicity of NO_2 must be recognized, weapon systems tested for unacceptable emissions of NO_2, armored vehicles, aircraft and naval vessels constructed to reduce the risk of fires, and safety measures such as ventilation systems and training of military personnel in the correct use of respirators must be instituted. As new research provides insights into the mechanisms of NO_2 lung injury, prophylactic and therapeutic interventions can be initiated to further reduce the risk of NO_2 toxicity.

ACKNOWLEDGEMENTS: The authors wish to thank all the investigators and technicians from the Department of Respiratory Research, Walter Reed Army Institute of Research and from the Pulmonary Biology-Toxicology Program, Los Alamos National Laboratory who contributed to this research over the years. Special thanks is given to Daniel S. Oh for graphics support.

The views, opinions and/or findings contained in this report are those of the authors and should not be construed as an official Department of the Army position, policy or decision unless so designated by other documentation.

In conducting research using animals, the investigators adhered to the "Guide for the Care and Use of Laboratory Animals", prepared by the Committee and Use of Laboratory Animals of the Institute of Laboratory Animal Resources, National Research Council (NIH) Publication NO. 87-23, Revised 1985).

LITERATURE CITED

1. Stavert, D.M., and Lehnert, B.E. Potentiation of the expression of nitrogen dioxide-induced lung injury by post exposure exercise. Environ Research, 1987; 42,1-13.
2. Januszkiewicz, A.J., Mayorga, M.A. Nitrogen dioxide-induced acute lung injury in sheep. Toxicol. 1994; 89, 279-300.
3. Stavert, D.M., Lehnert, B.E., Wilson, J.S. Exercise potentiates nitrogen dioxide toxicity. Toxicologist, 1987; 7, A46.
4. Gurley, L.R., Valdez, J.G., London, J.E., Dethloff, L.A., Lehnert, B.E. , An HPLC procedure for the lavage of proteins in lung lavage fluid. Anal. Biochem. 1988; 172,465-478.
5. Januszkiewicz, A.J., Snapper, J.R., Sturgis, J.W., Rayburn, D.B., Dodd, K.T., Phillips, Y.Y., Ripple, G.R., Sharpnack, D.D., Coulson, N.M. and Bley, J.A. Pathophysiologic response of sheep to brief high-level nitrogen dioxide exposure. Inhal. Toxicol. 1992; 4,359-372.

RECEIVED April 20, 1995

Chapter 23

Application of the Naval Medical Research Institute Toxicology Detachment Neurobehavioral Screening Battery to Combustion Toxicology

G. D. Ritchie[1], J. Rossi III[2], and D. A. Macys[2]

[1]Geo Centers, Inc. and [2]Naval Medical Research Institute Detachment (Toxicology), 2612 Fifth Street, Building 433, Wright-Patterson Air Force Base, OH 45433-7903

A comprehensive screening battery of neurobehavioral tests, applicable to combustion toxicology research, is being developed at the Naval Medical Research Institute Toxicology Detachment (NMRI/TD) of the Tri-Service Toxicology Consortium at Wright-Patterson AFB, OH. This screening battery, evaluating small animal responses, will be used to predict human neurobehavioral effects of non-lethal toxic exposures in a diversity of real world operational scenarios. While the scientific literature contains over 100,000 studies addressing effects of toxic substances on lethality or single behavioral endpoints, few have used an integrated test battery to predict toxicity effects in real world scenarios. Many jobs, occupations and human functions require attention and judgment, performance of precise actions in a specified sequence, fine motor control and integrated motivation. Regardless of the scenario, the critical factor in risk assessment is the capability for prediction of the individual's ability to make decisions and execute precise behaviors that will enable mission completion. The NMRI/TD test battery is designed to evaluate the impact of individual toxicants or the interaction of combined toxicants within a complex mixture on a specific operational performance. Current applications of the battery model human behavioral and performance deficits associated with acute exposure to fire gases, fire extinguishants and extinguishant byproducts. Validation of the battery involves comparisons of documented human deficits associated with exposure to acceptable levels of known toxicants or pharmaceutical drugs with animal responses to comparable levels of the same compounds.

Screening of potential toxicants encountered in real-world operational scenarios is important in both military and commercial sectors. Acute or repeated brief exposure to varying concentrations of individual toxicants or complex mixtures, in changing physical environments, provides a complex challenge to traditional risk assessment methodology. For military applications, knowledge of the behavioral and performance endpoints impacted by a toxic exposure may be as important as understanding the biological causes and physiological mechanisms underlying the response. Ability to predict performance deficit in the operation of aircraft, radar, sonar and weaponry control systems during combat or damage control conditions, for instance, is often critical to mission completion and personnel survival. The capacity to predict potential toxicity from complex mixtures, as inevitably exist in combat theatres, may be of significant value in protecting the short- and long-term health of military veterans. A neurobehavioral screening battery, using non-primate animal tests, may more reliably, rapidly and cost-effectively predict CNS and neuromuscular toxicity than traditional histopathological or descriptive chemical techniques. Behavioral toxicity screening can provide more immediate indication that alternatives should be considered including eliminating hazardous uses of a chemical, selection of the least toxic of alternative compounds, testing of methods for safe exposure to toxicants that cannot be eliminated or replaced, and avoidance of situations in which a toxicant can generate a significant hazard.

The scientific literature contains over 100,000 studies reporting the effects of specific toxicants on lethality (LD_{50}) or individual behavioral endpoints. Research programs that provide the methodology to identify the impact of individual toxic components in complex mixtures on specific performance criteria are few. Tests used to evaluate the impact of toxic substances on behavior can be divided into at least two types; in the first, chemical analyses of contaminants are conducted; deficits are predicted based on the known toxic effects. Exposure to complex mixtures, however, produces deficits that are often greater or lesser than the deficits expected from exposures to the individual components of the mixture. Furthermore, these deficits may be exacerbated or mitigated by environmental factors. The non-lethal behavioral toxicity of complex mixtures in general, and of combustion atmospheres in particular, cannot yet be adequately predicted simply from identification of their chemical composition.

The second type of test, addressed in this paper, is the quantitation of neurobehavioral responses to the toxic exposure. In these paradigms, subjects are exposed to atmospheric contaminants. Their toxicity is determined by assessing alterations in one or more measures to define and quantitate a specific behavior. These tests model progressive performance deficits through incapacitation and subsequent recovery, and can often be correlated with physiological changes. Examples of the single performance endpoints that have been used to assess incapacitation include: loss of equilibrium or righting, decreases in general activity, leg flexion shock avoidance, signaled or unsignaled avoidance versus escape, positively reinforced operant behavior and various endurance tests including rotorod and forced treadmill *(1-10)*. Neurobehavioral toxicity evaluation, through detailed analysis of single endpoints, is very useful assuming the toxicant is adequately understood or its application sufficiently defined to allow appropriate selection of the behavioral testing paradigm. The initial testing of a novel toxicant, however, requires use of a comprehensive screening battery of behavioral and performance tests to ensure that critical areas of deficiency are not ignored by the preference for specific individual tests. It is necessary to include tests that measure: (a) neurological activity; (b) motor system integrity and endurance; (c) physiological irritation; (d) sensory acuity and incapacitation; (e) social behavior and emotionality; (f) motivation level and frustration; (g) spatial and temporal discrimination; and (h) higher cognitive capacity. As more "molecularized" tests emphasizing cellular-level electrophysiological and biochemical markers are validated for inclusion in the battery they will be selected to replace existing neurobehavioral measures.

The balance of this chapter will discuss issues relative to the application of neurobehavioral screening batteries to combustion toxicology, including a brief history of the use of neurobehavioral batteries in general toxicology. An experimental research program ongoing in our laboratory is summarized to illustrate the applicability of neurobehavioral analysis to combustion toxicology risk assessment.

Problems with Complex Combustion Atmospheres

U.S. fire statistics show that 69% of all fire deaths are associated with post-flashover fires, with the majority of deaths due to smoke inhalation outside the room of fire origin. The relative toxicities of the common atmospheric contaminants produced by combustion, as measured by lethality and incapacitation endpoints, are well established. Analytical methods can be used to identify and quantify the major components of combustion atmospheres, yet these analyses can involve considerable complexity. Over 400 different thermal decomposition products were identified, for instance, during the thermal decomposition of seven plastics *(11)*. Even with identification of combustion gas components, it cannot be assumed that knowledge of the toxicities of individual contaminants is sufficient to predict accurately the toxicity of the mixture. The possibility of production of unique "supertoxicants" from the combustion or pyrolysis of previously untested combinations of materials is very real, especially in the military combat scenario.

Fortunately, if non-gas contaminants (particulate matter and aerosols), environmental factors and "supertoxicant" formations are eliminated from consideration, it appears that a relatively small number of fire gases are sufficient to account for the incapacitating and lethal effects induced by exposure to combustion atmospheres. Sakurai *(12)* used a murine model to test the incapacitating effects of brief (5-30 min.) exposure to pure fire gases (CO, HCl, HCN, NH_3, NO_2, N_2 and CO_2) and several of the interactive combinations of these gases. The results demonstrated conclusively that the expected additivity of component gases was insufficient to predict incapacitative endpoints. The N-Gas model of combustion gas toxicity *(13-14)* assumes the toxicity of fire smoke is determined mainly by a small number of gases (CO, CO_2, O_2, HCN, HCl, HBr, NH_3 and NO_x) which act additively, synergistically or antagonistically. [Additivity implies that if compounds are introduced simultaneously, the toxic response is equal to the sum of the effects of the compounds if introduced separately. A synergistic effect is an effect greater than the effects predicted by simple additivity. An antagonistic effect assumes effect mitigation by adding compounds.] This N-Gas model *(11, 13-14)* assumes that, given constant exposure times, fire temperature and atmospheric conditions, relative concentrations of each major gas generated under post-flashover conditions are sufficient to calculate potential for human lethality. Interestingly, the presence of low concentrations of irritants (such as HCl, HBr, NO_2, NH_3, *etc.*) in a complex combustion mixture were shown to reduce the debilitating and lethal effects of exposure to such gases as CO and HCN, while excessive CO_2 significantly exacerbated progress toward these endpoints. Additional factors (*i.e.*, visual obscuration), while insufficient to induce incapacitation or lethality, can impact escape from an enclosed space such that lethality from CO and HCN exposure can ensue.

While identification and quantification of the major fire gases in a complex combustion atmosphere are apparently sufficient to predict severe incapacitation (unconsciousness) and lethality, there have been few published models to predict specific, less severe behavioral endpoints. The neurobehavioral screening battery has the potential to predict reliably the potential of specific combustion atmospheres (regardless of composition and possible interactive effects between component gases) that will induce specific behavioral and performance deficits. The neurobehavioral deficits produced by exposure to varying concentrations of fire gases, extinguishants and extinguishant combustion byproducts are of extreme interest to the operational military. Aircraft, shipboard and vehicle fires, and the byproducts produced by use of appropriate extinguishants, jeopardize both personnel involved in fire control and, perhaps more importantly, the capacity of personnel exposed to even small (sometimes undetectable) toxicant concentrations to perform mission-critical functions.

Use of Animal Models in Behavioral Toxicology

The use of animals to predict human toxicological response is essential in military as well as private sector applications. Human exposures to high concentrations of individual toxicants or unknown complex mixtures, especially fire gases, is prohibited by safety considerations. Field military personnel may, for instance, be exposed to unknown toxicants, unresearched combinations of toxicants, byproducts of the use or destruction of equipment, or interactive toxicity from extinguishment of fires. It is important that these contingencies are as thoroughly explored as possible using animal testing before the opportunities for such exposures occur. When unexpected exposures do occur, validated animal models must be available for the rapid deployment for toxicological screening as well as for evaluating potential therapeutic treatments. In some military situations, highlighting combat scenarios, immediate knowledge of the potential toxicity of unique environments is required, necessitating decision-making based on use of animal models.

The direct validation of animal models for human application for cases of extreme toxic potential is difficult or impossible. The NMRI/TD Neurobehavioral Screening Battery contains tests that have been used to evaluate human, non-human primate and lower animal performance. For example, there is a wealth of literature on human and animal response to specific pharmaceuticals, drugs of abuse and low doses of common toxic gases (CO, certain CFC's, *etc.*). Carboxyhemoglobin equilibrium level curves for different CO exposures in humans and rats, for instance, has been shown to be nearly identical *(11)*. During testing of the relative behavioral toxicity of ozone depleting substances (ODS) and their proposed replacements at NMRI/TD, the data reflected remarkable similarity between rodent and humans in dose response. Rat behavioral response to Freon-12 and Halon-1211, two widely tested ozone depleting gases *(15-20)*, was found to be so similar to known human response that we are confidently using the model for preliminary risk assessment of new replacement candidates. We have attempted to select tests for inclusion in the NMRI/TD battery that typically reflect similar responses in both humans and non-human animals. Although it cannot be assumed that rodent and human behavioral response during exposure to an unknown compound will be similar, it is logical to eliminate, without further testing, the use of any compound that induces severe deficits in animal performance.

Use of Neurobehavioral Tests and Batteries in Combustion Toxicology

Klimsch et al. (21) provided an excellent review of a number of studies where single behavioral endpoints, including incapacitation, leg-flexion shock avoidance, activity wheel performance, pole-climb conditioned avoidance/escape response, rotorod performance, shuttlebox, operant performance deficits, *etc.* were used to assess behavioral toxicity. The balance of this section focuses on studies where single behavioral endpoints were utilized to investigate toxicity induced by exposure to major fire gases or combinations of gases. Highlighting CO-induced toxicity research, this short review emphasizes the considerable inconsistency in methodologies, experimental results and conclusions in this literature.

Goldberg and Chappell (22) published research indicating a performance enhancement in operant behavior following several exposures to CO. The authors utilized operant chambers housed within Wahmann-like inhalation exposure chambers to evaluate the effects of low doses of CO on behavior in the Lewis rat. CO administered for 55 min. over 3 successive days at doses of 250 or 500 ppm resulted in dose-related, progressive reduction in responding on a continuous reinforcement schedule. Testing in room air on day 4 resulted in a significant enhancement of performance over pretest baseline in the 500 ppm exposure group. No detectable effects were observed from a 2 hr. CO exposure below 200 ppm on operant responding to a VR-3 schedule.

Several studies report an effect of combined CO-CO_2 exposures that could be expected in terms of the increased respiration and tidal volume produced by excessive CO_2. Gaume et al. (23), using mice on an exercise wheel, reported an increase in "time of useful function" with CO-CO_2 mixtures as opposed to CO alone. Carter et al. (24) tested the effects of CO (1000 ppm) on operant behavior (FR-15) in the Sprague-Dawley rat with 2.5%, 5.0% or 7.5% CO_2. While concentration of CO_2 (to 7.5%) alone had no effect on operant performance, atmospheric levels of CO_2 of 5-7.5% partially protected animals against severe deficits in operant performance produced by CO exposure.

Kaplan et al. (25-27) investigated the effects of physiological irritancy on CO-induced toxicity in rats and a non-human primate. The authors established a carbon monoxide EC_{50} of 6850 ppm for escape/avoidance from a shock chamber by the juvenile African Savannah baboon. Acrolein (12-2780 ppm) or HCl (190-17,200 ppm) was shown to neither prevent nor hinder escape, despite mild to severe irritant effects, including eventual lethality. An EC_{50} of 6780 ppm of CO was established for rats escaping shock in a shuttlebox. Five minute exposures to HCl (11,800-76,730 ppm) did not prevent escape, despite severe post-exposure respiratory effects or lethality. In both species, escape time was not effected by HCl, although intertrial responding increased. The similarity in dose response to CO behavioral toxicity between rodent and non-human primate models increases the confidence in direct application of these animal model data to certain human risk assessments. Hartzell et al. (4), in contrast, reported that the guinea pig, as compared to the rat, was found to be (1) three times more sensitive to HCl due to greater bronchoconstriction; (2) less sensitive to CO than the rat; and (3) equally sensitive to HCN.

Rowan and Fountain (28) attempted to factor effects on brain reward systems and hypoxia-induced fatigue in CO-induced deficits in operant responding for brain-stimulation reward in Long-Evans hooded rats. The authors reported that CO exposure, resulting in 65% COHb in blood, resulted in no significant reduction in responding rate (fatigue, hypoxia, ataxia, dizziness, nausea, *etc.*) in a single lever task, but rather a significant elevation in the stimulus duration threshold (SDT) of rewarding brain stimulation required to support responding. When rats exposed to the highest CO level tested were required to alternate between two levers, requiring additional physical exertion, more severe deficits occurred. The authors concluded that CO exposure influences both brain reward systems and other systems impacted by hypoxia-induced fatigue.

Benignus *et al.* (29) developed dose effect curves for animal and human models as a function of COHb levels. The authors reviewed the literature in compiling a list of human COHb-induced deficits including: impaired visual dark adaptation sensitivity, lower critical flicker fusion, reduced temporal discrimination accuracy, vigilance and tracking. Animal studies reviewed were limited to those measuring deficits in operant behavior in rodents and monkeys, and included CRF, multiple FR/FI, DRL, ratio straining, and consecutive number schedules. The authors concluded that while substantial variability and inconsistency in the reporting of results was shown to exist in the literature: (a) CO exposures resulting in less than 20% COHb were generally without significant effects; (b) COHb levels above 20% produced deficits in a non-linear manner, with a progressively steeper slope consistent with increasing COHb levels; and (c) rodent, non-human primate and human curves for comparable tasks and COHb levels were directly comparable.

Numerous researchers have utilized from one to several behavioral endpoints for toxicity testing, but few have attempted to develop and validate a screening battery sufficiently comprehensive to identify the factors underlying specific behavioral and performance effects. Several underlying neurobehavioral systems will be impacted in any toxicity-induced performance deficit or enhancement. A simple operant discrimination task may, for example, reflect induced changes in sensory acuity, physiological irritancy, neurological transmission and processing, motivation, cognitive processing, decision-making, motor response and endurance. Thorough understanding of induced effects is difficult, if not impossible, without a battery sufficient to both identify deficits or enhancements and factor the specific component systems underlying the observed changes.

Guidelines for conducting neurobehavioral tests of motor activity, schedule-controlled operant performance, and a functional observational battery (FOB) were published (30) by the U.S. EPA Office of Toxic Substances (1985). Efforts have infrequently focused on the development and validation of a neurobehavioral screening battery in combustion toxicology in compliance with these guidelines. Mosher and MacPhail (31-32) used these guidelines to evaluate the behavioral toxicity of six compounds that produce different syndromes of intoxication in the rat. The battery consisted of an FOB (home cage and open-field observations, tests of reflexes, reactivity, neuromuscular function and tone, and physiological parameters), measure of motor activity in a modified figure-8 maze, and performance on a Fixed Interval (FI 3-min.) operant schedule for milk reinforcement.

Purser (33-34) developed a battery limited to physiological tests, such as respiration, cardiac function, respiratory blood gases, EEG, auditory cortical evoked potentials, peripheral nerve conductivity and muscle strength, to test the Cynomolgus monkey in combustion product environments (CO, HCN and CO_2). There was reduced nerve conduction velocity and severe CNS depression at 1000 ppm CO exposure. At 60 ppm HCN there was slight CNS depression, and at 80-150 ppm there was severe CNS depression and incapacitation. Hypoxia (10% O_2) resulted in muscle weakness, decreased nervous system conductivity, abnormal cardiac function, lowered blood pressure, and CNS depression. Five percent CO_2 exposure tripled respiratory minute volume.

The most sophisticated existing test battery used with animal models, measuring cognitive capacity, sensory system integrity and motivation, was developed by Merle Paule and his colleagues *(35-39)* at the National Center for Toxicological Research, Jefferson, AK. Used primarily to evaluate the effects of injectable drugs on cognitive capacity in rhesus monkeys, the battery has not yet been used in combustion toxicology or modified for use in rodent studies. The operant test battery contains five behavioral tasks thought to model different CNS functions. The tasks and the CNS function that each is presumed to model are: delayed matching-to-sample (short-term memory and attention); conditioned position responding (color and position discrimination); progressive ratio (motivation); temporal response discrimination (time perception); and incremental repeated acquisition (learning). The use of non-human primate models for predicting human cognitive deficit in behavioral toxicology provides a very useful tool for isolating the specific CNS systems impacted; their use for behavioral toxicity screening, however, presents ethical and practical problems sufficient to preclude their further consideration.

This brief review highlights the inconsistencies in methodology and results in the literature reporting the neurobehavioral consequences of exposure to CO, a single and reasonably well understood fire gas. The study of the toxic effects of complex mixtures of fire gases presents substantial additional challenges. A validated comprehensive neurobehavioral screening battery should provide a consistent standard for non-lethal combustion toxicology research, and allow direct comparison among studies.

The NMRI/TD Neurobehavioral Screening Battery

The NMRI/TD neurobehavioral research group is involved in a long-term project to define and validate a battery of performance tests. These tests, using non-primate animal subjects, will reliably predict toxicant-induced performance deficits in similarly exposed human populations. This battery will provide the capacity to predict rapidly the potential of virtually any substance to induce performance deficits. This battery can also be used in conjunction with more traditional toxicological methods to provide risk assessment for potential unavoidable toxic exposures or to define prophylactic and therapeutic medical solutions to existing toxic exposures. Ultimately, results from the neurobehavioral screening battery research will be correlated with emerging cellular-level markers to provide a highly comprehensive risk assessment tool.

The NMRI/TD Neurobehavioral Screening Battery consists of at least three different tests for each sub-area of performance outlined in guidelines for conducting neurobehavioral tests published by the U.S. EPA Office of Toxic Substances *(30)*. Many tests are applicable to more than one area of neurological function or performance, allowing factoring of the specific sub-areas within a given deficit. A complex regression equation is being developed to predict neurobehavioral deficits based upon results from tests appropriate to the operational scenario being investigated. The NMRI/TD Neurobehavioral Screening Battery will provide the tools for: (a) rapid screening to identify a wide range of potential deficits when the nature of a novel toxicant is unknown; (b) identification of deficits within a specific category of behavior or performance; (c) delineation of the exact neurological, behavioral or performance components that contribute to the observed deficit; and (d) ultimately the ability to predict human toxicologically-induced deficits in specific operational scenarios based on regression analysis from the results of a limited number of carefully selected tests. The current research program is evaluating the following areas of neurological and performance integrity testing:

NMRI/TD Neurobehavioral Screening Battery

(1) *Measures of Neurological Activity*

(a) **Electroencephalogram (EEG):** Discrete change in the simultaneous, summed electrical activity of cortical cells.

(b) **Evoked Potential:** Measurement of specific event-related electrical potentials in response to discrete sensory stimuli.

(c) **P-300 Wave:** Electrical measurement of CNS recognition of a change in continuous sensory stimulation (novel stimulus effect).

(d) **Startle Response:** Measurement of neural transmission time to observable startles following auditory stimulus or air puff.

(e) **Tail Flick:** Measurement of nerve conductivity, based on the time required to flick the tail away from a source of mild electroshock.

(f) **Rating of Convulsive Behaviors:** Identification of observed motor seizures, from stereotypies to generalized motor seizure, using the NMRI/TD Behavioral Seizure Identification Scale (Buring *et al.*, NMRI/TD, in preparation).

(2) *Measures of Motor System Integrity & Endurance*

(a) **NMRI/TD Roto-Wheel Performance:** Measurement of resistance to: (1) loss of equilibrium (fall from rotorod); (2) hindlimb dysfunction (in activity wheel); (3) forelimb dysfunction; (4) loss of righting (incapacitation); (5) recovery of righting and locomotory function (Rossi *et al.*, NMRI/TD, in preparation).

(b) **Self-Driven Treadmill**: Measurement of locomotion in self-powered treadmill.

(c) **Grip Strength Response**: Tail-pull force required to detach front paw gripping of horizontal strain bar.

(d) **Forced Swim Test:** Time to submersion of rat in a forced swim.

(3) Physiological Irritation Measures

(a) **Sensory and Respiratory System Irritancy:** Eye tearing or opaqueness, nasal discharge, salivatory drooling and pulmonary distress.

(b) **Emotionality Rating Scales:** Rating of emotionality response to a series of sensory events *(e.g.,* handling, looming stimulus, *etc.)*

(c) **Rat and Pigeon Avoidance or Escape Conditioning:** Animals are trained in an operant chamber to barpress or keypeck in response to a cue to avoid footshock, or escape from foot shock if initiated. Measures of impairment include: (1) number of shocks avoided versus escapes; (2) latency to response; (3) number of errors on signaled lever discrimination; and (4) number of shocks avoided and escaped versus non-responsive trials.

(4) Measures of Sensory Acuity, Including Color Vision

(a) **Startle Response:** See Section 1(d).

(b) **Conditioned Position Responding (Hue Discrimination):** Pigeon must peck center key, illuminated blue or green, or red or yellow, to initiate trial. After a variable period of time (0-20 sec.), the left and right keys are illuminated white. If center key had been blue or green, pigeon must peck right key; if center key had been red or yellow pigeon must peck left key.

(c) **Social Contact (or Play) Behavior:** Juvenile or adult male rats, deprived of social contact for up to 48-hr., are rated for total activity, contacts, play pins, and separations during a 5-min. conspecific exposure in a modified open field.

(d) **Delayed Matching-to-Sample (Pattern Vision):** Pigeon is trained to peck center key in response to the appearance of a geometric shape. After a variable period of blackout (0-20 sec.), three different geometric shapes appear on the right, center and left keys. Pigeon must peck the key containing the same shape as was initially presented to earn a food reinforcement.

(e) **Tail Flick:** See Section 1(e)

(5) Social Behavior and Emotionality

(a) **Social Contact (or Play) Behavior:** See Section 4(c).

(b) **Emotionality Rating Scales:** See Section 3(b).

(c) **Rat and Pigeon Avoidance/Escape Conditioning:** See Section 3(c).

(6) Measures of Motivational Level & Frustration

(a) **Intracranial Self-Stimulation Thresholds (ICSS):** Measure of minimum current required to reinforce operant responding for electrical stimulation through electrodes implanted in brain reward centers.

(b) **Negative Reinforcement Paradigms:** Animal must press the bar under the illuminated left or right light within 2 sec. to avoid a foot shock for 10-sec. Incorrect or no response is punished by brief footshock. Motivation level is measured by the minimum foot shock level required to elicit bar pressing.

(c) **Operant Progressive Ratio:** Rat is required to increase the amount of work required for each food reinforcement. Rats trained to barpress three times to receive one pellet (FR-3) must barpress 5 times, then 7 times, then 9 times, then 11 times, etc., to earn one pellet, until extinction occurs. Motivation level is measured by the ratio (breakpoint) at which barpressing ceases for 5 min., latency to collect pellets, latency to resume lever pressing after reinforcement, number of entries to the food magazine between reinforcements (expectancy), and number of presses on an unreinforced bar (general arousal or frustration).

(d) **Schedule Induced Polydipsia:** A reliable measure of frustration; total ingestion of water available *ad lib* to rats working on a fixed-interval schedule increases as a function of time and the length of the interval separating food pellet deliveries.

(7) Spatial & Temporal Discrimination

(a) **Forced Swim Test:** Also, see Section 2(d). Rat, in a large circular swim tank, is trained to swim in the direction of an environmental position cue to locate a partially submerged platform. Time to reach platform and patterns of swimming are measures of spatial discrimination integrity. The tank can be rotated relative to fixed environmental cues.

(b) **Rat Complex Operant Discrimination:** Rat is trained to respond on FR-3 (three barpresses for one food pellet) appetitive schedule when left chamber light is illuminated. Barpressing when right chamber light is illuminated results in timeout. Cognitive integrity, including response inhibition, is measured by total reinforcements earned, slope of cumulative record and error to reinforcement ratio.

(c) **Differential Reinforcement of Low Rates of Responding (DRL):** Rat is reinforced with a food pellet for the first response after a specified Interresponse Time (IRT). Incorrect barpresses reset the IRT.

(d) **Conditioned Position Responding:** See Section 4(b).

(8) *Measures of Higher Cognitive Decrement*

    (a) **Delayed Matching-to-Sample:** See Section 4(d).

    (b) **Operant Progressive Ratio**: See Section 6(c).

    (c) **Rat Complex Operant Discrimination**: See Section 7(b).

    (d) **Differential Reinforcement of Low Rates of Responding (DRL)**: See Section 7(c)

    (e) **Conditioned Position Responding:** See Section 4(b).

    (f) **Modified Forced Swim Test:** Rat is allowed to swim in an inescapable bell jar partially filled with water (during or after toxic exposure) until an immobility (floating) response is emitted. Rat is reintroduced to the bell jar after 24 hours. Time to exhibit the immobility response is a measure of memory consolidation.

Application of NMRI/TD Neurobehavioral Battery to Combustion Toxicology

Examples of interest to the military include sonar and radar operation, weaponry guidance, and firefighting/damage control during exposure to multiple fire gases of varying concentrations. Specific applications of the program under development at the Naval Medical Research Institute Toxicology Detachment emphasize human behavioral and performance deficits associated with acute exposure to fire gases, fire extinguishants and extinguishant byproducts. Operations on ships and submarines, for example, require that the effects of combustion gases on personnel in the immediate areas of the fire, and in areas remote from the fire scene be addressed. In aircraft and land vehicles, small smoldering fires may produce combustion gases in such small concentration that performance impairments may occur before awareness of the exposure initiates protective response. We are including a brief discussion of an ongoing research project at NMRI/TD to demonstrate use of behavioral screening tests to evaluate relative toxicity of compounds of military interest. The methodology presented in the study is very simple relative to the substantial complexity of results evaluation from use of the entire battery, but demonstrates the applicability of behavioral battery testing to real-world scenarios.

Behavioral and performance effects of brief exposure to the ozone-depleting substance replacement R-134a

Ritchie, G.D., Rossi J. III and Buring M.S.

In compliance with the Montreal Protocol, as amended in London in 1990 and in Copenhagen in 1992, and the U.S. Clean Air Act Amendments, the United States will soon eliminate the use of fully-halogenated chlorofluorocarbons. Substantial research is in progress to evaluate the relative toxicity of a number of EPA-approved ozone-depleting substance replacement candidates. R-134a, a nonozone-depleting alternate air conditioning coolant, refrigerant and pharmaceutical propellant is undergoing toxicological evaluation for widespread international application. Possible extinguishant applications include Naval ships and submarines, military aircraft and flight operations, and combat vehicles, emphasizing the release of relatively high concentrations of the extinguishant into enclosed areas. The LC_{50} (567,000 ppm for four hours in rats) *[16]* for R-134a is relatively high, but little research has investigated non-lethal toxic effects on behavior and performance as a result of brief exposure to the lower concentrations expected in real world scenarios.

Methods

Subjects. Sixty (60) male adult Wistar rats were briefly exposed to increasing concentrations of R-134a, Freon-12, Halon-1211 or nitrogen control atmospheres in a rotorod/activity wheel (NMRI/TD Roto-Wheel) apparatus or operant box enclosed within a modified Hinners-type chamber.

Materials.
R-134A (1,1,1,2 tetrafluoroethane) - The leading non-ozone depleting candidate to replace Freon-12 coolant and refrigerant applications; proposed replacement for Halon-1211 and Halon-1301 airfield and military vessel or vehicle extinguishant uses.

Freon-12 (dichlorodifluoromethane) - Ozone depleting gas; the most common military coolant and refrigerant.

Halon-1211 (bromochlorodifluoromethane) - Ozone depleting gas; a common military flightline extinguishant with limited applications for fire control in enclosed areas.

Gas Exposure Groups.
NC: Nitrogen Control - A control atmosphere was utilized where oxygen was progressively depleted from 21% to 8% through the introduction of nitrogen. This atmosphere was used to control for possible hypoxic effects associated with exposure to increasing concentration of the test gases.

R: R-134a WITHOUT OXYGEN REPLACEMENT GROUP- Exposure to incapacitation, with R-134a concentration increased from 0 to 37%; oxygen content decreased as R-134a volume increased.

R+O_2: R-134a WITH OXYGEN REPLACEMENT GROUP-Exposure to incapacitation, with R-134a increased from 0 to 37%; oxygen content was maintained at 21%.

F: FREON-12 WITHOUT OXYGEN REPLACEMENT GROUP- Exposure to incapacitation, with Freon-12 increased from 0 to 47%; oxygen content decreased as Freon-12 volume increased.

F+O$_2$: FREON-12 WITH OXYGEN REPLACEMENT GROUP-Exposure to incapacitation, with Freon-12 increased from 0 to 47%; oxygen content was maintained at 21%.

H: HALON-1211 WITHOUT OXYGEN REPLACEMENT GROUP - Exposure to incapacitation, with Halon-1211 increased from 0-28%; oxygen content decreased as Halon-1211 concentration increased.

Apparatus & Training Procedures.

Roto-Wheel: A specially designed apparatus, combining the functions of a traditional motorized activity wheel and rotorod, was housed within a Hinners-type chamber. The 240 L chamber allowed for highly precise exposure to test gases or control atmospheres at a dynamic flow rate of 65 L/min. After four days of conditioning, animals were placed on the rotorod turning at 11 rpm suspended in the center of the activity wheel. Experimental gas (R-134a, Freon-12 or Halon-1211) or a control atmosphere (NC) was then introduced into the chamber. Experimental gas concentrations were gradually increased from 0-47% (at the same rate), until the Loss of Righting (LOR) endpoint was achieved. Test gas concentrations were continuously monitored using infrared spectrometry (Miran 1A, Foxboro Co., Eastbridgewater, MA), and oxygen level was monitored with Hudson RCI units. For the R+O$_2$ and F+O$_2$ groups, atmospheric oxygen was maintained at 21% through the addition of pure oxygen to the chamber atmosphere; for the R, F and H groups, oxygen content was allowed to decrease to as low as 11% in response to induction of the experimental gas. For the Nitrogen Control (NC) group, atmospheric oxygen was gradually replaced by pure nitrogen gas until oxygen concentration was as low as 8%. When the animal fell from the rotorod (LOE) to the inside surface of the activity wheel it was permitted to walk until Loss of Righting (LOR) was observed. Time to occurrence and experimental gas concentration at the LOE, HLD, FLD and LOR behavioral endpoints were recorded. When the LOR endpoint was observed, experimental gas flow was immediately terminated and the chamber was flushed with room air. Time to Recovery (REC) was recorded. Animals were then removed from the chamber and inspected for sensory system or pulmonary irritancy.

Neurobehavioral Endpoints.

LOE: LOSS OF EQUILIBRIUM-Rat loses contact with rotorod surface (four paws) and falls to surface of activity wheel.

HLD: HINDLIMB DYSFUNCTION-Rat fails to utilize hindlimbs for forward locomotion in activity wheel for 15 sec.

FLD: FORELIMB DYSFUNCTION (GRIP STRENGTH) - Rat fails to utilize forelimbs in forward locomotion or gripping activity in activity wheel for 15 sec.

LOR: LOSS OF RIGHTING [INCAPACITATION] -Rat fails to right for 15 sec. during rotation of activity wheel.

REC: RECOVERY-Rat regains righting and uses all limbs for locomotion during rotation of activity wheel.

Operant Chamber: A standard operant chamber was modified to maximize gas flow and was housed within a Hinners-type chamber almost identical to the one described above. The chamber contained left and right white stimulus lights, a single bar below the left light and a feeder equidistant between the lights. Rats were thoroughly conditioned on a variety of operant tasks (CRF through FR-15) in room air atmospheres for 45-mg food pellet reinforcement until performance was extremely stable. The final pretraining schedule required barpressing (FR-3) when the left light was illuminated, and inhibition of barpressing when the right light was illuminated. The first barpress when the right light was illuminated resulted in a warning tone and dimming of the houselight. Subsequent barpresses resulted in an additional time-out. The lengths of the barpressing and inhibitory periods were randomly varied, with an average length of 9.8 sec. Three measures were recorded: (a) total reinforcements (three barpresses for one reinforcement); (b) errors; and (c) slope of cumulative record. After four stable 30-min. baseline performances, groups of four rats each were tested tri-weekly while exposed to Freon-12, R-134a or Halon-1211 atmospheres. Rats were exposed to a room air atmosphere for 15-min., and a test gas atmosphere for 15 min. of each 30 min. test session. The order of presentation of control or test atmospheres was counterbalanced, so that animals were equally likely to experience a control or test gas atmosphere during the first or second 15 min. subsessions. Freon-12 and R-134a atmospheres were progressively increased (by increments of 2%) from 2% to 18% (20,000-180,000 ppm), and Halon-1211 atmospheres were progressively increased from 0.5-4% (5,000-40,000 ppm) by 0.5 or 1% increases. Animals experienced each level of the assigned test gas for four successive daily sessions. Animals that were exposed to the assigned test gas for up to 36 subsessions were examined daily for any observable physiological deficits (*i.e.,* tremors, weight loss, chronic sensory system irritancy, *etc.*).

Results

Effects of Oxygen Replacement. Introduction of a test gas (0-47%) into the Hinners-type chamber resulted in reduced atmospheric oxygen levels. It was important to determine if the incapacitative effects observed were due primarily to the direct effects of test gas exposures, or simply reflected hypoxia-induced fatigue. While oxygen levels never fell below 11% in any test gas group, atmospheric O_2 was reduced (through nitrogen replacement) to as low as 8% in the NC group. One-way ANOVA's comparing mean time to each behavioral endpoint (LOE, HLD, FLD and LOR) among the R, R+O_2, F, F+O_2, H and NC groups were significant ($p < .001$). Values of 1800 sec. were assigned to animals in the NC group because FLD and LOR were never observed. Scheffe tests indicated a significant difference ($p < .01$) between the means for the NC group and each test gas group at each behavioral endpoint. No other comparisons were significant at the .01 level of confidence. This result indicated conclusively that those animals in the NC group, exercising vigorously for up to 30 min. as atmospheric oxygen was reduced from

21% to as low as 8%, were significantly less impaired on all behavioral endpoints tested than animals in any of the test gas groups.

Behavioral Incapacitation - Roto-Wheel Studies. One-way ANOVA's comparing mean times to behavioral endpoints and mean gas concentrations at behavioral endpoints among the test gas groups (R, R+O_2, F, F+O_2 and H) were significant (p < .001) for LOE, HLD, FLD and LOR functions. Results are summarized in Tables I & II. Behavioral endpoint comparisons between groups, discussed below, were conducted using Scheffe tests and were considered significant when p < .01.

Table I
Time to behavioral endpoints for test gas and control groups with 1800 sec. exposure

	LOE	HLD	FLD	LOR	REC
NC	1410	1620	1800+	1800+	XX
F	209	388	498	981	48
F+O_2	192	507	639	1062	47
R	285	400	453	850	37
R+O_2	228	402	476	691	43
H	57	115	138	246	79

Times (sec.) to observation of Loss of Equilibrium (LOE), Hindlimb Dysfunction (HLD), Forelimb Dysfunction (FLD), Loss of Righting (LOR) and subsequent Recovery (REC) are expressed as a function of the nitrogen control atmosphere (NC) or test gas [freon-12 (F), Freon-12 with oxygen replacement (F+O_2), R-134a (R), R-134 with oxygen replacement (R+O_2) or Halon-1211 (H)] exposure.

Table II
Mean test gas concentration (%) at each behavioral endpoint

	LOE	HLD	FLD	LOR
F	20%	31%	35%	45%
F+O_2	17%	34%	38%	45%
R	24%	30%	32%	39%
R+O_2	20%	29%	32%	37%
H	4%	8%	10%	16%

Mean test gas concentrations [Freon-12 (F), Freon-12 with oxygen replacement (F+O_2), R-134a (R), R-134 with oxygen replacement (R+O_2) or Halon-1211 (H)] within the Hinners-type chamber at the observation of Loss of Equilibrium (LOE), Hindlimb Dysfunction (HLD), Forelimb Dysfunction (FLD) and Loss of Righting (LOR) endpoints.

Loss of Equilibrium (LOE). Halon-1211 without oxygen replacement (H) effected loss of equilibrium (LOE) significantly more rapidly, and at a significantly lower gas concentration, than any other test gas atmosphere. R-134a without oxygen replacement (R) significantly protected rats against loss of equilibrium compared to exposure to approximately equal concentrations of R+O$_2$. Time to LOE was significantly greater for the R group than for the R+O$_2$, F, F+O$_2$ or H groups.

Hindlimb Dysfunction (HLD). The H atmosphere induced inability to use the hindlimbs for forward locomotion (HLD) significantly more rapidly and at a lower gas concentration than comparable R+O$_2$, R, F+O$_2$ or F atmospheres. No other group means were significantly different for the HLD endpoint.

Forelimb Dysfunction (FLD). Inability to locomote using the forelimbs or grip the bars of the rotorod (FLD) was significantly more impaired by H than other test gas atmospheres. Time to endpoint and gas concentration at behavioral endpoint comparisons were both significantly different ($p < .01$). R-134a with oxygen replacement (R+O$_2$) produced FLD significantly more rapidly than any other atmospheres, except H, and at significantly lower concentrations than either F+O$_2$ or F.

Loss of Righting. Loss of righting (LOR), or incapacitation, occurred more quickly and at a lower gas concentration in the H atmosphere than in F+O$_2$, F, R+O$_2$ or R atmospheres. R+O$_2$ produced LOR more rapidly than R, F+O$_2$ or F, and at significantly lower concentrations than in F+O$_2$ or F atmospheres.

Recovery from Incapacitation. Forward locomotion, using all four limbs, in response to the turning of the roto-wheel was recovered in all groups in less than 80 sec. (mean). Although the R group recovered more slowly than the other gas exposure groups, there were no significant differences at the .01 level of confidence between groups for time to REC. Data summarizing time to REC following gas-induced loss of righting data is presented in Table I.

Physiological Symptoms. Animals were carefully examined for evidence of acute sensory system and pulmonary irritancy immediately following each exposure. There was no evidence of tearing, nasal discharge or pulmonary congestion observed with exposure to any gas tested. The animals consistently exhibited shallow, rapid breathing and a very rapid heart rate during and immediately following recovery from incapacitation. No other physical symptoms were detected. Animals were observed for one month following the exposures, and no long-term problems were detected.

Epileptic Seizures. Many animals, particularly those exposed to Halon-1211, experienced epileptic seizures between the FLD and REC endpoints. All animals experiencing seizures, with one exception, were in groups (H, F or R) where atmospheric oxygen content was not controlled. Table III indicates the percentage of animals experiencing each seizure identification category while exposed to the control or test gas indicated.

Table III
Percentage of animals experiencing gas-induced seizures during NRMI/TD Roto-Wheel training

	NC	F	F+O2	R	R+O2	H
Sub-Clinical Seizures	0%	20%	0%	20%	0%	100%
Tonic-Clonic Seizures	0%	10%	0%	10%	0%	50%

Percentages of rats experiencing sub-clinical (nontonic-clonic) or generalized tonic-clonic seizures during exposure to various test gas atmospheres [Freon-12 (F), Freon-12 with oxygen replacement (F+O_2), R-134a (R), R-134 with oxygen replacement (R+O_2) or Halon-1211 (H)].

Operant Conditioning

Exposure to 20,000 - 140,000 ppm R-134a. Exposure to atmospheres containing from 2-10% R-134a, compared to room air atmospheres, resulted in reduced (but not significantly different) rates of responding and a larger number of errors. Rats exposed to R-134a atmospheres (2-14%) for as many as twenty-eight 15-min. sessions showed no differences in daily baseline responding over time, nor any observable evidence of physiological deficits. Significant deficits in rate of responding and percentage of errors to correct responses (errors/reinforcements) were observed for both the 12% ($p < .01$) and 14% ($p < .001$) R-134a concentration exposures compared to room air sub-sessions. Rats whose rates of responding were significantly depressed during the test gas subsessions showed almost immediate recovery to baseline levels when the test atmosphere was flushed and replaced with room air. Error ratio data is presented in Table IV.

Exposure to 20,000 - 180,000 ppm Freon-12. Exposure to atmospheres containing from 2-16% Freon-12, compared to room air, resulted in reduced (but not significantly different) rates of responding and a larger number of errors. Rats exposed to Freon-12 atmospheres (2-16%) for as many as twenty-eight 15-min. sessions showed no differences in daily baseline responding ($p < .001$) over time, nor any observable evidence of physiological deficits. Significant deficits in rate of responding and percentage of errors to correct responses were observed for the 18% ($p < .001$) concentration exposures compared to responding during the control subsessions. The order of presentation of the control and test gas subsessions did not influence rate of responding or error ratio. Rats whose rates of responding were significantly depressed during the test gas subsessions showed almost immediate recovery to baseline levels when the test atmosphere was flushed and replaced with room air. Error ratio results are presented in Table IV.

Exposure to 5,000-40,0000 ppm Halon-1211. Exposure to atmospheres containing from 0.5-3% Halon-1211, compared to room air atmospheres, resulted in reduced (but not significantly different) rates of responding and a larger number of errors. Significant deficits in rate of responding and percentage of errors to correct responses were observed for the 4% ($p < .001$) Halon-1211 concentration exposures compared to responding during the control sub-sessions. The order of presentation of the control and test gas subsessions did not influence rate of responding or error ratio. Rats whose rates of responding were significantly depressed during the test gas subsessions showed almost immediate recovery to baseline levels when the test atmosphere was flushed and replaced with room air. Error ratio results are presented in Table IV.

Table IV
Percentage of errors versus reinforced responses on operant task for test gas exposures

	R-134a	FREON-12	HALON-1211
5,000 ppm	XX	XX	9.1%
10,000 ppm	XX	XX	9.0%
20,000 ppm	XX	XX	15.2%
30,000 ppm	XX	XX	17.2%
40,000 ppm	9.0%	9.9%	38.3%
60,000 ppm	9.0%	8.3%	XX
80,000 ppm	7.4%	7.1%	XX
100,000 ppm	7.5%	5.6%	XX
120,000 ppm	11.4%	13.0%	XX
140,000 ppm	29.5%	15.0%	XX
160,000 ppm	XX	13.1%	XX
180,000 ppm	XX	17.5%	XX

Error ratio percentages (errors/number of reinforcements) exhibited by rats during exposure to various concentrations of Freon-12, R-134a or Halon-1211.

Discussion

Brief exposure to R-134a, Freon-12 or Halon-1211 can result in a diversity of mild to severe behavioral and performance deficits in rats, including deficits in operant response and induction of motor seizures. For the purposes of military application, the study indicates (in terms of exposure to equal concentrations of the test gases for equal periods) that R-134a is slightly more toxic than Freon-12, but substantially less toxic than Halon-1211. While the LC_{50} endpoints for both freon-12 and R-134a are relatively high *(15-16)*, significant behavior and performance deficits were shown to occur after brief exposure to much lower concentrations. Operant response data clearly indicated that deficits observed in 140,000 ppm atmospheres of R-134a, 180,000 ppm atmospheres of Freon-12 and 40,000 ppm atmospheres of Halon-1211 consisted of both motivational and higher cognitive components. It is of concern to the military, because these gases are colorless and with little detectable odor, that inadvertent exposure to low concentrations of R-134a could result in non-lethal incapacitation, or at least deficits in cognitive function or motivation. Military exposures might be expected to include rupture of cooling or refrigerant systems within ships or submarines, damage to cooling or fire extinguishant systems within military vehicles, or flightline firefighting scenarios. In the private sector, R-134a is currently utilized in the air conditioning systems of many new automobiles, where release of all coolant into the enclosed passenger compartment could result in concentrations calculated at 8-12% (General Motors dealership service management, Dayton, OH, personal communications, 1994). The maintenance of 21% oxygen or in the test gas atmospheres, or reduction to as low as 11% with test gas introduction, was shown to have little effect on the deficits observed. In the NC group, where oxygen levels reached 8-10% during 30-min of strenuous exercise, no loss of forelimb function or incapacitation was ever observed. All of the seizures observed occurred in groups where oxygen was maintained at 21%.

Recovery from incapacitation (loss of righting) induced by test gas exposure was unexpectedly rapid, averaging less than 80 sec. across all groups. There were no significant differences in time to recovery among groups, although times to incapacitation were significantly different. Animals recovered from loss of righting, and apparent anesthesia, before flushing of the chamber was complete and showed little or no observable residual deficit. Animals exposed to Freon-12 or R-134a as many as 36 times showed no sensory or pulmonary system irritancy, gained weight and appeared normal in all behavioral parameters.

It appears, then, that brief exposure to Freon-12, R-134a or Halon-1211, in that order, at concentrations well below the LC_{50}, can result in behavioral and performance deficits. The deficits demonstrated in the animal model, expected to be equally or more severe in human subjects at equal concentrations and times of exposure, could be detrimental to mission accomplishment.

Summary

The development, validation and application of a comprehensive neurobehavioral screening battery, using small animal subjects, should provide a valuable tool to several scientific fields, including combustion toxicology. The ability to evaluate rapidly, reliably and cost-effectively the relative toxicity of unique individual toxicants or combinations of known and unknown potential toxicants should be useful to both military and private sector concerns. The lethality and incapacitation effect of many toxicants is well documented, but very little information exists concerning the neurobehavioral effects of brief, non-lethal exposures as can occur in numerous real world operational scenarios. The capacity to factor a complex neurobehavioral deficit into components effects on specific systems (*i.e.*, sensory versus motivational versus higher cognitive systems) is important in all phases of human risk assessment. The use of animal models for human risk assessment can be supported by a comparison of human versus animal responses to similar tasks at dose levels safe for human exposure, thus allowing prediction of human response to toxic levels of identical or similar toxicants.

Acknowledgments

The opinions and assertions contained herein are those of the authors and are not to be construed as official or reflecting the views of the Navy Department or the Naval Service at large. The experiments conducted herein were performed according to the principles set forth in the current edition of the *Guide for the Care and Use of Laboratory Animals*, Institute of Laboratory Animal Resources, National Research Council. The work described herein was supported by the Office of Naval Research under Naval Medical Research and Development Command work units 0602233N MM33I30.008-1408, 1409 and 1420, and work unit 63706NM0096.004-1314. Experimental data reported in this manuscript were obtained in collaboration with Dr. T.K. Narayanan, Mr. L.E. Bowen, Mr. J. Reboulet and Dr. E.C. Kimmel. Additionally, the study reported herein would not have been possible without the hard work of HM1 Michael S. Buring, Ms. Cynthia Ademujohn, HM3 Jay Smith, HM3 Tracy Morton, Ms. Mandy Langley, Ms. Heather Hedges and Mr. Bill Binole. We wish to additionally thank Ms. Ademujohn for her technical editing of the document.

Literature Cited

1. Farrar, D.G.; Galster, W.A. Biological endpoints for the assessment of the toxicity of the products of combustion materials. *Fire Mater.*, 1980, *5*, 50-58.
2. Fitzgerald, W.E.; Mitchell, D.S.; Packham, S.C. Effects of ethanol on two measures of behavioral incapacitation of rats exposed to CO. *J. Combust. Toxicol.*, 1978, *5*, 64-74.
3. Hartung, R.; Ball, G. L.; Boettner, E. A.; Rosenbaum, R.; Hollingsworth, Z.R. The performance of rats on a rotorod during exposure to combustion products of rigid polyurethane foams and wood. *J. Combust. Toxicol.*, 1977, *4*, 506-522.
4. Hartzell, G. E.; Grand, A. F.; Switzer, W. G. Modeling of toxicological effects of fire gases: VII. Studies on evaluation of animal models in combustion toxicology. *J. Fire Sci.*, 1988, *6*, 411-442.
5. Hilago, C.J; Gall, L.A. Relative toxicity of pyrolysis products of some wood samples. *J. Combust. Toxicol.*, 1977, *4*, 193-199.
6. Kaplan, H.L.; Grand, A.F.; Hartzell, G.E. Toxicity and the smoke problem. *J. Combust. Toxicol.*, 1982, *9*, 121-138.
7. McGuire, P.S.; Annau, E. Behavioral effects of exposure to the combustion products of flexible polyurethane foam. *Neurobehav. Toxicol.*, 1980, *2*, 355-362.
8. O'Mara, M.M. The combustion products from synthetic and natural products. Part 1: Wood. *Fire Flammability*, 1974, *5*, 34-53.
9. Spurgeon, J. The correlation of animal response data with the yields of selected thermal decomposition products of typical aircraft interior materials. *FAA-NA-78-45*. National Aviation Facilities Experimental Center, Atlantic City, N.J., 1978.

10. Weiss, B. and Laties, V. (eds.). *Behavioral Toxicology.* New York : Plenum. 1975.
11. Levin, B.C. A summary of the NBS literature reviews on the chemical nature and toxicity of the pyrolysis and combustion products from seven plastics: Acrylonitrile-butadiene-styrenes (ABS), nylons, polyesters, polyethylenes, polystyrenes, poly(vinyl chlorides) and rigid polyurethane foams (NBSIR 85-3267) [U.S.] Natl. Bur. Stand., 1986. Also *Fire and Materials, 11*, 1987, 143-157.
12. Sakurai, T. Toxic gas tests with several pure and mixed gases using mice. *J. Fire Sci., 1989, 7(1),* 22-77.
13. Levin, B. C.; Paabo, M.; Gurman, J. L.; Harris, S. E.; Braun, E. Toxicological interactions between carbon monoxide and carbon dioxide. *Toxicology.* 1987, *44,* 135-164.
14. Babrauskas, V.; Levin, B.C.; Gann, R.C.; Paabo, M.; Harris, Jr., R.H.; Peacock, R.D.; Yusa, S. Toxic potency measurement for fire hazard analysis. *NIST Special Publication 827,* U.S. Dept. of Commerce Press, December 1991.
15. Brock, W. J. Material safety data sheet - "Freon" 12. Du Pont Chemicals. 1993, *2022,* FR:1-5.
16. Brown, K. P. Material safety data sheet - HFC-134a. Du Pont Company. *E-94938-2*:1-5; 1989.
17. Edling, C.; Ohlson, C. G.; Ljungkvist, G.; Oliv, A.; Soderholm, B. Cardiac arrhythmia in refrigerator repairmen exposed to fluorocarbons. *Brit. J. Ind. Med.* 1990, *47,* 207-212.
18. Lam, C.; Weir, F. W.; Williams-Cavender, K.; Tan, M. N.; Galen, T. J.; Pierson, D.L. Toxicokinetics of inhaled bromotrifluoromethane (Halon 1301) in human subjects. *Fund. Appl. Toxicol.* 1993, *20,* 231-239.
19. Lerman, Y.; Winkler, E.; Tirosh, M.S.; Danon, Y.; Almong, S. Fatal accidental inhalation of bromochlorodifluoromethane (Halon-1211). *Human Exper. Technol.* 1991, *10,* 125-128.
20. Silber, L. S.; Kennedy, G. L. Acute inhalation toxicity study of tetrafluoroethane (R-134a). *Haskell Laboratory Report No. 422-79,* 1979.
21. Klimsch, H.J.; Doe, J.E.; Hartzell, G.E.; Packham, S.C.; Pauluhn, J.; Purser, D.A. Bioassay procedures for fire effluents: Basic principles, criteria, and methodologies. *J. Fire Sci.,* 1987, *5,* 73-104.
22. Goldberg, H.D.; Chappell, M.N. Behavioral measure of the effects of carbon monoxide on rats. *Arch. Environ. Health,* 1967, *14,* 671-677.
23. Gaume, G.; Bartek, P.; Rostami, H.J. Experimental results on time of useful function (TUF) after exposure to mixtures of serious contaminants. *Aerospace Med.,* 1971, *42,* 987-990.
24. Carter, V.L.; Jr.; Schultz, G.W.; Lizotte, L.L.; Harris, E.S.; Feddersen, W.E. The effects of carbon monoxide-carbon dioxide mixtures on operant behavior in the rat. *Toxicol. Appl. Pharmacol.*, 1973, *26,* 282-287.
25. Russell, R.W.; Flattau, P.E.; Pope, A.M. (eds.). *Behavioral Measures of Neurotoxicity: Report of a Symposium.* Washington, D.C.: National Academy Press, 1990.
26. Kaplan, H.L.; Grand, A.F.; Switzer, W.G.; Mitchell, D.S.; Rogers, W.R.; Hartzell, G.E. Effects of combustion gases on escape performance of baboon and rat. *J. Fire Sci.,* 1985, *3,* 228-244.

27. Kaplan, H.L. Effect of irritant gases on avoidance/escape performance and respiratory response of the baboon. *Toxicology,* 1987, *47,* 165-179.
28. Rowan, J.D.; Fountain, S.B. Carbon monoxide exposure reduces the rewarding quality of brain stimulation reward in rats. *Neurotoxicol. and Teratol.,* 1991, *12,* 175-179.
29. Benignus, V.A.; Muller, K.E.; Malott, C.M. Dose-effects functions for carboxyhemoglobin and behavior. *Neurotoxicol. Teratol.,* 1989, *12,* 112-118.
30. U.S. Environmental Protection Agency. *Toxic Substances Control Act Testing Guidelines.* 40 CFR 798, subpart G section 798.6050, Federal Register 50:39458-39460, 1985.
31. Moser, V.C. Screening approaches to neurotoxicity: A functional observational battery. *J. Am. Coll. Toxicol.,* 1989, *8,* 85-93.
32. Moser V.C.; MacPhail, R.C. Comparative sensitivity of neurobehavioral tests for chemical screening. *Neurotoxicol.,* 1990, *11(2),* 335-344.
33. Purser, D.A. A bioassay model for testing the incapacitating effects of exposure to combustion product atmospheres using Cynomolgus monkeys. *J. Fire Sci.,* 1984, *2(1),* 30-36.
34. Purser, D.A.; Grimshaw, P. The incapacitive effects of exposure to thermal decomposition products of polyurethane foam. *Fire Mater.,* 1984, *8,* 10-16.
35. Ferguson, S.A.; Paule, M.G. Acute effects of chlorpromazine in a monkey operant behavioral test battery. *Pharmacol. Biochem. Behav.,* 1992, *42,* 331-341.
36. Ferguson, S.A.; Paule, M.G. Acute effects of pentobarbital in a monkey operant behavioral test batter. *Pharmacol. Biochem Behav.,* 1993, *45,* 107-116.
37. Paule, M.G. Use of the NCTR operant test battery in nonhuman primates. *Neurotoxicol. Teratol.,* 1990, *12,* 413-418.
38. Paule, M.G.; Schultze, G.E.; Slikker, W., Jr. Complex brain function in monkeys as a baseline for studying the effects of exogenous compounds. *Neurotoxicol.,* 1988, *9,* 463-470.
39. Schultze, G.E.; McMillan, D.E.; Bailey, J.R.; Scallet, A.C.; Ali, S.F.; Slikker, W., Jr.; Paule, M.G. Acute effects of delta-9-tetrahydrocannabinol in rhesus monkey as measured by performance in a battery of complex operant tests. *J. Pharmacol Exper. Therapeutics, 245(1),* 178-186.

RECEIVED January 31, 1995

Chapter 24

Smoke Production from Advanced Composite Materials

D. J. Caldwell[1], K. J. Kuhlmann[2], and J. A. Roop[3]

[1]U.S. Army Medical Research Detachment, Wright-Patterson Air Force Base, OH 45433−7400
[2]ManTech Environment, Inc., Dayton, OH 45431−0009
[3]Air Force Institute of Technology, Wright-Patterson Air Force Base, OH 45433−7765

The use of advanced composite materials (ACM) in the B-2 bomber, composite armored vehicle, and F-22 advanced tactical fighter has rekindled interest concerning the health risks of burned or burning ACM. The objective of this work was to determine smoke production from burning ACM. A commercial version of the UPITT II combustion toxicity method developed at the University of Pittsburgh, and refined through a US Army-funded basic research project, was used to establish controlled combustion conditions which were selected to evaluate real-world exposure scenarios. Production and yield of toxic species varied with the combustion conditions. Previous work with this method showed that the combustion conditions directly influenced the toxicity of the decomposition products from a variety of materials.

Introduced in the 1960s, advanced composite materials (ACM) are expected to compose 40-60 percent of future airframes. Figure 1 illustrates the increased use of ACM in US Air Force aircraft. During the 1990s, several events focused attention on the human and environmental consequences resulting from fabrication and incidental combustion of ACM. In addition, although the fibers and epoxy resins of advanced composites appear to be safe in their original state, the chemical transformation to a hazardous substance during combustion is not well characterized. These resins, such as epoxies, polymides, phenolics, thermosets, and thermoplastics, may release potentially lethal gases, vapors, or particles into the atmosphere when burned.

As the uses of composites increase, so do the potential risks to the environment and those exposed to the smoke and combustion gases during aircraft mishaps. The objective of this work was to determine smoke production from burning ACM and predict its toxicity.

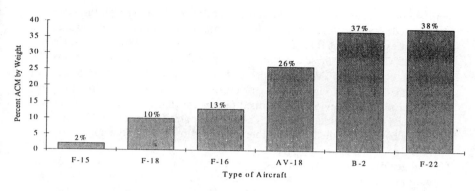

Figure 1. Percent of ACM (by weight) for selected U.S. Air Force aircraft.

The apparatus used to establish controlled combustion conditions is a commercially available version of the cone heater combustion module of the UPITT II method developed at the University of Pittsburgh (1). Previous work with this method showed that the combustion conditions directly influenced the toxicity of the smoke from a variety of materials (2). The toxic potency of the ACM smoke can be estimated from experiments conducted over a variety of combustion conditions selected to enable evaluation of real-world exposure scenarios. The smoke concentration can be calculated and used to predict toxicity (2). This prediction can be subsequently validated by bioassay to detect the presence not only of unusual or uncommon toxicants but also of biological interactions between common gases. If a further refinement is desired, the time-to-toxic effect can also be determined (2,3).

Experimental

Materials. A graphite fiber/modified bismaleimide resin ACM (approximate 2:1 ratio by weight) was used in these studies. Coupons were fabricated with epoxy/graphite skins; woven graphite/epoxy, MS-240; Hercules AG380-5H/8552 Prepreg; Chemmat 4011, 8-ply; thickness: 0.125 inches. The ACM coupons were 110 mm square (approximately $0.01 m^2$) by 2.5 mm thick. The average mass of sixteen coupons was 54.60 ± 0.18 g.

Combustion Module. A commercial version of the UPITT II combustion toxicity apparatus (1) was used to establish controlled combustion conditions. This apparatus consists of a truncated cone heater, as used in the cone calorimeter, to irradiate specimens at selected, controlled heat flux levels; a balance to determine the mass loss rate of the burning specimen, and an enclosure so the ventilation (airflow) can be accurately measured and controlled. This is the first combustion system to combine the three essential elements necessary to describe the burning conditions to which a specimen is submitted: the heat flux, ventilation level, and mass loss rate. With this apparatus, the radiant heat flux level and ventilation level can be varied, and thus, well defined burning conditions can be investigated over a wide scale. A post-crash conflagration would have a maximum external heat flux, from a JP-4 fuel fire, of approximately 84 kW/m^2 (i.e., 873.5°C) (4). For these experiments the sample was horizontal, heat flux (Q) was set at 38, 57, or 84 kW/m^2 and the airflow ($\dot{V}$) was 19.6, 28.7, 35.6, 41.5, or 51.3 L/min. The time to ignition (T_{ign}), percent mass lost, and mass loss rate ($\dot{m}$) were determined as previously described (1) except that a 10-minute period was used instead of a 30-minute period. The resulting smoke concentration (SC) was calculated by dividing the mass loss rate by the airflow through the apparatus.

Thermogravimetric Analysis. A Perkin-Elmer TGA7 was used to conduct thermogravimetric analysis of ACM specimens isothermally at 650°C, 770°C and 950°C.

Combustion Product Identification. A Perkin-Elmer Model 1650 FT-IR spectrometer was used to obtain transmission spectra of the filtered smoke produced by the burning ACM coupon. A Perkin-Elmer Q-Mass 910 GC/MS was used to analyze extracts prepared from cold-trapped vapor and the soot residue.

Screening Electron Microscopy. Burned material was allowed to directly settle on SEM stubs that were prepared with adhesive tabs. SEM stubs were dried overnight in a vacuum desiccator, coated with a 100-A layer of gold and examined in an Amray 1000B scanning the electron microscope equipped with a standard tungsten filament at 30 kV accelerating voltage. The working distance was 8 to 18mm.

Results

Time To Ignition And Mass Loss Rate. Results from sixteen experiments conducted under flaming conditions are presented in Tables I through III. The T_{ign} decreased as Q increased, while the mass loss rate increased with increasing Q.

Table I. Time to ignition, percent mass lost, mass loss rate, and smoke concentration for 0.01m^2 ACM specimens irradiated for 10 minutes at 38.5 kW/m^2 and indicated airflow

Sample No.	$\dot{V}$ (L/min)	T_{ign} (sec)	Mass Lost %	$\dot{m}$ (g/min)	SC (g/L)
22	19.6	120	21.5	1.180	0.060
10	28.7	110	24.2	1.323	0.046
14	35.6	94	23.4	1.273	0.036
17	41.5	69	23.2	1.262	0.030
27	51.3	97	23.5	1.290	0.025
Mean	-	98.0	23.16	1.266	-
Std. Dev.	-	23.16	1.00	0.053	-

Table II. Time to ignition, percent mass lost, mass loss rate, and smoke concentration for 0.01m² ACM specimens irradiated for 10 minutes at 57.2 kW/m² and indicated airflow

Sample No.	$\dot{V}$ (L/min)	T_{ign} (sec)	Mass Lost %	$\dot{m}$ (g/min)	SC (g/L)
9	19.6	45	30.0	1.632	0.083
5	28.7	50	28.1	1.536	0.054
12	28.7	47	27.7	1.514	0.053
15	35.6	39	26.6	1.459	0.041
18	41.5	34	28.3	1.541	0.037
28	51.3	35	29.0	1.595	0.031
Mean	-	41.7	28.28	1.546	-
Std. Dev.	-	6.6	1.15	0.061	-

Table III. Time to ignition, percent mass lost, mass loss rate, and smoke concentration for 0.01m² ACM specimens irradiated for 10 minutes at 84.2 kW/m² and indicated airflow

Sample No.	$\dot{V}$ (L/min)	T_{ign} (sec)	Mass Lost %	$\dot{m}$ (g/min)	SC (g/L)
3	19.6	30	30.8	1.676	0.086
13	28.7	34	29.6	1.609	0.056
16	35.6	32	30.5	1.666	0.047
19	41.5	18	31.7	1.728	0.042
29	51.3	16	32.0	1.751	0.034
Mean	-	26.0	30.92	1.686	-
Std. Dev.	-	8.4	0.96	0.056	-

Consolidated results found the average mass loss of the ACM coupons to be 27.5 ± 3.4%. Further review of the results from these experiments under controlled conditions demonstrated that the mass loss rate increased with heat flux but was independent of airflow. The graphical presentation of these data are found in Figures 2 and 3.

TGA Data. The TGA has the capability to heat a sample in a nitrogen atmosphere or in air. The atmosphere made a significant difference in the mass loss characteristics of the ACM.

TGA In Nitrogen: After the initial mass loss due to the polymer pyrolysis, the mass stabilized at a little over 75% of the initial mass, and stayed there for the rest of the thirty minute test run. There was no significant change with extended time.

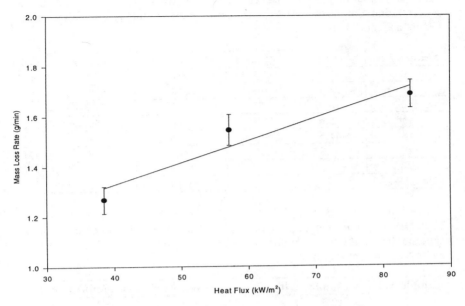

Figure 2. Coupon mass loss rate vs. external heat flux

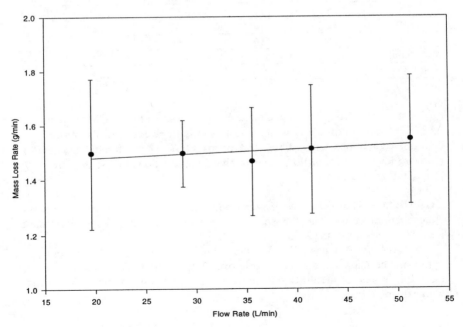

Figure 3. Coupon mass loss rate vs. airflow

TGA In Air: The specimen mass did not stabilized after the initial pyrolysis mass loss (due to the polymer resin loss). The mass loss curve changed with temperature, the slope of which increased as temperature increased. Given enough time, the graphite fibers completely disappeared, i.e., at 950°C all mass was lost within 25 minutes, while at 650°C the time required increased to over 60 minutes. Mass versus time curves for data presented in Table IV are illustrated in Figure 4.

Smoke And Aerosol Characterization. The composition of the smoke and properties of the aerosol particles were evaluated. Initial evaluation indicated the smoke was composed of phenol groups, aniline groups, carbon monoxide, and carbon dioxide. Major spectrum peaks from an FT-IR spectrum of smoke from an experiment conducted at 57 kW/m^2 and 28.7L/min are identified in Table V.

Table IV. Isothermal TGA of ACM specimens under nitrogen or air atmosphere at indicated temperature

Sample Number	Atmosphere	Temperature (°C)	Initial Mass (g)
9	N_2	770°	0.0422
11	N_2	770°	0.0632
20	Air	950°	0.0811
28	Air	770°	0.0765

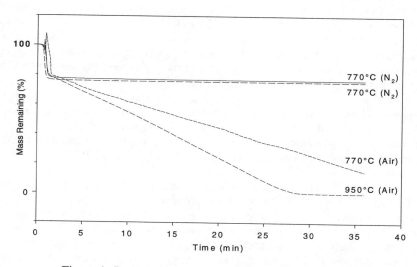

Figure 4. Percent mass remaining vs time for TGA in air and nitrogen at indicated temperature

Table V. Peak Identification From Representative FT-IR Spectrum

cm-1	Height	Identity
3708	17.64	Aniline
3628	19.23	Aniline
3596	25.35	Aniline
3566	53.31	Aniline
3324	53.36	Phenol
2510	81.09	Carbon Dioxide
2174	50.68	Carbon Monoxide
2116	56.36	Carbon Monoxide
1526	10.45	Aniline
1304	29.53	Aniline
1164	47.28	Phenol
1138	53.65	Phenol
730	10.25	Phenol

The particle density was determined using standard laboratory practice and found to be 0.2879 mg/mL. Air samples analyzed by electron microscopy identified a range of aerosol diameters from 0.5 to 1.5μm. Given this density and the observed range of particle diameters the gravitational settling velocity was calculated to be from 6.5×10^{-6} to 3.8×10^{-4} m/sec.

GC/MS Data. Two samples were analyzed by GC/MS; the first being combustion vapor condensed in a cold trap, and the second being a solvent extract of the soot. The quantitation amounts were obtained by using the Response Factor = 1 approximation as specified in the USEPA Contract Laboratory Program Statement of Work for Organics Analysis; Multimedia, Multiconcentration (5). The normal method of quantification used in GC/MS analysis is based on comparison of the area of the most prominent ion in a known concentration of the material to that of the most prominent ion of an internal standard and developing a ratio known as the response factor. As over 90 compounds were tentatively identified in the extract from the soot, it can be seen that preparation of this many standards is impractical for several reasons, including time and expense. An alternative method of quantification, namely that specified in the EPA Contract Laboratory Program Statement of Work, was chosen.

This method of quantification assumes that the response factor is 1 and the total areas, not just those for the most prominent ions, are used. While not extremely accurate, this quantitation will give a general feel for the amount of material present in the sample. Based on over eight years of environmental laboratory experience in a production setting, the numbers given by this technique are probably low (K.J.K., personal observation). Most of the internal standards are deuterated PAHs with strong, sharp responses. In general, most other compounds do not respond as well, which would give a lower concentration by this method than if one were basing calculations on actual, measured response factors.

In the first sample a vapor aliquot was collected using a cold trap with the results summarized below in Table VI. The experimental conditions were heat flux of 57 kW/m^2 and airflow of 28.7L/min.

Table VI. Quantitation of Identified Compounds by GC/MS

Compound	Air Conc μg/m3
Aniline	0.57
Phenol	1.60
4-Methylphenol (o-cresol)	0.11
2-Methylphenol (p-cresol)	0.12
3-Methyl-1-isocyanobenzene	0.01
Quinoline	0.04
Biphenyl	0.01
Diphenyl Ether (diphenyl oxide)	0.19

In the second sample approximately 1.4g of the soot was extracted with 50:50 Methylene Chloride:Acetone solution. It was apparent, upon examination of the injection port liner, that many of the extracted compounds were not suitable for analysis by GC/MS as there was obvious evidence of pyrolysis and deposition in the injection port liner. We identified over 90 different compounds in the extract. The major compounds (i.e., >1g/kg) that were identified are shown in Table VII below.

The compounds were identified by GC/MS, using the most recent version of the NIST mass spectral library. All matches were inspected by an experienced analyst, and the best matches were selected by the analyst rather than only relying on the computer algorithm. GC/IR was performed on the same samples, and the data was used in a confirmatory way. There are very few vapors phase libraries for IR currently available, and most of the compounds were not locatable.

Several other comments need to be made. As these identifications were made by mass spectroscopy, it is quite probable that some of the compounds identified may be a different isomer. That is one of the weaknesses of the technique. There were a number of PAH peaks in the soot extract which were of too low an intensity to characterize properly, and were not included in Table VII. Additionally, as mentioned above, there are probably many compounds which either did not extract in the first place or did not make it out of the GC injection port. The key point is that several of these fourteen compounds, those annotated with (C) in Table VII, are known to be carcinogens (6). Cancer biassays have not been performed for most of the remaining identified compounds.

Table VII. Identification and Quantitation by GS/MS of Major Compounds Extracted from Soot

Compound	Conc. in Soot g/kg
Aniline (C)	2.99
Phenol	2.18
2- and 3-Methylaniline	1.20
Quinoline	3.48
5-Methylquinoline	1.20
Diphenylether	1.05
2-Methoxyethoxybenzene	1.66
1,2-Dihydro-2,2,4-trimethylquinoline	2.21
1-Isocyanonaphthalene	2.21
Dibenzofuran (C)	1.36
1-Isocyanonaphthalene	1.66
Anthracene (C)	1.70
N-Hydroxymethylcarbazole	1.29
Fluoranthene (C)	2.13

Discussion

ACM Mass Loss Rate. The primary objective of these combustion experiments was to obtain a mass loss rate for the ACM to be used to model atmospheric dispersion and deposition of smoke from a burning aircraft. However, one significant limitation to this study was the lack of research on the heat transfer properties of this ACM. Therefore, we assumed that the flame spread characteristics demonstrated by this bench-scale combustion equipment accurately simulated those of a full-scale aircraft. To test this hypothesis, experiments were repeated with ACM coupons of different surface area. The results, under controlled heat flux and airflow conditions, identified a linear relationship between the mass loss rate and the area of the burning composite (7). (The results from experiments conducted with ACM coupons of different surface areas are not presented here.)

Regression of the mass loss rate data with the sample coupon area was performed. The equation for the regression line is found below in Equation 1.

$$\dot{m} = \beta_1 (Area) + \beta_2 (HeatFlux) - 0.01 \qquad (1)$$
where:
$$\beta_1 = 1.98$$
$$\beta_2 = 1.86 \times 10^{-4}$$

The regression results provided a linear equation ($R^2 = 0.999$) that allowed accurate prediction of an emission rate for a full-scale aircraft. These findings enable regression analysis of a linear equation for the emission rate given constant heat flux, airflow, and area conditions. Aerosol properties were identified which enabled calculation of the gravitational settling velocity. This, in turn, will serve to better estimate the downwind plume characteristics. The combined results allow for accurately modeling the smoke and aerosol smoke plum generated during the combustion of composite material aircraft.

Comparison of UPITT II with TGA. The manner in which energy was applied to the samples in the TGA was different than that in the UPITT II apparatus. The TGA uses a cup design with the sample in the center of a small furnace. It wasn't possible to shield the sample from the heat during the heatup cycle, as was possible in the UPITT II apparatus. Despite this difference, and the much smaller sample size used in the TGA, the mass loss measured by the two units was remarkably similar. During the first two to three minutes of TGA, the sample lost approximately 25% of its mass. What happened in the TGA after this time period was a direct result of the differences between the TGA and the UPITT II combustion paradigms.

We suspect that the graphite fiber was being "eroded" by the oxygen in the air. Unpublished work on diamond showed a molecular surface effect (R. E. Langford, personal communication). Apparently, when the material absorbs enough energy (heated), the impact of an oxygen molecule on the surface is enough to pull off a carbon atom and form CO_2 or CO. This hypothesis is supported by infrared spectroscopy data, which show continuous evolution of these gases until the mass goes to zero.

Other Observations From The TGA Experiments. Once the resin was pyrolysed off the graphite fiber matrix, the fibers separated and puffed to several times the original volume and lost any cohesion or tendency to group together. After an experiment where the graphite was not completely consumed, there was considerable difficulty in cleaning the instrument. This phenomenon happened whether the experiment was conducted in nitrogen or air.

The fibers are extremely fluffy, and potentially electrically conductive. Therefore, they could travel a significant distance in a mild breeze, and have the potential to short out electrical equipment from computers to power lines.

In general, even after a long experiment, the fibers remained visible and were therefore not respirable. It is, however, possible that some fibers were being eroded to the point where they could be respirable. At the present time, the answer is unknown, however, there are clear hazards associated with what is known to be contained in the soot particles and that these present the greater hazard to life and property than does the physical shape (i.e., particle or fiber). It is critical that measures be taken in fighting a fire involving these materials to reduce dust and aerosols.

Predicting Smoke Toxicity. As described above, the smoke concentration (in g/L) can be calculated by dividing the mass loss rate (in g/min) by the airflow through UPITT II apparatus (in L/min). For the experiment conducted at a heat flux of 57 kW/m^2 with an airflow of 28.7 L/min, taking the average mass loss rate at that $\dot{V}$ of 1.525g/min from Table II, the resulting smoke concentration was calculated to be 0.053 g/L (53 mg/L). Exposure to this concentration of smoke is expected to be lethal since it is nearly twice the LC_{50} previously determined for a variety of synthetic materials (e.g., rigid polyurethane foam, plasticized polyvinyl chloride, etc.) (4). This statement is supported by the FT-IR identification in smoke and the GC/MS quantifications of toxic species in the cold trapped vapor from this experiment as identified in Tables VI and VII.

Conclusions

The mass loss rate increased with heat flux, but changes in airflow had no effect on mass loss rate. The total mass loss determined by TGA was comparable to that determined by the UPITT II apparatus. The particle density and aerosol diameters of the soot particles were measured. Combustion products quantitated by GC/MS included oxygenated compounds. Fluffy graphite fibers were observed in TGA studies, but none were collected on filter media during testing with the UPITT II apparatus. Small specimen size may have contributed to this phenomenon for the TGA findings. Smoke production and yield of toxic species from ACM varied with the combustion conditions. This finding is consistent with previous work with this method which showed that the combustion conditions directly influenced the yield and toxicity of smoke produced by a variety of materials (2).

Future work will incorporate animal exposures to determine the toxic potency of the smoke and evaluate alternate non-lethal endpoints such as incapacitation. In addition to toxic potency in terms of smoke concentration, we can also determine the time to effect, i.e. lethality or incapacitation. This approach will result in the selection of safer advanced composite materials for new and existing weapons systems.

References

1. D. J. Caldwell and Y. C. Alarie, *J. Fire Science*, **8**, 23-62 (1990).
2. D. J. Caldwell and Y. C. Alarie, *J. Fire Science*, **9**, 470-518 (1991).
3. D. J. Caldwell, *Toxicology Letters*, **68**, 241-249 (1993).
4. D. D. Drysdale, *An Introduction to Fire Dynamics* (1985).
5. USEPA, *Document No. OLMO1.8*, (Aug, 1991).
6. N. I. Sax, *Dangerous Properties of Industrial Materials* (1989).
7. J. A. Roop, *Modeling Aerosol Dispersion from Combustion of Advanced Composite Materials During an Aircraft Mishap,* Master's Degree Thesis (AFIT/GEEM/ENV/94S-21), Air Force Institute of Technology (1994).

RECEIVED November 28, 1994

Chapter 25

Formation of Polybrominated Dibenzodioxins and Dibenzofurans in Laboratory Combustion Processes of Brominated Flame Retardants

Dieter Lenoir and Kathrin Kampke-Thiel

Institute of Ecological Chemistry, GSF Research Center for Environment and Health, 85758 Oberschleissheim, Germany

Oxidative thermal degradation of a collection of polymers with 10 different kinds of brominated flame retardants has been studied under standardized laboratory conditions using varying parameters including temperature and air flow. Polybrominated diphenyl ethers like the deca-, octa-, and pentabromo compounds yield a mixture of brominated dibenzofurans while burning in polymeric matrices. Besides cyclization, debromination/hydrogenation is observed. Influence of matrix effects, presence of various metals, and burning conditions on product pattern have been studied, the relevant mechanisms have been proposed and the toxicological relevance is discussed.

Hazards identification and prevention in oxidative thermal degradation of polymers has become an important issue in the research of European countries (1) and in the US (1, 2). Great concern is given to the problem of acute toxicity of gaseous hazards like carbon monoxide, hydrogen cyanide and hydrogen halides during various kinds of accidental fires (3). Somewhat less concern is given to long-term toxic effects on humans caused by other pollutants emitted by fires like polycyclic aromatic hydrocarbons (4). Therefore, we have studied the formation of polybrominated dibenzodioxins and -furans (PBDD/F) from various types of brominated flame retardants under laboratory conditions. Some related results of PBDD/F formation during accidental burning of these materials in tunnels and in houses will be discussed.

Experimental

The following 10 bromine compounds were investigated, in general within their polymeric matrices (see Scheme 1):

decabromobiphenyl ether (1)
octabromobiphenyl ether (2)
pentabromobiphenyl ether (3)
1,2-Bis (2,4,6,tribromphenoxy) ethane (4)
tetrabromobisphenol A (5)
tetrabromophthalic anhydride (6)
dibromopropyldian (7)
1,2-Bis (tetrabromophthalimide) (8)
polybrominated styrene (9)
hexabromocyclododecane (10)

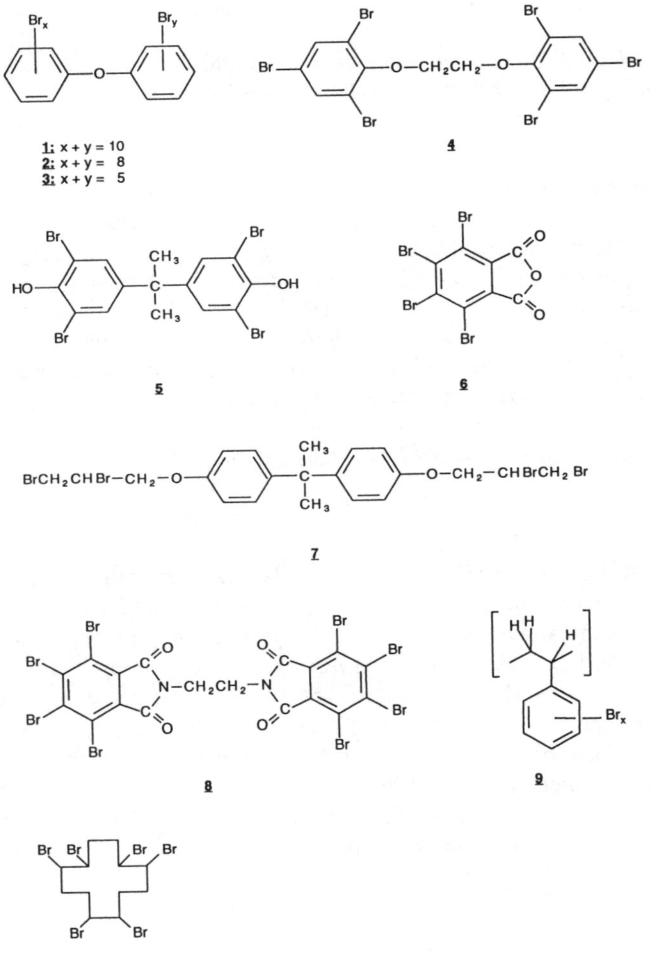
1: x + y = 10
2: x + y = 8
3: x + y = 5
4
5
6
7
8
9
10

Scheme 1. Structure of investigated Brominated Flame Retardants.

decabromobiphenyl ether (**1**)
octabromobiphenyl ether (**2**)
tetrabromobisphenol A (**5**)
tetrabromophthalic anhydride (**6**)
dibromopropyldian (**7**)

pentabromobiphenyl ether (**3**)
1,2-Bis (2,4,6,tribromphenoxy) ethane (**4**)
1,2-Bis (tetrabromophthalimide) (**8**)
polybrominated styrene (**9**)
hexabromocyclododecane (**10**)

The following polymeric materials (commercial products) were investigated:

a) polystyrene with 10 % of **1** and 4 % Sb_2O_3
b) polystyrene with 12,5 % of **1** and 4 % Sb_2O_3
c) ABS with 14 % of **2** and 6 % Sb_2O_3
d) polyurethane with 15,4 % of **3**;
e) ABS with 18 % of 4 and 7 % Sb_2O_3
f) epoxilaminate with **5**;
g) epoxilaminate with **5** (other product)
h) epoxilaminate copper-laminated with **5**;

j) polycarbonate with 12 % of **5**;
k) polyurethane with 33 % of **6**;
l) polyurethane with 6,4 % of **6**;
m) polyester with **7**;
n) polyester with **7** (other sample);
o) polypropylene with 5,9 % of **8**;
p) ABS with 11 % of **9**:
q) polystyrene with 3 % of **10**;

Laboratory combustion processes were performed with the three different furnaces: The German VCI-Apparatus and the German BIS-oven (BIS: <u>B</u>ayer, <u>I</u>CI, <u>S</u>hell) are shown in Figures 1 and 2. Besides these furnaces the German DIN-oven was also used, details can be found in the literature (5, 6, 7).

These furnaces are designed to model the situation of a fire under laboratory conditions. Therefore we prefer the term „oxidative thermal degradation" compared to the term „pyrolysis", since the term „pyrolysis" is generally used for thermolysis without oxygen.

By variation of temperature and air flow rate burning conditions ranging from a smoldering fire to an open fire can be modeled. Details can be found in the literature (5, 6, 7). The furnaces are complementary to each other. In general, similar results are obtained.

Special clean-up procedures have been developed for PBDD/F in the thermal degradation products formed by either furnace. One of these clean-up procedures is outlined below (see Scheme 2); details will be found in the literature (7, 8).
For the relevant structures of PBDD/F, see formula below:

$x + y = 1$ to 8
(135 Isomers)

$x + y = 1$ to 8
(75 Isomers)

Identification and quantification of PBDD/F was performed by GC/MS techniques (5-9). This was done for all brominated PBDD and PBDF from mono-

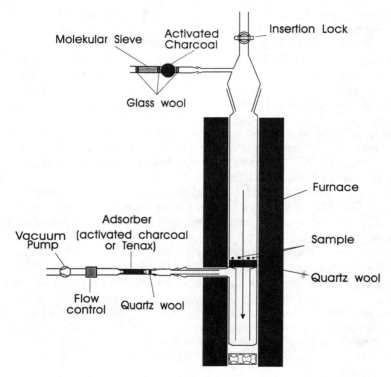

Figure 1. VCI-Apparatus (Furnace of Verband der Chemischen Industrie; German Chemical Industry Association).

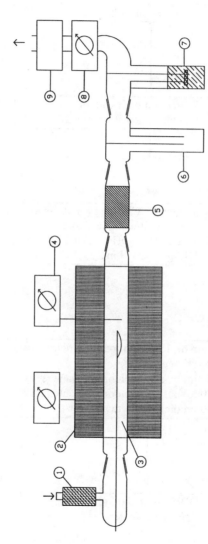

Figure 2. Scheme of BIS Apparatus (**B**ayer, **I**CI, **S**hell); **1:** inlet with filter, **2:** oven, **3:** quartz tube with spoon, **4:** control, **5:** filter (quartz wool), **6:** impinger with hexane/dichloromethane 1:1, **8:** flow meter, **9:** control.

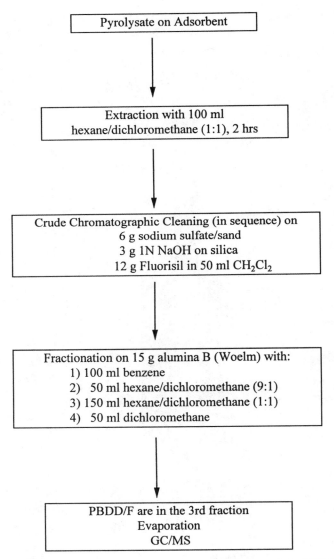

Scheme 2. Clean-up Procedure for PBDD/F in Pyrolysate of the VCI-Furnace.

through octabromo compounds using external standards which were either prepared or purchased. A total of 210 brominated compounds of PBDD/F exists. Since not all isomers are available a complete isomer-specific determination could not be performed. For further experimental details see also (9).

Results and Discussion. The following results have been obtained in laboratory combustion studies. Polymers containing bromine compounds 4 and 5 yield PBDD/F in the ppm range. Oxidative thermal degradation of compound 5 gives PBDD isomers but no toxic isomers. Polymers containing brominated diphenyl ethers 1-3 however, yield PBDF in very high yields (6-10); depending on the applied conditions the conversion can be nearly quantitative. Therefore, this class of compounds has been investigated more thoroughly by different groups (8, 10-12). Compounds 6-10 do not yield any detectable amounts of PBDD/F even in the ppm range.

First, thermal behaviour of decabromobiphenyl ether 1 will be described. The thermal reactivity of this compound depends on the applied conditions; the pure compound reacts completely different in comparison to its reaction in polymeric matrices. Thermolysis of the pure compound gives a good yield (60 %) of hexabromobenzene. The main products obtained by laboratory combustion in the DIN-oven at three temperatures for pure 1 and 1 within a polypropylene matrix are shown in Table 1 and Figures 3 and 4.

Table 1: Main Products obtained by Oxidative Thermal Degradation of Decabromobiphenyl Ether 1 at three different Temperatures; Yields are in Percent by Weight

Product	400°C	600°C	800°C
1	22,3^a ; nd^b	nd^a ; nd^b	nd^a ; nd^b
Hexabromobenzene	12,3^a ; nd^b	56,8^a ; nd^b	0,6^a ; nd^b
PBDD/F	0,08^a ; 25,5^b	0,3^a ; 12,4^b	0,2^a ; 5,5^b
CO$_2$	4,5^a ; 3,0^b	9,9^a ; 3,5^b	16,3^a ; 4,3^b
HBr	nd^a ; nd^b	nd^a ; 1,0^b	nd^a ; 2,2^b
Br$_2$	nd^a ; nd^b	9,4^a ; nd^b	10,9^a ; nd^b

^{a}: from pure compound 1
^{b}: from compound 1 with the polymeric matrix (polypropylene and 4% Sb$_2$O$_3$)
nd: not detected within the detection limit (<0,1 %)

The determined products do not add to 100 % since insoluble tarry products are also obtained. These results show that the thermochemistry of 1 depends strongly on whether it is performed within pure compound 1 or if 1 is heated within a polymeric matrix. A mechanistic explanation has been given, see Figure 6 (8).

Thermochemistry in the polymeric matrix follows a different pattern compared to the pure compound. Decabromobiphenyl ether 1 does cyclize according to the following reaction yielding brominated dibenzofurans (PBDF):

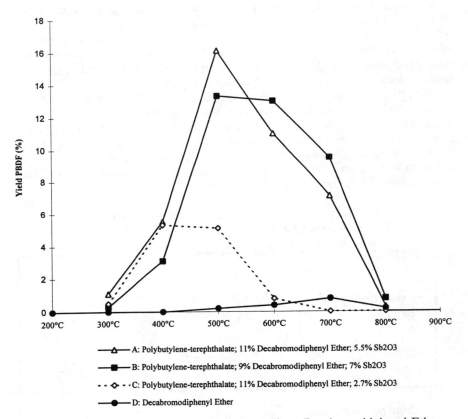

Figure 3. Yields of Brominated Dibenzofurans from Decabromobiphenyl Ether as Dependent on Temperature and Amount of Sb_2O_3 (Synergist) in the Polymeric Matrix (Polybutyleneterephthalate), Air flow rate 100ml/min.

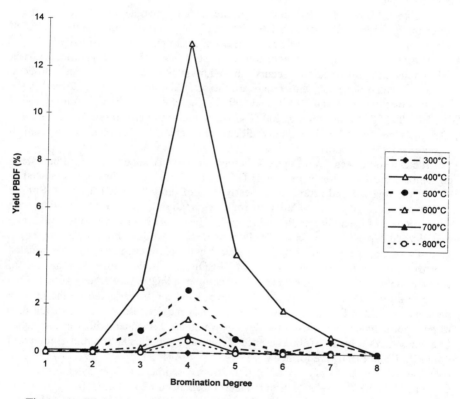

Figure 4. Yields and Distribution Pattern of PBDF formed from Thermal Degradation of Decabromodiphenyl Ether in Polybutyleneterephthalate with 5.5% Sb_2O_3 at 6 Different Temperatures.

[Reaction scheme: diphenyl ether with Br_{1-10} → (with [T], $-Br_2$) → dibenzofuran with Br_{1-8}]

The optimal yield for PBDF depends on the applied burning temperature (8, 10); this itself depends on the kind of polymeric matix, which is shown below for exposures in the DIN-oven.

Figure 3 shows that the amount of the added synergist Sb_2O_3 to the flame retardant strongly effects the PBDF yield and the optimal temperature of PBDF formation. The kind of polymeric matrix itself does not effect yields of PBDF.

During laboratory combustion of 1 in the polymeric matrix debromination/hydrogenation occurs in addition to the cyclization process. Tetrabrominated dibenzofuran isomers are the most abundant products formed in the temperature range between 300 °C and 400 °C. Figure 4 shows PBDF-composition at 300 °C - 800 °C. Combustion at 400 °C gives tetrabromo-benzofurans in yields up to 13 %. Besides of PBDF, brominated dibenzodioxins are also formed, but to a much lesser extent (30-90 ppm).

It can also be seen from Figure 4 that tetrabrominated dibenzofurans dominate at all temperatures used for incineration. The toxic 2,3,7,8-isomer has been detected as part of the mixture and estimated at nearly 2 % of the total PBDF amount. Similar results are obtained from oxidative thermal degradation of polymeric materials with octabromo- and pentabromodiphenyl ether. The temperature with the maximum PBDF-yield depends on the kind of polymeric matrix. All three brominated ethers 1-3 give the same isomer distribution pattern with preference for tetrabrominated dibenzofurans. The overall yield of PBDF is lower for degradation of pentabromobiphenyl ether 3, 4 % at 700 °C compared to 29 % for ether 1 at 500 °C.

The influence of metal species like copper has been investigated on the product pattern and yield of PBDD/F (Figure 5). This study may be relevant to accidental fires of polymeric materials of electronic devices which are associated with various metals like copper. As a result of the presence of the metal species substantial amounts of both PBDF and PBDD are formed.

The additional formation of PBDD from brominated diphenylether can be explained by a SET (single electron transfer) mechanism, see Figure 6 (8).

Important is the formation of PBDD by Ullmann reaction of brominated phenoxy radicals. Influences of added water, air content and other factors on PBDD/F yield and pattern have also been studied in detail (8).

The presence of metal oxides has a different influence on PBDD/F concentration and pattern, see Figure 7. Oxides of zinc and copper reduce the PBDD/F yields due to debromination reactions. The two oxides of copper show a distinctively different effect, while CuO leads to a strong reduction of PBD/F yield Cu_2O_2 enhances the PBDD concentration by the SET mechanism.

Besides these main products, formed in laboratory combustion of 1 in polymeric matrices, complex isomeric mixtures of brominated methyl-dibenzofurans and

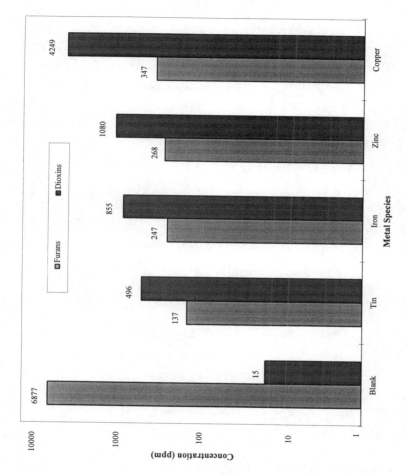

Figure 5. Influence of Various Metal Species on PBDD/F Formation during Oxidative Thermal Degradation of Decabromobiphenyl ether in Polybutyleneterephtalate at 500 °C (BIS-Furnace).

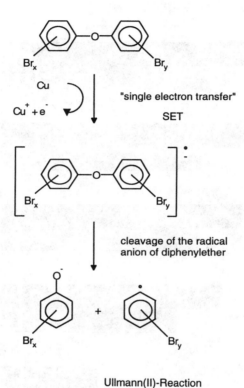

Figure 6. Formation Mechanism of PBDD from Brominated Diphenylether.

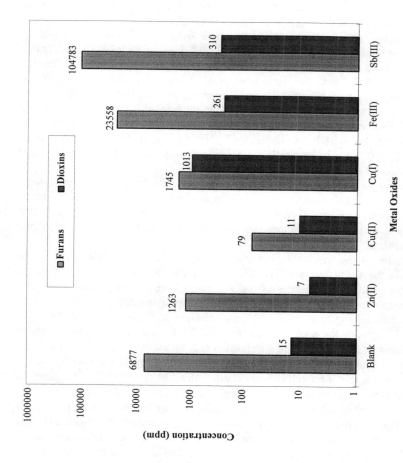

Figure 7. Influence of Various Metal Oxides on PBDD/F Formation during Thermal Degradation of Decabromobiphenyl ether in Polybutyleneterephthalate at 500°C (BIS Furnace).

brominated condensed systems like benzo[b]naphto[2,3,-d]furan have been identified by GC/MS (8).

It is possible that brominated dibenzodioxins and -furans are formed in trace levels during the preparation of technical bromoether products 1-3. Dumler (13) and others (14) have found PBDD/F in the ppm range in commercial samples of decabromodiphenyl ether. A more recent and advanced study has been performed on currently manufactured technical products of bromoethers 1-3; none of the 16 toxic PBDD/F isomers with the 2,3,7,8-pattern (the octabromo compounds were excluded) were detected within the detection limit (0,1 ng/kg) (15).

It was shown above that formation of PBDF from decabromobiphenyl ether 1 occurs even in the temperature range of 250-300 °C. Therefore it exists a chance that formation of these compounds starts during the technical extrusion process (mixing the polymeric material with the flame retardant). The concentration of PBDF in the plastic materials has been measured at three temperatures in a pilot plant test. PBDF were found in the resin blend in the range of 1-560 ppb as well as in the surrounding ambient air of the working place. High concentrations of PBDF were measured in the air; e.g. 34 ng/m^3 for T_3BDF; 143 ng/m^3 for T_4BDF 143 ng/m^3 for PBDF; 554 ng/m^3 for H_6BDF; 200 ng/m^3 for H_7BDF, besides smaller amounts of brominated dibenzodioxins (16).

While most studies of brominated flame retardants have been done under realistic condensed phase conditions, one study has been performed under gas phase oxidative and nonoxidative conditions (17). Brominated dibenzodioxins and -furans have also been observed in thermal decomposition of brominated diphenyl ethers under the applied conditions.

In addition to the laboratory combustion experiments described above some studies about the formation of PBDD/F in accidental fires of plastic materials in private houses and in vehicle fires of traffic tunnels have been performed (18, 19). The main source for PBDD/F in fires of private houses is the plastic TV-case, which contains brominated diphenyl ether. The concentrations of PBDD/F were in the range from 1 to 173 µg/kg in the fire residues (18). PBDD/F were spread from the source to all other rooms of the house. It was suggested that these special residues should be treated as hazardous waste (18). The potential hazards of a fire of a burning car and a subway waggon within a tunnel was modelled. Collectors for PBDD/F using special cotton surfaces modelling a passive and an active collection were developed (19).

The toxicology of PBDD/F has been investigated by several groups (20-22). It was shown that they exhibit similar toxicity studies in rats and mice compared to their chlorinated analogues. These results have lead to proposals for legislative actions for use of polybrominated diphenyl ethers in Germany and other countries (23). But it remains a controversial issue (24). In a special German program called "product controlled protection of the environment" new duroplastic materials for the electronic use with halogen free flame retardants have been developed (25, 26).

Summary. The hazardous potential of various brominated flame retardants during oxidative thermal degradation has been evaluated. Brominated diphenyl ethers form very high yields of toxic brominated dibenzofurans while burning in the polymeric matrix. The laboratory results are consistant with the reality of fire accidents in private houses and vehicle fire studies in tunnels.

Acknowledgements. We thank Mr. Helmar Liess for preparing some figures.

Literature cited

1. Paulatin, J. Fire Sciences, Vol. 11, 109-130 (1993);
   Engler, A.; Pieler, J.; Einbrodt, H. J. Wissenschaft und Umwelt, 1990, Vol. 3; p. 163-167; Vol. 4; 191-204

2. Fire and Polymers, Hazards Identification and Prevention, Nelson, G. L., Ed. ACS Symposium, Series No. 425, Washington 1989

3. Nelson, G. L. "Effects of Carbon Monoxide in Man: Exposure Fatality Studies", in Hirschler, M. M.; Debanne, S. M.; Larsen, J. B.; Nelson, G. L. Carbon Monoxide and Human Lethality: Fire and Non-Fire Studies, Elsevier Applied Sciences (Essex, England), 1993, pp. 33-60.

4. Kelley, M. E. Sources and Emissions of Polycyclic Organic Matter, EPA-450/5-83-010b (NTIS PB 84 - 144153); US Environmental Protection Agency: Research Triangle Park, NC, 1983

5. Dumler, R.; Teufl, C.; Lenoir, D.; Hutzinger, O. VDI-Report No 634, 287-292 (1987)

6. Hutzinger, O.; Dumler, R.; Lenoir, D.; Teufl, C.; Thoma, H. Chemosphere, 18, 1235-1242 (1989)

7. Klusmeier, W.; Vögler, P.; Ohrbach, K.-H.; Weber, H.; Kettrup, A. J. Anal. Appl. Pyrolysis, 13, 277-285 (1988)

8. Lenoir, D.; Zier, B.; Bieniek, D.; Kettrup, A. Chemosphere, 28, 1921-1928 (1994)

9. Ball, M.; Päpke, O.; Lis, A. German UBA-Report No. 10403364/01, Berlin 1992

10. Dumler, R.; Lenoir, D.; Thoma, H.; Hutzinger, O. J. Anal. Appl. Pyrolysis, 16, 153-158 (1989)

11. Luijk, R.; Wever, H.; Olic, K.; Govers, H. A. J. Organohalogen Compounds, Vol. 2, 335-338, Bayreuth 1990

12. Buser, H.-R. Chemosphere, 17, 889-902 (1988)

13. Dumler, R. Dissertation, University of Bayreuth, Germany, 1989

14. Wu, J. A Qualitative Method for Evaluation the Potential of Chemicals for Dioxin Contamination, Richover Science Institute, 1985

15. Freiberg, M.; McAllister, D. L.; Mazac, C. J.; Ranken, P. Analysis of Trace Levels of Polybrominated Dibenzo-p-dioxins and Dibenzofurans in Brominated Flame Retardants, Presentation on June 30, 1993 at Orgabrom in Jerusalem

16. Brenner, K. S.; Knies, H. Organohalogen Compounds, Vol. 2, 319-324, Bayreuth 1990.

17. Striebich, R. C.; Ruby, W. A.; Tirey, D. A.; Dellinger, B. Chemosphere, 23, 1197-1201 (1991)

18. Zelinski, V.; Lorenz, W.; Bahadir, M. Chemosphere, 27, 1519-1528 (1993)

19. Wichmann, H.; Scholz-Böttcher, B.; Zelinski, V.; Lorenz, W.; Bahadir, M. Chemosphere, 26, 1159-1166 (1993)

20. Löser, E., Ives, I. Chemosphere, 19, 759 (1989);
Nagao, T.; Neubert, D.; Löser, E. Chemosphere, 20, 1189-1192 (1990)

21. Buckley Kedderis, L.; Diliberto, J. J.; Linko, P.; Goldstein, J. A.; Birnbaum, L. S. Toxicology and Applied Pharmacology, 111, 163-172 (1991)

22. Schulz, T.; Golor, G.; Körner, W.; Hagenmaier, H.; Neubert, D. Organohalogen Compounds, Vol. 13, 145-148, Wien 1993

23. Meyer, H.; Neupert, M.; Pump, W. "Flammschutzmittel entscheiden über die Wiederverwertbarkeit", in: Kunststoffe, Vol. 83, 253-257 (1993).

24. Hardy, M. L. Status of Regulations Effecting Brominated Flame Retardants in Europe and the United States, in: Recent Advances in Flame Retardancy of Polymeric Materials, Vol. 5, in print, Lewin, M., Editor, Stamford, US, 1994

25. Gentzkow, v., W. Halogen Free Flame Retardants Thermosets for Electronics, in: Recent Advances in Flame Retardancy of Polymeric Materials, Vol. 5, in print, Lewin, M., Editor, Stamford, US, 1994

26. Lenoir, D.; Becker, L.; Thumm, W.; Kettrup, A.; Hauk, A.; Sklorz, M.; Hutzinger, O. Evaluation of Ecotoxicological Properties from Incineration of New Duroplastic Materials without Halogen as Flame Retardants, in: Recent Advances in Flame Retardancy of Polymeric Materials, Vol. 5, in print, Lewin, M., Editor, Stamford, US, 1994

RECEIVED March 22, 1995

Chapter 26

Analysis of Soot Produced from the Combustion of Polymeric Materials

Kent J. Voorhees

Department of Chemistry and Geochemistry, Colorado School of Mines, Golden, CO 80401

> Soot is composed of a carbonaceous matrix and often condensed products of incomplete combustion. Liquid chromatography, GC/MS, and pyrolysis-GC/MS have been used to analyze soot for the primary polymer degradation products, degraded polymers similar to the original structure, plus many of the additives of polymer systems. Both intact soot and extractable materials from the soot have been utilized. Specific examples of soot analyses from fires are discussed along with the success of applying the results to fuel source identification, fire growth, fire models, and environmental contamination.

The combustion of a polymer is a complex process that involves many parallel and sequential reactions. A simplified mechanism for this process can be represented in two primary steps:

polymer → polymer fragments + pyrolysates
pyrolysates → stable products (low and high molecular weight) + soot.

The initial reaction takes place as a pyrolysis process while the second reaction occurs in the combustion zone. Therefore, soot is composed of a highly condensed carbon material, an amorphous polymeric fraction, and condensed volatile substances. Early studies showed that for polyurethanes, the amorphous fraction was a degraded (lower molecular weight) material that had a similar repeating unit to the original polymer (1,2). In addition, many commercial polymer systems contain additives for stabilizing, processing, or property modifications (3). When these polymer systems are exposed to combustion conditions, many additives are distilled and eventually are condensed onto a soot particle (4,5).

    The thermal degradation products of many polymer systems have been extensively studied (6-9). The portion of these compounds that survive the combustion process as pyrolysates is usually a small percentage of the total isolated material. Some compounds in the pyrolysate fraction can be very diagnostic for establishing a particular polymer system as a fuel source. For example, polyvinyl chloride is known to quantitatively liberate hydrogen chloride at about 250°C (10) and has been identified in PVC soot (11,12). Other polymers such as polymethylmethacrylate "unzip" during pyrolysis to produce high percentages of monomer (4). Styrene monomer and dimer are observed from the degradation of polystyrene (13).

Several papers have been published describing the use of volatile compounds trapped in the soot matrix to identify a hydrocarbon fuel source (14-17). Takatsu and Yamamoto have examined the solvent extracts from laboratory soot samples produced from the combustion of benzene, toluene, xylenes, ethyl benzene, styrene, and cumene (14-17). The various chromatograms showed specific changes that varied according to the original fuel. Tsao and Voorhees (18) examined the volatile compounds in soot produced from the combustion of five polymeric materials with gasoline present as an accelerant. Pyrolysis-mass spectrometry was used with charcoal glued to the Curie-point wire (19) to fingerprint the volatile compounds. The spectra were successfully classified using factor analysis and the presence of gasoline established.

Characterization of the aerosol material which had been heated under vacuum to remove the volatile compounds has been studied by Voorhees and Tsao for both flaming and non-flaming combustion using both natural and synthetic polymers (12,20). These researchers found, using pattern recognition procedures on pyrolysis-mass spectrometry data, that the non-volatile material in soot could be used to identify the fuel producing the aerosol. Successful identification of 12 polymers in mixtures containing up to three of the individual polymers was above 70%.

The following paper summarizes the application of some of the previously described laboratory studies to the analysis of soot from actual fire sites. Applications to fuel source identification, fire growth, fire models, and environmental contamination are highlighted.

Experimental

Soot Collection

Soot samples were collected by scraping a soot coated surface with a razor blade followed by carefully placing the soot into a glass vial equipped with an aluminum lined cap. Except for Formica® surfaces, non-organic surfaces such as glass were chosen for scraping.

Gas Chromatography/Mass Spectrometry (GC/MS) Analysis

Extraction of the soot for GC/MS analysis was done using 20 to 65 mg of soot suspended in 0.15 mL of hexane. Suspensions were sonicated for 10 min, allowed to partition over a two-hour period, and then an aliquot of the hexane layer injected into the gas chromatograph. One-microliter injections were typically used for most of the samples.

A DB-5 fused capillary column (0.25 mm X 30 m) temperature programmed from 50° to 280°C at 10°C per minute was used in the GC/MS analysis. Detection of the eluents was accomplished using a Hewlett Packard MSD or a Finnigan TSQ-70 mass spectrometer in the electron ionization mode scanned from 40 to 450 Da. All chromatographic peaks were searched against the EPA/NIST mass spectral library. Homologous series of compounds were represented as selected ion plots. Both mass spectrometers were tuned with fluorinated standards to EPA specifications. Blank runs of the GC/MS system and the solvents were made before the initiation of the analysis. Hydrocarbon standards were run to establish a hydrocarbon number scale for the retention times.

Pyrolysis-Gas Chromatography/Mass Spectrometry Analysis (Py-GC/MS)

The pyrolysis of soot was performed using a Curie-point Pyrolyzer (University of Utah) with a Fisher Power supply (1.5 kW, 1.1 MHz) connected to a Finnigan TSQ-70 GC/MS-MS system. Samples (~ 10 µg of material) were coated as methanol suspensions onto 510°C Curie-point wires. The gas chromatography and mass spectrometer operating parameters were identical to those described for the GC/MS extract analysis.

Ion Chromatographic Analysis

Both soot and wipe samples have been analyzed by ion chromatography. The collection of soot has been previously described. The wipe samples were taken by wiping a 100 cm^2 area with a Whatman 40 filter paper wetted with 10 mL of isopropyl alcohol. Following collection, the filter paper was sealed in a glass vial using an aluminum lined cap. Appropriate field blanks, where the filter paper was wetted and then handled and placed in the container, were also collected.

Soot samples weighing approximately 20 mg were dissolved with sonication (10 min) in 10 ml of doubly distilled deionized water. Plastic laboratory equipment was exclusively used for the analysis. A Dionex ion chromatograph equipped with a 15 cm HPIC A54A analytical column was used for the chloride analysis with a mobile phase (flow rate = 2 mL/min) containing 1.8 mmoles Na_2CO_3 and 1.7 mmoles $NaHCO_3$. A 50 µL sample, previously filtered through a 0.45 µm filter, was injected through a 1.0 µL sample loop. The regenerate contained 25 mmoles of H_2SO_4. The retention time and the mass response for the chloride ion peak were determined by injecting standard sodium chloride solutions.

Sodium ion concentrations were determined for the same solutions by ion chromatography using a 15 cm HPIC CS3 analytical column. The regenerate was a 70 mmole solution of tetrabutylammonium hydroxide. Appropriate standards were used for retention times and mass response. The sodium and chloride levels for the soot samples are reported as weight percent.

The wipe sample filters were individually extracted using deionized water volumes from 0.5 mL to 15 mL. The ion chromatographic analysis scheme was identical to that previously described for the soot analysis. Because the actual amount of total material on the wipe samples is not known, the sodium and the chloride levels are reported as micrograms per wipe sample.

Polychlorinated Biphenyls (PCB) Analysis

The method used for the PCB wipe sampling was similar to that described by Ness (22). Gauze pads 3" X 3" (7.6 cm X 7.6 cm) saturated with 8 mL of Nanograde hexane were used to wipe a 100 cm^2 surface area. Following exposure, the pads were placed in a glass jar sealed with a metal foil lined screw cap. Blank samples were also taken where the gauze was saturated with hexane, handled and then returned to the glass jar.

Extraction of the gauze pad was done by sonicating for 5 min with 25 mL of Nanograde hexane followed by vacuum distillation of the solvent to a 1 mL final volume. The extracts were immediately analyzed on a gas chromatograph equipped with a Hall electrolytic conductivity detector. A 1 µL sample was injected onto a 30 m SPB-5 megabore column maintained at 180°C for one minute followed by a temperature program from 180 - 250°C at 10°C/min. The final temperature was held for 35 min. Other parameters for the analysis were: injector temperature - 220°C, detector temperature - 910°C, reaction gas - hydrogen, and carrier gas - helium at 17 ml/min. All chromatograms were electronically integrated and were compared to the appropriate PCB standards. Samples with the highest PCB levels were run by gas chromatography/mass spectrometry for verification. Chromatographic conditions for the GC/MS analysis were: column- 0.25 mm X 12 m, temperature programmed from 70 - 270°C at 10°C/min and scan range m/z 170 to 450. Mass spectral identifications were made by comparing the spectra to PCB mass spectra.

Results and Discussion

Ion Chromatography

Polyvinyl chloride is extensively used in commerce and has been involved in many structural fires. The following discussion presents the historical events and the chemical analyses of two fires that involved PVC combustion with the release of hydrogen chloride.

A fire occurred in a multistory office building that had been undergoing a refinishing project on the wood paneling in the executive offices located on one floor. The construction occurred on a Saturday and was finished for the day in the late afternoon. Oil, rags and other supplies were left in an office so that work could begin the next morning.

A single company occupied the entire floor of origin. The floor was laid out such that a series of executive offices surrounded a secretarial pool which was located in the center of the complex. The secretarial area was subdivided into stations and was nicely furnished. Polyvinyl chloride wall covering was present on the secretarial pool side of the divider wall between the executive offices and the secretarial stations.

A fire alarm from this floor was received on a central alarm board on the first floor and was observed by two security officers. One of the security people took an elevator to the fire floor where he was met by thick smoke. The guard later stated that he was overcome by the smoke and had to radio the other security person to override the elevator controls and return the elevator car to the first floor.

Several fire sources were proposed. The best scenario that fit the eye witnesses' accounts who were outside the building and the security guard, placed the fire origin in an executive office where the refinishing supplies had been left. The ignition source was speculated as occurring from spontaneous combustion of linseed oil soaked rags left in the office. Burned rags were recovered in this office during the post-fire investigation.

The PVC wall covering, known from purchase records to be in the secretarial pool, provided a probe to determine whether or not the fire had spread into this area at the time the security guard arrived on the floor. A smoke coating was visible on the inside surfaces of the elevator car that was used by the security guard. Wipe samples were collected from the inside of the car when scrapping provided insufficient quantities of soot. Table I summarizes the results of the ion chromatography analysis for the elevator wipe samples.

The major background contributor to the chloride level would primarily be sodium chloride. To gauge this contribution, it has been assumed that any sodium ion present was associated with sodium chloride. The measured sodium level was multiplied by a factor of 1.54 to obtain the chloride level that would be present if all sodium ions were in the form of sodium chloride. If large chloride levels and low sodium levels were obtained for a sample, it was assumed that the chloride was associated with hydrogen chloride released from the PVC combustion. Two soot samples collected from the fire floor during the post-fire investigation had the following sodium and chloride levels in weight percent:

| Sample #1 | Na-0.28% | Cl-1.7% | Excess Cl-1.3% |
| Sample #2 | Na-0.23% | Cl-3.8% | Excess Cl-3.4% |

Both samples show the presence of large quantities of chloride ion and low levels of sodium; therefore, significant quantities of hydrogen chloride were present.

The chloride values listed in Table I, except for sample #1, are near the background levels. Enough sodium is present to account for all chloride as sodium chloride. Sample #1 has excess chloride ion. Based on the results from a

Table I. Ion Chromatography Results from Elevator Wipe Samples

Sample	Na^+ Levels[1]	Cl^- Levels[1]
Deionized water blank	13.7	31.1
Field Blank #1 (Wetted)	26.3	33.0
Field Blank #2	15.3	29.3
Field Blank #3	5.0	24.3
Elevator Sample #1	43.5	149
Elevator Sample #2	44.0	39.6
Elevator Sample #3	31.8	45.6
Elevator Sample #4	20.3	36.1
Elevator Sample #5	25.8	33.7

[1] All levels for sodium and chloride ions are ug/wipe.

substantial number of samples analyzed in our laboratory, this value is very small and in most analyses would be considered as a negative for hydrogen chloride.

The conclusion from this analysis was that the PVC in the secretarial pool was not involved in the fire at the time the smoke was deposited on the elevator car's interior surfaces. Furthermore, these results support the hypothesis that the linseed rags in the office provided the combustion source for the fire.

Another application of ion chromatographic analysis has been to assess smoke movement associated with PVC combustion. A vinyl wallpaper was the major source of PVC in a recent hotel fire. This wall became involved in the fire immediately after flashover from the room of origin. The smoke from the burning PVC along with a portion of the smoke from the combustibles in the room of origin, was vented through the hotel casino and lobby where multiple deaths occurred.

Soot was collected by scraping non-organic surfaces throughout the lobby, casino, shops, and guest rooms. The results of the ion chromatographic analysis showed chloride levels from 4.6 wt % to a high of 32.7 wt %. Since this hotel was near the ocean, the question of sea spray was an important factor. In all but one sample, the ratio of sodium to chlorine clearly defined the high chloride levels to be from hydrogen chloride. The highest chloride level was found in a room off the casino area where very little fire or heat damage occurred. The lower chloride levels were in the casino and lobby areas that had suffered high heat damage. It became clear in correlating chloride levels to fire and heat damage that high heat fluxes cause thermal desorption of the adsorbed hydrogen chloride on deposited soot.

The other major application of chloride analysis we have used has been for directing cleanup after a PVC fire. Both soot and wipe sampling methods have been employed in these investigations.

Gas Chromatography/Mass Spectrometry Analysis of Soot

The development of GC/MS has provided the capability to analyze complex organic mixtures such as those obtained from extraction of soot. Figure 1 illustrates a total ion chromatogram of a soot solvent extract from the same hotel fire described previously. In general, the two compound classes in the greatest concentration in most samples are phthalate plasticizers (marked with an O) and polynuclear aromatic hydrocarbons (PAH) (marked with a X). Depending upon the characteristics of a fire, these two compound classes can dominate the chromatogram. However, even when this occurs as illustrated in Figure 1, marker compounds are often observed that can be used to identify a particular fuel source. For example, some of the minor peaks in this chromatogram were identified as phosphate plasticizers and fatty acids commonly used in PVC. The following example illustrates and expands the concept of using marker compounds to identify a commercial product present in a horse stable fire.

The manager of a horse training facility awoke early one morning to see an orange glow near his large horse barn. He saw that he could not extinguish the fire and immediately removed a tractor and then went into the stable portion of the barn and started to release the horses. While he was removing his prized Arabian stallion, the fire rapidly spread across the ceiling of the stable which resulted in the manager being overcome by the smoke and the horse eventually being killed. In total, 32 horses were killed in the fire.

The vapor barrier system associated with the ceiling appeared to provide the fuel for the rapid spread of the fire. Because of the legal aspects of the fire, it became important to identify the manufacturer of the barrier system. The legal discovery process provided composition information on several potential manufacturers' products and served as a database for the following investigation.

Figure 2 shows the total ion chromatogram of a hexane extract from a soot sample removed from a window in an area of the stable where only minor fire damage had occurred. Because this window had only minor fire impact, it had served as a cool surface for collection of the soot and distilled products. Table II summarizes the product identification associated with Figure 2. Polynuclear aromatic hydrocarbons in the extract are common combustion products; however, in this case, an asphalt binder used in the vapor barrier system was also a potential contributor. It should be noted that almost all soot sample extracts studied in our laboratory have contained some polynuclear aromatic hydrocarbons. The *normal* hydrocarbons observed in the stable soot sample are not a result of combustion, but of a distillation from one of the fuels.

After reviewing product information from the various manufacturers of vapor barrier materials that could have been in the barn, one was found that contained an asphalt material and a hydrocarbon wax. Figure 3 represents the total ion chromatogram of the hexane extract of the asphaltic-wax materials. Table III lists the compounds identified in the study. Note that a similar *normal*-hydrocarbon pattern of evenly spaced peaks with the same retention times exists in both Figures 2 and 3. The change in relative intensity occurs during distillation because of the differing vapor pressures and boiling points. A survey of the materials involved in this fire and the chemical formulations allowed for an identification of this product as being present in the stable barn.

Polymer additives have also been observed in soot extracts from other fires studied. Table IV lists the various types of organic additives that have been identified. Based on the complexity of the fire, select additives can be traced to a particular polymer system. The fire retardants used in polyurethanes are quite specific in identifying these materials as fuels.

Compounds of environmental interest are often found in soot extracts. We have observed polychlorinated biphenyls (PCB), chlorinated phenols, and insecticides in various fire samples. The primary sources of PCBs are light ballasts and paint. Frequently, the type of PCB (i.e., Araclor types) can be determined by comparing the soot extract data against PCB standards. Pentachlorophenol has been used as a wood preservative and is effectively distilled from treated wood prior to its combustion. A surprising class of compounds that have been detected in soot extracts is pesticides. These compounds were not the result of stored pesticide, but from the extended use of pesticides in an area later impacted by fire.

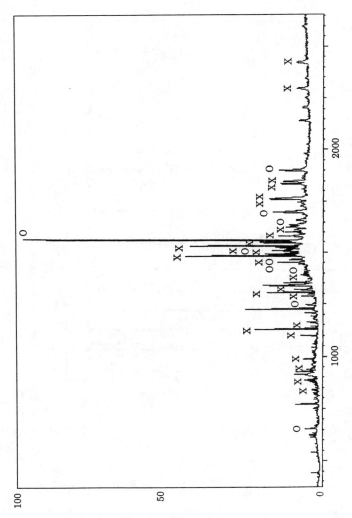

Figure 1. A Total Ion Chromatogram of a Hexane Extract of Soot Removed from a Hotel Fire Scene (O= plasticizer compounds, X = polynuclear aromatic hydrocarbons).

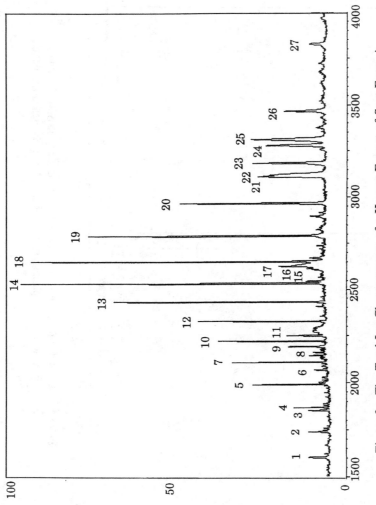

Figure 2. The Total Ion Chromatogram of a Hexane Extract of Soot From A Horse Barn Fire.

Table II. A Summary of Products Identified in a Hexane Soot Extract

Peak Number	Compound
1	n-Hexadecane
2	n-Heptadecane
3	Phenanthrene or anthracene
4	n-Octadecane
5	n-Nonadecane
6	Phthalate ester
7	n-Icosane
8	Dimethylphenanthrene or anthracene
9	Fluoranthene
10	n-Henicosane
11	not identified
12	n-Docosane
13	n-Tricosane
14	n-Tetracosane
15	Benzofluoranthene
16	Triphenylene
17	Chrysene
18	n-Pentacosane
19	n-Hexacosane
20	n-Heptacosane
21	Benzopyrene
22	Benzofluoranthene
23	n-Octacosane
24	Benzofluoranthene
25	Benzopyrene
26	n-Nonacosane
27	n-Triacontane

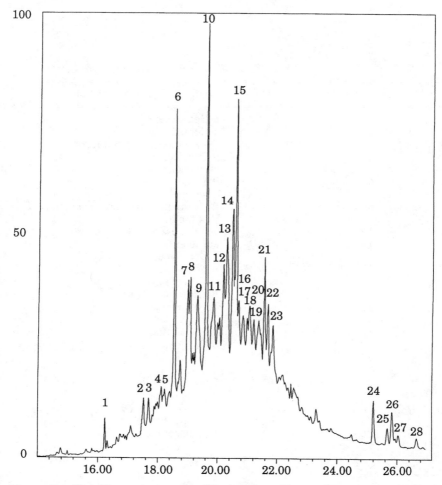

Figure 3. The Chromatogram Produced from a Hexane Extract of a Vapor Barrier System used in The Horse Barn.

Table III. Compounds Identified in the Hexane Extract of the Asphaltic Materials

Peak Number	Compound
1	n-Heptadecane
2	n-octadecane
3	Phenanthrene or anthracene
4	C_{18} hydrocarbon
5	C_{18} hydrocarbon
6	n-nonadecane
7	Methoxyphenylmethylbenzene
8	Methylphenanthrene or anthracene
9	Methylphenanthrene or anthracene
10	n-Icosane
11	Hexahydropyrene
12	Dimethylphenanthrene or anthracene
13	Dimethylphenanthrene or anthracene
14	Dimethylphenanthrene or anthracene
15	n-Henicosane
16	Dimethylphenanthrene or anthracene
17	C_{21} hydrocarbon
18	unknown
19	unknown
20	Trimethylphenanthrene or anthracene
21	n-Docosane
22	Trimethylphenanthrene or anthracene
23	Trimethylphenanthrene or anthracene
24	Phthalate
25	Hexacosane
26	Silicone peak
27	Silicone peak
28	Heptacosane

Table IV. Organic Additives Identified in Soot Extracts

Additive Type	Examples
Fire Retardants	Chlorinated phosphates
Plasticizers	Phthalate Esters
	Trialkyl Phosphates
	Triaryl Phosphates
	Fatty Acid Esters
	Glycerides
Antioxidants	Hindered Phenols
Lubricants	Fatty Acids

Direct Pyrolysis-GC/MS Analysis of Soot

Direct Curie-point pyrolysis-GC/MS (Py-GC/MS) from many of the soot samples that were subjected to extraction has also been conducted. Figure 4 represents a pyrolysis total ion chromatogram of the soot that was extracted to produce the chromatogram represented in Figure 2. In addition to the n-alkane series, many new peaks were observed in the pyrolysis chromatogram (Table V) when compared to the

Table V. A Summary of Compounds Identified from the Py-GC/MS of Soot

Peak Number	Compound
1	Acetic Acid
2	Phenol
3	Benzyl Alcohol
4	Isopropylbenzene
5	Hydroxymethoxybenzene
6	1,2-Dimethyl-2-pyrrolidine Carboxylic Acid
7	2-Ethoxynaphthalene
8	Trimethylnaphthalene
9	2,5-dimethylphenylbutanoic acid
10	Benzophenone
11	Pentachlorophenol
12	Phenanthracene
13	n-Nonadecane
14	Oxacycloheptadecanone
15	Hexadecanoic Acid
16	n-Icosane
17	Fluoranthene
18	n-Henicosane
19	Octadecanoic Acid
20	n-Docosane
21	n-Tricosane
22	n-Tetracosane
23	n-Pentacosane
24	n-Hexacosane
25	n-Heptacosane
26	Benzopyrene

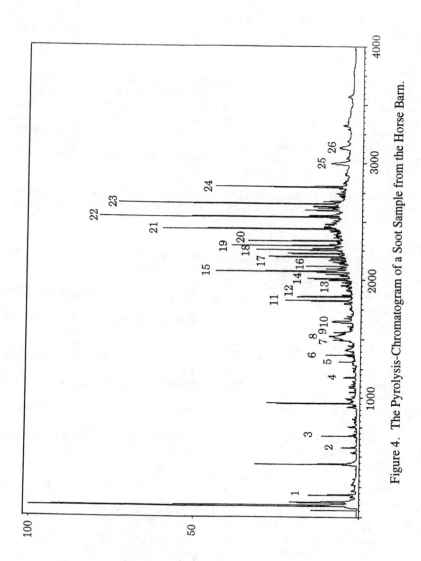

Figure 4. The Pyrolysis-Chromatogram of a Soot Sample from the Horse Barn.

extract. The small peaks near baseline were mostly polynuclear aromatics. In general, pyrolysis-GC/MS analysis has not been as successful as the extraction approach. This results from the added complexity that occurs from the degradation of the amorphous polymeric material remaining in the soot. It had been postulated that the polymeric material could be used to identify the presence of degraded polymer (12), but this has not occurred in most samples analyzed. The solvent extraction procedure has definitely been the more successful of the two approaches.

Conclusions
Soot has been shown to be an effective adsorbent material for both thermal polymer degradation products and polymer additives that are distilled from a polymer during combustion. These studies have shown that both organic and inorganic compounds in extracts can be detected using a variety of analytical procedures. The use of the data has been applied to solve a number of different problems associated with actual fire cases. For volatile organic analyses, data obtained from soot pyrolysis-GC/MS were more complex than the data obtained from soot extracts and generally contained less information about the fuel source of a fire.

The author would like to thank David N. Osborne for his assistance in conducting the analyses and Dr. Steven C. Packham for his interest in soot analysis applied to combustion toxicity.

Literature Cited
1. Hileman, F.D.; Voorhees, K.J.; Wojcik, L.H.; Birky, M.M.; Ryan, P.W.; Einhorn, I.N. J. Polym. Sci. Chem. Ed., 1975, 13, 571.
2. Wooley, W.D. Br. Polym. J., 1972, 44, 27.
3. Gachter, R.; Muller, H. In Plastic Additives, Hanser Publishers: New York and others, 1987.
4. Mordechai, P.; Zinn B.T.; Browner, R.F. Comb. Sci. and Tech., 28, 263 (1982).
5. Takatsu, M.; Yamamoto, T. Bunslei Kagaku, 1993, 42, 543.
6. Grassie, N. In Chemistry in High Polymer Degradation Processes, Interscience Publications: New York, 1964.
7. Madorsky, S.L. In Thermal Degradation of Organic Polymers, Interscience: New York, 1964.
8. Montaudo, G.; Puglisi, C. In Comprehensive Polymer Science, Pergamon Press: New York, 1992, p. 227.
9. Bryk, M.T. In General Degradation of Filled Polymers at High Temperatures and Thermal Oxidation Processes, Ellis Harwood: New York and others, 1991.
10. Boettner, E.A.; Weiss, B. J. Appl. Polym. Sci., 1969, 13, 337.
11. Stone, J.P.; Hazlett, R.N.; Johnson, J.E.; Carhart, H.W. J. Fire and Flam., 1973, 4, 42.
12. Tsao, R.; Voorhees, K.J. Anal. Chem., 1984, 56, 368.
13. Wu, B.; Yang, M.; Liu, G.;Pan X.; Han S. Sepu., 1993, 223.
14. Takatsu, M.; Yamamoto, T. Nippon Kagaku Kaishi, 1990, 1749.
15. Takatsu, M.; Yamamoto, T. Nippon Kagaku Kaishi, 1990, 880.
16. Takatsu, M.; Yamamoto, T. Nippon Kagaku Kaishi, 1991, 235.
17. Takatsu, M.; Yamamoto, T. J. Anal. Appl. Pyrol., 1993, 26, 53.
18. Tsao, R. Voorhees, K.J. Anal. Chem., 1984, 56, 1339.
19. Colenutt, B.A.; Thornburn, S. Chromatographia, 1979, 12, 12.
20. Voorhees, K.J.; Tsao, R. Anal. Chem., 1985, 57, 1630.

RECEIVED November 28, 1994

TOOLS FOR FIRE SCIENCE

Fire chemistry is difficult to investigate because reactions and reaction mixtures are complex. Key products such as char and char precursors are not soluble, and this feature makes more traditional chemical analyses difficult.

Fire scientists must be able to follow high-temperature complex chemistry, chemistry concentrated at surfaces, and interfaces. Bench-scale tools are necessary to access ignition, flame spread, and heat-release phenomena as they relate to definable real fires. The goal is to use bench-scale data in a mathematical model to understand the role of changes in material parameters on system fire performance. Those models need to be validated in the real world. Material goals for science-based specification or regulations can then be identified.

The chapters by Israel et al. and Wang discuss two powerful techniques for elucidating chemical reactions relevant to fire. Pyrolysis—chemical ionization mass spectrometry allows pyrolysis directly in the source of a double-focusing mass spectrometer. Rapid heating rates to high temperatures allow simulation of conditions more nearly akin to fire. Key volatile species can be identified as a function of material composition. For the solid phase, Wang applies X-ray photoelectron spectroscopy to several sample systems. Because XPS is a surface-sensitive tool, structures and chairing processes in the condensed phase can be studied as a function of temperature.

Over the past decade, heat-release calorimetry has become an important tool. However, bench-scale tests tend be overventilated, that is, the heat scenario is that of a fully developed fire while the oxygen availability is that of a developing fire. The development of a controlled-atmosphere cone calorimeter (Christy et al.) allows the investigator to more carefully control conditions and to study the effects of those conditions on material performance. Tewarson contrasts three heat-release rate apparatuses and their capabilities.

The need for lateral flame-spread data in addition to heat-release data for fire modeling has focused attention on the development of suitable bench-scale apparatus. Dietenberger discusses the evolution of the lateral ignition and flame-spread test (LIFT).

With suitable bench-scale assessment tools mathematical models can be used for engineering analyses of fire potential for real applications. Ohlemiller and Cleary discuss how three models of upward flame spread tested against intermediate-scale fire experiments. Data characterization for input into the models and model limitations are also discussed.

Janssens' chapter discusses a new single-compartment, single-vent generic fire model. Primary- and secondary-state variables are calculated as a function of fire. Temperature and composition of the two gas layers are defined, as are wall temperatures, mass flows through the vent, and entrainment into the fire. One or multiple fires can be specified; thus the model can be used for different types of fires.

In the future advances in analytical techniques, bench-scale evaluation, and models will permit sophisticated engineering analysis of materials and systems for fire performance at a level not possible today.

Chapter 27

The Computer Program Roomfire
A Compartment Fire Model Shell

Marc L. Janssens

American Forest and Paper Association, 1111 Nineteenth Street, N.W., Suite 800, Washington, DC 20036

This paper summarizes the physical basis for ROOMFIRE, a new single-compartment single-vent generic fire model. ROOMFIRE solves the conservation equations of mass and energy for the lower and upper gas layers in the compartment. Various auxiliary equations are solved simultaneously. Thus, ROOMFIRE calculates primary and secondary state variables as a function of time. The former define the geometry, temperature, and composition of the two gas layers. The latter include wall temperatures, mass flows through the vent, and entrainment into the fire. One or multiple fires are specified in separate modules that are external to ROOMFIRE. Therefore, ROOMFIRE can be used for different types of fires, by using it in combination with the appropriate fire modules.

Over the past few years, the North American wood industry has worked on the development of a single-compartment mathematical fire model. The main applications of the model are
- prediction of the fire risk of wall linings and their contribution to a growing compartment fire, relative to that from contents; and
- determination of natural fire exposure conditions for structural wood members, and wood-frame assemblies.

Therefore, the model needs to cover both the pre-flashover and post-flashover regimes.

There are numerous compartment fire models available, so why not adopt an existing one? CFAST is probably the most comprehensive pre-flashover fire model (*1*). However, this program is rather extensive, partly because it is a multi-room model. This makes customizing difficult and tedious. Moreover, a less complex and quicker program is desirable for fire risk assessment, which requires a large number of simulations for one particular case. COMPF2 is an excellent post-flashover compartment fire model (*2*). However, this model does not allow the user to specify different materials for walls, floor and ceiling. Various other models were examined, but all were found unsuitable for adoption due to specific problems or missing features. In addition, previous attempts to build on an existing model were not successful (*3*). For these reasons, it was decided to develop a new model.

The new model is modular. It consists of a generic compartment fire model, and one or several sub-models that simulate the fire. The same generic model is used for different types of fires (corner fires, furniture fires, crib fires etc.), by adding appropriate fire sub-model(s) or module(s). The generic model is named ROOMFIRE. It acts as a shell for the fire sub-model(s), and is the subject of this paper.

Overview of ROOMFIRE

ROOMFIRE is a zone fire model for a single compartment with a single vertical ventilation opening. The gas volume inside the compartment is subdivided into a lower layer of cold air, and an upper layer of hot gases. Given the fire (specified by sub-models), ROOMFIRE calculates primary and auxiliary state variables as a function of time. The primary variables define the two gas layers inside the compartment. The primary variables used in ROOMFIRE are location of the interface between lower and upper gas layer, z_i; neutral plane height, z_n; and lower and upper gas layer temperatures, T_l and T_u. The auxiliary variables are mass fractions of major species in the upper layer, $Y_{i,u}$; wall, floor, and ceiling temperatures, $T_1 \ldots T_{10}$; and mass flows through the ventilation opening, $\dot{m}_i$ and $\dot{m}_o$.

ROOMFIRE assumes fires are point sources of mass and heat. A certain fraction of the latter is released as radiation. To calculate heat transfer, the interior area of the compartment is subdivided in 10 rectangular sections, as shown in Figure 1. Since mixing between the layers is neglected, the lower gas layer is assumed to be transparent to thermal radiation. However, the lower layer gas temperature can exceed ambient temperature due to convective heat transfer with the floor and lower wall sections. The upper layer absorbs and emits radiation due to the presence of CO_2, H_2O, and soot. Radiative heat transfer calculations are performed according to Forney (*4*). For heat conduction, the walls, floor, and ceiling are approximated as semi-infinite solid slabs.

ROOMFIRE calculates the primary variables by solving the conservation equations of mass and energy for the two gas layers. Source terms in these equations are partly determined from the heat transfer calculations, and partly specified by the fire module(s). Since pressure equilibrates much more rapidly than the other variables, it is assumed constant and equal to ambient. This is an acceptable assumption for a single-compartment model. It avoids numerical problems due to the stiffness of the conservation equations (*5*).

Flashover is assumed when T_l reaches 600°C. At that time, the option exists to switch to a single gas layer geometry.

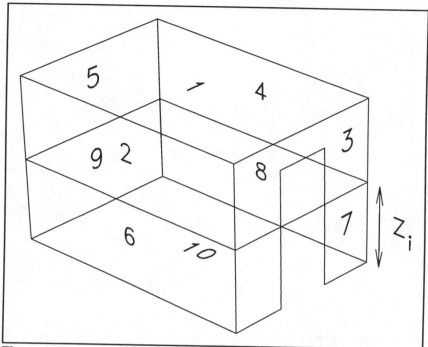

Figure 1   Wall sections for heat transfer calculations

Gas Layer Conservation Equations

Mass flows are shown in Figure 2. Mixing between the layers is neglected. The conservation equation of mass for the lower layer is

$$\frac{dm_l}{dt} = \dot{m}_i - \dot{m}_e. \tag{1}$$

Furthermore, to relate density to temperature for the lower and upper layer gases, the equation of state for dry air is used. Consequently, the mass of the lower layer can be expressed as

$$m_l \approx LWz_i \frac{\rho_\infty T_\infty}{T_l}. \tag{2}$$

Substitution of equation (2) in equation (1) leads to

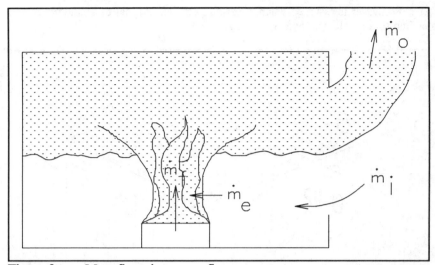

Figure 2 Mass flows in a room fire

$$\frac{dz_i}{dt} = \frac{T_l(\dot{m}_i - \dot{m}_e)}{LW\rho_\infty T_\infty} + \frac{z_i}{T_l}\frac{dT_l}{dt}. \qquad (3)$$

Mass flows and the temperature derivative on the right hand side are shown below to be a function of the primary variables. Hence, the lower layer mass conservation equation is of the following form

$$\frac{dz_i}{dt} = f_1(z_i, z_n, T_l, T_u). \qquad (4)$$

Conservation of mass for the upper layer is expressed by

$$\frac{dm_u}{dt} = \dot{m}_e + \dot{m}_f - \dot{m}_o. \qquad (5)$$

This can be transformed in an analogous way to

$$\frac{dz_i}{dt} + \frac{T_u(\dot{m}_e + \dot{m}_f - \dot{m}_o)}{LW\rho_\infty T_\infty} + \frac{H - z_i}{T_u}\frac{dT_u}{dt} = 0, \qquad (6)$$

or

$$f_2(z_i, z_n, T_l, T_u) = 0. \qquad (7)$$

Because the time derivative of pressure is neglected, upper layer mass conservation leads to an algebraic rather than a differential equation.

Since layer mixing is ignored, composition of the lower layer is equal to that of the incoming air. Composition of the upper layer is determined from the solution of species conservation equations

$$\frac{dY_{i,u}}{dt} = \frac{\dot{m}_e(Y_{i,l} - Y_{i,u}) + \dot{m}_f(\omega_i - Y_{i,u})}{m_u}. \tag{8}$$

Concentrations are tracked for N_2, O_2, CO_2, H_2O, soot and fuel. It is assumed that combustion is complete (no CO), except for soot production. Equation (8) is solved for the mass fraction of O_2, which is not permitted to be lower than a starvation limit. The concentrations of other species are determined on the basis of combustion stoichiometry.

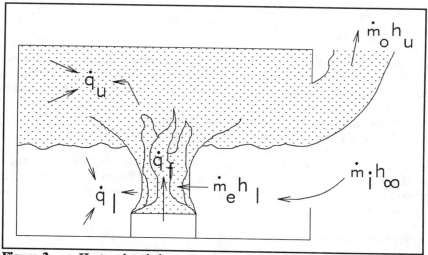

Figure 3 Heat and enthalpy terms in a room fire

Figure 3 shows the heat and enthalpy flows pertinent to the energy conservation of the gas layers. Application of the first law of thermodynamics to the lower layer results in the following energy conservation equation. More details can be found in reference (6).

$$\frac{dT_l}{dt} = \frac{\dot{m}_i[c_\infty(T_\infty - T_o) - c_l(T_l - T_o)] + \dot{q}_l}{m_l c_l + m_l(T_l - T_o)\left[\frac{dc_l}{dT}\right]_{T=T_l}}, \tag{9}$$

or

$$\frac{dT_l}{dt} = f_3(z_i, z_n, T_l, T_u). \tag{10}$$

The upper layer equation is obtained in a similar way. Neglecting the sensible enthalpy of $\dot{m}_f$, upper layer energy conservation is expressed as

$$\frac{dT_u}{dt} = \frac{\dot{m}_e[c_l(T_l-T_o)-c_u(T_u-T_o)]+\dot{q}_u+\dot{q}_f}{m_u c_u + m_u(T_u-T_o)\left[\frac{dc_u}{dT}\right]_{T=T_u}}, \tag{11}$$

or

$$\frac{dT_u}{dt} = f_4(z_i, z_n, T_l, T_u). \tag{12}$$

In equations (9) and (11), c_x is the average specific heat at constant pressure between temperatures T_o and T_x for gas mixture x. The heat absorbed by the lower layer, $\dot{q}_l$, and the upper layer, $\dot{q}_u$, are obtained from the heat transfer calculations discussed below. The generation rate of fuel volatiles, $\dot{m}_f$, is determined by the fire module(s). Heat release rate, $\dot{q}_f$, is obtained by multiplying $\dot{m}_f$ with the effective heat of combustion of the fuel volatiles, possibly adjusted for oxygen starvation.

In conclusion, the conservation of mass and energy for the two gas layers is expressed in the form of three ordinary differential equations (4), (10), (12); and one non-linear algebraic equation (7). The differential equations are solved via a 4th order Runge-Kutta method with stepsize control. The algebraic equation is solved using a bisection technique.

Auxiliary Equations

Additional equations are needed for mass flows and energy source terms in the conservation equations as a function of the primary variables. These auxiliary equations are described below.

Mass Flows through the Doorway. Several flow regimes have to be considered:
A. Shortly after ignition, combustion products accumulate underneath the ceiling. As the smoke layer descends, cold air is pushed out of the vent. The neutral plane is located below the door sill ($z_n < z_s$).
B. At a certain time, the upper smoke layer reaches the soffit. However, the upper layer volume still expands at a rate faster than the entrainment of air into the fire. So, both upper and lower layer gases leave the compartment, and z_n is still below z_s.

C. Eventually, a quasi-steady state establishes with cold air flow into the compartment at the bottom of the vent and upper layer gases leaving the room at the top. z_n is between z_s and z_d.
D. If the fire grows beyond a certain limit (usually after flashover), the inflow of air into the compartment may no longer be controlled by the entrainment rate, but is restricted by the ventilation opening size.
E. Eventually, when the fire dies down, a regime develops that is the opposite from B.

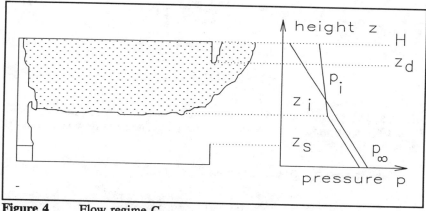

Figure 4   Flow regime C

In most compartment fires, regimes A and B prevail only for a short time (less than 30 seconds). Regime C is the most important regime during the remainder of the pre-flashover fire period and beyond.

The flow in and out of the compartment is driven by the hydrostatic pressure differences across the vent, which vary with height as shown in Figure 4 for flow regime C. For all regimes, flows are obtained from integration over height between z_s and z_d of velocity which is a function of pressure difference (or height) according to Bernouilli's equation. The resulting equations for the most important flow regime are given below. Details can be found in reference (7). The equations are rather complex. To simplify the notation, some new symbols are defined:

$$\theta_{\infty u} \equiv \frac{1}{T_\infty}\left[\frac{1}{T_\infty} - \frac{1}{T_u}\right], \tag{14}$$

$$\Theta_{ll} \equiv \frac{1}{T_l}\left[\frac{1}{T_\infty} - \frac{1}{T_l}\right], \tag{15}$$

$$\Theta_{ul} \equiv \frac{1}{T_u}\left[\frac{1}{T_\infty} - \frac{1}{T_l}\right], \tag{16}$$

$$\Theta_{uu} \equiv \frac{1}{T_u}\left[\frac{1}{T_\infty} - \frac{1}{T_u}\right], \tag{17}$$

$$\gamma_i \equiv 1042 C_i W_d, \tag{18}$$

$$\gamma_o \equiv 1042 C_o W_d, \tag{19}$$

where C_i and C_o are orifice coefficients for inflow and outflow respectively (both $\approx$ 0.7), and W_d is the width of the ventilation opening. The mass flow equations differ for $z_i < z_n$ and $z_i \geq z_n$.

Case 1: $z_i < z_n$. The equation for $\dot{m}_i$ is

$$\dot{m}_i = \frac{\gamma_i}{\Theta_{\infty l}}\left\{\left[\Theta_{\infty l}(z_i - z_s) + \Theta_{\infty u}(z_n - z_i)\right]^{3/2} \right. \\ \left. + \left(\Theta_{\infty u}^{1/2}\Theta_{\infty l} - \Theta_{\infty u}^{3/2}\right)(z_n - z_i)^{3/2}\right\}. \tag{20}$$

T_l should always be higher than T_∞, so that there is no problem in the evaluation of equation (20). In the limiting case of $T_l = T_\infty$, application of de l'Hôpital's rule leads to

$$\dot{m}_i = \gamma_i \Theta_{\infty u}^{1/2}(z_n - z_i)^{1/2}\left[z_n + \frac{z_i - 3z_s}{2}\right]. \tag{21}$$

An equation for the outflow is obtained in a similar way as for the inflow:

$$\dot{m}_o = \gamma_o \Theta_{uu}^{1/2}(z_d - z_n)^{3/2}. \tag{22}$$

Case 2: $z_i \geq z_n$.

$$\dot{m}_i = \gamma_i \Theta_{\infty l}^{1/2}(z_n - z_s)^{3/2}. \tag{23}$$

$$\dot{m}_o = \frac{\gamma_o}{\Theta_{uu}}\left\{\left[\Theta_{uu}(z_d-z_i)+\Theta_{ul}(z_i-z_n)\right]^{3/2}\right.$$
$$\left.+\left(\Theta_{ll}^{1/2}\Theta_{uu}-\Theta_{ul}^{3/2}\right)(z_i-z_n)^{3/2}\right\}. \qquad (24)$$

Mass Flow Entrained in the Fire. ROOMFIRE allows for fires located in the middle of the room, against a wall, or in a corner. The location is specified by the fire module(s). For a fire in the middle of the room, $\dot{m}_e$ is calculated as a function of fire strength and entrainment height below z_i from McCaffrey's correlations (8). For a fire in a corner, correlations obtained by Tran and Janssens are used (9). For a fire against a wall, an average between the center and corner entrainment values is used.

Radiative Heat Transfer. Heat transfer to the layers has to be determined as part of a more general calculation that also includes the fire(s), floor, vertical walls, and ceiling. The inside surface of the enclosure is subdivided into 10 smaller surfaces, each with a uniform temperature T_j ($j=1...10$). The enclosure is also partially filled with an absorbing-emitting gas at temperature T_u. 5 of the 10 surfaces ($j=1...5$) are in direct contact with the upper layer. The remaining 5 surfaces ($j=6...10$) are in contact with the transparent lower layer gas.

In addition to the 10 surfaces, radiation from fires also has to be considered. Fires are treated as point sources that emit, but do not absorb radiation. Therefore, no explicit radiation transfer equation has to be considered for fires. Each fire releases a certain fraction χ_r of its heat release rate, $\dot{q}_f$, in the form of thermal radiation, where χ_r is specified by the fire module(s). Consequently, radiative heat transfer between the 10 surfaces is described by the following set of transfer equations ($k=1...10$):

$$\sum_{j=1}^{10}\left[\frac{\delta_{kj}}{\epsilon_j}-F_{kj}\tau_{kj}\frac{1-\epsilon_j}{\epsilon_j}\right]\dot{q}''_{j,r} = \sum_{j=1}^{10}\left[(\delta_{kj}-F_{kj}\tau_{kj})\sigma T_j^4 - F_{kj}\alpha_{kj}T_u^4\right]$$
$$-\sum_{fires}\frac{F_{fk}\tau_{fk}\chi_r \dot{q}_f}{A_k}-F_{kf}\alpha_{kf}\sigma T_u^4. \qquad (25)$$

This set is obtained by integrating similar equations for monochromatic radiation derived by Siegel and Howell (10) over all wavelengths, assuming that the gas and all surfaces are grey, and taking flame radiation into account. Forney's algorithms are used to determine the net radiation to each surface, and the radiation absorbed by the upper layer as a function of surface temperatures (4).

If neither of the surfaces j or k are in contact with the upper layer, τ_{kj} and α_{kj} are equal to 1 and 0 respectively. Otherwise, they are calculated from

$$\alpha_{kj} = 1-\tau_{kj} = \epsilon_u, \qquad (26)$$

where the emissivity of the upper layer, ϵ_u, follows from an equation suggested by Tien et.al. (11):

$$\epsilon_u = \epsilon_s + \epsilon_g(1-\epsilon_s). \tag{27}$$

ϵ_s is the soot emissivity, which is a function of soot concentration in the upper layer and T_u according to equation (17-143) in reference (10). The gas emissivity, ϵ_g, is obtained as a function of $(X_{H_2O}+X_{CO_2})S$ and T_u as proposed by Hadvig (12). The mean beam length, S, for the upper layer gas volume, V_u, is estimated as $3.6\, V_u/A_u$, where A_u is the total enveloping surface area of V_u. Equation (26) is also valid for τ_{fk}, provided the fire is located in the upper layer or surface j is in contact with the upper layer.

Convective Heat Transfer. The convective heat flux to each wall surface, $\dot{q}''_{j,c}$, can be written as

$$\dot{q}''_{j,c} = h_j(T_{fluid} - T_j), \tag{28}$$

where T_{fluid} is the temperature of the gas layer in contact with surface j. The convection coefficient, h_j, is obtained from empirical correlations for natural convection over an isothermal flat plate (13). Such correlations have the following non-dimensional form:

$$Nu_j = \frac{h_j l_j}{k_j} = C(Gr_j Pr_j)^n, \tag{29}$$

Both C and n depend on the orientation of the plate, the flow regime (laminar or turbulent) and whether the plate is cooled or heated by the fluid. Appropriate values are taken from reference (13). The Nusselt, Grashof, and Prandtl numbers in equation (29) are calculated using properties of air evaluated at the film temperature, i.e., the average of fluid and surface temperature. The length scale, l_j, is equal to $(L \cdot W)^{1/2}$ for horizontal surfaces, and to the height for vertical surfaces.

Heat Conduction. The net heat flux to a surface, $\dot{q}''_{j,net}$, is the sum of net radiative and convective heat fluxes, which are calculated as described above. Given the net heat flux to a surface at time t, surface temperature is updated to the next time step via a numerical solution of the heat conduction equation. It is assumed floor, walls, and ceiling behave as semi-infinite solids. A numerical implementation of Duhamel's superposition theorem then results in the following equation:

$$T_j(t+\Delta t) = T_\infty + \frac{1}{\sqrt{\frac{1}{(T_j(t)-T_\infty)^2} + \frac{\Delta t \pi k_j \rho_j c_j}{E_j^2(t)}}} + \dot{q}''_{j,net}(t)\sqrt{\frac{4\Delta t}{\pi k_j \rho_j c_j}}, \quad (30)$$

where $E_j(t)$ is the total absorbed energy prior to t:

$$E_j(t) \approx E_j(t-\Delta t) + \frac{\Delta t \left(\dot{q}''_{j,net}(t-\Delta t) + \dot{q}''_{j,net}(t)\right)}{\sqrt{4\pi k_j \rho_j c_j}}. \quad (31)$$

However, as long as T_j remains close to the initial temperature, T_∞, and E_j is very small, the following equation is used instead of equation (30):

$$T_j(t+\Delta t) = T_j(t) + \dot{q}''_{j,net}(t)\sqrt{\frac{4\Delta t}{\pi k_j \rho_j c_j}}. \quad (32)$$

Values must be chosen for thermal inertia, $k_j \rho_j c_j$, that are representative for the range of temperatures T_j is expected to cover. Suitable values can be found in the literature. For combustible materials, such values can also be obtained from analysis of piloted ignition data (14).

Sequence of Calculations

This section outlines the sequence of calculations followed by ROOMFIRE. If values are available for the primary and secondary variables at time t, the following steps are taken to obtain values at $t+\Delta t$:
- calculate net radiative fluxes to the surfaces $j=1...10$, and radiation absorbed by the upper layer at time t (eq. 25-27);
- calculate convective fluxes to the 10 surfaces at time t (eq. 28-29);
- calculate surface temperatures at $t+\Delta t$ (eq. 30-32);
- calculate total heat transfer to lower ($\Sigma \dot{q}_{j,c}$, $j=6...10$) and upper ($\Sigma \dot{q}_{j,c}$, $j=1...5$ and $\Sigma \dot{q}_{j,r}$, $j=1...10$) layers;
- solve conservation equations by iteration (eq. 4, 7, 10, and 12), and determine updated mass flows in the process; and
- pass values at $t+\Delta t$ to fire module(s) and obtain updated fire information.

This sequence is repeated every time step, Δt (≈ 1 second), until the user-specified simulation time is reached.

Conclusions

This paper summarizes the physical basis for ROOMFIRE, a new single-compartment single-vent generic fire model. One or multiple fires are specified in separate modules that are external to ROOMFIRE. Therefore, ROOMFIRE can be used for different types of fires, by using it in combination with the appropriate fire modules. The first application of ROOMFIRE, to be completed in the near future, will be the simulation of an extensive series of room/corner tests with various wood products used as wall linings (15)(16). For this purpose, a fire module will be constructed on the basis of room/corner flame spread algorithms developed by Quintiere (17).

Nomenclature

c	specific heat at constant pressure ($kJ \cdot kg^{-1} \cdot K^{-1}$);
F	configuration factor for radiative heat transfer;
f	function;
H	room height (m);
h	enthalpy ($kJ \cdot kg^{-1}$), convection coefficient ($kW \cdot m^{-2} \cdot K^{-1}$);
L	room length (m);
m	mass (kg);
$\dot{m}$	mass flow ($kg \cdot s^{-1}$);
p	pressure (Pa);
$\dot{q}$	heat flow (kW);
S	mean path length for radiation (m);
T	temperature (K);
t	time (s);
W	room width (m), width (m);
X	mole fraction;
Y	mass fraction ($kg \cdot kg^{-1}$); and
z	height (m).

Greek Symbols

α	absorptivity;
γ	auxiliary constant in door flow calculations ($kg \cdot K \cdot s^{-1} \cdot m^{-3/2}$);
δ	Kronecker delta;
ϵ	emissivity;
Θ	auxiliary variable in door flow calculations (K^{-2});
ρ	density ($kg \cdot m^{-3}$);
τ	transmittance;
χ	fraction; and
ω	species yield ($kg \cdot kg^{-1}$).

Subscripts

c	convection;
d	door;
e	entrained;
f	fuel;
i	interface, inflow, species #, interior;

j	surface #;
k	surface #;
l	lower layer;
n	neutral plane;
o	outflow, reference;
r	radiation;
s	sill;
u	upper; and
∞	ambient, initial.

Literature Cited

(1) R. Peacock, G. Forney, P. Reneke, R. Portier, and W. Jones, Technical Note 1299, National Institute of Standards and Technology, Gaithersburg, MD, 1993.
(2) V. Babrauskas, Technical Note 991, National Bureau of Standards, Gaithersburg, MD, 1979.
(3) M. Janssens, ASTM STP 1233, ASTM, Philadelphia, PA, 169-185, 1994.
(4) G. Forney, NISTIR 4709, National Institute of Standards and Technology, Gaithersburg, MD, 1991.
(5) R. Rehm and G. Forney, NISTIR 4906, National Institute of Standards and Technology, Gaithersburg, MD, 1992.
(6) M. Janssens, *Heat Release in Fires*, Ch. 6, Elsevier, London, 1992.
(7) M. Janssens and H. Tran, *J. of Fire Sciences*, (10), 528-555, 1992.
(8) B. McCaffrey, NBSIR 79-1910, National Bureau of Standards, Gaithersburg, MD, 1979.
(9) M. Janssens and H. Tran, *J. of Fire Protection Engineering*, (5), 53-66, 1993.
(10) R. Siegel & J. Howell, *Thermal Radiation Heat Transfer*. Hemisphere Publishing Co., New York, NY, 1981.
(11) Tien C., Lee K. and Stretton A., *SFPE Handbook of Fire Protection Engineering*, Ch.1-5, SFPE, Boston, MA, 1989.
(12) S. Hadvig, *Journal of the Institute of Fuel*, (1), 129-135, 1970.
(13) N. Özişik, *Heat Transfer - A Basic Approach*. McGraw-Hill Book Company, New York, NY, 1985.
(14) ASTM E1321, *Annual Book of Standards*, (04.07), 993-1008, 1993.
(15) H. Tran and M. Janssens, *J. of Fire Sciences*, (7), 217-237, 1989.
(16) H. Tran and M. Janssens, *J. of Fire Sciences*, (9), 106-124, 1991.
(17) J. Quintiere, *Fire Safety J.*, (20), 313-339, 1993.

RECEIVED November 21, 1994

Chapter 28

Upward Flame Spread on Composite Materials

T. J. Ohlemiller and T. G. Cleary

Building and Fire Research Laboratory, National Institute of Standards and Technology, Gaithersburg, MD 20899

Three existing models of upward flame spread were tested against intermediate-scale experiments on a vinyl-ester/glass composite. Characterization of rate of heat release per unit area, needed as input to the models, was obtained at external radiant fluxes below the minimum for ignition by adaptation of a method due to Kulkarni. There are several limitations on the accuracy of the material characterization when applied to composites. Each of the flame spread models has definite limitations as well. Nevertheless, all three models produced predictions of spread behavior in sufficiently quantitative agreement with the experiments that they should prove useful for engineering analyses of flame spread potential.

The U. S. Navy is investigating the use of composite materials for both ship and submarine compartment construction. Load-bearing compartment walls would consist of flat (or possibly more complex) panels of such material. The composite construction materials of interest consist of two components: long fibers of high tensile strength material which are the primary load bearing elements and a continuous organic resin which surrounds the fibers to protect them, hold them in place and transfer stresses between fibers (*1*). Typically the fibers are in several layers or plies, each of which may be woven into some fixed pattern and oriented to maximize resistance to load stresses in specific directions. Such composites offer a high strength-to-weight ratio and other advantages, such as corrosion resistance, which make them attractive for a wide variety of structural uses.

In any structural use of composites, one must be aware of two potential limitations brought on by the organic nature of the binder resin. Both are potential consequences of a fire in proximity to the composite. First, the key structural roles of the resin--to hold the fibers in place and transfer stresses among them--may be compromised as the temperature of the composite approaches the glass transition temperature of the resin. The result is that the composite loses its strength, especially toward compressive loads. This is particularly a problem for certain relatively inexpensive thermoplastic resins (2). Second, the resins are typically flammable, to varying degrees, and thus may become involved in and contribute to the spread of a fire.

It is the second problem area which is addressed here. In particular, we are seeking a consistent way in which to characterize composite materials in small-scale tests so as to obtain the necessary data to predict their full-scale flammability behavior. The focus here is on upward spread of flames on the surface of a composite because it typically represents the fastest mode of fire growth. Three existing models of this spread process are compared with data obtained on a non-flame retarded vinyl ester/glass roving composite in an intermediate scale facility. As will be seen, certain modifications were necessary in the normal procedure for small-scale characterization of a solid fuel by means of heat release rate calorimetry.

FLAME SPREAD MODELS

Thermophysical Processes. Experience with other materials indicates that surface flame spread can be treated with reasonable accuracy as the movement of an ignition front, with that front being represented by a fixed isotherm, the ignition temperature. As this front passes a given point the fuel pyrolysis rate goes from negligible to significant, i.e., the actual Arrhenius-like temperature dependence of the gasification process is approximated by a step function. The speed of this pyrolysis front movement or flame spread process can then be predicted by following the chemically-inert heating, to the ignition temperature, of successive fuel elements ahead of the front. In the case of upward flame spread, that heat-up process can be a result of two heat fluxes. The first is the convective/radiative flux from the flame which buoyancy causes to rise upward into contact with unignited portions of the fuel. The second, which is not always present, is the radiative flux from a nearby burning object or from hot combustion gases trapped by a compartment ceiling.

There are three key empirical elements to any model of upward flame spread based on the above idealization of the chemical behavior of the fuel: 1) the magnitude and spatial variation of the heat flux from the flame to the fuel surface as a function of distance above the pyrolysis front; this dictates how fast the inert fuel heats up; 2) the height of the flame as a function of the total heat release rate (kilowatts per unit width) below the flame front; this helps dictate the time over which the upward moving flame heats the fuel surface and thus it interacts with (1) to affect how rapidly the fuel heats-up to ignition; 3) the dependence of the heat release process from an element of ignited fuel (below the upward moving flame front) on both external flux and time; this also interacts with (1) and (2) to further complicate the time dependence of the heat flux which the flame provides to fuel surface above the pyrolysis front. The models examined here treat each of these differently.

It should be noted that when an external radiative flux is present (as it is in the experiments described here), that flux is typically below the minimum flux needed to cause ignition of the material. Ignition then occurs during the spread process as the flames provide additional heat. This poses a problem for item (3) above, the rate of heat release rate from unit area of burning fuel, since the data needed are for external radiant fluxes below that needed to ignite the material. The normal technique for measuring rate of heat release from a material (in the Cone Calorimeter, for example) employs radiative fluxes above that needed to ignite the material. The real requirement here is that the heat release data input to the model reproduce (or be able to reproduce) that which is evolved in the full-scale flame spread situation. This essentially means that the net heat flux into the sample surface during its burning be the same in the small-scale and large-scale situations. This is discussed further below.

Cleary and Quintiere Model. The simplest model of upward flame spread to be considered here, that of Cleary and Quintiere (*3*), addresses the three empirical input requirements listed above as follows. It assumes the flame heat flux above the pyrolysis front to be spatially constant up to the flame tips (and zero above this), the flame height to be a linear function of total heat release rate below the front and the fuel heat release rate per unit area to be constant in time until complete consumption occurs at any given height on the fuel surface. The constant heat release rate is an average of that obtained in Cone Calorimetry; however, the proper value for the incident radiant flux to be used for this small-scale measurement is not well-defined for some situations. These simplifying assumptions allow an analytical solution for pyrolysis front position versus time.

Mitler's Model. The model of Mitler (*4*) is numerical in nature. It uses separate sub-models, grounded in data from gas burner experiments, to calculate the radiative and convective fluxes from the flame versus height. The flame height vs. heat release rate correlation is also based on gas burner results. A transformation procedure based on a balance of the various heat fluxes crossing the fuel surface during burning is used to take Cone Calorimeter heat release data obtained at an arbitrary external heat flux and convert it to any external flux for which the flame spread process is being predicted. In principle, these data could then be measured at fluxes above or below the minimum heat flux needed to ignite the material.

Model of Brehob and Kulkarni. The model of Brehob (*4*) and Kulkarni, *et al* (*6*), also numerical, uses an experimentally-fitted exponential decay law for flame heat flux as a function of height above the pyrolysis front and an experimental flame height vs. heat release rate correlation, also based on gas burner behavior, but different from that used by Mitler. Finally, this model uses experimental rate of heat release data at the specific external flux of interest for flame spread prediction; no means for interpolating or extrapolating heat release data to other fluxes is provided. Kulkarni, *et al* (*6*) used mass loss rate data obtained for a small sample in the presence of a turbulent gas burner flame (plus an external radiant flux (*5*)) and the heat of combustion to infer the needed rate of heat release inputs.

Application to Composites. In all of these models the chemistry of the fuel is present only implicitly, in the heat release behavior and in the ignition temperature. Such models can handle fuel composition variations only through substitution of new experimental data on these properties. That is a limitation but also an advantage in dealing with real materials for which detailed chemical parameters would be nearly impossible to obtain.

Composite materials have two features not found in other materials for which the above models have been tested. First, they have a high content of inert material (the glass fibers in this study). If the degrading resin does not liquify and wick through this residual structure, the fiber mats will perform a function similar to a char in a burning material such as wood. That is, as the outer layers of resin gasify, deeper layers will be somewhat insulated from the flame heat flux by the glass left behind and thus gasify more slowly, weakening the flame. If wicking does occur, this self-protective feature is lessened. In any event, the small-scale measurements of rate of heat release should capture this effect in a manner equally pertinent to the full-scale test.

The second feature peculiar to composites is the tendency to delaminate during intense surface heating as a result of gas generation between plies. This has at least two consequences. First, it implies that the thermal properties (particularly thermal conductivity) of the delaminated layers are changed appreciably. Second, when performing measurements on small-scale samples, one notes a tendency for the evolved gases to come out the sample edges, especially late in the burning process. This edge burning issue was examined to a limited extent in a previous study (7). Suppressing edge burning leads to rather erratic heat release behavior that may not be relevant to full-scale flame spread testing; it was not suppressed in the present study. The consequences of delamination for model/experiment comparisons are believed to lie mainly in the thermal property effects but this issue may need further study.

EXPERIMENTAL

Flame Spread Experiments. The intermediate-scale experiments were performed with the facility sketched in Figure 1. A flat, 0.95 cm thick (13 ply) composite panel (vinyl ester/E-glass, plain weave, woven roving), 0.38 m wide by 1.22 m tall, was uniformly irradiated ($\pm 5\%$) on one surface by a pair of electrically-heated panels. The flux was varied from test to test by changing the panel temperature, typically in the 325-600 °C range. Irradiation was started at time zero by removal of a shutter; at the same time a methane-fueled line burner, spanning the sample width at its base, was ignited. This burner proceeded to ignite the composite across its base, initiating upward flame spread. The location of the pyrolysis front was noted by an observer at frequent intervals, to an accuracy of a few centimeters; it was made visible by the tendency of the fuel gases to emerge in small jets from the highest permeability locations in the woven glass roving.

Heat flux gages were inserted through the sample (sensing surface flush with the front face of the sample) at three heights (1/4, 1/2 and 3/4 of sample height) to aid in assessing flame fluxes and appropriate conditions for small-scale assessment of heat release rate; the gages were operated at a temperature of 100 °C or above to eliminate water condensation on their sensing surface. Flame spread rate was measured, in separate experiments, as a function of both incident radiant flux and igniter heat release rate. The hood above the apparatus is designed to catch all of the fire plume. Oxygen and gas flow rate measurements in the hood exhaust stream thus allow calculation of the time-dependent total rate of heat release from the burning sample (8).

Heat Release Experiments. A Cone Calorimeter was used to measure heat release rate per unit area on vertical, 100 mm square samples of the composite. Heat release rate is inferred from the same variable measurements as in the intermediate-scale experiments. As noted above, the desired data are for external radiant fluxes below that needed for sample ignition; these are the flux levels used in the larger-scale experiments. Recall that in the larger-scale tests, the flame spreads as a consequence of the flame heat flux and the external radiant flux, typically set at a value well below that needed to directly ignite the sample.

The Cone sample was pre-heated, typically 600 s, with an external flux comparable to that used in the above experiments; this mimicked the pre-heating the sample sees in the large-scale experiments before the flame arrives. (A pre-heat time of 400 s gave distinctly more erratic heat release behavior.) At the end of this interval a methane line burner below the sample was ignited. The function of this burner is not to produce upward flame spread but rather to produce a turbulent flame over the sample face resulting in a total heat input (flame flux plus external radiant flux from the Cone heater) to the sample surface comparable to the total in the above intermediate-scale experiments; in this way, the net heat input to the sample surface (flame plus radiative input minus surface radiative loss) should be the same in both situations. The heat release rate of the sample plus burner combination was recorded and the constant burner contribution subtracted to obtain the desired input data for the models. The accuracy was nominally $\pm$ 5% but the repeatability, especially at lower fluxes, was as poor as $\pm$ 25% in portions of the heat release history.

This rate of heat release measurement technique is unconventional; it is an adaptation of the mass loss procedure developed by Kulkarni, *et al* (6) to get heat release data at external radiant fluxes below that necessary to ignite a material. In previous work where the gas flame was not used (a high external radiant flux was used to yield ignition then reduced to a low value) the burning of the sample was erratic and unsatisfactory (7).

RESULTS AND DISCUSSION

Flame Heat Fluxes. Measurement of the flame heat fluxes for this composite proved to be somewhat problematical. As noted above, the flux gage was inserted from the rear of the sample through a hole just slightly larger than the 6 mm gage diameter itself. This hole proved to be an intermittent relief hole for fuel gas pressure build-up between plies of the composite. When this happened strong jets of flame in front of the gage falsified the flux data obtained. It was necessary to exclude the flux data obtained during jetting; videotapes of the tests made it possible to do this after the fact. Figure 2 shows the results of the flux gage measurements as a function of incident radiant flux in the intermediate-scale experiments. Note that these are the total fluxes (flame plus external radiation) seen after the pyrolysis front has moved above the gage location. Also included in this figure are data obtained with an unretarded polyester composite with the same type of fiber plies; this composite allowed flame spread in the absence of an external flux. It is apparent that the noise level in these data is high, in spite of exclusion of data from the times of visible jet flames around the gages. Apparently this is because of the inherently fluctuating nature of the flames emerging, as mentioned above, as small jets from high permeability points in the glass weave pattern. Extrapolation of the heat fluxes to zero external radiant flux indicates that the turbulent flame alone supplies a flux in the range of 25-40 kW/m^2; the polyester composite, which burns without an external flux, has a flame flux in this range, as well. Analogous flux measurements in the Cone Calorimeter indicate that the flame flux there, a result of the commingled gas burner and sample flames, is about 40 kW/m^2. This is at the high end of the full-scale range, probably because the gas burner flame in the Cone apparatus is not fully turbulent. (In order to make this flame fully turbulent, it would be necessary to increase its height (and thus its heat release rate); this in turn would decrease the accuracy of the measurement of the heat release contribution from the sample itself.) This comparison of small-scale and full-scale fluxes does indicate that the Cone data are being obtained at approximately the same flame flux conditions as exist in our intermediate-scale experiments. Thus this manner of obtaining heat release data at fluxes below that needed to directly ignite a material (in this case, below 15 kW/m^2) appears to be reasonably satisfactory.

Cone Calorimeter Measurements. Figure 3 shows Cone data obtained in this manner for the vinyl ester composite at three external radiant fluxes. The curves shown for the lowest and highest external fluxes are averages of two to three tests; the curve at the intermediate flux is the result of a single test and is clearly more noisy. The qualitative behavior as a function of time resembles that seen for char-forming, non-fiber-reinforced materials. That is, the rate of heat release shows an immediate decay from its initial value due to consumption of resin at and just below the heated surface coupled with the insulating effect of the non-volatile residue (here, the fibers). This suggests (but does not prove) that wicking of liquified resin degradation products to the front surface is minimal. It should be noted that the raw data for this peak were corrected upward over the first 100 seconds or so because video tapes of the test showed that ignition was not uniform over the sample surface; thus the heat release rate per unit area from the area actually burning was greater than the raw data showed. After this initial peak, one might expect the increasing insulating effect of the depleted glass plies would cause the heat release rate to continue to drop. However, the thermal

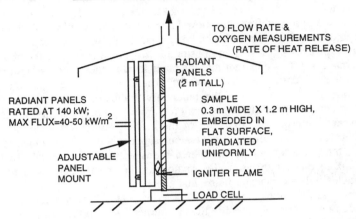

Figure 1. Radiant panel and heat release rate apparatus showing placement of sample in front of panels.

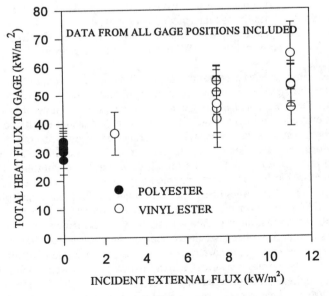

Figure 2. Total heat flux to gages embedded in sample surface.

wave in the sample soon reaches the insulated back surface and the sample becomes, in effect, increasingly pre-heated in depth. This causes the second, broad peak in the heat release curves. The effect of the external flux level is not very great in the range shown; data scatter nearly masks a rough upward trend of heat release rate with flux. This implies that one need not precisely match the net flux into the sample in small-scale and large-scale tests.

The source of the scatter in these tests appears to be largely the result of non-reproducibility in the ply-to-ply delamination during sample burning. Certainly this is evident during the pre-ignition heat-up and early burning process. Since these materials are the result of a hand lay-up process, there could also be local variations in the amount of resin present in a given ply layer. These types of small-scale variations tend to be averaged out in the behavior of the intermediate-scale samples cut from the same composite slabs.

These data were used in the models discussed above. Where the detailed heat release history was to be used, rather than a simple averaged value (as in the Cleary/Quintiere model), the data were fit with a polynomial or otherwise approximated by a smoothed curve to eliminate the noise seen in Figure 3.

Comparisons of Intermediate-Scale Data with Model Predictions. Figures 4, 5 and 6 show a comparison of typical upward flame spread data (circles) with the three spread models examined here. The data shown for each flux are the result of two replicate tests. The gas burner igniter was set at 6 kW for all of these tests; this provided a flame whose tips reached up to about 20-30 cm from the bottom edge of the sample. The upward spreading pyrolysis front was generally only roughly flat due to a distinct tendency for the buoyant flames to contract laterally as they moved upward on the sample face (in spite of vertical fins on the outer edges of the sample holder intended to inhibit lateral flow). Another contributing factor to non-two-dimensional behavior was occasional preferential ignition (by the spreading flames) of local regions along the front probably due to delamination bubbles which decreased the local thermal conductivity.

The rather ragged-looking plateau in the upward spread process was seen in both tests at an incident radiant flux of 2.5 kW/m^2; doubling of the igniter gas flow rate shrank the duration of this plateau to less than 100 s. At an intermediate incident flux of 7.5 kW/m^2, two tests (with the 6 kW igniter) showed behavior that was essentially a somewhat slowed version of that seen at 11.5 kW/m^2 (i.e., smoothly accelerating upward spread) but a third test again yielded a substantial plateau. These plateaus appear to be a result of delaminations which cause earlier than expected ignition on portions of the composite surface. When the resin burns out of the delaminated ply or plies the pyrolysis front stops and then gradually resumes a more normal upward pace for the given external flux.

The model predictions in Figures 4, 5 and 6 are shown as solid lines. The input data to the models, characterizing the composite, are the same for all cases. These data, particularly thermal properties, are problematical, however.

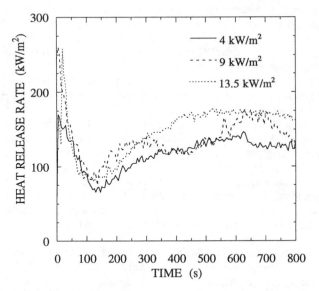

Figure 3. Rate of heat release per unit area of vinyl ester composite in Cone Calorimeter.

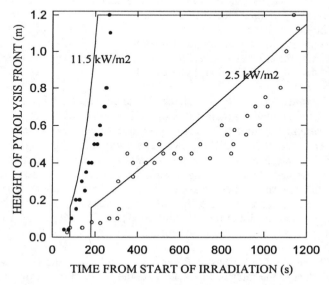

Figure 4. Measured pyrolysis front position as a function of time vs. prediction of Cleary/Quintiere model.

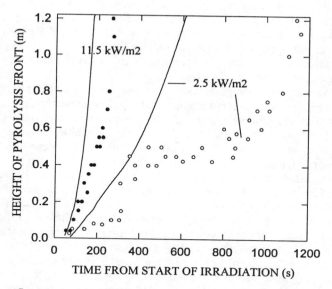

Figure 5. Measured pyrolysis front position as a function of time vs. prediction of Mitler's model.

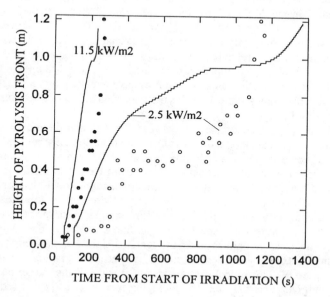

Figure 6. Measured pyrolysis front position as a function of time vs. prediction of Brehob and Kulkarni model.

The thermal conductivity was measured at about 100 °C for an undegraded composite sample and this is the value used here. One can readily argue that this value is approximate at best in this application. Ordinarily the conductivity is an increasing function of temperature but, as the resin begins to degrade and interply delamination occurs, the conductivity can be expected to drop. There are no quantitative data on these trends at present.

The effective, average heat capacity between room temperature and the ignition temperature also is not well-defined *a priori*; endothermic degradation reactions prior to ignition can increase it by an undetermined amount. Here this problem has been handled via an analysis of the ignitability behavior (ignition delay time vs. incident heat flux) of the composite as determined in the Cone Calorimeter. Ignition delay is dependent on the thermal inertia (product of conductivity, density and heat capacity). The effective value of thermal inertia is inferred by matching the experimental ignition data with a thermal model of the ignition process in the Cone Calorimeter. This model is based on constant thermal properties and a constant ignition temperature. It includes radiative and convective losses from the front surface and an adiabatic rear surface. The model is solved numerically to determine the time to reach the ignition temperature as a function of incident radiant flux; the results are compared to the experimental ignition data and the thermal inertia is adjusted to get a match.

The effective thermal inertia obtained in this manner depends somewhat on the heat flux at which the match is made; this probably reflects the fact that the ignition temperature for a real material increases with heat flux. Here a value of 1.4 times the nominal room temperature value of thermal inertia is inferred from a match in the 20-25 kW/m^2 range. This multiplier is applied entirely to the input value of the heat capacity. However, since the spread process is one of successive ignitions, only the thermal inertia matters. The effective thermal inertia inferred here probably incorporates both heat capacity and thermal conductivity changes. The impact of delamination on a small sample in the Cone may not be as great as that on a full-scale sample (interply pressure is relieved more readily in small samples) so the thermal conductivity effects on the two scales may not be equal.

Figure 4 shows a comparison between the data and the model of Cleary and Quintiere (*3*). This model deals with sample heat-up in a simplified way and thus cannot correctly calculate the pre-heating of the sample as a function of height due to the external radiant flux. Instead we have calculated the pre-heat effect at mid-height using the numerical heat conduction model mentioned above applied for a time up to the experimental arrival time of flame at that height. This initial temperature is applied to the entire sample. The model also simplifies the ignition process induced by the gas flame igniter at the sample base. The flux is taken to be uniform over the igniter height and thus this full height ignites simultaneously yielding the vertical lines at the beginning of each model trace in Figure 4. This model uses a constant average of the rate of heat release data for the given external flux; the averaging time interval is that required for the pyrolysis front to reach mid-height. Here that time was found iteratively by successive model solutions as follows. An initial guess for averaging time provided an average rate of heat release as the model input. The model was then solved and its predicted time for the flame to reach mid-height was used to obtain a corrected average heat release value. This cycle was repeated until it converged. In spite of these approximations (and others noted above), this model gives better quantitative agreement with the flamespread data than do the more mechanistically exact models discussed below. There is probably a fortuitous element to this.

Figure 5 shows a comparison of the data with the predictions of Mitler's model (*4*). With the thermal properties described above, this model clearly predicts somewhat faster upward spread than is seen experimentally. It should be noted that neither this nor the preceding model predicts the "plateau" behavior seen at the lowest flux. As noted above, the plateau appears to be associated with interply delamination and its effect on local thermal conductivity. Since this mechanism is not incorporated in any existing flame spread model, it is not surprising that no plateau is predicted (but see below).

Mitler's model starts out reasonably well for both fluxes but then over-predicts the speed of full upward spread. It uses heat release data from an intermediate external flux (4 kW/m^2) transformed to the actual flux of interest. The predictions are somewhat different if heat release data from another flux are used instead, implying that the transform method is only approximately applicable to this composite material. The transform is based on variations in the net heat flux to the sample surface, an approach that is plausible for a simple ablating material which leaves no residue on the surface. Here the increasing layer of glass fibers makes this method less accurate with increasing time.

Figure 6 shows a comparison with the model of Brehob (*5*) and Kulkarni, *et al* (*6*); note the extended time scale. The heat release data used as inputs are those measured at 4 kW/m^2 and 13 kW/m^2; data were not available at the run conditions of 2.5 and 11.5 kW/m^2. Such data would probably improve the agreement between model and experiment slightly. Also used as input data was an experimentally obtained correlation for this composite between total flame heat flux and distance above the pyrolysis front. The correlation is exponential in form as was that obtained by Kulkarni, *et al* (*6*) but the data were quite noisy; further work on this issue is being done.

This model does predict a plateau at the low flux though later than that seen experimentally. This model also predicts a peculiarly short plateau at the higher flux. Since the model does not account for delamination which appeared to be the actual cause of the experimental plateau, its plateau mechanism requires further study.

Conclusions. All of the models could probably be brought into better agreement with experiment by adjustments of the effective thermal inertia of the composite. This has not been done here because no further data on thermal properties of this composite exist now. In addition, one of the points of this study has been to see how accurate the models are with no adjustments to the inputs obtained from the small-scale tests in the manner described here. From the standpoint of an engineering assessment of the potential hazard of fire growth on composite walls made of these materials, all of the models appear to be adequate and somewhat conservative for the limited conditions examined here. For the vinyl ester composite studied her, for example, all of the models indicate that it will contribute to fire growth in the presence of fairly minor ignition sources coupled with weak external radiant heating. Thus it would not take a very large nearby burning object for this material to participate in fire growth.

More assessment of model accuracy is certainly desirable. None of the models should be used to make fine judgements about the speed of flame spread. For use in a dynamic compartment fire situation, model extensions are needed so as to accept changing ambient oxygen levels and, except fot Mitler's model which can already handle this, changing radiative inputs.

ACKNOWLEDGMENTS

The authors would like to acknowledge the helpful assistance of H. Mitler and P. Reneke. This work was sponsored by the U. S. Navy Surface Weapons Center.

LITERATURE CITED

1. Reinhart, T ; Clements, L., *Engineered Materials Handbook, Vol.1, Composites*, ASM International, Metals Park, OH, 1987, p. 28
2. Sorathia, U.; Beck, C.; Dapp, T., *J. Fire Sciences*, **1993**, 11, p. 255
3. Cleary, T.; Quintiere, J., *Fire Safety Science - Proc. Third Int'l Sympos.*, Elsevier, New York, NY, 1991, p. 647
4. Mitler, H., *Twenty-Third Sympos. (Int'l) on Combustion*, Combustion Institute, Pittsburgh, PA, 1990, p.1715
5. Brehob, E., Ph. D. Thesis, Penn State Univ., Mechanical Eng'g Dept., 1994
6. Kulkarni, A.; Kim, C.; Kuo, C., Nat'l Inst. Stds. Technol. NIST-GCR-91-597, May, 1991
7. Ohlemiller, T.; Cleary, T.; Brown, J.; Shields, J., *J. Fire Sciences*, 1993, 11, p. 308
8. Janssens, M.; Parker, W., "Oxygen Consumption Calorimetry" in *Heat Release in Fires*; Babrauskas, V. and Grayson, S., Editors, Elsevier Applied Science, New York, NY, 1992, p.31

RECEIVED November 28, 1994

Chapter 29

Protocol for Ignitability, Lateral Flame Spread, and Heat Release Rate Using Lift Apparatus

Mark A. Dietenberger

Forest Service, Forest Products Laboratory, U.S. Department of Agriculture, One Gifford Pinchot Drive, Madison, WI 53705–2398

> In this study, protocols in ASTM E 1321 for determining piloted ignition and flame spread properties and in E 1317 for deriving heat release rate are modified to improve generality, accuracy, and efficiency. We report on new methods that improve the calibration of radiative and convective heat flux profiles on the exposed material, such as Douglas Fir plywood. A simple direct measure of surface emissivity completes the boundary conditions for material thermal analysis. A crank-operated indicator, interfaced with data acquisition system, tracks the lateral flame front to the spread of flame limit. Features of our lateral flame spread model include a range of preheating the sample before flame spreading, the transitioning from thick to thin thermal behavior, and a limit to flame spread. Our improvement to E 1317 in measuring heat release rate profile is demonstrated.

Our initial intention was to implement ASTM E-1321 (*1*) and E-1317 (*2*) protocols and use these standards to obtain material and flammability properties for wood products. Specifically, the Forest Products Laboratory (FPL) wanted to assess ignitability of exterior siding materials and interior panels. Also, thermophysical material properties were needed for use in room fire growth models. Despite substantial progress made in understanding and measuring ignitability and creeping flame spread, inconsistencies remained between the various bench-scale fire tests and their suitability to derive thermophysical properties of test samples. Even with the Cone Calorimeter (ISO 5660, (*3*)), one finds great variation in time to ignition near the critical irradiance (which also varies greatly (*4*)) when effects of geometry and ignition mode are investigated (*5*). As a result of trying to implement E 1321, we needed to remedy deficiencies in calibrations of burner irradiances and convective heat transfer coefficients and in modeling piloted ignition. These are discussed in the ignitability section. Measurement and modeling of lateral flame spread also needed improvement, because test derived thermophysical constants were not consistent with those derived from

This chapter not subject to U.S. copyright
Published 1995 American Chemical Society

ignitability analysis, and changes in certain thermal behaviors during flame spreading were not accounted for in E 1321. These are addressed in the lateral flame spread section.

ASTM E-1317 provides a thermopile method to measure the heat release rate of a spreading lateral fire on the 150- by 800-mm specimen sample for a fixed panel burner setting. The Lateral Ignition and Flame-spread Test (LIFT) apparatus employed by ASTM E-1317 and E-1321 is not as accurate as the Cone Calorimeter for measuring heat release rates on a small square specimen because oxygen consumption calorimetry is not used in LIFT. However, with flame spreading on the larger specimen, the measurement of total heat and mass release is useful in validation of a fire growth model. Upon further investigation, we determined that improvement to the thermopile method for the LIFT apparatus could be made. As a result, new procedures to calibrate the heat release rate measurements and correlate heat release rate with the thermopile, compensator thermocouple, and pyrometer measurements are discussed in the heat release rate section.

Ignitability

Piloted ignition is difficult to assess for composite materials such as wood products. By using a finite difference model (FDM), Janssens (6) shows there is a transition behavior from thermally thick to thin as the ignition time gets very large. If the material also has a broad range of surface temperature in which char develops, such as wood, surface emissivity and temperature at ignition may deviate significantly so that the critical heat flux is elusive to measure. In his work with wood products tested in the Cone Calorimeter, Janssens suggested nominal values for the surface emissivity and convective heat transfer coefficient in analogy with the protocol described for the LIFT apparatus. We examined these physical processes in depth for the LIFT apparatus and found an alternative, more accurate approach for deriving thermophysical properties using Douglas Fir plywood as an example.

Calibrations. First, we calibrated the irradiance profile on the specimen by using a linear relationship between panel-burner heat-fluxes measured by the pyrometer and the water-cooled fluxmeters. These fluxmeters were inserted flush with the exposed surface and laterally every 100-mm along the 150- by 800-mm calibration board which is orientated 15 degrees laterally away from the panel-burner. The specimen holder was aligned such that the irradiance profile was within the limits required by E 1321. Nonlinearity between these two fluxes near the hot end of the specimen occurs because of (1) convective and radiative heat losses from the fluxmeters at ambient temperatures other than the optimum value of $T_a = 26$ °C, (2) flame impingement on the sample at high burner irradiances, and (3) orange-colored flames resulting from excessive carbon deposits on the burner grid. Thus, it has become our practice prior to testing to ramp slowly (in at least 10 minutes) the LIFT panel burner to the maximum and back to a minimum value while recording the signals from the pyrometer and at least two fluxmeters (The first at 50-mm location and the second at 350-mm location.). Then, these linear relationships are checked against the original calibration before proceeding with the daily testing. To better define the specimen's irradiance, we normalized its profile with respect to the pyrometer's irradiance rather than with respect to the fluxmeter at the 50-mm location as specified in E 1321.

Second, the convective heat transfer coefficient was calibrated for our LIFT apparatus. Since the values for convective heat flux were needed in an intermediate step, accurate values for the difference of net radiative heat flux and specimen's surface conductive flux were obtained. In addition to the panel irradiance profile already calibrated, the surface emissivity and temperature needed to be measured to derive the net radiative surface heat flux. The value for the surface conductive heat flux is obtained from the product of material thermal conductivity and temperature gradient at the surface. To maximize the convective heat losses and obtain fairly large temperature gradient in the material, we used insulative plugs inserted in place of the fluxmeters. These plugs were the same insulative material as the board holding the fluxmeters and the plugs, so that with our low-cost emmisometer we could directly measure the surface hemispherical emittance. Three thermocouples were inserted at fixed depths within the plugs and an interpolation function was used to derive the surface temperature and its gradient. Long before the temperatures reached steady-state during exposure to selected irradiances, the material conductive heat flux became small compared with other terms, and the derived convective heat transfer coefficient, h_c (convective heat flux over surface temperature rise) became constant.

The E-1321 recommendation of 0.015 kW/Km² for h_c was investigated using quite different assumptions than those in the previous paragraph. Surface emissivity, ε_s, was assumed to be unity. An FDM code together with a surface thermocouple were used to derive the surface conductive heat flux. Burner irradiance was equated to that measured with a water-cooled fluxmeter. The reported measurements seem limited to locations within 250 to 600 mm and at a flux of 50.5 kW/m² at the 50-mm location. There was a need to consider full lateral length of specimen, full range of panel burner irradiance, and full range of methane calibration burner.

Figure 1 shows the result using our method of deriving h_c and shows measurements for 50-, 350-, and 650-mm positions, for panel irradiances at 50 mm of about 22, 37, 51, and 71 kW/m². The derived values of h_c are much greater than can be expected from the air flow induced by the heated specimen (6). In Figure 1, data correlate best with one-fourth power of the panel irradiance, I_{50}, which suggests that the gas panel burner induce air flow over the specimen; h_c also has a significant linear decrease with lateral distance from the hot end,

$$h_c = (0.0139 - 0.0138x) I_{50}^{1/4} \quad kW/Km^2 \qquad (1)$$

that may be expected with a turbulent air flow within an angled wedge. The methane calibration T burner covered the lateral positions from 100 to 300 mm so that it was between two instrumented plugs. The objective was to see if flaming from the specimen location has an effect on h_c derived at instrumented plug locations. The results show specimen flaming has nil effect on h_c. This systematic study of the

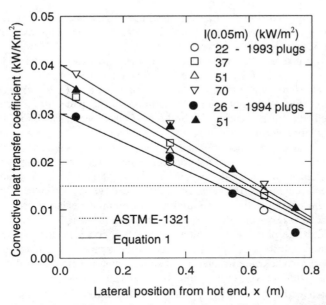

Figure 1 Correlation of convective heat transfer coefficient on specimen holder

convective heat transfer coefficient suggests that the value recommended by E-1321 is severely in error and should be replaced by a procedure to calibrate the convective heat transfer coefficient as outlined in this section.

Douglas Fir Plywood Example. The times to ignition of Douglas Fir plywood with varying thicknesses (6 to 15 mm) were measured over the full range of the gas panel irradiances up to 70 kW/m². The plywood samples were prepared in the test lab, where conditions were about 30% relative humidity and 23 °C. Emissivity measured an average of 0.86. For ignition time less than 30 seconds, a flip-out aluminum sheet was used to protect the specimen until the initiation of the test. Data obtained were initially plotted as time to ignition, t_{ig}, raised to the -0.547 power as function of irradiance following Janssens (6) recommendation. The thin plywood (6.55 mm and 12.7 mm) did not correlate well with the recommended straight line and had the largest disagreement near the critical flux level, q_{ig}, which corresponds to an infinite ignition time. With a finite difference analysis, Janssens shows a theoretical basis for this bias to be the transition behavior from thermally thick to thin for finitely thick materials. As an added difficulty, criteria for the transition use thermal properties not generally available. To remedy this situation, formulas based on our extensive finite element model solutions were developed for materials of finite thickness which was cooled convectively and radiatively on the exposed side and insulated on the unexposed side. Most accurate is the interpolation between thermally thick and thermally thin correlations with an adjustment factor including the Biot number, ***Bi***, and Fourier number, Fo, as follows:

$$\frac{q_{ig}}{q_e} = T = \left(T_{thick}^5 + \frac{T_{thin}^5}{1 + 0.574\, Bi^{1.31}} \right)^{1/5} \quad (2)$$

where

$$T_{thick} = 1 / (1 + 0.73\, (Bi^2 Fo)^{-0.547}) \quad (3)$$

$$T_{thin} = 1 - \exp(-Bi\, Fo) \quad (4)$$

$$Fo = \alpha\, t_{ig}/\delta^2 \quad (5)$$

$$Bi = h_{ig}\delta/\lambda \quad (6)$$

$$h_{ig} = h_c + \varepsilon_s \sigma (T_{ig}^2 + T_a^2)(T_{ig} + T_a) \quad (7)$$

$$q_{ig} = h_{ig}(T_{ig} - T_a)/\varepsilon_s \quad (8)$$

In examining equations 2 to 8, we note that material thickness, δ, should be measured in addition to the usual values of ε_s, h_c, I_{50}, t_{ig}, and T_a. There are three fitting parameters: surface temperature at ignition, T_{ig}, thermal diffusivity, α, and thermal conductivity, λ. In a test series, the variation of a step-function irradiance with time to ignition is measured. This is shown in Figure 2 for three different thicknesses of Douglas Fir plywood along with the curves given by the predicted irradiance, $q_e = q_{ig}/T$, fitted to the pyrometer measured values, I_{50}. Values of fitting parameters are $T_{ig} = 608.2$ K (a typical pyrolysis temperature of wood), $\alpha = 1.48 \times 10^{-7}$ m²/s (a known value for thermal diffusivity of Douglas Fir wood (6)), and $\lambda = 0.164$ W/km. Corresponding parameters also have the values $q_{ig} = 17$ kW/m², $h_{ig} = 0.048$ kW/Km², and $\rho C_p = 1108$ kJ/Km³.

Further examination of Figure 2 shows the effect that improving the functional correlations has on fitting the ignitability data. Janssens (6) demonstrated that the correlation used in the E-1321, seen with dot-dashed line legend in Figure 2, is strictly valid only where q_{ig}, given by the dotted line, is less than 30% of q_e. This means fluxes greater than 59 kW/m², instead of all data as recommended in E 1321, should only be used to determine the parameter, b, in the correlation. The result is the dot-

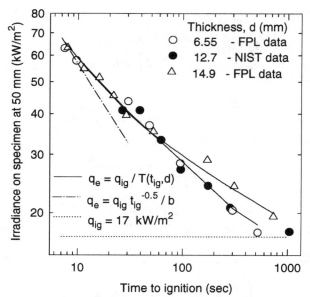

Figure 2 Correlation of irradiance versus time to ignition of Douglas Fir plywood

dashed line with $b = 0.095$. To validly incorporate more data, Janssens suggested using $q_e = q_{ig}/T_{thick}$ in the regime where the material is known to be thermally thick (upper solid curve in Figure 2). Thus, for the example of our 6.55-mm-thick plywood, this thermally thick regime is greater than 35 kW/m², just as the curves of equation 2 fan-out in Figure 2. In simpler correlations, the thermal diffusivity and conductivity could not be determined separately as can now be done with equation 2.

To clarify inconsistent critical heat flux values in recent studies of ignitability, let's take a value, $h_c = 0.0135$ kW/m²K, recommended by Janssens (6) for the Cone Calorimeter. Using this value and the calibrated values of emissivity and ignition temperature for Douglas Fir plywood in equations 7 and 8, the critical irradiance is computed as 12.1 kW/m². This is 5 kW/m² less than that for our LIFT apparatus! Thus, differing apparatus provide correct but differing critical fluxes and need a correction factor to extrapolate to other geometries.

Lateral Flame Spread

Since creeping flame spread can be an integral feature of a fire growth model, such as the furniture fire model (7), one tries to use flame spread properties derived from a flame spread apparatus, such as the LIFT. E-1321 prescribes flame front viewing rakes with pins spaced 50 mm apart. A pair of rakes, as observed through a floor level mirror, is used to line up pairs of pins to locations on the specimen. As the flame front reaches a lined-up pair of pins the observer writes down the time of occurrence. To derive flame spread properties for the model described in ASTM E-1321, the material sample is required to be heated to steady-state before igniting to get lateral flame spread. This preheating often causes surface properties to change significantly (such as charring on wood) and results in a lateral surface temperature profile not anticipated by the model. In addition, the imposed heat flux at ignition location is limited to 5 kW/m² above critical flux, because premature ignition can occur at greater fluxes before the material temperature reaches steady state, and flame spread rates need to be slow enough for manual observation and recording.

FPL Method. Our alternative indicator of the flame front position is a pair of pins perpendicular to each other mounted on a movable platform driven by a crank-operated, 1-m-long shaft brass screw. The shaft is connected to a potentiometer so that the lateral position of the movable platform is converted to millivolt signals. The operator cranks the device handle to keep the pair of pins in alignment with the flame front. The floor mirror is used so that the pins can be located away from the flame so as not to disturb the flame. With our data acquisition system, we can obtain data at short time intervals. This allows us to capture the initial high rates of flame spread and rapid decreases to low flame spread velocities. In addition, the spread of flame limit can be accurately pinpointed. Measuring the arrival times at 50-mm increments and performing a numerical differentiation to obtain the flame spread velocities causes significant errors during the initial rapid spread and near the spread of flame limit. Using the movable indicator should improve the required resolution.

This method of measuring creeping flame spread will likely show the limitations with simple correlations for flame spreading rate. A formula developed by Dietenberger (8) shows certain physical features such as (1) the transition from thermally thick to thermally thin behavior, both in preheat temperature and the flame spreading rate; (2) the transition from convective cooling by the induced air flow to convective heating from the flame edge; and (3) the transition to spread of flame limit at some minimum flux below the critical flux. This single formula was successfully applied to many existing creeping flame spread data for polymethylmethacrylate and paper and is simplified for Douglas Fir plywood to result in the following:

$$V_f = V_{f\infty} \left(\frac{1}{2} + \sqrt{\frac{1}{4} + (\delta^*/\delta)^{1.3}} \right)^{\frac{2}{1.3}} \tag{9}$$

$$V_{f\infty} = \frac{\delta_f}{\rho C_p \lambda} \left(\frac{q_{cf}}{T_{ig} - T_e} - (h_{ig} - h_c) \right)^2 \tag{10}$$

$$\delta^* = \lambda \left(\frac{q_{cf}}{T_{ig} - T_e} - (h_{ig} - h_c) \right)^{-1} \tag{11}$$

$$\delta_f = 2\lambda_a / \rho_a C_{pa} V_a \tag{12}$$

$$q_{cf} = \sqrt{2}\lambda_a(T_f - T_{ig})/\delta_f \tag{13}$$

$$V_a = V_f + \sqrt{2C(T_f/T_a - 1)g\delta_f} \tag{14}$$

The preheat temperature, T_e, is substituted for ignition temperature, T_{ig}, in equation 8 and is where the panel irradiance and convective cooling profiles affect the flame spread behavior as a function of time-to-ignition. The convective heat flux from the flame edge has the magnitude, q_{cf}, at the point of ignition and an exponential surface decay length of δ_f. Since the point of ignition moves at the flame spreading rate, V_f, the surface distance traveled was converted to the time domain so that the Laplace transform solution of the thermal diffusion equation predicts the surface temperature rise from preheat value to ignition value as a function of flame spreading velocity. A convenient interpolation between the exact thermally thick and thin solutions similar to equation 2 was rearranged to result in equations 9 to 11. Note that if the preheat temperature is low enough, then equation 10 approaches zero as a result of adjusted surface radiant flux equating to flame edge heat-flux. In addition, equation 11 will approach a large value, causing a thermally thin spreading behavior for equation 9. On the other hand, if the preheat temperature approaches ignition temperature, the flame spread rate will approach infinity and is the basis for piloted ignitability. Equations 12 to 14 were used because they represent creeping flame spread by the deRis solution and provide additional information about the flame edge. Finally, a better comparison with the data can be achieved by numerically integrating the predicted flame spread rate and plotting the result with the experimental flame position as a function of time. We did this for Douglas Fir plywood.

Flame Spread on Douglas Fir Plywood. Figure 3 shows the lateral position of the flame front in the application of our movable flame spread indicator. Minimal preheating, just enough to ignite the plywood on the hot end within a few seconds, was applied. The solid curve in Figure 3 is the numerical integration of the flame spread velocity given by equation 9 and with the constants $C = 1.3$ and $T_f = 1234$ K. Substituting these constants into equations 12 to 14, we obtain q_{cf} = **80** kW/m^2 and δ_f = **0.29** mm, which are reasonable. In Figure 3, four phases of lateral flame spread are represented. The first phase is characterized by initial rapid flame spread and decelerating rapidly as a result of the preheated temperature profile. Recall that the preheated temperature is affected significantly both by irradiance and convective cooling. In the next phase, a strip fire separates from the primary fire because an irradiance below a minimum level (about 13 kW/m^2 at the 370 mm location) is reached by which a heavily charred wood surface does not sustain burning. The third phase occurs at 200 seconds, when preheated temperature transitions to thermally thin

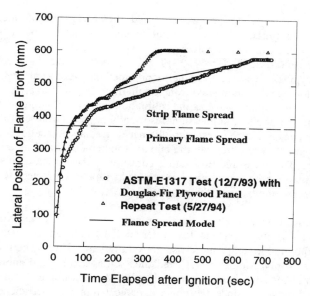

Figure 3 Flame spread data using new indicator and comparison with formula

behavior as it approaches equilibrium temperature. The final fourth phase occurs when flame spread rate transitions to a thermally thin behavior right near spread of flame limit (irradiance of 3.2 kW/m^2 at 580-mm position). Our repeat test of flame spread shown in Figure 3 deviates significantly from that of the initial test at long ignition times. Thus, uniformity and conditioning of the test material needs improvement.

Rate of Heat Release

Beyond construction details of E-1321, E-1317 describes the construction of the fume stack and the locations of the thermopile and compensator thermocouple on the stack. The idea is to sense the temperature rise of the air exiting the fume stack caused by flaming of the specimen and to convert the signals to calculations of the heat release rate. The purpose of E-1317 is to provide flammability ratings using the irradiance at 50 mm set at 50.5 kW/m^2. The flame front position and the heat release rate are measured as a function of time. This information is processed into different representations that suggest levels of flammability for the material. To gain additional information, we installed a fluxmeter about 40 mm below the compensator thermocouple on the fume stack.

For our methane calibration burner, we first used the E-1317 design. It prescribes a 2-m-length tube with holes near the hot end and was placed in the middle of the calibration holder. However, the tube covered our previously mentioned instrumented plugs and stuck precariously out from the apparatus. As a more convenient, but equivalent design, we modified the ASTM E-906 methane "T" calibration burner and attached it to a portable stand so that the burner head was located directly on the calibration holder.

The approach described in E-1317 for calibrating the heat release rate measurement is typical when using the thermopile method. As a first step, a fraction of the compensator signal is subtracted from the thermopile signal to arrive at a signal that is approximately a square wave as the calibration burner is flaming and then shut off. This correction to the thermopile signal is usually needed because of the thermal inertia of the fume stack. In the next step, the panel irradiance is set at 50.5 kW/m^2, and the methane gas flow rate through the burner is incremented and measured as the burner is flaming. This measured flow rate is converted to heat release rates using the heat of combustion for methane gas and is correlated with the "corrected" thermopile's signal. We examined the basis of this approach.

FPL Modifications. Our results with the fluxmeter show that the total heat flux from the heated air in the fume stack to the water-cooled fluxmeter is solely a power function of the thermopile temperature-rise above ambient. This striking correlation was also found to apply at other panel irradiance settings to a good approximation. Thus, the heat transfer coefficient is proportional to the thermopile temperature-rise raised to the 0.65 power. Since the convective heat transfer coefficient primarily varies as a function of the air mass flow rate near the exit of the fume stack, the air mass flow rate is then only a function of the thermopile temperature. Since the enthalpy heat rate of exiting air is the product of the mass flow rate and the temperature-rise, it is therefore only a function of the thermopile temperature rise. Examination of the ratio of fume stack temperature-rise to thermopile temperature-rise showed that it has a steady state value around 0.645, independent of the methane burner or panel irradiance setting. However, during the transient phase (when the methane burner is lit or turned off), the compensator lags behind the thermopile because of the heat capacitance of the fume stack. This means that the steady-state heat-loss to the fume stack is proportional to the heat flux on the water-cooled fluxmeter installed on the fume stack. Therefore, the steady heat loss rate from the fume stack to surroundings is ultimately only a power function of the thermopile temperature rise. Adding this heat loss rate back to the enthalpy rate of the exiting air flow gives the enthalpy rate of the air entrained into the fume stack and is ultimately only a function of the thermopile temperature-rise. As a result, the fume stack is designed optimally to derive the sensible heat release rate from the thermopile temperature-rise.

The signal from the compensator can be used to correct the lag in the thermopile signal by using deviations from their steady-state ratio. To calibrate the appropriate correction, we set the panel irradiance at 51 kW/m^2, imposed a stepping function of the heat release rate into the fume stack, and recorded signals from the thermopile and compensator. Results are shown in Figure 4. A compensation formula given by

$$T_{cp} = T_p + 0.5\,(0.645\,T_p - T_c) \tag{15}$$

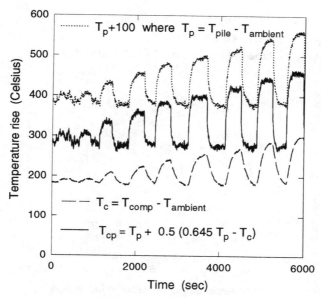

Figure 4 Compensation to thermopile temperature using fume temperature

was found adequate for flattening the peaks and valleys, as shown by the solid line in Figure 4, and yet retained steady-state values of the thermopile signal. This compensation formula was found applicable to other panel irradiance settings. In correlating data, we used nine levels of heat release rate. This is shown in Figure 5. Of significance is the open triangle data obtained with the gas panel burner turned off. The pyrometer sensed a small radiant reflection of methane flame by the gas panel burner and is included in the simple and invertible correlation,

$$T_{cp} = \frac{50\,\dot{H}^{0.8} + 11.5\,I_{50}}{1 + 0.021\,I_{50}} \tag{16}$$

Note that the correlation is not a smooth curve because the actual irradiance values were used in correlating the data, and the irradiances reported in Figure 5 are merely averaged values for the purposes of organizing the data. If the heat release rate, $\dot{H}$, is zero, the panel irradiance at 50 mm, I_{50}, will preheat the entrained air (the numerator term in equation 16) and affect air flow rate (the denominator term in equation 16) at entrance to the fume stack. As the heat release rate increases, its effect on heating and accelerating the entrained air is correlated by the power law correlation (the numerator term in equation 16). Since most of the flame is within the fume stack and not visible to the thermopiles because of the baffles, most radiant energy loss to

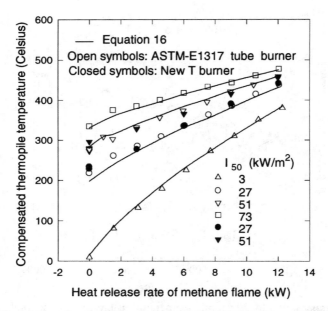

Figure 5 Correlation for the heat release rate using the methane calibration burner

the stack becomes recaptured by the entrained air flow. The error in heat release rate caused by radiant energy loss from the specimen will depend on how different the radiant emission of the methane flames is from that of the specimen. The T burner and the E-1317 tube burner gave equivalent calibration results.

Heat Release Rate of Douglas Fir Plywood. To calculate heat release rate using equation 16, our data acquisition system recorded the signals from the pyrometer, the thermocouple at ambient temperature, the thermopile, and the compensator. These data were processed to provide heat flux at 50 mm, thermopile temperature-rise, and compensator temperature rise as a function of time. Equation 16 was inverted to obtain heat release rate as a function of I_{50}, T_p, and T_c and is plotted as function of time (Figure 6). A heat release rate divided by burning area during the transition to strip fire spreading at time 150 seconds is computed as 85 kW/m². In the corresponding Cone Calorimeter data on a vertical Douglas Fir plywood specimen with irradiance at 50 kW/m² and at 150 seconds, the heat release flux is 87.4 kW/m², which is about the same. Although the irradiance on the LIFT at the 370-mm location is about 40% that at the 50-mm location, thus implying a lower overall heat release flux, it is compensated by a peak heat release flux as high as 200 kW/m² at the 370-mm location. To more accurately calculate this effect, we are modifying Dietenberger's (7) Furniture Fire Model for our wood specimen in the LIFT setup.

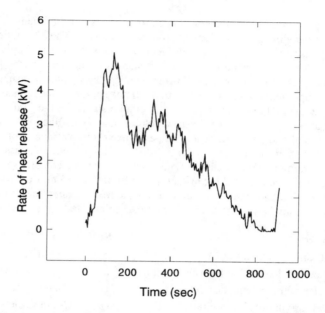

Figure 6 Heat release rate of DF plywood using FPL's thermopile method

Conclusions

Results of this research indicate that for studies of ignitability, surface emissivity and thickness of the specimen should be measured before testing. In addition, the irradiance profile and the convective heat transfer coefficient should be carefully calibrated and correlated. During testing, a thermal shutter should be used at high panel irradiances. A full range of achievable irradiances for piloted ignition is recommended for ignitability testing, because our empirical formula for materials of finite thickness can be used to derive reasonable values for the ignition temperature, thermal diffusivity, and thermal conductivity. If we had used "nominal" values of emissivity, 0.88, and convective heat transfer coefficient, 0.015 kW/Km^2, and applied Janssens's correlation to our thickest plywood, the derived ignition temperature would have been 68 °C greater than our value. Indeed, the same techniques developed in this work can be applied to the Cone Calorimeter or similar apparatus for ignitability. It is not surprising that the critical heat fluxes vary significantly among the different apparatuses, because convective cooling of the specimen is quite different.

If proposals to eliminate the preheating before lateral flame spreading are made part of E 1321, then there is a need to improve measurements of the lateral flame spread. A movable indicator linked to a potentiometer has been tested and proved adequate. An appropriate flame spread formula has been used that accounts for three transitions of heat transfer behavior occurring on a typical specimen during flame spreading. These transitions are thermally thick to thin transition, convective cooling to flame heating transition, and the transition to the spread of flame limit. Flame spread in various directions can benefit from this analysis. These new procedures for analyzing ignitability and flame spread data obtained with the LIFT apparatus should also result in a major revision of the E-1321 standard (and similar standards) and an increased use of the LIFT apparatus. Further studies that include measurements of the flame front properties should add insight to the flame spread mathematical formulation.

A new method to compensate the thermopile signal for the fume stack thermal inertia and a new calibration of the heat release rate as a function of compensated thermopile signal and panel irradiance are significant improvements for E-1317. The methane "T" burner similar to that in E-906 is a convenient substitute for E-1317 methane line burner. The methane flame does not affect convective heat transfer coefficients over instrumented calibration plugs located at 50 mm or further from the flame. Thus, simultaneous calibrations for heat release rates and convective heat transfer coefficients have been achieved and can be adopted for greater efficiency.

Nomenclature

Bi	Biot number (equation 6)
C	Factor for flame-edge buoyant velocity
Fo	Fourier number (equation 5)
g	Local gravitational acceleration (m/s^2)
h_c	Convective transfer coefficient (kW/Km^2)
h_{ig}	Linearized transfer coefficient (kW/Km^2)
$\dot{H}$	Heat release rate (kW)
I_{50}	Irradiance at 50 mm position (kW/m^2)
T	Temperature (K)
T	Normalized temperature
q	Surface heat flux (kW/m^2)
V	Lateral velocity (m/s)
x	Lateral position (m)
α	Thermal diffusivity (m^2/s)
δ	Thickness (m)
δ^*	Transitional thickness (m)
λ	Thermal conductivity (kW/Km)
ε_s	Surface emissivity
ρC_p	Heat capacitance (kJ/Km^3)
σ	Stefan-Boltzmann constant (5.67e-11 kW/K^{4}m^2)

Subscripts

a	Entrained air
c	Fume stack thermocouple
cf	Flame edge convection
cp	Compensated thermopile
e	Surface preheating
f	Lateral flame edge
$f\infty$	Infinitely thick material
ig	Material surface at ignition
p	Thermopile
-	No-subscript for material
Thick	Thermally thick material
Thin	Thermally thin material

Acknowledgements

The author is indebted to Dr. Robert White for his support and guidance for this internal study and to Anne Fuller for acquiring data with the LIFT apparatus.

Literature Cited

1. ASTM E 1321-90. Standard Test Method for Determining Material Ignition and Flame Spread Properties. American Society for Testing and Materials, Philadelphia (1990).
2. ASTM E 1317-90. Standard Test Method for Flammability of Marine Surface Finishes. American Society for Testing and Materials, Philadelphia (1990).
3. ISO 5660. International Standard for Fire Tests - Reaction to Fire - Rate of Heat Release from Building Products. International Organization for Standardization, Geneva (1991).
4. Whitely, R.H., Short Communication - Some Comments on the Measurement of Ignitability and on the Calculation of 'Critical Heat Flux'. *Fire Safety Journal* **21** (1993) 177-183.
5. Shields, T.J.,Silcock G.W., and Murray, J.J., The Effects of Geometry and Ignition Mode on Ignition Times Obtained Using a Cone Calorimeter and ISO Ignitability Apparatus. *Fire and Materials* **17** (1993) 25-32.
6. Janssens, M., Fundamental Thermophysical Characteristics of Wood and their Role in Enclosure Fire Growth. Ph.D. Thesis, University of Gent, Belgium, November 1991.
7. Dietenberger, M., Chapter 13: Upholstered Furniture - Detailed Model. Heat Release in Fire, V. Babrauskas and S. Grayson, eds., Elsevier Science Publishers Ltd., 1992.
8. Dietenberger, M.A., Piloted Ignition and Flame Spread on Composite Solid Fuels in Extreme Environments. Ph.D. Thesis, University of Dayton, Dayton Ohio, December 1991.

RECEIVED December 23, 1994

Chapter 30

Fire Properties of Materials for Model-Based Assessments for Hazards and Protection Needs

A. Tewarson

Factory Mutual Research Corporation, 1151 Boston-Providence Turnpike, Norwood, MA 02062

Fire properties for model based assessments of hazards and protection needs, and test apparatuses and methods used for their quantification are discussed.

The fire properties discussed are: 1) Critical Heat Flux (CHF), 2) Thermal Response Parameter (TRP), 3) surface re-radiation loss, 4) heat of gasification, 5) flame heat flux limit, 6) yields of products, 7) heats of combustion, 8) Corrosion Index (CI), 9) Flame Extinction Index, and 10) Fire Propagation Index (FPI).

The apparatuses discussed are: 1) the Ohio State University (OSU) Heat Release Apparatus, designed by the Ohio State University, 2) the Flammability Apparatus, designed by the Factory Mutual Research Corporation (FMRC), and 3) the Cone Calorimeter, designed by the National Institute of Standards and Technology.

The test methods discussed are identified as: 1) the Ignition Test Method, 2) the Combustion Test Method (non-flaming: pyrolysis/ smoldering and flaming), 3) the Fire Propagation Test Method , and 4) the Flame Extinction Test Method.

For better protection to life and property from fires, various design changes and modifications are performed on the materials, buildings and their furnishings, fire detection and fixed and mobile fire protection systems. The design changes and modifications are made based on the past fire loss experiences, small-scale "standard" and full-scale fire test results and/or model based assessments. Small-scale "standard" test results are unreliable as they cannot account for the variations in the fire scenarios. Large-scale tests are expensive to perform and results cannot account for the variations in the fire scenarios either. Fire model based assessments for hazards and protection needs for the fire scenarios of concern are becoming more reliable and less expensive and thus their use is increasing rapidly.

As a result of the rapid increase in the use of the fire models, there is a great demand for the fire property data. In order to satisfy this demand, four tests methods have been developed. The test methods are utilized in the following three most widely used apparatuses: 1) the Ohio State University (OSU) Heat Release Apparatus [1-3]; 2) the Flammability Apparatus at the Factory Mutual Research Corporation (FMRC) [4-7], and 3) the Cone Calorimeter [8-10].

FIRE PROPERTIES

The fire properties are associated with the processes of ignition, combustion, fire propagation and flame extinction.

Fire Properties Associated with the Ignition Process: the fire properties of a material associated with the ignition process are:

1) *Critical Heat Flux (CHF)*: it is the minimum heat flux at or below which there is no ignition. CHF is independent of the fire scale;

2) *Thermal Response Parameter (TRP)*: it represents resistance of a material to generate flammable vapor-air mixture. It consists of the density, specific heat, thickness, thermal conductivity, and ignition temperature above ambient. The TRP value is independent of the fire scale.

Fire Properties Associated with the Combustion Process: the combustion process can proceed with or without a flame. A combustion process without a flame, defined as *non-flaming combustion process*, proceeds as a result of the heat flux supplied to the material by external heat sources (*pyrolysis*) or by the heterogeneous reactions between the surface of the material and oxygen from air (*smoldering*). A combustion process with a flame, defined as *flaming combustion process*, proceeds as a result of the heat flux supplied by the flame of the burning material back to its own surface. The fire properties of a material associated with the combustion process are:

1) *Surface Re-Radiation Loss*: it is the heat loss to the environment by the hot material surface via radiation. Surface re-radiation loss is independent of the fire scale;

2) *Heat of Gasification:* it is the energy required to vaporize a unit mass of a material originally at the ambient temperature. Heat of gasification is independent of the fire scale;

3) *Flame Heat Flux*: it is the heat flux supplied by the flame of the burning material back to its own surface. Flame heat flux is dependent on the fire scale;

4) *Yield of a Product*: it is the mass of a product generated per unit mass of the material vaporized in non-flaming or flaming combustion process. The product yield is independent of the fire size for the buoyant turbulent diffusion flames, but depends on the fire ventilation;

5) *Heat of combustion*: it is the energy generated in the flaming combustion process per unit mass of the material vaporized. If the material burns completely with water as a gas, it is defined as the *net heat of complete combustion*. The energy generated in the actual combustion of the material is defined as the *chemical heat of combustion*[1]. The chemical heat of combustion has a convective and a radiative component. The *convective heat of combustion* is the energy carried away from the combustion zone by the flowing product-air mixture. The *radiative heat of combustion* is the energy emitted to the environment from the combustion zone.

The heat of combustion is independent of the fire size for the buoyant turbulent diffusion flames, but depends on the fire ventilation;

6) *Corrosion Index*: it is the rate of corrosion per unit mass concentration of the material vapors. It is independent of the fire size.

Fire Properties Associated with the Fire Propagation Process: fire propagation process is associated with the movement of a vaporizing area on the surface of a material. The rate of the movement of the vaporizing area on the surface is defined as the fire propagation rate. The fire property associated with the fire propagation process is the:

Fire Propagation Index: it is related to the rate of fire propagation beyond the ignition zone and is expressed as the ratio of the heat flux from the flame transferred back to the surface of the material to the Thermal Response Parameter of the material. It is independent of the fire size.

Fire Properties Associated with the Flame Extinction Process: flame extinction is achieved by applying the agent in the gas phase such as Halon or alternates or on to the surface such as water. The fire property associated with the flame extinction process is the :

Flame Extinction Index: it is the volume or mass fraction of an agent in the gas phase or on the surface of the burning material required for flame extinction. It is independent of the fire size for the buoyant turbulent diffusion flames, but depends on the fire ventilation.

---

[1]It is defined as the *effective heat of combustion* in the Cone Calorimeter

TEST APPARATUSES

The most widely used apparatuses to measure the fire properties are: 1) the OSU Heat release Rate Apparatus, 2) the Flammability Apparatus, and 3) the Cone Calorimeter.

The OSU Heat Release Rate Apparatus: the apparatus is shown in Fig. 1. It was designed for the flaming fires by the Ohio State University in the early 70's. It is an ASTM standard test apparatus [1-3].

The Flammability Apparatus: the apparatus is shown in Figs. 2A and 2B. It was designed by the Factory Mutual Research Corporation's (FMRC) in the early 70's for non-flaming, flaming and propagating fires [4-7]. It is the FMRC standard test apparatus for cables and wall and ceiling insulation materials (replacing the 25 ft-Corner Test). It is being proposed as an apparatus for the FMRC standard test for conveyor belts. It is used extensively for the evaluation of composite and packaging materials, storage commodities, nonthermal damage assessment for various occupancies, and minimum concentrations of agents (water and Halon alternates) required for flame extinction. Nonthermal damage is defined as the damage associated with the toxic and corrosive products, reduced visibility and smoke damage [11].

A Fire Growth and Spread (FSG) model and a Global Equivalence Ratio (GRE) model are available within the data analysis package of the Flammability Apparatus. The FSG model is used to assess the fire propagation behavior of the materials and the GRE model is used to predict the gas temperatures and concentrations of products at various ventilation conditions for the fire scenario of concern to assess the nonthermal damage. The FSG model and the GRE model operate concurrently with the test being performed and the results are available at the end of the test.

The Flammability Apparatus is not a "Consensus based Standard Test Apparatus", but has a potential of being an advanced apparatus with model prediction capabilities for adoption by the ASTM, ISO, IEC, etc. for the examination of: a) flaming and non-flaming fires; b) upward and downward fire propagation, c) nonthermal damage, and d) flame suppression and extinguishment by water, Halon alternates, inert, dry powders, foams, water mist, etc.

The Cone Calorimeter: the apparatus is shown in Fig. 3. It was designed by the National Institute of Standards and Technology's in the 90's [8-10]. It is an ASTM standard test apparatus. Several design features and testing principles used in the Cone Calorimeter are taken from the Flammability Apparatus at FMRC, thus there are many similarities between the two apparatuses and equivalency of the data for ignition and combustion.

TEST CONDITIONS IN THE APPARATUSES

The test conditions used in the apparatus are listed in Table 1.

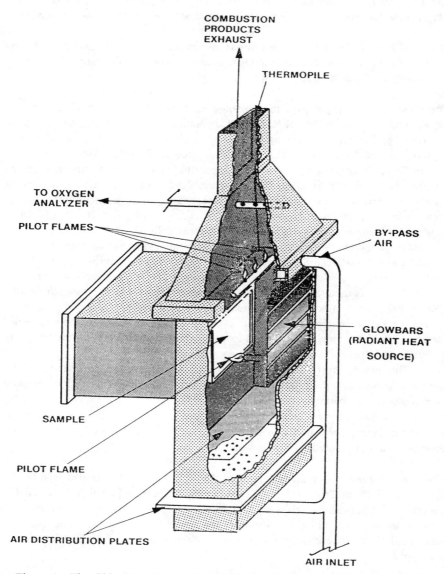

Figure 1. The Ohio State University (OSU) Heat Release Rate Apparatus [1-3].

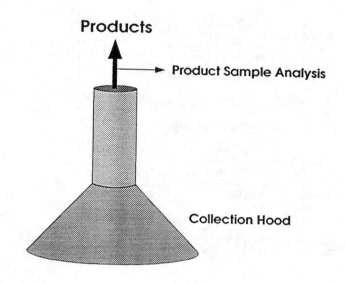
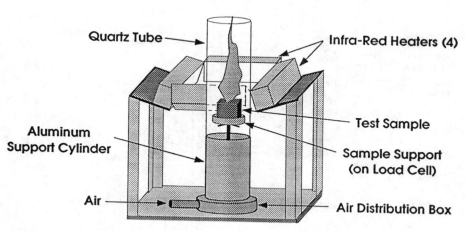

Figure 2A. The Flammability Apparatus for Horizontal Sample Configuration at the Factory Mutual Research Corporation (FMRC) [4-7].

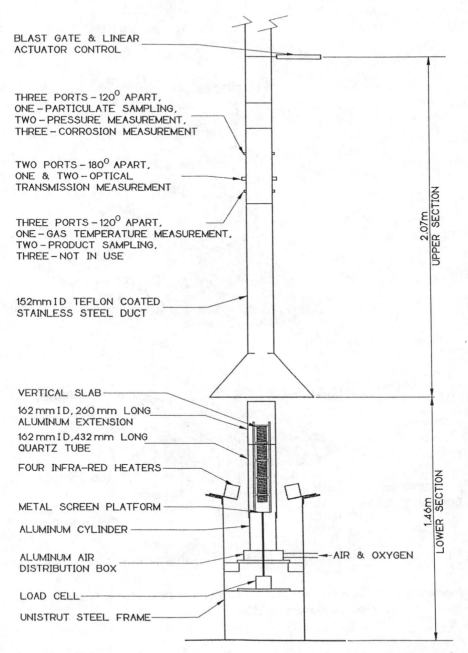

Figure 2B. The Flammability Apparatus for Vertical Sample Configuration at the Factory Mutual Research Corporation (FMRC) [4-7].

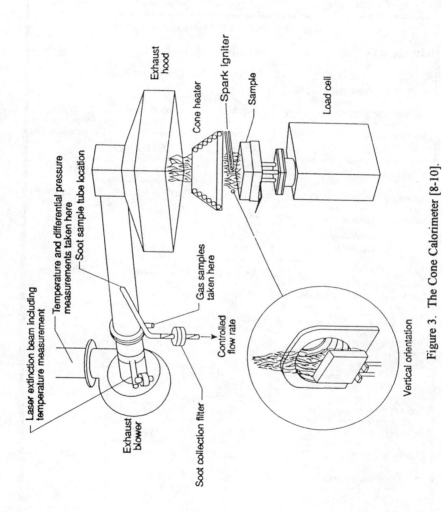

Figure 3. The Cone Calorimeter [8-10].

Table 1. Test Conditions, and Measurements in the OSU Apparatus, The Flammability Apparatus and the Cone Calorimeter

Design and Test Conditions	OSU Apparatus[a]	Flammability Apparatus	Cone Calorimeter[b]
Design Features and Test Conditions			
Inlet gas flow	Co-flow	Co-& natural flow	Natural
Oxygen concentration (%)	21	0 to 60	21
Gas velocity (m/s)	0.49	0 to 0.146	NA
External heaters	Silicon Carbide	Tungsten-Quartz	Electrical Coils
External flux (kW/m^2)	0 to 100	0 to 65	0 to 100
Exhaust gas flow (m^3/s)	0.04	0.035 to 0.364	0.012 to 0.035
Sample dimension (mm) horizontal vertical	110 x 150 150 x 150	100 x 100 100 x 600	100 x 100 100 x 100
Ignition	Pilot flame	Pilot flame	Spark plug
Heat release capacity (kW)	8	50	8
Measurements			
Time to ignition	yes	yes	yes
Mass loss rate	no	yes	yes
Fire propagation rate	no	yes	no
Product generation rates	yes	yes	yes
Light obscuration	yes	yes	yes
Smoke property	no	yes	no
Gas phase corrosion	no	yes	no
Chemical heat release rate	yes	yes	yes
Convective heat release rate	yes	yes	no
Radiative heat release rate	no	yes	no
Flame Extinction : water and halon alternates	no	yes	no

[a]: as specified in the ASTM E 906-83 [1-3]; [b]: as specified in the ASTM E 1354-90 [8-10].

Environment: the apparatuses are designed for co-flow and natural air flow conditions with normal air as well as with air-nitrogen or oxygen gas mixtures, and extinguishing agents such as water, Halon and alternates, inerts, water mist, and foams. The gas velocity under co-flow is similar to one expected under natural air flow condition for a buoyant turbulent diffusion flame. Thus, results under well-ventilated co-flow condition are very similar to the results under natural flow condition.

The OSU Apparatus uses the co-air flow, the Flammability Apparatus uses the co-air flow with 0 to 60 % oxygen and natural air flow, and the Cone Calorimeter uses the natural air flow. The co-flow is used to examine the effects of ventilation, flame radiation, and gaseous flame extinguishing agents on the combustion and flame spread behaviors of the materials.

External Heat Flux Capabilities of the Apparatuses: different types of external radiant heaters are used. The tungsten-quartz heaters used in the Flammability Apparatus expose the sample to a constant external heat flux value instantaneously. The sample can also be exposed to increasing rate of heat flux, as the controller for the tungsten-quartz lamps can be computer programmed for any desired rate of power input.

In the OSU Apparatus (silicone carbide heater) and the Cone Calorimeter (electrical rod heater), power to the heaters is switched on prior to testing to achieve an equilibrium condition. The sample is inserted after the equilibrium condition has been achieved.

The maximum external heat flux that can be applied to the sample surface in the OSU Apparatus and the Cone Calorimeter is 100 kW/m^2, whereas it is 65 kW/m^2 in the Flammability Apparatus.

The external heaters are adjusted in the OSU Apparatus and the Cone Calorimeter to obtain reasonably constant heat flux initially at the sample surface; no consideration is given for regression of the surface as the sample vaporizes with surface contraction or expansion. The external heaters in the Flammability Apparatus are adjusted to obtain reasonably constant heat flux within a 100 x 100 x 100 mm three dimensional space. Thus the flux remains reasonably constant at the regressing or expanding surface as the sample vaporizes. The adjustment of the heaters for the three dimensional space, however, results in the reduction of the maximum value of the external heat flux.

For the measurements of the fire properties, it is necessary to know the external heat flux value absorbed by the sample fairly accurately. The effects of in-depth adsorption of external heat flux and surface emissivity of the sample thus need to be considered. These two factors are not considered in the OSU Apparatus and the Cone Calorimeter; in the Flammability Apparatus, their effects are reduced or eliminated by coating the sample surfaces with thin layers of fine graphite powder or they are painted black.

Product Flow in the Apparatuses: the products generated in the tests are exhausted through the sampling ducts with forced flow. In the Flammability Apparatus and the Cone Calorimeter, products mixed with air are captured in a sampling duct, where measurements are made for the gas temperature, total flow, concentrations of fire products, optical transmission, etc. The maximum exhaust flow in the Flammability Apparatus is about 10 times the maximum flow in the Cone Calorimeter. The OSU Apparatus does not use the exhaust sampling duct, but measurements are made at the top of the sample exposure chamber.

Sample Configuration and Dimensions Used in the Apparatuses: samples in horizontal and vertical sheet (two dimensional) and box-like (three-dimensional) configurations are used. The sample dimensions used in the horizontal and vertical sheet configurations are: 1) about 110 x 150 mm and about 150 x 150 mm in the OSU Apparatus respectively, 2) about 100 x 100 mm in the Cone Calorimeter in both the configurations, and 3) about 100 x 100 mm in the horizontal configuration and about 100 mm wide and up to about 600 mm in the vertical configuration in the Flammability Apparatus.

The box-like (three-dimensional) configuration is used only in the Flammability Apparatus. One to eight, 100 mm cube box-like samples, in one to four layers with one to four boxes per layer are used. Each box and layer is separated by about 10 mm. The two- and three-dimensional vertical sample heights in the Flammability Apparatus are sufficient to perform the upward and downward fire propagation tests.

Measurements in the Apparatuses: measurements are made for the time to ignition, mass loss rate, heat release rate, generation rates of products, optical transmission through smoke, fire propagation rate, metal corrosion rate, smoke damage (color and odor) and flame extinction, utilizing the four test methods described in the next section.

THE IGNITION TEST METHOD

The Ignition Test Method is used to determine: 1) the *Critical Heat Flux (CHF)* and 2) the *Thermal Response Parameter (TRP)*. In the test, flammable vapor-air mixture is created by exposing the sample to various external heat flux values. The flammable mixture is ignited either by a small pilot flame (OSU and the Flammability Apparatuses) or by a spark plug (Cone Calorimeter).

Several tests are performed with variable external heat flux values and time-to-ignition is measured in each test. External heat flux value at which there is no ignition for 15 minutes, taken as the CHF value, is also determined, such as indicated in Fig.4.

The inverse of the time-to-ignition and its square-root are plotted against the external heat flux as shown in Fig. 4. The plot which shows a linear relationship, away from the CHF value, is used to obtain the TRP value. The TRP value is obtained from the inverse of the slope by performing a linear regression analysis. In Fig. 4, the square-root of the inverse of the time-to-ignition shows a linear relationship, which is a

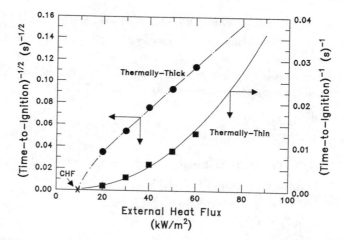

Figure 4. Time to Ignition Versus External Heat Flux for a Silicone Based Polymer. Data Measured in the Flammability Apparatus. Data Satisfy the *Thermally-Thick* Behavior but Not the *Thermally-Thin* Behavior Away from the Critical Heat Flux Value.

relationship for the *thermally-thick materials* [6,7,12,13]:

$$\sqrt{\frac{1}{t_{ig}}} = \frac{\sqrt{4/\pi}\ (\dot{q}_e'' - CHF)}{(TRP)_{thick}} \tag{1}$$

where t_{ig} is the time to ignition (sec), $\dot{q}_e''$ is the external heat flux (kW/m^2), CHF is the Critical Heat Flux (kW/m^2), and $(TRP)_{thick}$ is the Thermal Response Parameter for thermally thick material (kW-s$^{1/2}$/m^2). $(TRP)_{thick}$ is expressed as [6,7,12,13]:

$$(TRP)_{thick} = (k\rho c_p)^{1/2}(T_{ig} - T_a) \tag{2}$$

where k is the thermal conductivity (kW/m-K), c_p is the specific heat (kJ/kg-K), ρ is the density (kg/m^3), T_{ig} is the ignition temperature (K), and T_a is the ambient temperature (K).

For a sample behaving as a *thermally-thin material*, the relationship between inverse of time to ignition and external heat flux is linear [6,7,12,13]:

$$\frac{1}{t_{ig}} = \frac{(\pi/4)(\dot{q}_e'' - CHF)}{(TRP)_{thin}} \tag{3}$$

where $(TRP)_{thin}$ is the Thermal Response Parameter for thermally-thin material (kJ/m^2). $(TRP)_{thick}$ is expressed as [6,7,12,13]:

$$(TRP)_{thin} = \rho c_p \delta (T_{ig} - T_a) \tag{4}$$

where δ is the thickness of the material (m).

The Ignition Test Method is routinely used in the Flammability Apparatus at FMRC. We have also used it to determine the TRP values from the time-to-ignition data measured in the Cone Calorimeter [14,15]. The CHF and TRP values for numerous materials, determined from the time-to-ignition data from the Flammability Apparatus and the Cone Calorimeter have been reported [6,7], an example is shown in Table 2. There is a reasonable agreement between the TRP values obtained from the time-to- ignition data from the Flammability Apparatus and the Cone Calorimeter for the materials for which surface emissivities are close to unity.

In general, physically thick materials behave as thermally-thick materials and physically thin and expanded materials (foams) as thermally-thin materials. Materials with higher CHF and TRP values have higher resistance to ignition and fire propagation. The effectiveness of the passive fire protection to resist ignition and fire propagation through fire retardancy, chemical structural changes, coatings, etc. can be assessed by the magnitude of the increase in the CHF and TRP values. In general,

Table 2. Ignition Data for Materials

Material	CHF[a] (kW/m²)	(TRP)$_{thick}$ (kW-s$^{1/2}$/m²) Flamm. App.[a]	(TRP)$_{thick}$ (kW-s$^{1/2}$/m²) Cone[b]	(TRP)[a]$_{thin}$ (kJ/m²)
Thermally-Thin Materials				
100 % cellulose	13	-	-	159
Corrugated paper	13	-	-	385
News paper	11	-	-	175
Tissue paper	13	-	-	130
Polyester	11	-	-	161
Polypropylene	12	-	-	278
Rayon	17	-	-	227
Polyester & Rayon	11	-	-	286
Thermally-Thick Materials				
Polystyrene	13	162	-	-
Polypropylene	15	193	291	-
Styrene-butadiene	10	198	-	-
Polyvinyl ester	-	-	263	-
Polyoxymethylene	13	269	-	-
Nylon	15	270	-	-
Polymethylmethacrylate	11	274	380	-
High density polyethylene	15	321	364	-
Polycarbonate	15	331	-	-
Polyvinylchloride	10	194	285	-
Epoxy fiberglass	10	156	198	-
Phenolic fiberglass	33	105	172	-
Phenolic kevlar	20	185	258	-

a: from the Flammability Apparatus at FMRC; b: calculated from the data reported in Refs. 14 and 15.

materials with CHF value ≥ 20 kW/m^2 and (TRP)$_{thick}$ value ≥ 450 kW-s$^{1/2}$/m^2 have a higher resistance to ignition and fire propagation.

The CHF and TRP values are used in the assessment of the fire propagation behavior of the materials. For example, they are used in the FPI concept and the complementary FSG model predictions at FMRC. It would be useful if the Ignition Test Method is adopted in the ASTM E 906-83 (the OSU Apparatus) and ASTM E 1354-90 (the Cone Calorimeter) such that model based assessments can be made for fire hazards and protection needs.

THE COMBUSTION TEST METHOD

The Combustion Test Method is used to determine: 1) surface re-radiation loss, 2) heat of gasification, 3) flame heat flux, 4) yields of products, 5) heats of combustion, and 6) Corrsion Index. In the test, the sample is exposed to various external heat values in a co-flowing inert environment or air with 10 to 60 % oxygen concentration or in natural air flow.

In the tests, measurements are made for the mass loss rate, heat release rate, generation rates of products, corrosion rate, optical transmission through the products, and smoke color and odor as functions of time and external heat flux.

Mass Loss Rate (Generation Rate of Fuel Vapors) in the Non-Flaming Combustion Process

The mass loss rate and its integrated value are measured as functions of time at three to four external heat flux values. The mass loss rate is plotted against the external heat flux value. From the inverse of the slope, the heat of gasification is determined using the linear regression analysis, as suggested by the following relationship for the steady state condition [4,6]:

$$\Delta H_g = \frac{\dot{q}_e'' - \dot{q}_{rr}''}{\dot{m}_f''} \tag{5}$$

where $\dot{m}''$ is the mass loss rate (kg/m^2-s), $\dot{q}_e''$ is the external heat flux (kW/m^2), $\dot{q}_{rr}''$ is the surface re-radiation loss (kW/m^2) and ΔH_g is the heat of gasification (kJ/kg).

For the transient conditions, the linear regression analysis is performed using the relationship between the integrated values of the mass loss rate and the external heat flux value [4,6]:

$$\Delta H_g = \frac{E_{ex}(t)}{W_f(t)} \tag{6}$$

where E_{ex} is the net external energy (kJ) and W_f is the total mass lost (kg). In Eq. 6,

$E_{ex}(t)$ is determined from the following relationship:

$$E_{ex}(t) = A \sum_{n=t_0}^{n=t_n} (\dot{q}_e'' - \dot{q}_{rr}'')\Delta t_n \qquad (7)$$

where t_0 is the time at which the sample starts loosing its mass (s) and t_f is the time at which the sample stops loosing its mass (s). $\dot{q}_{rr}''$ value in Eqs. 5 and 7 is the external heat flux value at which there is no measurable mass loss rate. For higher accuracy, surface temperature is measured as a function of time to determine the $\dot{q}_{rr}''$ value. The $W_f(t)$ value in Eq. 6 is determined from the following relationship:

$$W_f(t) = A \sum_{n=t_0}^{n=t_f} \dot{m}_f''(t)\Delta t_n \qquad (8)$$

The heat of gasification and surface re-radiation loss values for numerous materials from the mass loss rate measurements for non-flaming fires in the Flammability Apparatus have been reported [4,6]; table 3 shows an example. The heat of gasification values from the mass loss rates for flaming fires at high external heat flux values in the Cone Calorimeter have also been calculated [4,6]. As can be noted in Table 3, the heat of gasification values from the data from the Flammability Apparatus, the Cone Calorimeter and the Differential Scanning Calorimeter show reasonable agreement.

Materials with higher values of surface re-radiation loss and heat of gasification have low intensity fires and generate lower amounts of products. The effectiveness of the passive fire protection through fire retardancy, chemical structural changes, coatings, and others can be assessed through the magnitude of the increase in the values of the surface re-radiation loss and heat of gasification.

The surface re-radiation loss and heat of gasification values are used as direct inputs to the FSG model in the Flammability Apparatus to predict the fire propagation rate. Alternately, the mass loss rate as a function of time at three external heat flux values is used directly by the FSG model in the Flammability Apparatus to predict the fire propagation rate.

It would be useful if the Combustion Test Method for the heat of gasification and surface re-radiation loss is adopted in the ASTM E 906-83 (the OSU Apparatus) and ASTM E 1354-90 (the Cone Calorimeter) such that model based assessments can be made for fire hazards and protection needs.

Mass Loss Rate (Generation Rate of Fuel Vapors) in the Flaming Combustion Process

The mass loss rate in the flaming combustion process is higher than in the non-flaming combustion process because of the additional heat flux from the flame:

Table 3. Surface Re-Radiation Loss, Heat of Gasification, and Flame Heat Flux Transferred Back to the Surface of the Burning Material in Large-Scale Fires

Material[a]	Surface Re-Radiation Loss (kW/m^2)	Heat of Gasification (MJ/kg)			Large-Scale Flame Heat Flux (kW/m^2)
		Flamm. App	Cone	DSC[b]	
Water	1.0	2.6	-	2.6	-
Wood (Douglas fir)	10	1.8	-	-	-
Particle board	-	-	3.9	-	-
Polypropylene	15	2.0	1.4	2.0	67
Polyethylene (ld)	15	1.8	-	1.9	61
Polyethylene (hd)	15	2.3	1.9	2.2	61
Polyethylene/25% Cl	12	2.1	-	-	-
Polyethylene/36% Cl	12	3.0	-	-	-
Polyethylene/48% Cl	10	3.1	-	-	-
Polyvinylchloride (PVC)	15	2.5	2.3	-	50
PVC, LOI = 0.20	10	2.5	2.4	-	-
PVC, LOI = 0.30	-	-	2.1	-	-
PVC, LOI = 0.35	-	-	2.4	-	-
PVC, LOI = 0.50	-	-	2.3	-	-
Polyoxymethylene	13	2.4	-	2.4	50
Polymethylmethacrylate	11	1.6	1.4	1.6	60
ABS	10	3.2	2.6	-	-
Polystyrene	13	1.7	2.2	1.8	75
PU foams (flexible)	16-19	1.2-2.7	2.4	1.4	64-76

a: ld: low denaity; hd: high density; Cl: chlorine; LOI: limiting oxygen index; ABS: acrylonitrile-butadiene-styrene; PU: polyurethane; **b**: Differential Scanning Calorimeter

$$\dot{m}'' = (\dot{q}''_e + \dot{q}''_{fr} + \dot{q}''_{fc} - \dot{q}''_{rr})/\Delta H_g \tag{9}$$

where $\dot{q}''_{fr}$ is the radiative heat flux and $\dot{q}''_{fc}$ is the convective heat flux from the flame of the burning material transferred back to its own surface (kW/m²).
In the absence of the external heat flux, Eq. 9 becomes:

$$\dot{m}'' = (\dot{q}''_{fr} + \dot{q}''_{fc} - \dot{q}''_{rr})/\Delta H_g \tag{10}$$

Results from numerous small- and large-scale fire tests show that as the surface area of the material increases, the radiative heat flux from the flame increases and reaches an asymptotic limit, whereas the convective heat flux from the flame decreases and becomes much smaller than the radiative heat flux at the asymptotic limit [6,16]. With increase in the surface area, however, there is an over all increase in the heat flux from the flame, resulting in the increase in the mass loss rate. In large fires, the flame heat flux and the mass loss rate per unit surface area both reach constant asymptotic values.

In small-scale fires, for a fixed area of the sample, the increase in the oxygen mass fraction (Y_o) results in an increase in the radiative heat flux from the flame and decrease in the convective heat flux [6,16]. For $Y_o \geq 0.30$, the radiative heat flux from the flame reaches an asymptotic limit comparable to the limit for large-scale fires burning in the open [6,16]. The convective heat flux from the flame becomes much smaller than the radiative heat flux for $Y_o \geq 0.30$, similar to the value for the large-scale fires [6,16].

The dependency of the radiative and convective flame heat fluxes on the mass fraction of oxygen in co-flowing air in a small scale fire is shown in Fig. 5 for 100-x 100-mm x 25-mm thick slab of polypropylene, where data are from the Flammability Apparatus [6,16]. The increase in the flame radiative heat flux with Y_o is explained as due to the increase in the flame temperature and soot formation and decrease in the residence time in the flame [16]. The technique of Y_o variations to simulate large-scale flame radiative heat flux conditions in small-scale flammability experiments is defined as the *Flame Radiation Scaling Technique* [6,16].

In the Flame Radiation Scaling Technique, the heat flux from the flame is determined by performing the combustion tests at Y_o values in the range of 0.233 to 0.600, without the external heat flux and measuring the mass loss rate in each test and substituting the data in Eq. 10. Table 3 lists examples of the data obtained by the Flame Radiation Scaling Technique in the Flammability Apparatus.

The asymptotic values of the heat flux from the flame using the Flame Radiation Scaling Technique in the Flammability Apparatus show good agreement with the values obtained directly from the large-scale fire tests [6,16]. The asymptotic values of the heat flux from the flame vary from 22 to 77 kW/m², dependent primarily on the pyrolysis mode rather than on the chemical structures of the materials. For examples, for liquids, which vaporize primarily as monomers or as very low molecular weight oligomer, the asymptotic values of the heat flux from the flame are in the range of 22 to 44 kW/m^2, irrespective of their chemical structures. For polymers, which vaporize as high molecular weight oligomer, the asymptotic values of the heat flux from the flame increase substantially to the range of 49 to 71 kW/m^2, irrespective of their chemical structures. The independence of the asymptotic flame heat flux value from the chemical structure is consistent with the dependence of the flame radiation on optical thickness, soot concentration and flame temperature. The flame heat flux is one of the pertinent fire properties of materials used in the models to assess hazards and protection needs.

It would be useful if the Combustion Test Method for the Flame Radiation Scaling Technique is adopted in the ASTM E 906-83 (the OSU Apparatus) and ASTM E 1354-90 (the Cone Calorimeter) such that model based assessments can be made for fire hazards and protection needs.

Generation Rate of Products

The generation rates of major products (CO, CO_2, smoke and hydrocarbons) and depletion rate of oxygen and their integrated values are measured as functions of time at several external heat flux values in the non-flaming and flaming fires in the OSU Apparatus, in the Flammability Apparatus, and the Cone Calorimeter.

There is a direct proportionality between generation rate of a product and the mass loss rate or the total mass of the product generated to the total mass of the material lost [6,7]:

$$\dot{G}_j'' = y_j \dot{m}'' \tag{11}$$

$$W_j = y_j W_f \tag{12}$$

where $\dot{G}_j''$ is the generation rate of product j (kg/m^2-s), $\dot{m}''$ is the mass loss rate (kg/m^2-s), W_j is the total mass of the product generated (kg), W_f is total mass of the material lost (kg), which is determined from Eq. 8 and the proportionality constant y_j is defined as the yield of the product (kg/kg). W_j is determined from the following relationship:

$$W_j = A \sum_{n=t_0}^{n=t_f} \dot{G}_j''(t_n) \Delta t_n \tag{13}$$

The generation rates of products depend on the chemical structure of the material and additives, fire size, and ventilation. The yields of products for buoyant turbulent diffusion flames are independent of the fire size but depend on the fire ventilation [6].

For the model based assessments of hazards and protection needs, average yields of products are reported for the non-flaming and flaming combustion tests. The average yield of each product is determined from the ratio of the total mass of the product generated to the total mass of the material lost (Eq. 12). Extensive data tabulation for the average yields of products for variety of materials have been reported in Ref. 6; Table 4 shows an example of the data.

The yields of products associated with complete combustion such as CO_2, are higher for materials with aliphatic, carbon-hydrogen-oxygen containing structures. The yields of products associated with incomplete combustion, such as CO and smoke, increase with increase in the chemical bond un-saturation, aromatic nature of the bonds, and introduction of the halogen atoms into the chemical structure of the material. The yields of products are used in fire models to assess hazards and protection needs in conjunction with the mass loss rate, such as the combination of Eqs. 9 and 11:

$$\dot{G}_j^{''} = (y_j/\Delta H_g)(\dot{q}_e^{''} + \dot{q}_{fr}^{''} + \dot{q}_{fc}^{''} - \dot{q}_{rr}^{''}) \tag{14}$$

where $y_j/\Delta H_g$ is defined as the *Product Generation Parameter (PGP)*. PGP is a property of the material, its value for buoyant turbulent diffusion flame is independent of the fire size, but depends on the fire ventilation.

The hazards in fires are due to generation of heat and products. With increase in the generation rates of the products associated with the complete combustion, such as CO_2, the heat release rate increases with enhancement of hazard due to heat (*thermal hazard*). On the other hand, with the increase in the generation rates of the products associated with incomplete combustion, such as CO and smoke, hazard due to toxicity, corrosivity, reduced visibility and smoke damage (nonthermal hazard) increases. Equation 14 shows that the generation rates of products can increase by changes in several factors, alone or in combination. The factors are: 1) yields of the products, 2) heat of gasification of the material; 3) surface re-radiation loss for the material, 4) radiative and convective heat fluxes from the flame of the burning material transferred back to its own surface, i.e., fire size, 5) external heat flux, and 6) extent of fire propagation, i.e., area.

The external and flame heat fluxes and extent of fire propagation are strongly dependent on the fire scenarios and are usually incorporated into the models through various heat flux correlations and/or heat flux data and fire propagation models, such as the FSG model at FMRC in the Flammability Apparatus. The yields of products and heats of gasification, separately or as ratios (Product Generation Parameter) are used as input parameters.

The PGP value of each product is quantified by measuring the generation rate of the product at three or four external heat flux values, plotting the generation rate against the external heat flux and determining the slope by the linear regression analysis. Figure 6 shows an example for the steady state condition. For fires, where steady state condition cannot be achieved, PGP values are determined by plotting the

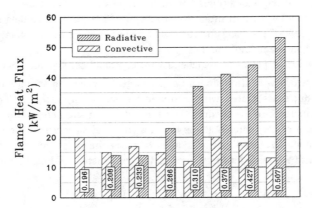

Figure 5. Flame Radiative and Convective Heat Fluxes at Various Oxygen Mass Fractions in the Co-Flowing Air for the Steady-State Combustion of 100 x 100 x 25 mm Thick Slab of Polypropylene. Data are from the Combustion Tests using the Flame Radiation Scaling Technique in the Flammability Apparatus at FMRC [6,16]. Numbers Within the Bars are Oxygen Mass Fractions.

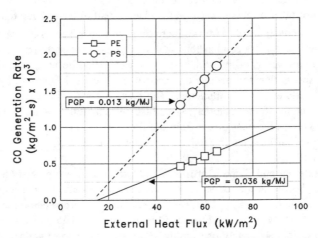

Figure 6. CO Generation Rate Versus the External Heat Flux. The Slopes of the Lines Represent the CO Generation Parameter for Polyethylene and Polystyrene. The Data are from the Flammability Apparatus at FMRC.

values of W_j calculated from Eq. 13 and E_{ex} calculated from Eq. 7. The technique is used routinely in the Flammability Apparatus and can also be used in the OSU Apparatus and the Cone Calorimeter. Table 5 shows an example of the data for PGP values for CO and smoke from the Flammability Apparatus (see Tables 3 and 4).

It would be useful if the Combustion Test Method for the determination of the Product Generation Parameter be adopted in the ASTM E 906-83 (the OSU Apparatus) and ASTM E 1354-90 (the Cone Calorimeter) such that model based assessments can be made for fire hazards associated with the fire products and protection needs.

Heat Release Rate

The chemical, convective, and radiative heat release rates and their integrated values are measured as functions of time at several external heat flux values in the flaming fires. Heat release rate in combustion reactions, within a flame, is defined as the *chemical heat release rate* [6]. The chemical heat released within the flame is carried away from the flame by flowing product-air mixture and is emitted to the environment as radiation. The component of the chemical heat release rate carried away by the flowing products-air mixture is defined as the *convective heat release rate* [6]. The component of the chemical heat release rate emitted to the environment is defined as the *radiative heat release rate* [6].

The chemical heat release rate is determined from the *Carbon Dioxide Generation (CDG)* [6] and *Oxygen Consumption (OC) Calorimetries* [6,8]. In the CDG Calorimetry, the chemical heat release rate is determined from the mass generation rate of CO_2 corrected for CO [6]. In the OC Calorimetry, the chemical heat release rate is determined from the mass consumption rate of O_2 [6,8]. The convective heat release rate is determined from the *Gas Temperature Rise (GTR) Calorimetry* by measuring the gas temperature above ambient and the total mass flow rate [1,6]. The radiative heat release rate is determined from the difference between the chemical and convective heat release rates [6].

The OSU Apparatus is designed to use the GTR calorimetry [1-3], but now also uses the OC calorimetry. The Flammability Apparatus is designed to use CDG, OC, and GTR calorimetries [6]. The Cone Calorimeter is designed to use the OC Calorimetry [8].

The chemical heat release rate follows the same relationships as the generation rates of products (Eqs. 11,12,14). The heat release rate relationship analogues to the relationships for the generation rates of products (Eqs. 11 and 12) are:

$$\dot{Q}_i = \Delta H_i \dot{m} \tag{15}$$

$$E_i = \Delta H_i W_f \tag{16}$$

Table 4. Yields of Major Products and Heats of Combustion of Materials[a]

Material	Yield (kg/kg)			Heat of Combustion[b] (MJ/kg)	
	CO_2	CO	Smoke	Chem	Con
Gases					
Ethylene	2.72	0.013	0.043	41.5	27.3
Propylene	2.74	0.017	0.095	40.5	25.6
1,3-Butadiene	2.46	0.048	0.125	33.6	15.4
Acetylene	2.60	0.042	0.096	36.7	18.7
Liquids					
Ethyl alcohol	1.77	0.001	0.008	25.6	19.0
Acetone	2.14	0.003	0.014	27.9	20.3
Heptane	2.85	0.010	0.037	41.2	27.6
Octane	2.84	0.011	0.038	41.0	27.3
Kerosene	2.83	0.012	0.042	40.3	26.2
Solids					
News paper	1.32	-	-	14.4	-
Wood (red oak)	1.27	0.004	0.015	12.4	7.8
Wood (Douglas fir)	1.31	0.004	-	13.0	8.1
Wood (pine)	1.33	0.005	-	12.4	8.7
Corrugated paper	1.22	-	-	13.2	-
Wood (hemlock)[c]	1.22	-	0.015	13.3	-
Wool 100 %[c]	1.79	-	0.008	19.5	-
ABS[e,d]	-	-	0.105	30.0	-
Polyoxymethylene	1.40	0.001	-	14.4	11.2
Polymethylmethcrylate	2.12	0.010	0.022	24.2	16.6
Polyethylene (PE)	2.76	0.024	0.060	38.4	21.8
Polypropylene	2.79	0.024	0.059	38.6	22.6

Table 4. Continued

Material	Yield (kg/kg)			Heat of Combustion[b] (MJ/kg)	
	CO_2	CO	Smoke	Chem	Con
Polystyrene	2.33	0.060	0.164	27.0	11.0
Silicone	0.96	0.021	0.065	10.6	7.3
PE + 25 % Chlorine	1.71	0.042	0.115	22.6	10.0
PE + 36 % Chlorine	0.83	0.051	0.139	10.6	6.4
PE + 48 % Chlorine	0.59	0.049	0.134	7.2	3.9
Polyvinylchloride (PVC)	0.46	0.063	0.172	5.7	3.1
PVC-1[c] (LOI=0.50)	0.64	-	0.098	7.7	-
PVC-2[c] (LOI=0.50)	0.69	-	0.076	8.3	-
PVC[c] (LOI = 0.20)	0.93	-	0.099	11.3	-
PVC[c] (LOI = 0.25)	0.81	-	0.078	9.8	-
PVC[c] (LOI = 0.30)	0.85	-	0.098	10.3	-
PVC[c] (LOI = 0.35)	0.89	-	0.088	10.8	-
PEEK-FG[c,d]	1.88	-	0.042	20.5	-
Polyester1- FG[c]	2.52	-	0.049	27.5	-
Polyester2- FG[c]	1.47	-	-	16.0	-
Polyester3- FG[c]	1.18	-	-	12.9	-
Polyester4-FG	1.74	-	-	19.0	-
Polyester5-FG	1.28	-	-	13.9	-
Polyester6-FG	1.47	0.055	0.070	17.9	10.7
Polyester7-FG	1.24	0.039	0.054	16.0	9.9
Polyester8-FG	0.71	0.102	0.068	9.3	6.5
Epoxy1-FG[c]	2.52	-	0.056	27.5	-
Epoxy2-FG	1.10	0.166	0.128	11.9	-
Epoxy3-FG	0.92	0.113	0.188	10.0	-
Epoxy4-FG	0.94	0.132	0.094	10.2	-

Continued on next page

Table 4. Continued

Material	Yield (kg/kg)			Heat of Combustion[b] (MJ/kg)	
	CO_2	CO	Smoke	Chem	Con
Epoxy5-FG	1.71	0.052	0.121	18.6	-
Epoxy-FG-paint	0.83	0.114	0.166	11.3	6.2
Phenolic1-FG	0.98	0.066	0.023	11.9	-
Phenolic2-FG[c]	2.02	-	0.016	22.0	-
Phenolic-FG-paint	1.49	0.027	0.059	22.9	11.5
Epoxy-FG-Phenolic	1.06	0.134	0.089	11.5	-

a: Data from the Flammability Apparatus; **b**: Chem: chemical; Con: convective; Rad: radiative; **c**: calculated from data reported in Refs. 14 and 15; **d**: ABS-acrylonitrile-butadiene-styrene; PEEK: polyether-ether ketone; PPS-polyphenylene sulfide; FG- fiber glass; LOI: limiting oxygen index.

Table 5. Carbon Monoxide and Smoke Generation Parameters

Material[a]	Product Generation Parameter (g/MJ)	
	CO	Smoke
Wood (red oak)	2.2	8.3
Polypropylene	12	30
Polyethylene (ld)	13	33
Polyethylene (hd)	10	26
Polyethylene/25% Cl	20	55
Polyethylene/36% Cl	17	46
Polyethylene/48% Cl	16	43
Polyvinylchloride (PVC)	25	69
Polyoxymethylene	0.42	0
Polymethylmethacrylate	6.3	14
Polystyrene	35	96

a: Cl- chlorine; ld- low density; hd- high density

where subscript i represents chemical, convective, or radiative, E_i is the chemical, convective, or radiative energy (MJ) and the proportionality constant ΔH_i is defined as the chemical, convective, or radiative heat of combustion (MJ/kg). E_i is determined from the following relationship:

$$E_i = A \sum_{n=t_{ig}}^{n=t_{ox}} \dot{Q}_i^{''}(t_n) \Delta t_n \tag{17}$$

The heat release rate depends on the chemical structure of the material and additives, fire size, and ventilation. The chemical, convective, and radiative heats of combustion for buoyant turbulent diffusion flames are independent of the fire size but depend on the fire ventilation [6]. For the model based assessments of hazards and protection needs, average heats of combustion are reported. The average heat of combustion is determined from the ratio of the energy to the total mass of the material lost (Eq. 16). Extensive data tabulation for the average chemical, convective, and radiative heats of combustion for variety of materials have been reported in Ref. 6; Table 4 shows an example of the data.

The chemical and convective heats of combustion are generally higher and the radiative heat of combustion is generally lower for materials with aliphatic, carbon-hydrogen-oxygen containing structures. The chemical and convective heats of combustion decrease with increase in the chemical bond un-saturation, aromatic nature of the bonds, and introduction of the halogen atoms into the chemical structure of the material. The heats of combustion are used in fire models to assess hazards and protection needs in conjunction with the mass loss rate, such as the combination of Eqs. 9 and 15:

$$\dot{Q}_i^{''} = (\Delta H_i / \Delta H_g)(\dot{q}_e^{''} + \dot{q}_{fr}^{''} + \dot{q}_{fc}^{''} - \dot{q}_{rr}^{''}) \tag{18}$$

where $\Delta H_i / \Delta H_g$ is defined as the *Heat Release Parameter (HRP)*. HRP is a property of the material, its value for buoyant turbulent diffusion flame is independent of the fire size, but depends on the fire ventilation.

The hazards in fires are due to generation of heat and products. With increase in the heat release rate, the thermal hazard increases. Equation 18 shows that the heat release rate increases by changes in several factors, alone or in combination. The factors are: 1) heat of combustion of the material, 2) heat of gasification of the material; 3) surface re-radiation loss for the material, 4) radiative and convective heat fluxes from the flame of the burning material transferred back to its own surface, i.e., fire size, 5) external heat flux, and 6) extent of fire propagation, i.e., area.

The external and flame heat fluxes and extent of fire propagation are strongly dependent on the fire scenarios and are usually incorporated into the models through various heat flux correlations and/or heat flux data and fire propagation models, such as the FSG model at FMRC in the Flammability Apparatus. The heat of combustion and heats of gasification, separately or as ratios (Heat Release Parameter) are used as input parameters.

The chemical, convective, and radiative HRP values are quantified by measuring the chemical and convective heat release rates (radiative heat release rate by difference) at three or four external heat flux values, plotting the heat release rate against the external heat flux and determining the slope by the linear regression analysis. Figure 7 shows an example for the steady state condition. For fires, where steady state condition cannot be achieved, such as shown in Fig. 8, the HRP values are determined by plotting the values of E_i calculated from Eq. 17 and E_{ex} calculated from Eq. 7. The technique is used routinely in the Flammability Apparatus and can also be used in the OSU Apparatus and the Cone Calorimeter. Table 6 shows an example of the data for the chemical and convective HRP values from the OSU Apparatus, the Flammability Apparatus and the Cone Calorimeter taken from Ref. 6. The HRP values from the OSU Apparatus, the Flammability Apparatus, and the Cone Calorimeter are in reasonable agreement.

It would be useful if the Combustion Test Method for the determination of the Heat Release Parameter be adopted in the ASTM E 906-83 (the OSU Apparatus) and ASTM E 1354-90 (the Cone Calorimeter) such that model based assessments can be made for fire hazards associated with the fire products and protection needs.

THE FIRE PROPAGATION TEST METHOD

Fire propagation process is associated with the movement of a vaporizing area of a material on its surface. The rate of the movement of the vaporizing area on the surface is defined as the fire propagation rate. The vaporizing area is defined as the *pyrolysis front*.

Fire propagation tests are performed in the Flammability Apparatus, but not in the OSU Apparatus and the Cone Calorimeter. The fire propagation test in the Flammability Apparatus can be considered as a larger version of the ASTM D-2863 Oxygen Index test, with an ignition zone provided by four external heaters (Fig. 2B). In the test, downward and upward fire propagation are examined under co-flowing air with the oxygen mass fraction (Y_o) in the range 0 to 0.60 [6,7,12,13]. Materials as vertical slabs and cylinders of up to 600 mm in length and up to about 25-mm in thickness, 100-mm in width or 50-mm in diameter are used. Pyrolysis front, flame height, chemical, convective, and radiative heat release rates, and generation rates of products are measured during fire propagation.

The Downward Fire Propagation

Figures 9 and 10 show the fire propagation data for 25-mm thick, 100-mm wide, and

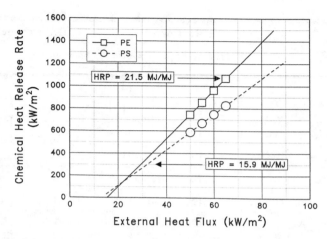

Figure 7. Chemical Heat Release Rate Versus the External Heat Flux. The Slopes of the Lines Represent the Chemical Heat Release Parameter for Polyethylene and Polystyrene. The Data are from the Flammability Apparatus at FMRC.

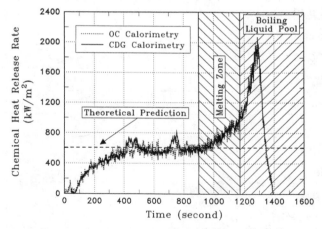

Figure 8. Chemical Heat Release Rate for 100-mm Diameter and 25 mm Thick Slab of Polypropylene Exposed to an External Heat Flux of 50 kW/m^2 for 0.09 m/s Co-Flowing Normal Air in the Flammability Apparatus at FMRC. The Theoretical Prediction is Based on Eq.18 and Data from Tables 3 and 4. OC: Oxygen Consumption Calorimetry; CDG: Carbon Dioxide Generation Calorimetry.

Table 6. Chemical and Convective Heat Release Parameters[a]

Materials	HRP-Chemical		HRP-Convective	
	Flamm.App.	Cone	Flamm.App.	OSU
ABS	-	14	-	-
Polyamide	21	21	-	-
Polypropylene	19	-	11	-
Polyethylene	17	21	12	-
Polystyrene	16	19	6	-
Polymethylmethacrylate	15	14	10	-
Nylon	12	-	7	-
Polyoxymethylene	6	-	5	-
Polyethylene /25 % Cl	11	-	5	-
PVC, LOI 0.25	-	5	-	-
PVC, LOI 0.30	-	5	-	-
Polyethylene/36 % Cl	4	-	-	-
PVC, LOI 0.50	-	3	-	-
Rigid PVC	2	3	1	-
Polyethylene/48 % Cl	2	-	-	-
ETFE (Tefzel)	6	-		
PFA (Teflon)	5	-		
FEP (Teflon)	2	-		
TFE (Teflon)	2	-		
Wood (Douglas fir)	7	-	5	-
Epoxy-fiber glass	4	5	2	1
Epoxy/kevlar	4	4	2	2
Phenolic-fiber glass	4	3	2	1

a: calculated from the data measured in the Flammability Apparatus and reported in Refs. 14 and 15

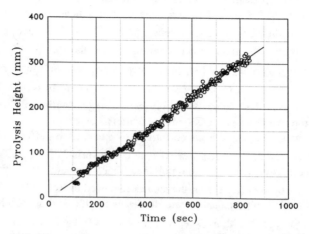

Figure 9. Pyrolysis Front Versus Time for the Downward Fire Propagation for 300-mm Long, 100-mm Wide and 25-mm Thick Polymethylmethacrylate Vertical Slab Under Opposed Air Flow Condition in the Flammability Apparatus at FMRC. Air Flow Velocity = 0.09 m/s. Oxygen Mass Fraction = 0.334 [12].

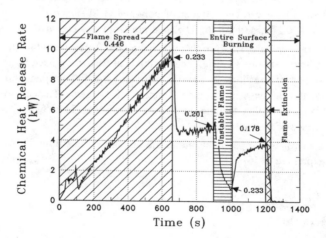

Figure 10. Chemical Heat Release Rates Versus Time for the Downward Fire Propagation, Combustion, and Flame Extinction for 300-mm Long, 100-mm Wide and 25-mm Thick Polymethylmethacrylate Vertical Slab Under Opposed Air Flow Condition in the Flammability Apparatus at FMRC. Air Flow Velocity = 0.09 m/s. Numbers in the Figure are the Oxygen Mass Fractions [12].

300-mm long vertical slab of polymethyl- methacrylate, taken from Ref.12. The slopes of the lines represent the fire propagation rate. Figure 9 shows the pyrolysis front for $Y_o = 0.334$. Figure 10 shows the chemical heat rate accompanying the pyrolysis front for the downward fire propagation for $Y_o = 0.446$ and for the burning of the entire slab after the flame reaches the bottom of the slab for $Y_o \leq 0.233$. The flame extinction occurs at $Y_o = 0.178$, in excellent agreement with the predicted value of 0.18 and values measured by other researchers for larger samples [12]. The Y_o value for flame extinction in the Flammability Apparatus (buoyant turbulent diffusion flame) is lower than from the Oxygen Index (laminar flame) as expected.

Numerous studies have been performed to examine the effects of Y_o on fire propagation (reviewed in Ref. 12). An example is shown in Fig. 11 for the downward fire propagation rates versus Y_o for the polymethylmethacrylate slab with width $\leq$ 25-mm and length $\leq$ 300 mm (data from the studies reviewed in Ref. 12). The data show that for $Y_o < 0.30$, fire propagation rate decreases rapidly and approaches the flame extinction zone for $Y_o = 0.178$, in excellent agreement with the flame extinction value in Fig. 10 and predicted value of $Y_o = 0.18$ [12].

The Upward Fire Propagation

Figure 12 shows the pyrolysis front for the upward fire propagation for 25-mm diameter, 600-mm long vertical cylinder of polymethyl- methacrylate in co-flowing air with Y_o = 0.233, 0.279, and 0.446. The data are from the Flammability Apparatus at FMRC [12]. The upward fire propagation is much faster than the downward fire propagation as expected due to differences in the heat flux transferred by the flame.

The slopes of the lines in Fig. 12 represent the fire propagation rate, which increases with the increase in the mass fraction of oxygen. This behavior is not a surprise as Flame Radiation Scaling Technique shows that flame heat flux transferred back to the surface of the material increase with Y_o.

The upward fire propagation rate in the direction of air flow for thermally thick materials is expressed as [6]:

$$u^{1/2} = \frac{\delta_f^{1/2} \dot{q}_f''}{\Delta T_{ig} \sqrt{(k\rho c_p}} \tag{18}$$

where **u** is the fire propagation rate in m/s; δ_f is an effective flame heat transfer distance (m) generally assumed to be a constant, $\dot{q}_f''$ is the heat flux transferred from the propagating flame ahead of the pyrolysis front (kW/m²), and $\Delta T_{ig} \sqrt{k\rho c_p}$ is the Thermal Response Parameter (TRP) for the thermally thick materials in kW-s$^{1/2}$/m² (Eq. 2).

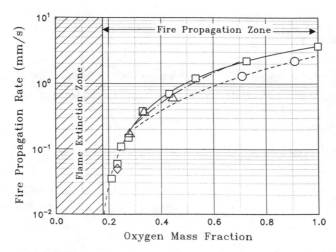

Figure 11. Downward Fire Propagation Rate Versus Oxygen Mass Fraction for Vertical Polymethylmethacrylate Slabs with Width ≤ 25 mm and Lengths ≤ 300 mm. Data are Taken from Various Studies Reported in the Literature (Reviewed in Ref. 12).

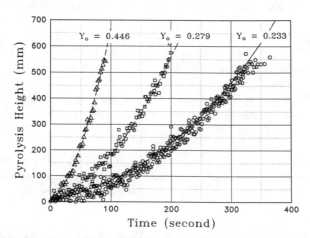

Figure 12. Pyrolysis Front Versus Time for the Upward Fire Propagation for 600-mm Long and 25-mm Diameter PMMA Cylinder Under Co-Air Flow Condition in the Flammability Apparatus at FMRC. Air Flow Velocity = 0.09 m/s. Numbers in the Figure are the Oxygen Mass Fractions [12].

Through data correlations, it has been shown that the heat flux transferred from the propagating flame ahead of the pyrolysis front satisfies the following relationship [6]:

$$\dot{q}_f'' \propto (0.42 \dot{Q}_{ch}')^{1/3} \tag{19}$$

where $\dot{Q}_{ch}'$ is the chemical heat release rate per unit width or circumference of the material as a slab or a cylinder respectively (kW/m). From Eqs. 18 and 19:

$$u^{1/2} \propto \frac{(0.42 \dot{Q}_{ch}')^{1/3}}{\Delta T_{ig} \sqrt{(k\rho c_p)}} \tag{20}$$

The right hand side of Eq. 20, with a proportionality constant assumed to be 1000, $\dot{Q}_{ch}'$ in kW/m and $\Delta T_{ig} \sqrt{k\rho c_p}$ in kW-s$^{1/2}$/m^2, is defined as the *Fire Propagation Index (FPI)* [6,7,17]:

$$FPI = 1000 \frac{(0.42 \dot{Q}_{ch}')^{1/3}}{TRP} \tag{21}$$

Classification of Materials Based on Their Fire Propagation Behavior

The following FPI values, based on the data from the Flammability Apparatus with validation in the large-scale fires, have been found to characterize the general fire propagation behavior of materials [6,7,17]:

1) **FPI** $\leq$ **7**: no fire propagation beyond the ignition zone. Materials are identified as *non-propagating Group N-1 materials*. Flame is at the critical extinction condition;

2) **7 < FPI < 10**: decelerating fire propagation beyond the ignition zone. Materials are identified as Group D-1 materials. Fire propagates beyond the ignition zone although in a decelerating fashion. Fire propagation beyond the ignition zone is limited;

3) **10 $\leq$ FPI < 20**: fire propagates slowly beyond the ignition zone. Materials are identified as *propagating Group P-2 materials*;

4) **FPI $\geq$ 20**: Fire propagates rapidly beyond the ignition zone. Materials are identified as *propagating Group P-3 materials*.

For the classification of material for their fire propagation behavior, the Fire Propagation Index Test is performed and materials are classified as Group 1, 2, or 3 materials.

The Fire Propagation Index (FPI) Test

Two sets of tests are performed:

1) Thermal Response Parameter Test : ignition test is performed for up to 100- x 100-mm or 100-mm diameter and up to 25-mm thick sample in the Flammability Apparatus (Fig. 2A) and the Thermal Response Parameter (TRP) value is determined from the time to ignition versus external heat flux relationship (Eqs. 1 or 3).

2) Upward Fire Propagation Test : fire propagation test is performed for vertical slabs, sheets, or cylinders in the Flammability Apparatus (Fig. 2B). About 25-mm thick, 100-mm wide slabs or 50-mm diameter cylinders with lengths of up to about 600-mm long are used. The bottom 120- to 200-mm of the sample is kept in the ignition zone, where it is exposed to 50 kW/m^2 of external heat flux in the presence of a pilot flame. Beyond the ignition zone, fire propagates by itself, under co-air flow condition with Y_o = 0.40. During upward fire propagation, measurements are made for the chemical heat release rate and generation rates of the fire products as functions of time.

The TRP value and the chemical heat release rate are used in Eq. 21 to calculate the Fire Propagation Index (FPI) value as a function of time and determine the propagating and non-propagating fire behavior of the material. An example is shown in Fig. 13 for five composite systems [17]. In the tests, there was no fire propagation beyond the ignition zone for all the systems, the FPI values are less than 5, and thus the composites are identified as non-propagating, Group 1N materials.

The FPI test procedure for electrical cables is the FMRC cable standard [18], where cables are classified as Group 1 (FPI< 10)- non-propagating or decelerating, Group 2 (10 ≤ FPI< 20) - slowly propagating, and Group 3 (FPI ≥ 20)- rapidly propagating. The FPI test procedure has also been adoped as a FMRC standard for conveyor belts [19, to be issued shortly]. It is also being used to classify wall and ceiling insulation panels with modifications [20,21], ducts, chutes, clean room materials, and others. Table 7 lists examples of the FPI values quantified in the Flammability Apparatus.

Fire Propagation Index (FPI) is one of the most important fire properties of materials to assess fire hazards and protection needs. Increasing the TRP value and decreasing the heat release rate for materials by various passive fire protection techniques would decrease the FPI value and change the fire propagation behavior from propagating to decelerating to non-propagating. Passive fire protection techniques

Table 7. Fire Propagation Index Values for Materials[a]

Materials[b]	Diameter/Thickness (mm)	FPI	Group	Fire Propagation[c]
Synthetic Polymers				
Polymethylmethacrylate	25	30	3	P
Fire retarded polypropylene	25	>>10	3	P
Electrical Cable Insulation and Jacket				
PVC/PVC (power)	4-13	11-28	2-3	P
PVC/PVC (communications)	4	36	3	P
PE/PVC (power)	11	16-23	3	P
PE/PVC (communications)	4	28	3	P
PVC/PE (power)	34	13	2	P
PVC/PVF (communications)	5	7	1	N
Silicone/PVC (power)	16	17	2	P
Silicone/XLPO (power)	55	6-8	1	N-D
Si/XLPO (communications)	28	8	1	D
EP/EP (power)	10-25	6-8	1	N-D
XLPE/XLPE (power)	10-12	9-17	1-2	D-P
XLPE/XLPO (communications)	22-23	6-9	1	N-D
XLPE/EVA (power)	12-22	8-9	1	D
XLPE/Neoprene (power)	15	9	1	D
XLPO/XLPO (power)	16-25	8-9	1	D
XLPO,PVF/XLPO (power)	14-17	6-8	1	N-D
EP/CLP (power)	4-19	8-13	1-2	D-P
EP,FR/None (power)	4-28	9	1	D
EP-FR/none (communications)	28	12	2	P
ETFE/EVA (communications)	10	8	1	D
FEP/FEP (communications)	8-10	4-5	1	N

Table 7. Continued

Materials[b]	Diameter/Thickness (mm)	FPI	Group	Fire Propagation[c]
Composite Systems				
Polyester 1-70%FG	4.8	13	2	P
Polyester 2-70%FG	4.8	10	2	P
	19	8	1	D
	45	7	1	D
Epoxy 1- 65%FG	4.4	9	1	D
Epoxy 2-65%FG	4.8	11	2	P
Epoxy 3-65%FG	4.4	10	2	P
Epoxy 4-76%FG	4.4	5	1	N
Phenolic-80%FG	3.2	3	1	N
Epoxy-82%FG-Phenolic	-	2	1	N
Phenolic-84%Kevlar	4.8	8	1	D
Cyanate-73%Graphite	4.4	4	1	N
PPS-84 %FG	4.4	2	1	N
Epoxy-71%FG	4.4	5	1	N
Conveyor Belts				
Styrene-Butadiene Rubber (SBR)	-	8-11	1-2	D-P
Chloroprene Rubber (CR)	-	5	1	N
CR/SBR	-	8	1	D
PVC		4-10	1-2	N-P

a: calculated from the data measured in the Flammability Apparatus; **b**: PVC-polyvinylchloride; PE-polyethylene; PVF-polyvinylidene fluoride; XLPO-cross-linked polyolefin; Si- silicone; EP-ethylene-propylene; XLPE-cross-linked polyethylene; EVA-ethylvinyl acetate; CLP- chlorosulfonated polyethylene; FEP-fluorinated ethylene-propylene; FG- fiber glass; PPS- polyphenylene sulfide; **c**: **P:** propagating; **D:** decelerating propagation; **N:** non-propagating.

could involve modifications of chemical structures, incorporation of fire retardants, and changes in the shape, size, and arrangements of the materials, use of coatings, inert barriers. Heat release rate could also be reduced by the application of active fire protection agents such as water, foam, inert, dry powders, Halon and alternates, etc.

Recently the Fire Spread and Growth (FSG) model has been incorporated into the Flammability Apparatus at FMRC to assess the fire propagation behavior of the materials under various fire scenarios, complementing the FPI based classification of materials. The FSG model operates concurrently with the test being performed and the FSG model results are available at the end of the test. The fire propagation rate predictions by the FSG model so far supports the FPI classification.

In the future, it is anticipated that the FPI test classification will be replaced by the Flammability Apparatus based test methodology using the FSG model assessments with risk profile predictions using a FMRC risk model currently under development.

The Flammability Apparatus with FSG and risk models has a potential of being considered as an apparatus with advance test procedures for adoption by the ASTM, ISO, IEC. and others.

NONTHERMAL DAMAGE TEST METHOD

Damage due to heat is defined as thermal damage and damage due to smoke, toxic, and corrosive products is defined as nonthermal damage [11]. Nonthermal damage depends on the chemical nature and deposition of products on the walls, ceilings, building furnishings, equipment, components, etc., and the environmental conditions. The severity of the nonthermal damage increases with time. Some examples of nonthermal damage to property are corrosion damage, electrical malfunctions, damage due to discoloration and odors, etc. Toxic effects of fire products on human body resulting in an injury or loss of life is an example of nonthermal damage of residential occupancies. The subject of toxicity has been discussed in detail in Ref. 22.

The subject of corrosion for commercial and industrial occupancies has been reviewed based on the knowledge derived from the telephone central office (TCO) experience for the deposition of atmospheric pollutants and fire products on equipment, severity of corrosion damage and ease of cleaning the equipment [23,24].
In TCO fires involving PVC based electrical cables, contamination levels in the range of about 5 to 900 microgram/cm^2 have been observed [23,24]. In general, an electronic switch would be expected to accumulate zinc chloride levels in the range of about 5 to 9 microgram/cm^2 from the interaction with the environment over its expected lifetime of 20 + years. A clean equipment is expected to have less than about 2 microgram/cm^2 of chloride contamination, whereas, contaminated equipment can have as high as 900 microgram/cm^2. Thus, equipment contamination levels due to chloride ions and ease of restoration have been classified into four levels [24], which are listed in Table 8.

Currently the nonthermal damage is assessed by toxicity tests [22] and by smoke and corrosion tests [11].

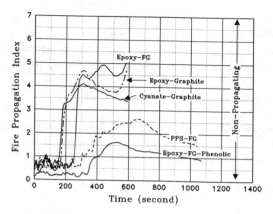

Figure 13. Fire Propagation Indices Versus Time for 4-mm Thick, 100-mm Wide, and 600-mm Long Vertical Slabs of Composite Systems in Co-Flowing Air with an Oxygen Mass Fraction of 0.40. The Sides and Back of the Slabs are Covered Tightly with Ceramic Paper and Heavy Duty Aluminum Foil. The Data are from the Flammability Apparatus at FMRC [17].

Table 8. Contamination Levels for the Surface Deposition of Chloride Ions for Electronic Equipment [a]

Chloride Ion (microgram/cm^2)	Level	Damage/Cleaning/Restoration
2	One	No damage expected. No cleaning and restoration required.
<30	Two	Equipment can be easily restored to service by cleaning without little impact on long-term reliability
30 to 90	Three	Equipment can also be restored to service by cleaning, as long as no unusual corrosion problems arise, and the environment is strictly controlled soon after the fire.
>90	Four	The effectiveness of cleaning the equipment dwindles and the cost of cleaning quickly approaches the replacement cost. Equipment contaminated with high chloride levels may require severe environmental controls even after cleaning in order to provide potentially long-term reliable operation

a: compiled from the information given in Ref. 24

within the data analysis package of the Flammability Apparatus can predict the concentrations of the products for various ventilation conditions for the fire scenarios of concern. The concentration predictions can be combined with the toxicity, smoke damage and corrosion assessments as overall hazard analysis package.

The Corrosion Test Method

The corrosion test method is used in the Flammability Apparatus. Corrosion damage is assessed in terms of rate of corrosion of a material exposed to a unit concentration of material (fuel) vapors, defined as the *Corrosion Index* (CI) [6]:

$$CI = \{\delta_{loss}/\Delta t_{exposure}\}/\{W_f/W_T \Delta t_{test}\} \tag{22}$$

where CI is in (Å/min)/(mg/g), δ_{loss} is the metal loss due to corrosion (Å), $\Delta t_{exposure}$ is the time the corrosive product deposit is left on the surface of the metal (min), W_f is the total mass of the mateiral lost (Eq.7) for the test duration (g), W_T is the total mass flow rate of the mixture of fire products and air (kg/s) and Δt_{test} is the test duration (s).

In the corrosion test method, a Rohrback Cosasco [RC] atmospheric metal corrosion probe, designed for the Flammability Apparatus, is placed inside the sampling duct of the Appartus (Fig. 2B). The probe consists of two metal strips (5,000 Å), embedded in epoxy- fiber glass plates. One metal strip is coated and acts as a reference and the other un-coated metal strip acts as a sensor. As the sensor strip corrodes and looses its thickness, its resistance changes. The change in the resistance, which represents the extent of corrosion of the metal, is measured as a function of time, by the difference in the resistance between the two strips. The probe readings remain reliable up to about half the thickness of the metal strip (2500 Å), the probe is thus identified as 2500 Å probe.

In the test, corrosion is measured every minute for first three hours and every hour after that up to a maximum of 16 hours. Figure 14 shows an example of the metal corrosion from the combustion products of the polyvinylchloride (PVC) homopolymer and PVC commercial materials, as measured in the Flammability Apparatus. The metal corrosion is faster in the initial stages and becomes slower in later stages due to protective oxide film formation on the surface, within the test duration of about 20 minutes. The data in Fig.14 show that the metal corrosion from the combustion products of the PVC homopolymer is significantly higher than the metal corrosion from the products of the PVC commercial materials, indicating dilution and/or partial neutralization of HCl by the pyrolysis products of non-halogenated additives in the commercial materials.

The metal corrosion in Fig. 14 is quite fast as it occurs within the test duration of 20 minutes. For less corrosive products, for the same test duration, the metal corrosion process takes about 12 to 14 hours to complete. Thus a maximum of 16 hour exposure is used for products showing slow corrosion.

In the corrosion test method, three types of measurements are made: 1) mass loss rate as a function of time and total mass of the material lost and its duration, 2) total mass flow rate of the mixture of the products with air as a function of time, and 3) metal corrosion as a function of time. The average corrosion rate is obtained from the difference between the initial and final corrosion values divided by the time duration. The data are used in Eq. 22 to calculate the CI value.

The CI values for various materials have been determined in the Flammability Apparatus; Table 9 lists values for some selected materials, as examples. The CI values for non-halogenated polypropylene and wood are negligibly small. For the highly halogenated materials, the CI value is high if hydrogen atoms are present in the structure (PVC) or halogenated fire retardant additive is present (PP/FR), and is low if there are no hydrogen atoms in the structure (Teflon® TFE). The difference in the CI values indicate the importance of water as a combustion product to generation acids for PVC and PP/FR. Teflon® (TFE) does not generate water as a product of combustion and thus the formation of an acid (HF) would depend on the efficiency of the hydrolysis process between the ambient water from air and Teflon® (TFE) vapors. The hydrolysis process appears to be inefficient. The CI value decreases with increase in the amount of non-halogenated additive (PVC-1 to -4).

The Smoke Damage Test

Smoke is a mixture of black carbon (soot) and aerosol. Smoke damage is considered in terms of reduction in the visibility, discoloration and odor of the property exposed to smoke, interference in the electric conduction path and corrosion of the parts exposed to smoke.

In the tests, measurements are made for the optical density in the sampling duct of the Flammability Apparatus and the Cone Calorimeter and above the sample exposure chamber in the OSU Apparatus. Tests are also performed to quantify the odor, color, and electrical properties of smoke in the Flammability Apparatus [6].

FLAME EXTINCTION TEST METHOD

Flame extinction process is associated with the interference with the chemical reactions and/or heat removal and/or dilution by liquids, gases, solid powders, or foams within the flame and/or on the surface of the burning material. The most commonly used liquid and gaseous chemical inhibition agents at the present time are: Halon- 1211 ($CBrClF_2$), 1301 ($CBrF_3$), and 2402 ($CBrF_2CBrF_2$). (The numbers represent: *First*- number of carbon atoms; *Second*- number of fluorine atoms; *Third*- number of chlorine atoms; *Fourth*- number of bromine atoms).

Because of the contribution of Halons to depletion of the stratospheric ozone

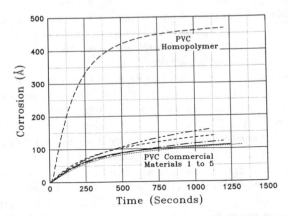

Figure 14. Copper Metal Corrosion Due to Exposure to Combustion Products of Polyvinylchloride Homopolymer and five Commercial Materials (1 to 5). Corrosion was Measured as a Function of Time in the Sampling Duct of the Flammability Apparatus at FMRC by a Rohrback Cosasco 2500 Å Copper Atmospheric Corrosion Probe Designed Specially for the Apparatus.

Table 9. Values for the Corrosion Index for Selected Materials[a]

Materials	Corrosion Index [(Å/min)/(mg of material vapors/g of air)]
Polyvinylchloride (PVC)[b]-1	1789
PVC-2	780
PVC-3	597
PVC-4	364
Polypropylene	74
Polypropylene/fire retardant	1678
Teflon® (TFE)	281
Wood	88

a: determined in the Flammability Apparatus at the Factory Mutual Research Corporation; b: amount of non-halogenated additive increasing from 1 to 4.

layer, they will, however, not be used in the future [25]. There is thus an intense effort underway to develop alternative fire suppressants to replace ozone layer depleting Halon [25]. The Halon alternatives belong to one of the following classes:

1) Hydrobromofluorocarbons (HBFC);
2) Chlorofluorocarbons (CFC);
3) Hydrochlorofluorocarbons (HCFC);
4) Perfluorocarbons (FC);
5) Hydrofluorocarbons (HFC);
6) Inert gases and vapors.

The Environmental Protection Agency (EPA) has provided the following information for the use of the Halon alternates [26]:

Acceptable Total Flooding Agents Feasible in Normally Occupied Areas
1) HFC-23: CHF_3 (Du Pont (FE13)
2) HFC-227ea: CF_3CHFCF_3 (Great Lakes FM 200)
3) FC-3-1-10: C_4F_{10} (3M PFC 410) {restricted use}
4) [HCFC Blend} A (NAF S III) (N.A.Fire Guardian)
5) [Inert Gas Blend] A (Inergen).

Other Acceptable Total Flooding Agents
1) HBFC-22B1: CHF_2Br (Great Lakes FM100)
2) HCFC-22: $CHClF_2$ (Du Pont FE 232)
3) HCFC-124: CF_3HClF
4) HFC-125: CF_3CHF_2 (Du Pont FE-25)
5) HFC-134a: CF_3CH_2F
6) Powdered Aerosol (Spectrex)
7) Solid Propellant Gas Generator (Rocket Research).

Streaming Agents:Commercial and Military Uses Only
1) [HCFC Blend]B (Halotron I)
2) HCFC-123: CF_3CHCl_2 (Du Pont FE-241)
3) FC-5-1-14: C_6F_{14} (3M PFC 614)-(restricted use)
4) HBFC-22B1: CHF_2Br (Great Lakes FM 100).

Total Flooding Agents (Pending)
1) Water Mist (Securiplex; Yates)
2) Powder Aerosols (Spectrex; Service)
3) Inert Gas Blends (Securiplex; Minimax)
4) SF_6 (Discharge test agent)

5) C_3F_8 (3M CEA-308; PFC-218)

6) Fluoroiodocarbons (CF3I)

Streaming Agents (Pending)

1) HCFC-124: CF_3HClF

2) HFC-134a: CF_3CH_2F

3) HFC-227ea: CF_3CHFCF_3 (Great Lakes FM-200)

4) HCFC/HFC Blewnd (NAF P III)

5) HCFC Blend (NAF Blitz III)

6) Powdered Aerosol/HFC or /HCFC Blend (Powsus).

The most common test to screen the Halon alternates is the "Cup Burner" test, where concentrations of Halons or alternates required for extinction of a small laminar diffusion flame are determined [25]. Table 10 lists the concentrations of Halon 1301 and alternates required for heptane flame extinction in the "Cup Burner" test, where the values are taken from Refs. 25 and 27. Acceptable total flooding agents in normally occupied areas are indicated in the table.

The Flammability Apparatus operates under principles very similar to the "Cup Burner", where both laminar and buoyant turbulent diffusion flames are examined [28]. An example of the flame extinction data for Halon 1301 is shown in Fig. 15, where initially there is a rapid decrease in the chemical heat release rate followed by an increase between 5.40 and 6.25 %, due to increase in the flame luminosity and flame heat flux transferred back to the fuel surface. Flame extinction occurs at 6.25 %.

Figure 16 shows a rapid increase in the generation efficiencies of CO, mixture of hydrocarbons, and smoke with increase in the Halon concentration. Generation efficiency is the ratio of the experimental yield of the product to the maximum possible stoichiometric yield of the product [6].

The effect of Halon on the generation efficiencies in Fig. 16 is strong for CO and the mixture of hydrocarbons and weak for smoke. This type of combustion behavior is similar to one found with the decreasing ventilation by decreasing the amount of oxygen. The behavior is postulated to be due to the increasing preference of fuel carbon atom to convert to CO and to the mixture of hydrocarbons rather than to smoke [29]. It thus appears that the chemical interruption processes in the oxidation zone for flame extinction are very similar with increasing amounts of Halon and decreasing amounts of oxygen. This experimental finding is consistent with the concept that a critical Damkohler number exists at the flame extinction condition [26].

The existence of the critical conditions at flame extinction has also been postulated by the "Fire Point Theory" [6] and supported by the experimental data for the critical mass pyrolysis and heat release rates [6].

The extinction test method in the Flammability Apparatus is performed in a fashion very similar to the Combustion Test Method, except that air with different

Table 10. Concentrations of Halon 1301 and Alternates Required for Flame Extinction in the "Cup Burner" Test[a]

Agent Name	Formula	Concentration (Volume %)	Relative Concentration
Halon 1301	CF_3Br	2.9	1.0
Trifluoromethyl Iodide 1311	CF_3I	3.0	1.03
FC-14	CF_4	13.8	4.76
HCFC-22 (Du Pont FE 232)	$CHClF_2$	11.6	4.00 [b]
HBFC-22B1 (Great Lakes FM100)	$CHBrF_2$	4.4	1.52
HFC-23 (Du Pont FE13)	CHF_3	12.4	4.28
HFC-32	CH_2F_2	8.8	3.03
FC-116	CF_3CF_3	7.8	2.69
HCFC-124	$CHClFCF_3$	8.2	2.83
HBFC-124B1	CF_3CHFBr_3	2.8	0.97
HFC-125 (Du Pont FE 25)	CF_3CHF_2	9.40	3.24
HFC-134	CHF_2CHF_2	11.2	3.86
HFC-134a	CF_3CH_2F	10.5	3.62
HFC-142b	$CClF_2CH_3$	11.0 (calc)	3.79
HFC-152a	CHF_2CH_3	27.0 (calc)	9.31
HFC-218	$CF_3CF_2CF_3$	6.1	2.10
HFC-227ea (Great Lakes FM 200)	CF_3CHFCF_3	6.1	2.10[b]
C318	C_4F_8	7.3	2.52
FC-5-1-14 (3M PFC 614)	C_4F_{10}	5.5	1.90[b]

a: compiled from the information given in Refs. 25 and 27; b: acceptable total flooding agents in normally occupied areas.

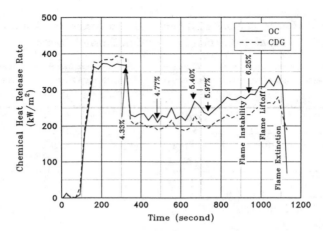

Figure 15. Chemical Heat Release Rate Versus Time for the Combustion of 100-mm x 100-mm x 25-mm Thick Horizontal Slab of Polymethylmethacrylate Exposed to 40 kW/m^2 in Co-Air Flow with Varying Halon 1301 Concentration and a Velocity of 90 mm/s in the Flammability Apparatus at FMRC. Numbers are Halon 1301 Concentrations in Volume Percents and Arrows Show Application Times. Times for Flame Instability, Liftoff, and Extinction are also Indicated.

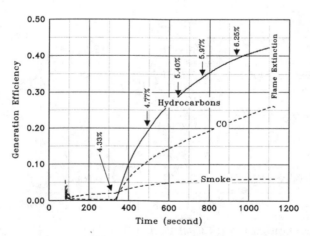

Figure 16. Generation Efficiencies of CO, Mixture of Hydrocarbons, and Smoke Versus Time for the Combustion of 100-mm x 100-mm x 25-mm Thick Horizontal Slab of Polymethylmethacrylate Exposed to 40 kW/m^2 in Co-Air Flow with Varying Halon 1301 Concentration and a Velocity of 90 mm/s in the Flammability Apparatus at FMRC. Numbers are Halon 1301 Concentrations in Volume Percents and Arrows Show Application Times. Times for Flame Instability, Liftoff, and Extinction are also Indicated.

amounts of the gaseous agent is used to determine the minimum concentration for flame extinction. Water is applied directly the surface as large drops and its application rate for flame extinction is determined. Nonthermal damage due to corrosive products and smoke is also determined.

Currently the Flammability Apparatus is being used quite extensively to determine the flame extinction concentrations of halon alternates and water, heat release rate, the types of products generated and the nonthermal damage potential for flame extinction conditions.

SUMMARY

1. The most widely used apparatuses are the OSU Apparatus, the Flammability Apparatus at FMRC, and the Cone Calorimeter. These apparatuses are capable of providing the necessary input combustion data needed for the model based assessments of fire hazards and protection needs;

2. The Flammability Apparatus is capable of providing additional input data to the models for the assessment of hazards due to fire propagation, corrosion and smoke damage, and concentrations of agents required for flame extinction;

3. A Fire Growth and Spread (FSG) model and a Global Equivalence Ratio (GRE) model are available within the data analysis package of the Flammability Apparatus. The FSG model is used to assess the fire propagation behavior of the materials and the GRE model is used to predict the gas temperatures and concentrations of products at various ventilation conditions for the fire scenario of concern to assess the nonthermal damage. The FSG model and the GRE model operate concurrently with the test being performed and the results are available at the end of the test. It is planned to combine this effort with a risk model currently under development at FMRC. It is anticipated that this tool (the Flammability Apparatus, the FSG model, the GRE model, and the risk model) would be a powerful tool for the assessment of hazards and protection needs in various types of fire scenarios. The tool will be available for adoption by the ASTM, ISO, IEC, and others in the very near future.

REFERENCES

1. ASTM E 906-83 "Standard Test Method for Heat and Visible Smoke Release Rates for Materials and Products", The American Society for Testing and Materials, Philadelphia, PA., 1984.
2. Smith, E.E., "Heat Release Rate of Building Materials", Ignition, Heat Release, and Non-combustibility of Materials, ASTM STP 502, p. 119. The American Society for Testing and Materials, Philadelphia, PA., 1972.

3. Smith, E.E., "Measuring Rate of Heat, Smoke, and Toxic Gas Release", *Fire Tech*, **8**, 237, August 1972.
4. Tewarson, A., and Pion, R.F., " Flammability of Plastics. I. Burning Intensity", *Combustion and Flame*, **26**: 85, 1976.
5. Tewarson, A, "Heat Release Rate in Fires", *J.Fire & Materials*, **8**, 151, 1977.
6. Tewarson, A. "Generation of Heat and Chemical Compounds in Fires", Revised Chapter 13 in The SFPE Handbook of Fire Protection Engineering National Fire Protection Association Press, Quincy, MA (in print).
7. Tewarson, A., "Flammability Parameters of Materials: Ignition, Combustion, and Fire Propagation", *J.Fire Sciences*, **12**, 329, 1994.
8. ASTM E 1354-90 "Standard Test Method for Heat and Visible Smoke Release Rates for Materials and Products Using Oxygen Consumption Calorimeter", The American Society for Testing and Materials, Philadelphia, PA , 1984.
9. Babrauskas, V.,"Release Rate Apparatus Based on Oxygen Consumption", *Fire and Materials*, **8**, 81, 1984.
10. Babrauskas, V., "Effective Measurement Techniques for Heat, Smoke, and Toxic Fire Gases", *Fire Safety Journal*, **17**, 13, 1991.
11. Tewarson, A. "Non-thermal Damage", *J.Fire Science*, **10**, 188, 1992.
12. Tewarson, A., and Ogden, S.D., "Fire Behavior of Polymethylmethacrylate", *Combustion and Flame*, **89**, 237, 1992.
13. Tewarson, A., and Khan, M.M., "Flame Propagation for Polymers in Cylindrical Configuration and Vertical Orientation", *Twenty-Second Symposium (International) on Combustion*, p. 1231. The Combustion Institute, Pittsburgh, PA, 1988.
14. Scudamore, M.J., Briggs, P.J., and Prager, F.H., "Cone Calorimetry- A Review of Tests Carried Out on Plastics for the Association of Plastics Manufacturers in Europe", *Fire and Materials*, **15**, 65, 1991.
15. Hirschler, M.M., "Fire Hazard and Toxic Potency of the smoke from Burning Materials", *J.Fire Sciences*, **5**, 289, 1987.
16. Tewarson, A., Lee, J. L., and Pion, R.F., "The Influence of Oxygen Concentration on Fuel Parameters for Fire Modeling" Eighteenth Symposium (International) on Combustion, p. 563. The Combustion Institute, Pittsburgh, PA, 1981.
17. Tewarson, A., "Fire Hardening Assessment (FHA) Technology for Composite Systems", Technical Report ARL-CR-178, Prepared for the Army Research Laboratory, Watertown, MA Under Contract DAAL01-93-M-S403 by the Factory Mutual Research Corporation, Norwood, MA., November, 1994.
18. *Specification Standard for Cable Fire Propagation, Class No.3972*, Factory Mutual Research Corporation, Norwood, MA., 1989.
19. Khan, M.M., "Classification of Conveyor Belts Using Fire Propagation Index", Technical Report J.I. OT1E2.RC, Factory Mutual Research Corporation, Norwood, MA., 1991.

20. Newman, J.S., and Tewarson, A., "Flame Spread Behavior of Char-Forming Wall/Ceiling Insulations", *Fire Safety Science-Proceedings of the Third International Symposium*, Elsvier Applied Science, New York, 679-, 1991.
21. *Approval Standard for Class 1A) Insulated Wall or Wall and Roof/Ceiling Panels, B) Plastic Interior Finish Materials, C) Plastic Exterior Building Panels, D) Wall/Ceiling Coating Systems, E) Interior or Exterior Finish Systems, Class Number 4880*, Factory Mutual Research Corporation, Norwood, MA., March 1993.
22. Fire and Smoke: Understanding the Hazards. Committee on Fire Toxicology, Board on Environmental Studies and Toxicology, Commission on Life Sciences, National Research Council, 1986, National Academy Press, Washington, D.C.
23. Network Reliability: A Report to the Nation, Compendium of Technical Papers, Section G, Presented by the Federal Communications Commission's Network Reliability Council, National Engineering Consortium, Chicago, Ill. June 1993.
24. Reagor, B.T., "Smoke Corrosivity: Generation, Impact, Detection, and Protection", *J. Fire Sciences*, **10**, 169, 1992.
25. "Evaluation of Alternative In-Flight Fire Suppressants for Full-Scale Testing in Simulated Aircraft Engine Nacelles and Dry Bays", W.L Grosshandler, R.G.Gann, W.M Pitt, Editors. National Institute of Standard and Technology, Gaithersburgh, MD., Special Publication 861, May 1994. Superintendent of Documents, U.S.Government Printing Office, Washington, D.C. 20402.
26. Metchis, K., "The Regulation of Halon and Halon Substitutes", *Proceedings of the Halon Options Technical Working Conference 1994*, pp.7-30 The University of New Mexico, New Mexico Engineering Research Institute, Center for Global Environmental Technologies, Albuquerque, New Mexico, May, 1994.
27. Heinonen, E.W., and Skaggs, S.R., "Fire Suppression and Inertion Testing of Halon 1301 Replacement Agents", *Proceedings -Halon Alternates Technical Working Conference 1992*, pp. 213-223. The University of New Mexico, New Mexico Engineering Research Institute, Center for Global Environmental Technologies, Albuquerque, New Mexico, May, 1992.
28. Tewarson, A., and Khan, M.M., "Extinguishment of Diffusion Flames of Polymeric Materials by Halon 1301", *J.Fire Sciences*, **11**, 407, 1993.
29. Tewarson, A., Jiang, F.H., and Morikawa, T., "Ventilation-Controlled Combustion of Polymers", *Combustion and Flame*, **95**, 151, 1993.

RECEIVED April 17, 1995

Chapter 31

Controlled-Atmosphere Cone Calorimeter

M. Robert Christy, Ronald V. Petrella[1], and John J. Penkala

Fire Science Technology Center, Dow Chemical Company, Midland, MI 48667

A new and unique cone calorimeter, the Controlled Atmosphere Cone Calorimeter, which can vary both the oxygen concentration and ventilation rate of its combustion gas is introduced. Three plastic materials, polymethylmethacrylate, and two polyisocyanurate rigid foams, one HCFC-141b blown and the other carbon dioxide blown, were evaluated in the Controlled Atmosphere Cone Calorimeter under nine different conditions of varying oxygen concentrations of the combustion gas (15 to 21%) and rates of ventilation (9 to 24 l/s). The combustion performance properties for each material are presented and discussed. This study demonstrates that the Controlled Atmosphere Cone Calorimeter can effectively model well-ventilated, fully developed fires as well as other types of fires, which standard cone calorimeters can not.

In an effort to address the short comings of existing small scale flammability tests, new laboratory scale tests are constantly being developed and evaluated. One such test is the cone calorimeter. The cone calorimeter was developed in the early 1980's, as a research test by Babrauskas at NBS (*1*). It derives its name from its radiant heat source which is in the shape of a truncated cone. The cone calorimeter can measure mass loss rate, rate of heat release, smoke obscuration, and decomposition products, specifically carbon dioxide and carbon monoxide. Within ten short years, the use of the cone calorimeter has quickly become the acceptance test for standards such as ASTM E 1354 (*2*) or ISO 5660 (*3*).

The cone calorimeter is one of the most sophisticated laboratory fire tests. It is designed to evaluate the combustion performance of laboratory samples (10 cm^2) under apparent fully developed (large scale) fire heat scenarios. This is achieved by igniting the sample under a pre-set radiant heat source. During a test, the actual heat flux gradually increases as a function of the heat released by the sample being evaluated. The ASTM method is performed under conditions of ambient air and at a ventilation rate of 24 l/s.

---

[1]Current address: Flamcon Associates, 3712 Hillgrove Court, Midland, MI 48642

Full scale fire properties are affected by both rate of ventilation and oxygen concentration. One of the main criticisms of the cone calorimeter is that it does not model real fire scenarios because it is over-ventilated. Under standard operating conditions, the cone calorimeter can not model a well ventilated, fully developed fire. Fully developed fires typically are oxygen depleted, or fuel-rich and oxygen lean. Under standard ASTM conditions, the fuel to air ratio in the cone calorimeter tends to simulate the very early stages of a developing fire, or oxygen-rich. Therefore, the cone simultaneously simulates the heat scenario of a fully developed fire while resembling the oxygen availability of an early stage, developing fire.

Cone calorimeters capable of evaluating the fire properties of materials under conditions of incomplete or partial combustion (by varying the oxygen concentration in the combustion gas) have been developed and described (4-5). This study describes a Controlled Atmosphere Cone Calorimeter (CACC) that has been developed which can vary both the oxygen concentration of the combustion gas as well as the rate of ventilation. The objective of this study was to demonstrate the capabilities of the CACC by evaluating three different plastics under varying oxygen concentrations and rates of ventilation. The dependence of combustion performance on oxygen concentrations and ventilation rates in the CACC will be discussed. This is the initial stage of a more comprehensive study to evaluate the potential value of the CACC towards predicting large scale fire performance.

Background

PMMA was chosen as a reference material as it is the standard for calibrating cone calorimeters. Two rigid polyurethane foams were chosen as representative of the emerging new polyisocyanurate insulation and spray foam technologies which have developed as a result of the mandated switch from the blowing agent CFC-11 (trichlorofluoro-methane) (6) to alternative blowing agents HCFC-141b (1, 1-dichloro-1-fluoroethane) or carbon dioxide. The effects of formulation changes on the processing, physical and flammability properties of such polyisocyanurate and spray foams have been reported elsewhere (7). Typically, carbon dioxide blown foams demonstrate higher heat release, smoke, and weight loss as compared to similar HCFC-141b blown foams (8).

The objective of developing laboratory or small scale flammability tests has been to predict the flammability performance of a material under actual fire conditions. Laboratory fire tests fall in to two main categories: 1. Those which are used to understand flammability fundamentals, or research tests and 2. Those which are used to set standards, or acceptance tests. All acceptance tests were once research tests. Ideally, as new research tests emerge, the best ones will become acceptance tests (9).

The ASTM E 84 Steiner Tunnel Test is the standard acceptance test for comparative surface burning characteristics of building materials such as these polyurethane foams. The ASTM E 84 test requires a 7.3 m x 51 cm test sample. The preparation of such samples is costly and time consuming, a major obstacle in the product development cycle for new polyurethane foam formulations. Dowling and Feske recently reported some excellent work enabling predictions of ASTM E 84 Tunnel performance for polyurethane foams based on cone calorimeter test results, thus utilizing a 10 cm^2 sample versus the much larger 7.3 m x 51 cm sample required in the Tunnel (10) It is reported that the gas flame source and limited ventilation rate lead to a reduced oxygen concentration in the Steiner Tunnel versus the cone. Limited evaluations at a lower ventilation rate (16 l/s) led them to conclude that under the conditions where their correlations were derived, burning was insensitive to the availability of oxygen but, they also commented that they were unable to make conclusions about the differences in combustion resulting from depleted oxygen concentrations.

Experimental

Controlled Atmosphere Cone Calorimeter. The Controlled Atmosphere Cone Calorimeter (CACC) utilized in this study was manufactured by Dark Star Research Ltd., Park Lane, UK, to specifications set by The Dow Chemical Company. In addition to maintaining the precision and reproducibility of a standard cone, it was designed to be able to model a fully developed, oxygen deprived, fire. The same principals employed by Babrauskas and Hshieh of controlling the oxygen concentration of the combustion gas were applied. The cone heater, sample holder, and associated ignition system are enclosed in a combustion chamber.

Rather than utilizing glass or metal doors as in previous designs, the CACC chamber is constructed from square double-walled stainless steel panels measuring 50 cm^2. To enhance heat absorption, thus minimizing heat build-up which could effect test results, the inside of the panels are painted flat black and cooled by water channels within the walls. The back of the chamber is fitted with a pressure relief flap. Additionally, to minimize exposure of the sample to the radiant heat source while the chamber is equilibrating, the sample holder sub-system, with load cell, is on a electronically operated retractable mechanical arm. The sample is rotated completely away from the cone rather than shielded in place from the cone by a water shutter. Once the chamber has equilibrated, the sample holder sub-system is rotated so that the sample is positioned under the cone. The feature which makes the CACC unique is the ability to regulate the rate of ventilation through the combustion chamber. This is achieved by pairing the exhaust fan with an air inlet fan and nitrogen delivery system. The fans are installed in a remote location to minimize any vibrational interference with the load cell (Figure 1).

The CACC can operate in the standard air mode or in the controlled atmosphere mode of reduced oxygen and/or ventilation. Operation at standard oxygen concentration (21%) with the CACC can be achieved by one of three conditions: 1. Running with the door open; 2. Removing lower panels on each side of the chamber; or 3. By adjusting the differential between the exhaust and inlet fans to the desired ventilation rate with the chamber sealed. The materials in this study were tested under the latter condition at ventilation rates from 9 to 24 l/s.

The reduced oxygen concentration of the combustion gas is achieved by mixing nitrogen with air. The mixture is delivered to the bottom of the chamber, which is lined with 75 mm of glass beads, through a perforated 6 mm stainless steel tube. The nitrogen concentration is manually adjusted. The resulting oxygen concentration is determined by an oxygen analyzer near the combustion gas inlet in the combustion chamber. The oxygen depletion, from which the heat release rate is determined, is calculated as the difference between the oxygen concentration analyzed in the bottom of the combustion chamber and oxygen level in the exhaust. The materials in this study were evaluated from 15 to 21% oxygen concentrations.

Materials. Three materials were studied, black polymethylmethacrylate from ICI, London, UK, and two flame-retarded polyisocyanurate rigid foams, one HCFC-141b blown (PU/H) and the other water-blown (carbon dioxide--PU/C), both foams were supplied by Foam Enterprises, Inc., Houston, TX (formulations proprietary). The PU foam samples are typical of formulation commercially available for building panels (PU/H) and spray foam insulation (PU/C). The PMMA samples measured 1 cm x 10 cm^2 and weighed 155 g. The PU foam samples measures 2.5 cm x 10 cm^2 and averaged 8.5 g (PU/H) and 14 g (PU/C). Samples were equilibrated at 23° C and 50% relative humidity for a minimum of two days prior to testing.

Procedure. The test procedures were similar to those described in ASTM E 1354 with the exceptions that: the samples were run in an enclosed chamber; ventilation rate and/or oxygen concentration of the combustion gas varied; and for the foams,

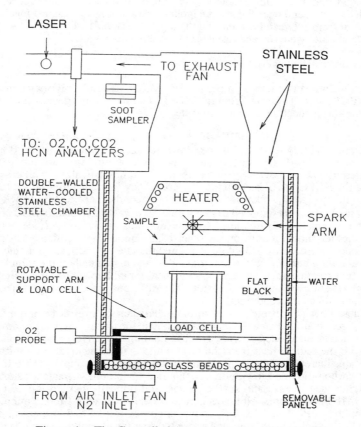

Figure 1. The Controlled Atmosphere Cone Calorimeter

combustion was usually complete within two to four minutes. The CACC tests were conducted at the following three ventilation rates: 9, 15, and 24 l/s, ±0.2 l/s and the following three oxygen concentrations: 15, 18, and 20.9%, ±0.3% for each sample. All samples were tested in an unrestrained, horizontal position, at a radiant heat flux of 50 kW/m^2, and data was collected at 5 s intervals. Once the chamber was equilibrated, the following steps were followed: the chamber door was quickly opened; the sample placed in the sample holder; door closed; the chamber allowed to re-equilibrate (< 60 s); a spark igniter electro-mechanically moved in place under the cone and started; the sample holder sub-system electro-mechanically rotated in place beneath the cone; and once the sample ignited, the spark igniter electro-mechanically rotated out from between the cone and the top of the sample. Typically, two to three replicate samples were run and results are reported as averages.

Results and Discussion

The combustion performance properties for PMMA, and two polyisocyanurate rigid foams, PU/H and PU/C, are listed in Table I as a function of oxygen concentration and Table II as a function of ventilation rate.

Burning Characteristics of Rigid Polyisocyanurate Foams. To maintain dimensional stability, carbon dioxide blown foams are typically 10 to 20% higher density than HCFC-141b blown foams. The physical burning characteristics of the two different types of polyisocyanurate foams differ dramatically. In all cases, upon rotating the HCFC-141b foams under the cone, they immediately foamed to 100 to 150% of their original thickness. After several seconds, they collapsed back to their original thickness and continued to maintain their dimensional stability (10 cm^2 size) through the remainder of the test. Combustion was on the top surface only. The carbon dioxide blown foams, on the other hand, immediately collapsed after being rotated under the cone to 50% of their original thickness. Then, all four sides would curl up allowing burning on the top, four sides, and underneath of the sample. Even though the experimental procedure was identical for both materials, the difference in physical reaction of the foams resulted in different effective heat fluxes, thus dissimilar combustion conditions.

These physical differences in burning characteristics result from the foaming action of residual HCFC-141b and may account for the difference in combustion performance as mentioned previously. The foaming effectively insulates the PU/H foam from the radiant heat from the cone while the dimensional instability of the PU/C foams increased the effective combustion area of the sample. Table III shows that the carbon dioxide blown polyisocyanurate foam demonstrates greater weight loss, higher heat release and smoke values, and typically burned longer than the HCFC-141b blown foam. All foams burned for approximately 1 to 1.5 minutes with the exception of PU/C at 21% oxygen which burned from 4 to 5 minutes. For these samples, a small blue flame persisted above the char for several minutes, it is believed that these last few minutes did not add appreciably to either mass loss or heat generation.

In an actual pre-flashover fire (fully developed/highly ventilated) both oxygen concentration and ventilation are depleted. One of the most surprising observations in this study was that neither foam would support combustion at the lowest oxygen level (15%) at the two higher ventilation rates (15 and 24 l/s) but would at the lowest ventilation rate (9 l/s). This demonstrates that the combustion performance of the foams is a function of *both* the oxygen content of the combustion gas and the rate of ventilation. Either of these foams investigated in a previously modified cone calorimeter, which could vary oxygen concentration only, would have led to an erroneous conclusion that they would not support combustion at 15% oxygen--because the test would have been run at 24 l/s.

The ability to sustain combustion at low oxygen levels and low ventilation rates was believed to be a result of one or both of the following: 1. Lower actual heat flux of the sample due to the cooling effect at higher ventilation rates of the air flow at the samples surface and/or 2. Air turbulence raising the flame from the sample surface at the higher ventilation rates thus, reducing the burning velocity near the surface. The cooling effect was discounted based on earlier work by Penkala (Dow Report, 1993), which showed that the CACC carefully controls the surface temperature of samples, even at high ventilation rates. In a diffusion type flame, at reduced oxygen levels, increasing combustion gas flow (higher ventilation rates) may lift the flame such that sufficient diffusion of secondary air at the surface can not occur (*11*). Thus, it is believed that turbulence must be the contributing factor which resulted in the polyisocyanurate foams not burning at high ventilation rates at a reduced oxygen level.

Mass Loss. The PMMA was virtually consumed under all 9 conditions. There is a direct correlation of mass loss rate (MLR) versus oxygen concentration, the lower the oxygen the lower the MLR (Figure 2) but, longer the burn times. There is no apparent dependence on ventilation.

Char yield is defined as the percent of the sample mass remaining after the completion of the run. It is a convenient way to describe the char forming characteristics of a material (Lukas, C., Stobby, W., and Boukami, E., Dow Report, 1993). For the foams, there is a direct correlation of char formation versus oxygen concentration (Figure 3), the lower the oxygen concentration, the shorter the burn times, and the greater the char yield. There is no apparent dependence of char yield versus the rate of ventilation. No apparent MLR correlation exists for the foams versus either oxygen concentration or ventilation.

Heat Release. Plots of average rate of heat release (RHR) for PMMA versus ventilation and oxygen concentration are presented in Figure 4. RHR for PMMA appears independent of rate of ventilation but is directly proportional to the oxygen concentration. More time is required to reach the peak rate of heat release (PRHR), the duration of combustion increases, and the PRHR decreased as the oxygen concentration decreased, Figure 5. Generally, total heat release (THR-Table I, column 8) is the same with increasing levels of oxygen.

Like PMMA, the PRHR for both polyisocyanurate foams increases with oxygen level, Figure 5. For both foams, there is higher THR at higher oxygen levels. There is little dependency for either foam versus ventilation rate at 21% oxygen. There is a slight dependence at 18% oxygen. The average RHR is the quotient of THR divided by the combustion duration time (ignition time minus flame-out time) and reflects a direct (PU/H) and inverse (PU/C) dependence on oxygen concentration for each foam, respectively, Figure 6. The inverse dependence for PU/C may be an artificial artifact of the unusually long burn times for the PU/C at 21% oxygen, as mentioned above. There does not appear to be a dependence on ventilation rate.

The total potential heat release available from a material due to complete oxidation represents the heat of combustion. Incomplete oxidation of materials actually occurs in real combustion processes. It is described as the effective heat of combustion and is the quotient of the RHR divided by the MLR. The effective heat of combustion for PMMA is fairly independent of oxygen levels or ventilation rates, although samples had slightly lower RHR at lower oxygen levels, they burned for longer times resulting in similar heats of combustion. The foams showed little dependence on ventilation at 21% oxygen. At 18% oxygen, the effective heat of combustion for the PU/H showed a dependence to ventilation, decreasing with lower ventilation rates. While the values for effective heat of combustion decreased for PU/C at 18 % oxygen, there was no dependence on ventilation rate.

Table I. Controlled Atmosphere Cone Calorimeter Combustion Performance vs Oxygen Concentration--PMMA and PUR Foams

Sample	Volume Fraction O2 %	Flow Rate l/s	Blowing Agent	Char Yield %	Peak Rate of Heat Release kW/m2	Average Rate of Heat Release kW/m2	Total Heat Released MJ/m2	Average Heat of Combustion MJ/kg	Peak Extinction Area m2/kg	Average Extinction Area m2/kg	Peak Mass Loss Rate g/s-m2	Average Mass Loss Rate g/s-m2	Average CO Production kg/kg	Average CO2 Production kg/kg	CO2/CO	Time to Ignition sec
1-PMMA	21	24	NA	0	815.50	501.02	298.07	25.01	398.18	176.37	31.44	22.58	0.017	2.69	158.24	50
4-PMMA	18	24	NA	0	793.20	454.00	290.51	24.88	342.83	201.34	30.77	20.74	0.020	2.67	133.50	56
7-PMMA	15	24	NA	0.003	689.80	393.99	291.5	24.82	300.18	179.88	26.34	18.24	0.039	2.64	67.69	58
2-PMMA	21	15	NA	0	799	498.24	291.42	25.49	352.7	156.66	30.73	22.38	0.016	2.68	167.50	50
5-PMMA	18	15	NA	0	783.90	440.10	283.83	24.4	363.6	209.08	30.53	20.94	0.017	2.70	158.82	38
8-PMMA	15	15	NA	0.003	691.80	404.72	293.38	24.55	308.05	179.96	27.27	18.83	0.033	2.57	77.88	55
3-PMMA	21	9	NA	0	790.8	483.19	282.63	24.51	337.53	151.84	30.79	22.64	0.015	2.62	174.67	53
6-PMMA	18	9	NA	0	726.20	444.88	277.97	24.73	394.39	208.72	28.24	20.59	0.017	2.68	157.65	41
9-PMMA	15	9	NA	0	636	386.84	266.88	23.57	322.47	181.79	26.83	18.79	0.030	2.66	88.67	43
1-PU/H	21	24	HCFC	46.9	202.65	100.40	8.44	9.49	540.75	314.37	14.74	10.75	0.150	1.12	7.73	5
4-PU/H	18	24	HCFC	55.3	153.00	78.44	7.45	9.82	694.97	385.63	12.24	8.75	0.180	1.16	6.44	4
7-PU/H	15	24	HCFC	Sample would not support combustion												
2-PU/H	21	15	HCFC	45.8	193.05	94.46	8.64	9.58	577.59	319.72	13.21	10.03	0.160	1.16	7.47	5
5-PU/H	18	15	HCFC	52.4	113.5	59.14	2.92	6.25	574	336.97	10.49	9.37	0.140	1.19	8.50	5
8-PU/H	15	15	HCFC	Sample burned for 10 seconds--insufficient data collected												
3-PU/H	21	9	HCFC	46.4	169.20	87.41	8.68	9.49	528.31	305.71	13.57	9.20	0.140	1.09	7.79	5
6-PU/H	18	9	HCFC	67	72.05	43.67	1.98	3.55	571.05	296.13	11.53	9.96	0.070	1.02	19.07	5
9-PU/H	15	9	HCFC	67.1	46.70	31.44	1.87	3.28	387.07	233.22	12.11	9.87	0.100	1.04	10.40	5

	21	24	H2O	19.1	707.70	167.88	36.70	15.91	995.13	843.67	44.89	25.23	0.130	1.54	12.47	5
1-PU/C																
4-PU/C	18	24	H2O	34.6	687.00	312.70	20.26	11.30	1082.59	1121.80	40.9	36.53	0.140	1.33	9.50	5
7-PU/C	15	24	H2O	Sample would not support combustion												
2-PU/C	21	15	H2O	17.3	640.85	135.79	41.53	16.96	931.91	722.15	38.57	13.95	0.160	1.65	10.86	5
5-PU/C	18	15	H2O	38.2	470.10	248.40	17.52	9.27	905.86	956.23	37.77	33.36	0.080	1.53	19.13	5
8-PU/C	15	15	H2O	Sample burned for 25 seconds--insufficient data collected												
3-PU/C	21	9	H2O	19.5	582.35	139.38	39.83	16.09	991.05	697.79	38.35	16.65	0.120	1.54	13.37	5
6-PU/C	18	9	H2O	34.9	389.90	277.90	16.99	8.94	889.66	913.00	38.90	34.08	0.110	1.38	12.55	5
9-PU/C	15	9	H2O	38.3	402.90	254.70	21.62	12.08	657.42	571.12	27.20	24.51	0.120	1.50	12.65	7

Table II. Controlled Atmosphere Cone Calorimeter Combustion Performance vs Ventilation Rate--PMMA and PUR Foams

Sample	Flow Rate l/s	Volume Fraction O2 %	Blowing Agent	Char Yield %	Peak Rate of Heat Release kW/m2	Average Rate of Heat Release kW/m2	Total Heat Released MJ/m2	Average Heat of Combustion MJ/kg	Peak Extinction Area m2/kg	Average Specific Extinction Area m2/kg	Peak Mass Loss Rate g/s-m2	Average Mass Loss Rate g/s-m2	Average CO Production kg/kg	Average CO2 Production kg/kg	CO2/CO	Time to Ignition sec	Flameout Time sec
1-PMMA	24	21	NA	0	815.50	501.02	298.07	25.01	398.18	176.37	31.44	22.58	0.017	2.69	158.24	50	654
2-PMMA	15	21	NA	0	799	498.24	291.42	25.49	352.7	156.66	30.73	22.38	0.016	2.68	167.50	50	640
3-PMMA	9	21	NA	0	790.8	483.19	282.63	24.51	337.53	151.84	30.79	22.64	0.015	2.62	174.67	53	644
4-PMMA	24	18	NA	0	793.20	454.00	290.51	24.88	342.83	201.34	30.77	20.74	0.020	2.67	133.50	56	700
5-PMMA	15	18	NA	0	783.90	440.10	283.83	24.4	363.6	209.08	30.53	20.94	0.017	2.70	158.82	38	688
6-PMMA	9	18	NA	0	726.20	444.88	277.97	24.73	394.39	208.72	28.24	20.59	0.017	2.68	157.65	41	674
7-PMMA	24	15	NA	0.003	689.80	393.99	291.5	24.82	300.18	179.88	26.34	18.24	0.039	2.64	67.69	58	801
8-PMMA	15	15	NA	0.003	691.80	404.72	293.38	24.55	308.05	179.96	27.27	18.83	0.033	2.57	77.88	55	783
9-PMMA	9	15	NA	0	636	386.84	266.88	23.57	322.47	181.79	26.83	18.79	0.030	2.66	88.67	43	737
1-PU/H	24	21	HCFC	46.9	202.65	100.40	8.44	9.49	540.75	314.37	14.74	10.75	0.150	1.12	7.73	5	92
2-PU/H	15	21	HCFC	45.8	193.05	94.46	8.64	9.58	577.59	319.72	13.21	10.03	0.160	1.16	7.47	5	101
3-PU/H	9	21	HCFC	46.4	169.20	87.41	8.68	9.49	528.31	305.71	13.57	9.20	0.140	1.09	7.79	5	105
4-PU/H	24	18	HCFC	55.3	153.00	78.44	7.45	9.82	694.97	385.63	12.24	8.75	0.180	1.16	6.44	4	101
5-PU/H	15	18	HCFC	52.4	113.5	59.14	2.92	6.25	574	336.97	10.49	9.37	0.140	1.19	8.50	5	56
6-PU/H	9	18	HCFC	67	72.05	43.67	1.98	3.55	571.05	296.13	11.53	9.96	0.070	1.02	19.07	5	58
7-PU/H	24	15	HCFC	Sample would not support combustion													
8-PU/H	15	15	HCFC	Sample burned for 10 seconds--insufficient data collected													
9-PU/H	9	15	HCFC	67.1	46.70	31.44	1.87	3.28	387.07	233.22	12.11	9.87	0.100	1.04	10.40	5	67

1-PU/C	24	21	H2O	19.1	707.70	167.88	36.70	15.91	995.13	843.67	44.89	25.23	0.130	1.54	12.47	5	243
2-PU/C	15	21	H2O	17.3	640.85	135.79	41.53	16.96	931.91	722.15	38.57	13.95	0.160	1.65	10.86	5	321
3-PU/C	9	21	H2O	19.5	582.35	139.38	39.83	16.09	991.05	697.79	38.35	16.65	0.120	1.54	13.37	5	299
4-PU/C	24	18	H2O	34.6	687.00	312.70	20.26	11.30	1082.59	1121.80	40.9	36.53	0.140	1.33	9.50	5	71
5-PU/C	15	18	H2O	38.2	470.10	248.40	17.52	9.27	905.86	956.23	37.77	33.36	0.080	1.53	19.13	5	83
6-PU/C	9	18	H2O	34.9	389.90	277.90	16.99	8.94	889.66	913.00	38.90	34.08	0.110	1.38	12.55	5	82
7-PU/C	24	15	H2O	Sample would not support combustion													
8-PU/C	15	15	H2O	Sample burned for 25 seconds–insufficient data collected													
9-PU/C	9	15	H2O	38.3	402.90	254.70	21.62	12.08	657.42	571.12	27.20	24.51	0.120	1.50	12.65	7	97

Table III. Mass Loss, Heat Release, and Smoke Comparisons for PUR Foams: HCFC-141b vs Carbon Dioxide Blown

Sample	Volume Fraction O2 %	Flow Rate l/s	Blowing Agent	Char Yield %	Peak Mass Loss Rate g/s-m2	Average Mass Loss Rate g/s-m2	Peak Rate of Heat Release kW/m2	Average Rate of Heat Release kW/m2	Total Heat Released MJ/m2	Peak Extinction Area m2/kg	Average Specific Extinction Area m2/kg	Flameout Time sec
1-PU/H	21	24	HCFC	46.9	14.74	10.75	202.65	100.40	8.44	540.75	314.37	92
2-PU/H	21	15	HCFC	45.8	13.21	10.03	193.05	94.46	8.64	577.59	319.72	101
3-PU/H	21	9	HCFC	46.4	13.57	9.20	169.20	87.41	8.68	528.31	305.71	105
4-PU/H	18	24	HCFC	55.3	12.24	8.75	153.00	78.44	7.45	694.97	385.63	101
5-PU/H	18	15	HCFC	52.4	10.49	9.37	113.5	59.14	2.92	574	336.97	56
6-PU/H	18	9	HCFC	67	11.53	9.96	72.05	43.67	1.98	571.05	296.13	58
7-PU/H	15	24	HCFC	Sample would not support combustion								
8-PU/H	15	15	HCFC	Sample burned for 10 seconds--insufficient data collected								
9-PU/H	15	9	HCFC	67.1	12.11	9.87	46.70	31.44	1.87	387.07	233.22	67

1-PU/C	21	24	H2O	19.1	44.89	25.23	707.70	167.88	36.70	995.13	843.67	243
2-PU/C	21	15	H2O	17.3	38.57	13.95	640.85	135.79	41.53	931.91	722.15	321
3-PU/C	21	9	H2O	19.5	38.35	16.65	582.35	139.38	39.83	991.05	697.79	299
4-PU/C	18	24	H2O	34.6	40.9	36.53	687.00	312.70	20.26	1082.59	1121.80	71
5-PU/C	18	15	H2O	38.2	37.77	33.36	470.10	248.40	17.52	905.86	956.23	83
6-PU/C	18	9	H2O	34.9	38.90	34.08	389.90	277.90	16.99	889.66	913.00	82
7-PU/C	15	24	H2O	Sample would not support combustion								
8-PU/C	15	15	H2O	Sample burned for 25 seconds--insufficient data collected								
9-PU/C	15	9	H2O	38.3	27.20	24.51	402.90	254.70	21.62	657.42	571.12	97

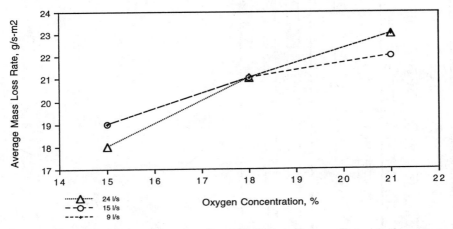

Figure 2. Average Mass Loss Rate: PMMA vs Oxygen Concentration

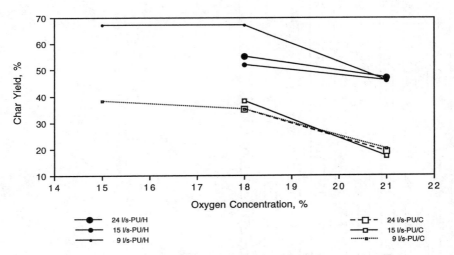

Figure 3. Char Yield vs Oxygen Concentration: Polyisocyanurate Foams

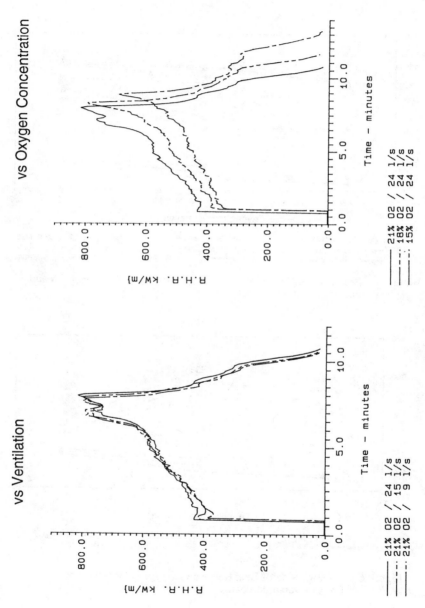

Figure 4. Average Rate of Heat Release: PMMA

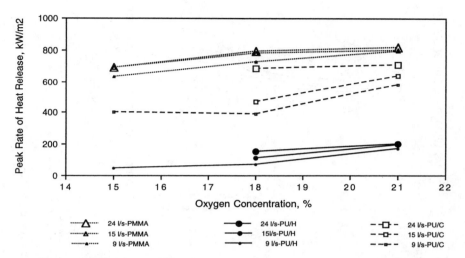

Figure 5. Peak Rate of Heat Release vs Oxygen Concentration: PMMA and Polyisocyanurate Foams

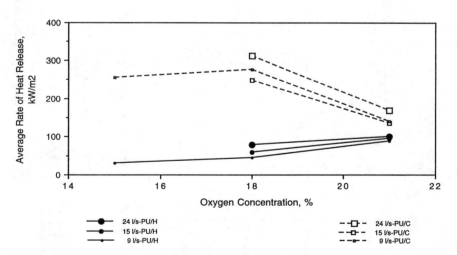

Figure 6. Average Rate of Heat Release vs Oxygen Concentration: Polyisocyanurate Foams

Smoke. The specific extinction area is the area that smoke particles would obscure if one kilogram of fuel was combusted, higher areas represent greater smoke production. For PMMA, there is not a correlation versus ventilation. An interesting trend is seen for PMMA and both polyisocyanurate foams: smoke production reaches a maximum at 18% oxygen, Figures 7 and 8. Unlike PMMA, there appears to be a correlation to ventilation for the foams, most pronounced at 18 % oxygen, the lower the ventilation the lower the smoke. The following observations can be made for the range of oxygen concentrations investigated: 1. There is a point where maximum smoke is generated in an oxygen depleted environment, above or below that concentration, less smoke is produced and 2. Lower ventilation at optimal smoke producing conditions (oxygen depleted) favors less smoke.

Carbon Dioxide / Carbon Monoxide. For PMMA, Figure 9, at a constant rate of ventilation (24 l/s) carbon dioxide (CO_2) development is relatively constant versus oxygen levels. But, there is a direct dependence for carbon monoxide (CO) development. There is not much change from 21% to 18% oxygen but from 21% to 15% oxygen the CO yield nearly doubles. There is a slight dependence on ventilation rate for CO, the lower the air flow, the lower the CO yield.

Purser (*12*) refers to ISO/TC 92 which utilizes the CO_2/CO ratio as one component in defining three major categories of fires: 1. Developing Fire (100-200); 2. Fully developed/highly ventilated (10-100); and 3. Fully developed/poorly ventilated (<10). He goes on to describe the effectiveness of small scale flammability tests in modeling full scale fire toxicity hazards. He wrote that the cone calorimeter can model decomposition conditions of a very early, very well ventilated fire (ratios 100-200). But, attempts to modify the combustion process and decrease the combustion efficiency to model other stages of fire have not proved successful (ratios < 100). The CO_2/CO ratios for PMMA versus oxygen concentration are shown in Figure 10. At lower oxygen levels, CO_2/CO ratios < 100 are achieved. Thus demonstrating that the CACC can effectively model a fully developed/well ventilated fire. A slight dependence on ventilation rate is also shown, the higher the ventilation rate, the lower the CO_2/CO ratio. As mentioned above, the CO_2/CO ratio is only one component and is not necessarily indicative of defining the stage of fire, alone.

For the polyisocyanurate foams, although there is a slightly broader scatter of values, there does not appear to be a dependence on CO_2 generation versus oxygen level. There is no apparent correlation to CO development and thus CO_2/CO ratios.

Conclusions

1. The Controlled Atmosphere Cone Calorimeter (CACC) introduced in this study, by regulating the oxygen concentration and rate of ventilation of the combustion gas, demonstrated that it can model well ventilated, fully developed fires, which standard cone calorimeters can not.

2. The ability of the CACC to model well ventilated, fully developed fires may make it a more useful laboratory tool to correlate small scale burn tests with actual fire performance.

3. The combustion characteristics of different plastic materials are affected differently by changes in oxygen level and ventilation rate in the CACC.

4. The mass loss rate for materials which under go almost complete combustion (PMMA) decreases linearly with oxygen concentration in the combustion gas. For those materials that produce char (polyisocyanurate foams), the char yields (percent material remaining after combustion) increase and burn times decrease with decreasing oxygen levels.

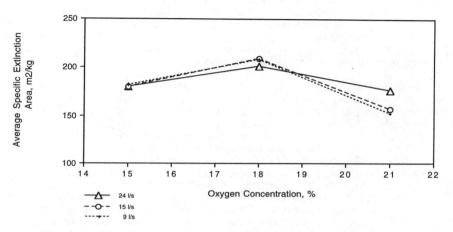

Figure 7. Specific Extinction Area vs Oxygen Concentration: PMMA

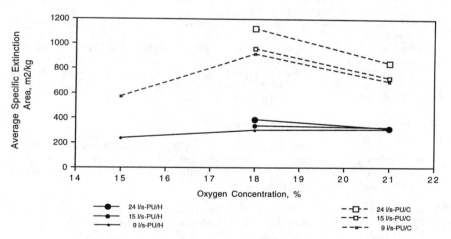

Figure 8. Specific Extinction Area vs Oxygen Concentration: Polyisocyanurate Foams

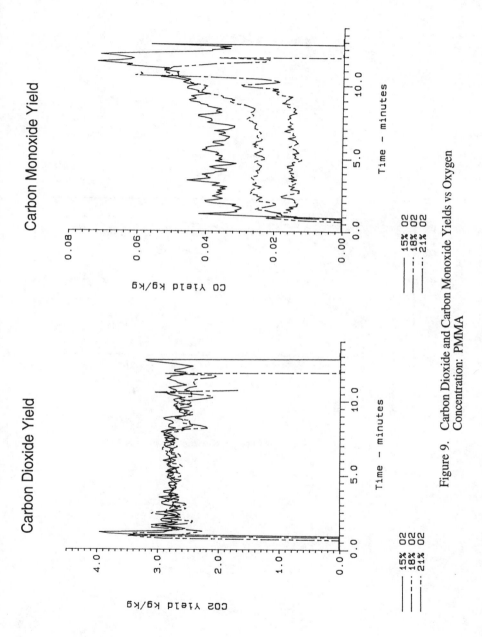

Figure 9. Carbon Dioxide and Carbon Monoxide Yields vs Oxygen Concentration: PMMA

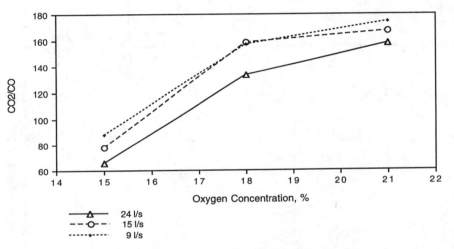

Figure 10. CO_2/CO vs Oxygen Concentration: PMMA

5. The peak rate of heat release (PRHR) for all materials studied, decreases with oxygen concentration.

6. No trend was determined for the effective heat of combustion versus oxygen concentration.

7. For all materials tested, a maximum smoke development was observed at 18% oxygen, above and below that concentration less smoke is produced.

8. Generally, for the three materials evaluated, the CO_2/CO yield ratios decreases with decreasing oxygen levels.

Acknowledgments

The authors would like to acknowledge Christine Lukas and Bill Stobby for their insights into interpreting some of the results reported here and to Deb Bhattacharjee for supplying the polyisocyanurate foam samples and related pertinent information.

Literature Cited

1. Babrauskas, V., "Development of the Cone Calorimeter: A Bench-Scale Heat Release Rate Apparatus Based on Oxygen Consumption" (NBSIR 82-2611). [U.S.] Natl. Bur. Stand., Gaithersburg, MD, **1982**.
2. Standard Test Method for Heat and Visible Smoke Release Rates for Materials and Products Using Oxygen Consumption Calorimeter (E 1354), American Society for Testing and Materials, Philadelphia, PA.
3. International Standard - Fire Tests - Reaction to Fire - Rate of Heat Release from Building Products (ISO 5660). International Organization for Standards, Geneva.
4. Mulholland, G., Janssens, M., Yusa, S., and Babrauskas, V., *Fire Safety Science Proceedings of the Third International Symposium,* **1991**, pp. 585-594.
5. Hshieh, F., Motto, S., Hirsch, D., and Beeson, H., *Proceedings of the International Conference on Fire Safety,* **1993**, vol 18, pp. 299-325.
6. Polyisocyanurate Insulation Manufacturers Association news release presented at the CSI meeting, Houston, TX, June 25, **1993**.
7. Schiff, D., Koehler, C., Parsley, K., and Bhattacharjee, D., *Proceedings of the Society of Plastics Industries Polyurethane Division* , **1994**, pp. 599-608.
8. Moore, S., Bhattacharjee, D., and Dressel, D., *Proceedings of the Society of Plastics Industries Polyurethane Division* , Boston, **1994**, pp. 418-427.
9. Hilado, C.J., *Flammability Handbook for Plastics,* Third Edition, Technomic Publishing, PA, **1982**, p. 97.
10. Dowling, K. and Feske, E., *UTECH Proceedings*, The Netherlands, **1994**.
11. Gaydon, A.G. and Wolfhard, H.G., *Flames,* Third Edition, Chapman and Hall Ltd., London, **1970**, pp. 14-18.
12. Purser, D.A., "The Relationship of Small Scale Toxicity Test Data to Full Scale Fire Hazard", *Proceedings of the Fire Retardant Chemicals Association*, Coronado, CA, **1991**, pp. 151-157.

RECEIVED December 5, 1994

Chapter 32

X-ray Photoelectron Spectroscopy (Electron Spectroscopy for Chemical Analysis) Studies in Flame Retardancy of Polymers

Jianqi Wang

National Laboratory of Flame Retarded Materials, College of Chemical Engineering and Materials Science, Beijing Institute of Technology, Beijing 100081, China

Thermal degradation of polymers and subsequent events (e.g. burning) occur rationally first from the surface exposed to the heat flux from surroundings and then go further in depth. This article aims at introducing and exemplifying the potential applications of XPS, a surface sensitive tool, in flame retardancy of polymers, such as, thermal degradation, structural identification, synergisms, and charring processes in condensed phase. Being coupled with other techniques C1s and O1s data about charring which can hardly be dealt with by conventional techniques were particularly provided and described.

The amazing growth and diversification of surface analysis techniques began in the 1960 with the development of electron spectroscopy -- first Auger spectroscopy (AES), then followed by X-ray Photoelectron Spectroscopy (XPS, or ESCA, Electron Spectroscopy for Chemical Analysis). Recently, these techniques have rapidly matured with the generation of a large volume of literature and are now routinely used in industry and universities providing a wealth of information, e.g. elemental composition, detailed knowledge of chemical bonding states, and depth profiles of the outermost atomic layers of solids.

Application of XPS to the Study of Flame Retardancy of Polymers

Polymers are increasingly pervasive in material science. The impact of XPS on materials characterization of polymers has been twofold : firstly, through its ability to analyze relatively intractable materials without the need for special sample preparation; secondly, through its surface sensitivity. It must be emphasized as well that with XPS are abstracted not only the surface properties as known, but also the bulk properties sometimes unavoidable depending upon the sample preparation. As more is learned about how the surface chemistry of polymers can play an important role in the studies of flame retardant systems, the more individual problems can be tailored to maximize better understanding of flame retardancy. The goal of this article is trying to abstract more experimental data of likely processes in flame retardancy of polymers at some level by XPS technique coupling with thermal analysis techniques.

32. WANG   XPS (ESCA) Studies in Flame Retardancy of Polymers

Brief Description of XPS Test. The spectra (MgK_α) were recorded on PHI 5300 ESCA SYSTEM (PERKIN-ELMER) at 250 W (12.5 KV x 20 mA) under a vacuum better than 10^{-6} Pa (10^{-8} Torr) calibrated by assuming the binding energy of the adventitious carbon to be 284.6 eV. Pass energy and step length were selected to be 35.75 eV and 0.05 eV respectively for better results. Specimens in thin films through spreading a droplet of dilute solution on aluminum foil (for example, chloroform was adopted for EVA copolymers) were preferred, if possible. For polyethylenes thin films of 0.1 mm thickness can be prepared by a mixing roll and parallel platen curing machine on aluminum foil.

Results and Discussion

Thermal Degradation of Polymers upon Heating. The basic approach in the past to the mechanistic studies of thermal degradation of polymers and polymeric composites generally invokes gas analysis primarily of the decomposition products with techniques, such as IR, NMR, MS etc. to identify the structural changes of the bulk properties. Here, we have made use of XPS to identify the surface chemistry of the solid materials, instead.

Wool, a Natural Polymer. It was chosen as one example to show its thermal behavior followed by XPS. Wool is a keratin protein composed of amino acids linked by peptide bonds, of which the units are crosslinked by disulfide bonds and have attached to them reactive groups such as amino, hydroxyl, sulfhydryl and carboxyl. The XPS investigation (*1*) of the S2p, N1s, C1s and O1s spectra as function of temperature gives further quantitative awareness into thermal behavior, as shown in Table I. Table I

Table I Quantitative analysis of S2p, N1s, C1s and O1s spectra as function of temperature in air[a]

Temperature,°C	S/N	N/C	O/C
Room temp.	1.00	0.03	0.14
150	1.10	0.03	0.14
200	0.96	0.03	0.15
245	0.18	0.09	0.16
300	0.074	0.10	0.08
350	0.040	0.11	0.06

([a] Reproduced with permission from ref.1. Copyright 1994)

shows an abrupt decrease in S/N at about 200°C and an inflection in the range of 230° - 240°C in fairly good consonance with data reported by Felix,et.al. (*2*). The changes in N/C and S/N ratio (Table I) imply the evolution of carbon monoxide, carbon dioxide, hydrogen sulfide, hydrogen cyanide, nitric oxide and acrylonitrile as combustion products from wool on heating reported by Koroskys (*3*). A larger decrease in S than in N was noted and ascribed by the disruption of the disulphide linkage, i.e. cystine decomposition (*1*). There are four types of sulfur-containing groups in the surface of wool, namely, -SH-, -S-S-, -SO_3H and SO_4^{2-} (see Table II). Among these, the first two are received from the complex wool structure, SO_4^{2-} is from washing detergent needed for XPS test, and $-SO_3H$ group located in the surface layers is due to oxidation of -S-S-. Upon heating, say from ambient temperature to 350°C, organic sulfur-containing compounds will dominate the surface at the expense of inorganic sulfur compounds till their complete disappearance. NH_3 evolved through bond scissions of terminal amine, side-chain amine, amide groups of lysine, arginine, asparagine and glutamine is likely to be the main gaseous product within this temperature range, as observed by Ingham (4). The emergence of -C=N- and -C≡N groups can also be observed in N1s spectra at about 250°C and 350°C, respectively, (Table II).

Table II Quantitative analysis of S2p and N1s spectra as function of temperature[a]

Temperature, °C	Atomic concentration, %							
	S2p				N1s			
	-SH-	-S-S-	$-SO_3H$	SO_4^{2-}	-NH-CO-	$-NH_2$	-C=N-[b]	-C≡N[b]
Room temp.	21.0	10.6	28.1	40.4	77.3	22.7	---	---
150	17.8	14.5	51.4	16.3	77.0	23.0	---	---
200	15.0	14.2	52.9	17.9	58.9	41.1	---	---
245	37.0	17.8	31.0	14.2	52.5	22.7	24.8	---
300	51.0	39.0	10.0	0.0	39.2	26.2	34.6	---
350	60.9	39.1	0.0	0.0	35.0	25.2	25.9	13.9

[a] Reproduced with permission from ref.1. Copyright 1994
[b] Evidenced by IR

The Study of Flame Retardant Mechanism in Gas Phase. Polypropylene (PP), Dechlorane Plus 25 (De-25) and Sb_2O_3 of commercial grade were used as a model system for the study of synergy between chlorine-containing flame retardant and antimony trioxide in gas phase with the aid of XPS (5). For comparison with De-25, BCPD (a chlorine-containing flame retardant with cyclic structure) was synthesized in the laboratory. The formulations were processed at 170 °C by blending as shown in Table III. A drastic change in the electronic state of the chlorine atom in systems K-1 to K-4 was carefully explored as function of temperature.

Table III Formulations of flame retarded systems

No.	Formulation	Parts by weight
K-1	PP + BCPD	31 : 14.5
K-2	PP + BCPD + Sb_2O_3	31 : 14.5 : 4.5
K-3	PP + DE-25	31 : 14.5
K-4	PP + De-25 + Sb_2O_3	31 : 14.5 : 4.5

([a] Reproduced with permission from ref.1. Copyright 1994)

Taking note of the change in electronic state of Cl atom blended in polymeric matrix Cl2p spectra have been recorded as function of temperature. The ionic component, Cl^{1-} % observed from the spectrum can be used as an indicator in halogen-containing systems to follow the reaction. The dependence of atomic concentration (AC) Cl^{1-}% in systems K-1 through K-4 upon temperature is shown in Figure 1. As seen, the Cl^{1-}% first decreases to a minimum (for BCPD (K-1, K-2) at ca. 290°C ; for De-25 (K-3, K-4) at ca. 320°C), then rises again due to the decomposition of both BCPD and De-25 during heating. It follows that the higher the temperature at the minimum occurs, the better the thermal stability of the flame retardant. The significance is twofold : firstly, the emergence of Cl[1]·links up with the interaction between PP and the radical Cl· which induces again the evolution of HCl giving a subsequent impetus to the catalytic degradation of PP, secondly, on exposure to heat in a fire environment the combination between HCl and Sb_2O_3 leads to the formation of antimony oxychlorides and then $SbCl_3$, both of which have proved to be powerful flame retardants in gas phase (5). It is also worth noting that the separation between the curves of K-3 and K-4 is much larger than those between K-1 and K-2, indicative of the fact that the system of Sb_2O_3/De-25 exhibits stronger synergistic effect than that of Sb_2O_3/BCPD system in good conformity with our thermal data (1). Through XPS analysis we have the ratio Cl/Sb ≅ 4-7 in the temperature of 260°-320°C (see, Table IV) comparable to the result of Costa, et al.(6), if the huge difference in the experimental conditions between TGA and XPS is taken into account. Another point of interest is that the ratio Cl[1]·/Sb of the intermediate antimony oxychlorides, which is possibly a catalyst for charring in the condensed phase, lies inbetween the range of 0.7-1.2, implying formulae something between SbOCl and $Sb_4O_5Cl_2$.

Table IV Total atomic concentrations of Cl, Cl[1]· and Sb in the surface within the range of 260°-320°C for system K-4

Temperature, °C	Total Cl %	Cl[1]·%	Total Sb %	Cl/Sb	Cl[1]·/Sb
260	0.77	0.14	0.19	4.1	0.74
320	1.24	0.22	0.18	6.9	1.20

(ª Reproduced with permission from ref.5. Copyright 1991)

The Study of Flame Retardant Mechanism in Condensed Phase. Zirpro process (zirconium and titanium treatment) is an effective and durable flame resistance technique with keratinous fibres such as wool, mohair, silk, cashmere and speciality animal fibres. Many patents have appeared since 1970s; however, few is known of its mechanism, because the zirconyl (or titanyl) fluorides can not easily be analyzed with reasonable accuracy. The TGA-DTG curves of the Zirpro flame retardant prepared by Zirpro process are shown in Figure 2 (1), where there are three peaks with T_{max} at 250°C, 350°C and 500°C, respectively. To ascertain the structure of Zirpro flame retardant, the stoichiometry has been derived by XPS to be F/N : N/Zr : F/Zr : O/Zr : C/Zr = 1.8 : 1.1 : 2.0 : 3.6 : 6.8. Among those, O/Zr and C/Zr do contain contribution from contamination. In the presence of fluorides, the structure of Zirpro flame retardant may be presumed as shown in Structure 1. The oxidation state of Zr(IV) with coodination number of five is expected to be d^0 configuration. Based on this structure we have tried to study the changes in the C/Zr, O/Zr, F/N, N/Zr and F/Zr ratios as functions of temperature (see Figures 3 and 4). Some conclusions can be readily drawn as follows : (1) C1s spectra of Zirpro flame retardant show that organic moiety begins to decompose at 150°C, with the evolution of water and carbon dioxide; (2) Both F/N and N/Zr change inversely as a function of temperature, resulting in a constant ratio of F/Zr (close to 2.0), indicative of a fact that the NH_4^+ moiety starts decomposing at a temperature ranging from 300°C to 350°C. Items 1 and 2 are in excellent agreement with TGA-DTG and DTA data (omitted for simplicity); (3) The fluorozirconate gradually loses fluoride during heating above 350°C. It is clear that there are drastic

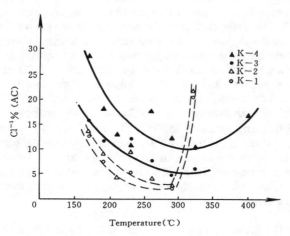

Figure 1 The dependence of Cl^{1}% in systems K-1, K-2, K-3 and K-4 on temperature
(Reproduced with permission from ref.5. Copyright 1991 Elsevier Science Ltd)

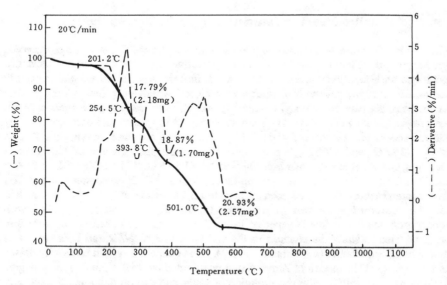

Figure 2 TGA-DTA curves for Zirpro flame retardant
(Reproduced with permission from ref.1. Copyright 1994 Elsevier Science Ltd)

32. WANG   *XPS (ESCA) Studies in Flame Retardancy of Polymers*   523

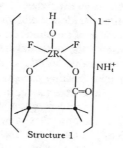

Structure 1

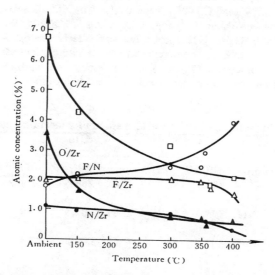

Figure 3   Degradation curves of Zirpro flame retardant on heating (XPS)

changes in percentages of -COOH and NH_4^+ starting at about 150°C and 300°C, characteristic of decarboxylation (150°-280°C) and deamination (300°- 380°C), respectively; (4) Of great significance is the ratio of F/Zr which remains constant (2.0) over a wide range of temperature from ambient temperature to 350°C, confirming that $ZrOF_2$ (or $TiOF_2$) may be the reactive intermediate responsible for the char formation, evidenced by chemical shift of 0.6 eV in C1s spectra.

The Structural Analysis of Intractable Intermediates. Melamine and its salts are widely used in the formulation of fire retardant additive systems for polymeric ma terials. For understanding the mechanism of melamine decomposition upon heating the problem can hardly be of access without the structural information of condensates formed on heating at temperature above 300°C. As a matter of fact, little is known about its mechanism of action upon heating except undergoing condensation with elimination of NH_3 and formation of insoluble products, of which the very sparse solubility of these products is a major drawback encountered in their structural investigation. Discrepancies arose in literature that they have been given the conventional names, "melamine", "melam", "melem" and "melon" before reaching wide agreement on their structures (7). The structures of melamine and melam have been well established to be s-triazine-like and dimer, of which the empirical formulae were determined to be $C_3H_6N_6$ and $C_6H_9N_{11}$, respectively. The XPS data give us the apparent quantitative results, see Table V (Wang, J. Q., Costa, L., Xiang, W., Camino, G., et. al., to be submitted). Using N1s spectra one can clearly see the stoichiometries of functional groups within both compounds which are in complete agreement with the well established structures. From the view-point of surface chemistry the stoichiometries seem not to be dependent too much on oxidation in case of both compounds.

Table V Curve-fittings of N1s and C1s spectra for melamine and melam

Compound	N1s			C1s			Stoichiometry in N1s		
	group	BE, eV	%	group	BE, eV	%	-C=$\underline{N}$	-$\underline{N}H_2$	
Melamine ($C_3H_6N_6$)	-C=$\underline{N}$	398.1	52.0	$\underline{C}$-C	284.6	41.4			
	-$\underline{N}H_2$	398.9	48.2	$\underline{C}$=N	287.5	52.9	1	1	
Melam ($C_6H_9N_{11}$)	-C=$\underline{N}$	398.0	54.6	$\underline{C}$-C	284.6	48.4	-C=$\underline{N}$-	-$\underline{N}$H$_2$-	$\underline{N}$H
	-$\underline{N}H_2$	399.0	36.4	$\underline{C}$=N	287.5	46.3	6	4	1
	-$\underline{N}$H	400.1	9.0	$\underline{C}$-O	288.0	5.3			

Presumed Structures for "Melem" and "Melon" Based on Theoretical Analysis. The correlation between binding energy and net charge of nitrogen atom requires substantiation based on theoretical calculation (CNDO-2) before getting to the heart of the structural argument of both "melem"and "melon". The extrapolated binding energy for $\underline{N}$ group was found to be 400.7 eV by the author, based upon which curve-fittings of N1s spectra of "melem"and "melon" were cautiously performed (see Table VI). The presumed structures for "melem"and "melon" is likely   as proposed in Table VII (Wang, J.Q., Xiang, W., Costa, L. and Camino, G., to be submitted).

Table VI Curve-fitting results of "melem"and "melon"

Compound	N1s			C1s			Stoichiometry in N1s
	Group	BE, eV	%	group	BE, eV	%	$C=\underline{N} : \underline{N}H_2 : \underline{N}H : \underline{N}$
"melem"	-C=$\underline{N}$	398.2	65.2	$\underline{C}$-C	284.6	50.1	
	-$\underline{N}H_2$	398.8	11.2	$\underline{C}$=N	287.6	49.9	
							6 : 1 : 1 : 1
	-$\underline{N}H$	399.9	11.7				
	$\underline{N}$	400.5	11.9				
"melon"	-C=$\underline{N}$	398.3	72.8	$\underline{C}$-C	284.6	38.7	
	-$\underline{N}H_2$	398.9	6.0	$\underline{C}$=N	287.6	61.3	
							12 : 1 : 2.5 : 1
	-$\underline{N}H$	400.0	15.2				
	$\underline{N}$	400.8	6.0				

Charring Process in Condensed Phase — Transition Temperature, (T_{gr}). In the wake of the rapid development in science and technology of polymer flame retardation   an insight into the mechanisms of flame retardancy in condensed phase becomes one of the focussing points at issue. How to characterize and trace experimentally the charring process of polymers upon heating is, therefore, a problem of prior importance. Krevelen (8) found that there is a very significant relation between the pyrolysis residue (%) and the Oxygen Index (OI) of a polymer, i.e. the higher the char residue, the lower the flammability. Ill-defined is the char,   of arbitrary definition. Therefore, an alternative parameter T_{gr} has been supplementarily defined in the laboratory (9) as a temperature at which transition between the mesophase and the graphitic phase of the char takes place.

EVA Copolymers as Model System for Charring Study. Compounds based on ethylene-vinyl acetate (EVA) copolymers can meet stringent requirements in the area, for instance, of halogen-free flame retardant cable jackets. To begin with, thermal data for EVA copolymers derived from TGA are shown first in Table VIII. The C1s spectra and valence bands for EVA with 15 % VA, for example, are shown in Figures 5 and 6 at different temperatures. The small peak appeared at the higher binding energy relative to the main peak in C1s spectrum (Figure 5) at room temperature was assigned to Ac (CH_3COO) group attached to the backbone. It disappears at 360°C (cf. T_{max}=359.1°C for the 1st peak), meaning the loss of the Ac group through scission of C-O bond. The C1s spectrum at 500°C displays a stupendous change in shape, i.e. tailing at left side on the one hand and shifting of the main peak towards right by ca. 2.6 eV on the other hand. An obvious change can also be found in the valence band where the two bands positioned within the ranges of 0-20 eV and 22-29 eV (see, Figure 6) are the fingerprints of the polyethylenic backbone and the acetate group, repectively. Similar to the C1s spectrum, the band corresponding to the Ac group in valence band disappears at 360°C as well. Both valence bands and C1s spectra at

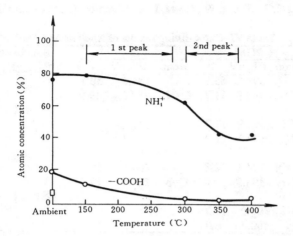

Figure 4 Atomic concentration of NH_4^+ and -COOH groups in Zirpro FR vs temperature (XPS)
(Reproduced with permission from ref.1. Copyright 1994 Elsevier Science Ltd)

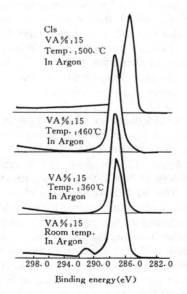

Figure 5 C1s spectra of EVA (15%VA) at different temperatures

Table VII The presumed structures for "melem" and "melon" based upon XPS data

Compound	Suggested structure	Stoichiometry -C=$\underline{N}$: -$\underline{N}$H$_2$: -$\underline{N}$H : $\underline{N}$				Comment
Melamine		1	1	0	0	triazine-ring
Melam		6	4	1	0	triazine-ring
"Melem"		6	1	1	1	cymeluric structure
"Melon"		12	1	2.5	1	cymeluric + triazine

Table VIII TGA data for EVA copolymers in 99.5 % N_2

VA %	Peak 1		Peak 2	
	Onset, °C	T_{max}, °C	Onset, °C	T_{max}, °C
15	275	359.1 (2.361%/min)	370	474.5 (26.38%/min)
45	260	354.8 (6.310%/min)	375	471.6 (15.99%/min)
60	220	351.9 (8.629%/min)	370	473.1 (12.33%/min)
Assignment (TGA/FTIR)	Evolution of HAc		Charring	

500°C are similar to the graphite-like structure (omitted for simplicity). The T_{gr} for EVA was carefully determined to be 470°C under Ar (9). Comparison of the C1s spectra and valence bands of EVAs at three levels of VA % was performed at T_{gr} (470°C) under Ar protection. In Figures 7 and 8 we see that as the sample of EVA (15 % VA) just happens to start converting by all appearances, the conversion of EVA with 60% VA is readily accomplished to the graphite-like structure; and that of EVA with 45% VA lies somewhere inbetween. In contrast with TGA data one conclude that the scission of Ac group from backbone leads to the formation of HAc, which does catalytically help hold the transition earlier, i.e. causing charring process easier. Consequently, EVA copolymers possess of higher flame retardancy than polyethylene. The OI of EVA increases linearly with the increase of VA content, i.e. the slope = +0.22% OI / 1% VA (10). Rationalization was carried out by Meisenheimer (11) in terms of enthalpy argument on the basis of a linear relationship between combustion enthalpy and VA content. Looking over our XPS data the problem does not seem to be so simple as expected before. Keeping TGA data in mind (the decrease rate of T_{max} vs VA% = -0.16°C / 1% VA; weight loss rate vs VA% = +0.14% / min / 1% VA) our XPS study has shown that the Ac group attached to the polyethylenic backbone expedites the formation of the graphite-like structure with highly crosslinked network, which plays a very important role in charring process in many cases.

The Role of Oxygen. Although all polymers degrade at high temperatures in the absence of air, degradation is almost always faster in the presence of oxygen. Table IX (Tu, H. B. & Wang, J. Q., to be submitted) gives thermal data of EVA copolymers in absence and presence of oxygen, respectively. The char residues at 500°C were measured by TGA for systems EVA, EVA/P_x and EVA/APP, see Table X (Jiang, Q. J., & Wang, J. Q., to be submitted).

Promoting Charring. It is evident that EVA copolymers themselves alone do not form char at all at 500°C (see Table X) within the detectable limit in TGA unlike that observed in XPS under inert gas. In XPS experiments run at 500°C a graphite-like structure of the residues even in very little amount can be clearly observed without any difficulties. The presence of oxygen enhances char formation through initiating the charring reaction at a lower temperature, as indicated by Δ_2 = -60°C in the case of EVA (15 % VA), for instance. Even a tiny amount (e.g. 0.5%) of oxygen in air can be sufficient for a substantial depression in the average transition temperature $\overline{T}_{gr}$ of

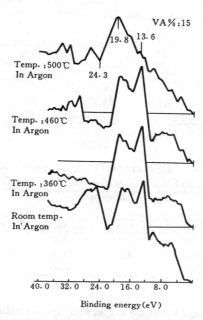

Figure 6 Valence bands of EVA at different temperatures

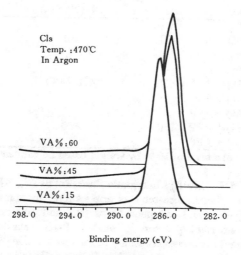

Figure 7 C1s spectra of EVA at 470°C

the char unambiguously observed by XPS (cf. Table XI). This fact can be seen more pronounced in oxygen-free polymers, for example, polyethylenes (see below) than oxygen-containing polymers, for example, EVA copolymers.

Table IX TGA data of EVA copolymers under N_2 and Air atmospheres

VA % by weight	m/n [a]	$(T_{max})_1$, °C		Δ_1 [b]	$(T_{max})_2$, °C		Δ_2 [b]
		Air	N_2		Air	N_2	
15	17.4	336.1	359.1	-23.0	415.4	474.5	-59.1
45	3.8	336.1	354.8	-18.7	438.5	471.6	-33.1
60	2.0	327.4	351.9	-24.5	450.0	473.1	-23.1

[a] EVA formula : $-(CH_2CH_2)_m-(CH_2-CH)_n-$; [b] $\Delta = (T_{max})_{air} - (T_{max})_{N_2}$
$\qquad\qquad\qquad\qquad\qquad\quad\ \ |$
$\qquad\qquad\qquad\qquad\qquad\ \ OOCCH_3$

Table X Char residues at 500°C for systems EVA, EVA/P_x and EVA/APP under (99.99%) N_2 and Air

System	Atmosphere					
	N_2			Air		
	15 % VA	45 % VA	60 % VA	15 % VA	45 % VA	60 % VA
EVA	0.0	0.0	0.0	8.0	10.0	9.0
EVA/P_x	5.0	6.0	8.0	20.0	19.0	23.0
EVA/APP	21.0	26.0	18.0	32.0	35.0	30.0

Table XI Average transition temperature $\overline{T}_{gr}$ of graphite-like structure for EVA, EVA/P_x, EVA/APP [a]

Copolymer	Average transition temperature, °C		
	Ar	N_2 (99.5 %)	Air
EVA (15%)			
EVA (45%)	470	430	400
EVA (60%)			
EVA (15%) + P_x			
EVA (45%) + P_x	> 480	470 - 480	460
EVA (60%) + P_x			
EVA (15%) + APP			
EVA (45%) + APP	460	460	400
EVA (60%) + APP			

[a] P_x, APP : at a level of 10 % by weight based on elemental P for each

Polyethylene (HDPE, LDPE) / Inorganic Phosphorus (P_x, APP). These systems merit our attention whether in research or application. Inorganic phosphorus flame retardants including elemental red phosphorus (P_x) and ammonium polyphosphate (APP) occupy an important position in modern flame retardation. A wealth of literature had appeared in succession. Many of the ideas relating to the mechanism of flame retardation by phosphorus compounds still lack definite experimental proof. All of our data derived from TGA of PE, PE/P_x and PE/APP are listed in Table XII. Figures 9 and 10 give C1s spectra of HDPE at different temperatures under N_2 and air atmosphere. The similar spectra of LDPE, LDPE/P_x, LDPE/APP were omitted for sim-

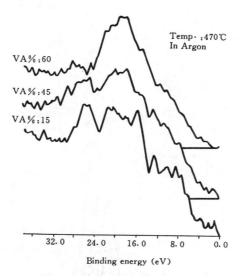

Figure 8 Valence bands of EVA at 470°C

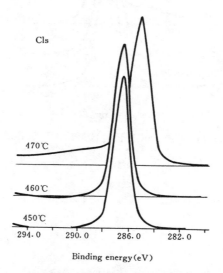

Figure 9 C1s spectra of HDPE, HDPE/P$_x$, HDPE/APP at different temperatures under N$_2$

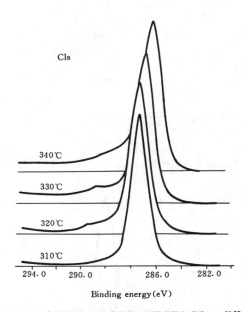

Figure 10 C1s spectra of HDPE, HDPE/P_x, HDPE/APP at different temperatures under air

Table XII TGA data of polyethylene/inorganic phosphorus systems [a]

Sample	N$_2$ T$_{max}$ °C	N$_2$ ΔW$_{440}$ %	Air T$_{max}$ °C	Air ΔW$_{400}$ %	CR[b], % N$_2$	CR[b], % Air
HDPE	468.0	5.5	339.2	90.0	0.4	0.0
HDPE/P$_x$	464.1	7.5	395.1	21.5	1.9	6.8
HDPE/APP	470.3	8.5	428.8	10.5	16.3	4.5
LDPE	455.9	14.0	337.2	78.0	0.4	0.0
LDPE/P$_x$	464.0	10.0	379.3	41.5	0.0	4.1
LDPE/APP	465.8	15.5	412.5	17.0	17.1	4.0

[a] T$_{max}$: maximum temperature at the first peak; ΔW$_t$: weight loss % at temperature t.
[b] Char residue

plicity. TGA data (cf. Table XII) disclosed that in the presence of oxygen both P$_x$ and APP exert an obvious stabilized effect compatible with data reported by Peters (12) for P$_x$. According to the method used in literature the transition temperature T$_{gr}$ and Δ(T$_{gr}$) = (T$_{gr}$)air - (T$_{gr}$)N$_2$ were acquired for systems of HDPE, HDPE/P$_x$, HDPE/APP, LDPE, LDPE/P$_x$, and LDPE/APP as shown in Table XIII. In harmony with TGA results it indicated that adding P$_x$ into PE matrix can cause a change in ΔT$_{gr}$ of around +20°C relative to (T$_{gr}$)N$_2$ (cf. Table XIII). In other words, char formation was postponed by 20°C relative to HDPE or LDPE In the opposite direction, however, it happened to APP by a change in ΔT$_{gr}$ of about 10° to 20°C, although APP can make PE systems more stable and larger reduction in Δ(T$_{gr}$) (12). It was exceptionally surprised to learn that for polyethylenes without P$_x$ or APP the reduction in T$_{gr}$ can be as large as 140°C, if compare this with polyethylenes containing P$_x$ or APP. For the latter two only a small additional change in ΔT$_{gr}$, say 20°C, can be observed. The influence of oxygen obviously plays much stronger role in charring process than P$_x$ or APP; even APP behaves more powerful in charring process than P$_x$ (see Table XIII). Another thing is the flame retardant activity of elemental phosphorus, which may appear somewhat unanticipated because of its highly negative oxidation enthalpy. To gain further information one needs to know the interaction occurred between polymer and oxygen at the surface. Curve-fittings of C1s spectra for this purpose have been carried out (cf. Table XIV). Should emphasized is the fact that the oxygen-containing functional groups at the surface are remarkably reduced by the presence of P$_x$.

Table XIII Influence of atmosphere on T$_{gr}$ of PE/inorganic phosphorus systems

System	HDPE	HDPE/P$_x$	HDPE/APP	LDPE	LDPE/P$_x$	LDPE/APP
Δ(T$_{gr}$) [a], °C	-140	-120	-160	-140	-120	-150

[a] Δ(T$_{gr}$) = (T$_{gr}$)air - (T$_{gr}$)N$_2$; Experimental error for Δ(T$_{gr}$) : ± 1.0°C

Table XIV Atomic concentration (%) of functional groups of various systems at 360°C

Group	HDPE	HDPE/P$_x$	HDPE/APP	LDPE	LDPE/P$_x$	LDPE/APP
-$\underline{C}$OO-	2.5	0.3	2.3	3.6	0.3	3.1
$\underline{C}$=O	7.0	2.4	4.5	10.0	1.5	6.1
$\underline{C}$-O	10.6	5.7	11.7	11.8	7.2	13.1
$\underline{C}$-C, $\underline{C}$-H	79.9	91.5	81.5	74.6	91.0	77.8

This is direct confirmation for the spectulation that P_X acts as a scavenger of oxygen containing radicals, isolating the polymeric surface from O_2 attack. From binding energies of P2p spectra at the surface (Table XV) and infrared absorption of C-O-P vibrational modes at 1090, 1190 cm^{-1} one can conclude that once the oxygen containing radicals are captured by P_X, it will be gradually oxidized to derivatives of phosphorous acid (P^{+3}) and phosphoric acid (P^{+5}). The derivatives of phosphorus containing compounds do cause an intense dehydration and thus, subsequent carbonization. Besides the above mechanism some other factors must be taken into account as well, e.g. oxidative reactions occurring at the surface in case of APP must have tremendous effect on cross-linking with higher speed than P_X. It is well known that highly cross-linked network does impart flame retardance to the polymers (*13*).

Table XV P2p and O1s spectra of phosphorus containing compounds

Char or compound	Temperature, °C	BE, eV	O/P	Structural assignment
EVA (15 % VA)/P_X (in Ar)	500	134.5	2.3	inbetween HPO_2 (metaphosphite) and H_3PO_3 (phosphorous acid)
EVA (45 % VA)/APP (in Ar)	500	135.5	3.3	inbetween P_2O_5 and H_3PO_3 (phosphoric acid)
P_X	room temp.	129.7	0.0	elemental phosphorus
APP	room temp.	133.9	3.0	ammonium polyphosphate
P_2O_5	room temp.	135.0	2.5	phosphoric anhydride

Cross-linking and Flame Retardancy. Cross-linking is recognized as an effective means for flame retardation of polymers. A simple means for determining the cross-linking has been expected for long. It turns out that XPS may prove to be an appropriate approach to fulfill the demand. As might have been expected, a clear-sighted description of thermal degradation of EVA copolymers under Ar atmosphere had been achieved (Wang, J. Q. & Tu, H. B., to be submitted). The absolute intensity of C1s peak must be used for heating from ambient through 500°C. Presenting a striking contrast to TGA data the XPS results divided into three regions, more or less, corresponding to the three regions in TGA can simply be stated as follows:

Decontamination. It begins from ambient to 200°C, without weight loss detected in TGA. Contamination of both carbon and oxygen, a phenomenon specific in surface analysis.

Weak Cross-linking. This stage starts at 200°C up to 450°C, with elimination of HAc in the range of 300°C to 400°C, nearly, one hundred degrees below the onset of the first peak in TGA. There is no weight loss detectable from 200°-300°C within the experimental error of TGA technique, indicative of some reactions taking place in the condensed phase. The minor cross-linking may originate from the C-O bond scission taken place at a very slightly rate, at least, in the outermost layers. Within the range of 300°-400°C occurs an elimiation of HAc, which should cause a weight loss and result in a sequential reduction of intensities in C1s and O1s spectra. In fact considerable cross-linking must have happened to the polymer within this range.

High Cross-linking. A sudden growth of intensity in C1s spectra around $T_{gr} \cong (T_{max})_2$ above 450°C (charring process) induces a highly cross-linked network of the char residue (Wang, J.Q. & Tu H.B., to be submitted). In the highly cross-linking region the surface can be abundant in free radicals, active enough to recombine with oxygen atoms from the residual fragments of Ac group retained in Ar atmosphere. This is why a broad scattering of the experimental points results.

In summarizing. The last two decades have seen a major growth in the use of surface analytical tools in materials science. XPS (or, ESCA), core spectra and valence band, were used quite common to characterize polymers, capable of accommodating intractable materials and little or no detrimental effect to the specimens. However, very few is known in its application to the study of flame retardancy of polymers. Based on XPS data in the laboratory exemplications are described in the paper in order to call attention to the potential utilization of the surface sensitive technique in areas relevant to flame retardnacy of polymers. Among those, charring process in condensed phase seems to be one of the most promising fields in view of the current demand for novel techniques to provide informative data in the study of the black char or non-black residual ash. It is shown indeed that coupled to some other techniques XPS may manifest itself to be a means of great worth in research and application of flame retardant polymers, if used properly.

Literature Cited. (*1*) Wang, J.Q., Feng, D.M. and Tu, H.B., *Polym. Degrad. Stab.*, **1994**,43, 93-99; (*2*) Felix, W.D., McDowell, M.A. & Eyring, M., *Text. Res. J.*, **1963**,33, 465; (*3*) Koroskys, M. J., *Am. Dyest. Rep.*, **1969**, 58 (6), 15; (*4*) Ingham, P., *J. Appl. Polym. Sci.*, **1971**, 15, 3025; (*5*) Wang, J.Q., Feng, D.M., Wu, W.H., Zeng, M.X. and Li, Y., *Polym. Degrad. Stab.*,**1991**,31, 129-140; (*6*) Costa, L., Goberti, P., Paganetto, G., Camino, G., & Sgarzi, P., *Third Meeting on Fire Retardant Polymers*, Torino, Italy, Sept. 1989, pp. 19, 21-22; (*7*) Costa, L., and Camino, G., *J. Thermal Analysis*, **1988**,34, 423-429; (*8*) Van Krevelen, D.W., *Chimia*, **1974**, 28, 504; (*9*) Wang, J.Q. & Tu, H.B., in *Proceedings of the 2nd Beijing International Symposium/Exhibition on Flame Retardants*, Oct.11-15, Beijing, China, 272-279, 1993; (*10*) Ford, J.A., in *Fire Resistant Hose, Cables and Belting* (International Conference) Plastics and Rubber Institute, Science and Technology Publishers Ltd September, 1986; (*11*) Meisenheimer, H., *Proceedings of the 138th ACS Meeting, Rubber Division*, October, 9-12, 1990; (*12*) Peters, E.N., *J. Appl. Polym. Sci.*, **1982**, 27, 1457; (*13*) Wilkie, C.A., in *Fire and Polymers*, Ed. Nelson, G.L., *ACS Symposium Series*, **425**, Chapter 13, 1990.

RECEIVED January 24, 1995

Chapter 33

Thermal Analysis of Fire-Retardant Poly(vinyl chloride) Using Pyrolysis—Chemical Ionization Mass Spectrometry

Sunit Shah[1,3], Vipul Davé[1,4], and Stanley C. Israel[2]

[1]Department of Plastics Engineering and [2]Department of Chemistry, University of Massachusetts, Lowell, MA 01854

The fire retardant effect of additives on poly (vinyl chloride) [PVC] was studied by pyrolysis directly in the ion source of a double focusing, magnetic sector mass spectrometer operating in the chemical ionization mode. By using a high temperature, fast heating rate, high temperature pyrolysis (HTP) probe, pyrolysis temperatures of 1200°C at controlled heating rates of up to 20,000°C/sec could be achieved. Pyrolysis of PVC/antimony oxide formulations suggest that the additive interacts with PVC in the gas phase to impart fire retardancy by the formation of $SbCl_2$ as the major volatile species. Pyrolysis of PVC/molybdenum oxide samples indicate a condensed phase interaction, with the additive acting as a smoke suppressant by decreasing the yields of benzene and other aromatic compounds.

Due to its inherent flame retardancy and low production costs, poly (vinyl chloride) [PVC] is extensively used by the electrical and communications industries in applications where fire hazards are present. Currently, it is consumed in excess of 8 billion pounds per year or 20% of the total thermoplastics market. More than half the PVC used is unplasticized (i.e. rigid) and is combustible, albeit of low flammability, since it contains significant amount of carbon. Since PVC is a brittle glassy polymer within the normal temperature range of use, it is necessary to add plasticizers and stabilizers in order to improve its melt processability. However, the resultant plasticized PVC is usually combustible and therefore a flame retardant is added to reduce its flammability. It should be noted that there is more plasticized PVC in use that is not flame retarded than PVC materials that contain flame retardants, and that unplasticized PVC is often flame retarded.

[3]Current address: D&S Plastics International, Grand Prairie, TX 75050
[4]Current address: Consumer Products Research and Development, Warner Lambert, 175 Tabor Road, Morris Plains, NJ 07950

It is well known that certain metal oxides can act as effective flame retardants in many polymer systems (*1*). Antimony oxide (Sb_2O_3) is most extensively used with PVC since the halogen/Sb_2O_3 synergism has been found to be very effective in this and many other polymer systems (*2-4*). In spite of the fact that antimony oxide imparts a degree of opacity to PVC, almost 80% of the antimony oxide consumed in the plastics industry is used in fire retardant PVC formulations (*5*). The fire retardance provided by the antimony oxide/halogen combination is a result of the interaction of the halogen source, usually an organic chloride or bromide, with antimony oxide to produce largely SbX_3 (X = chlorine or bromine) which in turn inhibits the flame (*4*). This synergistic flame retardancy effect is observed when antimony oxide is incorporated into PVC (*6*). The interaction mechanism between PVC/antimony oxide has been studied by laser-probe pyrolysis mass spectrometry (*7*). Volatile antimony trichloride was detected in the mass spectra along with a smaller amount of $SbCl_2$. The mechanism proposed for this synergism was the reaction of hydrogen chloride gas evolved from the thermal decomposition of PVC with antimony oxide resulting in the *in situ* formation of an intermediate compound, namely antimony oxychloride [SbOCl] (*7*). This intermediate compound further decomposes thermally to yield the actual flame retardant species, antimony trichloride [$SbCl_3$] (*7,8*). The hydrogen chloride gas evolved during the thermal decomposition of PVC was also found to react with antimony oxychloride to yield antimony trichloride.

PVC is one of the many materials which produces smoke on combustion. Several of the commonly used fire retardant additives, although effective in controlling flaming combustion, actually increase the amount of smoke produced when the material is incinerated in a fire. Antimony oxide exhibits this type of behavior. It is reported to perform well as a flame retardant for PVC, but the production of smoke is greatly increased due to its presence (*9*). A large number of chemical compounds have been reported as smoke retarders for PVC in both open and patent literature (*10*). The first mechanistic studies regarding smoke retarders in PVC were conducted with ferrocene (*11,12*). It was found that ferrocene increased char formation in PVC but there was no clear correlation between char formation and smoke. Molybdenum-based compounds, molybdenum oxide in particular, have been previously reported to be effective smoke suppressants and fire retardants as well (*13,14*). Molybdenum oxide functions as a smoke suppressant for PVC within the polymer matrix by suppressing the production of volatile aromatics, especially benzene, which are the principal source of smoke (*15-18*). In particular, the oxide was found to be a potent catalyst for the dehydrochlorination of the polymer and it also caused a dramatic reduction in the amount of benzene produced. Furthermore, the possibility of a gas phase mechanism for molybdenum oxide was doubtful as the compound yielded no volatile metallic species during the process of combustion (*19*). The smoke suppressant action of molybdenum oxide has been studied by various techniques such as NBS Smoke Chamber, Goodrich Smoke Char tests, pyrolysis-gas chromatography, pyrolysis-gas chromatography-mass spectrometry and dynamic mass spectrometry (*10,20,21*). It is evident that the mechanism of volatile pyrolyzate formation from the thermal decomposition of PVC plays a key role in determining the effectiveness of smoke retarder additives. Benzene is the principal aromatic substance formed during the burning of PVC and the smoke suppressant action of molybdenum oxide should be related to the ability of this oxide to reduce the benzene yield (*4,12*). During the thermal decomposition of PVC, dehydrochlorination step leads to the formation of a conjugated polyene structure and benzene formation evidently occurs via a cyclohexadiene intermediate (*20,22-25*). Intramolecular pathways to the formation of benzene have also been suggested by several workers (*26*). There are two conflicting theories which are proposed to explain the action of molybdenum oxide as a smoke retardant for PVC, namely the Lewis Acid Theory (*21,27*) and the Reductive Coupling Mechanism (*20*). In essence, both the theories explain the decrease in the yield of benzene formation and subsequent smoke suppression in PVC. Recently, more work has been done to explain the formation of the thermally labile allylic structures in PVC by hydrogen abstraction by macroradicals (*28,29*). Also, the use of activated copper-promoted reductive coupling mechanism has been investigated as a potential method of smoke suppression in PVC (*30*).

A relatively new technique to analyze the pyrolysis products of polymers is Chemical Ionization Mass Spectrometry in which the pyrolyzates are analyzed immediately on production without having to conduct a gas chromatographic separation. It was first developed by Munson and Field (*31*). Chemical ionization is a technique of ionizing molecules without imparting excess energy to them. It is essentially a gas phase acid-base reaction in which the acidic species is formed by subjecting a reagent gas, such as methane, to bombardment by a beam of high energy electrons (*32*). The sample molecule (M) gets ionized through the gain of a proton forming species represented by MH^+. This process can be controlled to avoid imparting excess energy to the molecular ion which in turn prevents secondary fragmentation. The chemical ionization mass spectrum is relatively easy to interpret due to little or no secondary fragmentation and each peak can be ascribed to a particular component in the mixture. It is a very convenient technique to characterize polymers as the pyrolysis and analysis of the pyrolyzates is completed in a very short period of time, and only a few milligrams of sample is required for each pyrolysis run.

The technique of pyrolysis-chemical ionization-mass spectrometry can be used to study the fire retardant and smoke suppressant characteristics of additives on different polymer systems. Polymer samples containing different concentrations of the additive can be pyrolyzed and the extent of interaction of the additive can be evaluated by comparing the mass spectra of different formulations with that of pure polymer for the identification of new products. We have previously reported the application of this technique to elucidate the mechanism of flame retardancy for Nylon 66 containing poly (pentabromobenzyl acrylate) and antimony oxide (*33-35*). The main objective of this paper is to apply the same technique to investigate the mechanisms of fire retardancy and smoke suppression of PVC by antimony oxide and molybdenum oxide, respectively.

Experimental

Materials. PVC and antimony oxide were obtained from B. F. Goodrich Chemical Co., and molybdenum oxide was obtained from Aldrich Chemical Co. All these materials were in powder form.

Sample Preparation. Different formulations of PVC samples containing 2%, 3%, 5%, 10% and 20% (w/w) antimony oxide were prepared by intimate mixing with agate mortar and pestle. PVC formulations containing 1%, 3%, 5%, 10%, 20%, 30% and 50% (w/w) molybdenum oxide were similarly prepared. Since the formulations were physical mixtures of powders, care was taken to obtain a homogenous blend in each case.

Instrumentation. Pyrolysis of all the polymer formulations was carried out directly in the ion source of a MAT Model 112S double focusing magnetic sector mass spectrometer interfaced to a Varian Model V-76 computer for data acquisition and reduction. The high temperature pyrolysis of all the samples was done using a high temperature, fast heating rate pyrolysis probe (HTP probe) controlled by a CDS pyroprobe 120 control unit (Chemical Data Systems, Oxford, PA) (36). The unit allowed controlled pyrolysis temperatures up to 1200°C at heating rates in excess of 20,000°C/sec. All determinations were made with the mass spectrometer operating in the chemical ionization (CI) mode using methane as the reagent gas.

Results and Discussion

Pyrolysis of PVC. The CI mass pyrogram of PVC obtained at 1000°C is shown in Figure 1. Methane was used as the reagent gas and the spectral scanning was started from mass 33 up to mass 1000 to detect the low molecular mass fragments. The first step in the thermal degradation of PVC is dehydrochlorination and the corresponding peak for hydrogen chloride (HCl) gas is observed at m/z=37. This is in agreement with previously reported studies (5,37). A smaller isotope peak is also observed at m/z=39 which corresponds to the relative abundance of Cl^{37}. The m/z=37:m/z=39 peak intensity ratios exhibit the expected 3:1 chlorine isotopic ratio. A methane reagent gas peak is also observed at m/z=41 which belongs to the $C_3H_5^+$ species. The loss of HCl leaves behind a conjugated polyene structure which undergoes pyrolysis to yield numerous hydrocarbon products that are observed in the peaks above m/z=37 and is in agreement with other investigators (5,38).

The most abundant aromatic pyrolyzate detected during the pyrolysis of PVC was benzene observed at m/z=79. Benzene is formed as a result of the intramolecular cyclization of the cis polyene structure remaining after the dehydrochlorination step. The peaks at m/z=93 and 129 are attributed to the presence of protonated toluene and naphthalene, respectively. The peak at m/z=107 may indicate the presence of either dimethyl benzene or ethyl benzene, both of which have been previously reported (25). The pair of peaks at m/z=117 and 119 are attributed to the protonated forms of indene which were also observed by Lattimer and Kroenke (25). Traces of methyl naphthalene and dimethyl naphthalene are also observed at m/z=143 and 157, respectively. No volatile chlorine containing species are observed in the mass pyrogram and is in agreement with previous studies (5).

Antimony Oxide Flame Retardance of PVC. The pyrolysis of PVC formulations containing antimony oxide were conducted in order to determine the mode of action of the fire retardant additive. The pyrolysis of pure PVC as well as the PVC/antimony oxide formulations were performed while maintaining the same conditions. Figure 2 show the mass pyrogram obtained at 1000°C on the pyrolysis of PVC sample containing 2% Sb_2O_3. A peak is observed at m/z=192 and another pair of peaks is seen at m/z=208 and 210. These peaks were not observed in the pyrolysis of pure PVC. The peak at 192 corresponds to the formation of antimony chloride, $SbCl_2^+$, which results from the interaction of Sb_2O_3 and HCl. Lum in his studies of antimony oxide fire retardance also observed a similar peak at mass 191 (7). However, he also observed the formation of antimony trichloride ($SbCl_3^+$) at mass 226 as the parent ion, which is the principle volatile fire inhibitor. No peak corresponding to $SbCl_3$ is seen in Figure 2. The peak at m/z=208 also indicates the presence of $SbCl_2$. It appears as a result of the formation of the adduct species ($SbCl_2+CH_5^+$) which is commonly found in methane CI mass spectrometry. An isotope peak is also observed at m/z=210. This peak is attributed to the formation of the same adduct species ($SbCl_2+CH_5^+$), corresponding to the relative abundance of Sb^{123}. However, it can also represent the $Sb^{121}Cl^{37}Cl^{35}$ species attached to the CH_5^+ radical as has been reported earlier (7).

The mass pyrogram of PVC samples containing 10% and 20% Sb_2O_3 is represented in Figures 3 and 4, respectively. Two major peaks are again observed at m/z=192 and 210. The intensity of these peaks increases as the amount of antimony oxide increases in the formulations. A smaller isotope peak is also seen at m/z=194 which corresponds to the relative abundance of $Sb^{121}Cl^{35}Cl^{37}$ as has been reported earlier (7). It is interesting to note that the intensity of the peak at m/z=37 (which represents HCl) reduces as the amount of Sb_2O_3 increases. This indicates that the amount of HCl which is evolved from PVC is reacting almost completely with Sb_2O_3 to form $SbCl_2^+$ (i.e., m/z=192). Again, no peak is detected at m/z=227 which would correspond to the evolution of $SbCl_3$.

Mass pyrograms were also obtained during the same pyrolysis run for each of the formulations of PVC and Sb_2O_3 after a delay of 10 seconds (not shown). No peaks were detected at m/z=192 and 210, and the spectra were identical to that of pure PVC. This shows that Sb_2O_3 does not enter into secondary reactions with other thermal decomposition products. It should be noted that the species that are found in the mass spectrometer and those that can be expected to be found in actual atmospheres may differ to a certain extent.

During the pyrolysis of the PVC samples, the ion intensity of the major pyrolyzates was monitored in an attempt to determine their total yield. The yield of

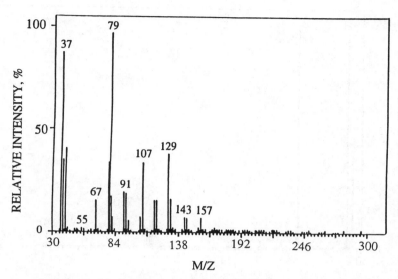

Figure 1. Methane-CI Mass Pyrogram of PVC at 1000°C.

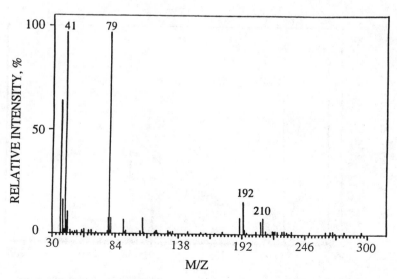

Figure 2. Methane-CI Mass Pyrogram of PVC/2% Sb_2O_3 at 1000°C.

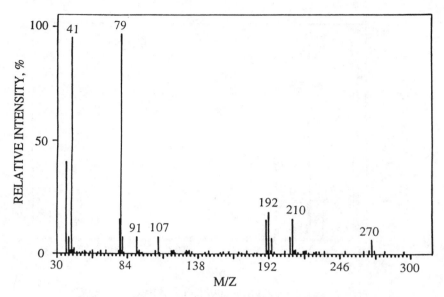

Figure 3. Methane-CI Mass Pyrogram of PVC/10% Sb_2O_3 at 1000°C.

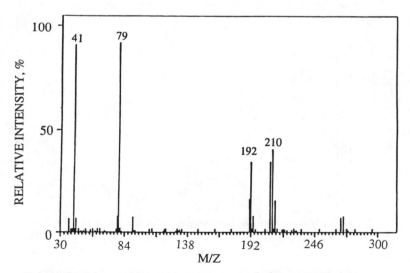

Figure 4. Methane-CI Mass Pyrogram of PVC/20% Sb_2O_3 at 1000°C.

each pyrolyzate was calculated from the area under the ion intensity profile for each sample. The yield of the PVC/Sb_2O_3 interaction products (m/z=192 and 210) were calculated relative to a yield of 10 for the m/z=210 peak produced in the 20% Sb_2O_3 sample. The relative abundance of the PVC/Sb_2O_3 interaction products as a function of Sb_2O_3 content in PVC is plotted in Figure 5. It is clearly seen that none of the pyrolyzates are observed in pure PVC and the abundance of both the products increases as the amount of Sb_2O_3 increases. Maximum of both pyrolyzates is noted for the 20% Sb_2O_3 sample. The total production of aromatic pyrolyzates including benzene was monitored in a similar fashion with the yields calculated from the ion profile data. The yield from each sample is reduced to a yield of 100 for benzene evolved from the 5% Sb_2O_3 sample. The yields of the other major aromatic pyrolyzates are presented as the sum of all mass fragments between 91 and 131. These include toluene, dimethyl and ethyl benzene, and naphthalene which accounts for the major peaks in the pure PVC spectra. The relative abundance of benzene and other aromatic pyrolyzates as a function of antimony oxide content in PVC is plotted in Figure 6. It can be seen that benzene evolution decreases to a small degree on the addition of 2% antimony oxide, but then increases again in the presence of 3%, 5% and 10% antimony oxide. Overall it appears that there is no appreciable decrease in the yield of benzene and this has been reported in earlier studies (15,39). This shows that while antimony oxide is a good fire retardant for PVC mainly by $SbCl_2$, it is unsuccessful as a smoke suppressant as it does not decrease the yield of benzene which is the principle source of smoke during PVC combustion. The evolution of the major aromatic pyrolyzates apart from benzene is also not decreased to any great extent by the increasing amount of antimony oxide.

Molybdenum Oxide Smoke Suppression of PVC. The pyrolysis of PVC samples containing molybdenum oxide (MoO_3) was conducted under identical conditions as that of pure PVC. Figure 7 shows the mass spectrum for PVC sample containing 10% molybdenum oxide. The major peaks observed were the same as those in pure PVC and no new interaction products were formed during the pyrolysis due to the increased presence of MoO_3 in PVC. The absence of volatile molybdenum containing species in the mass pyrogram is in agreement with that reported by Edelson et al. (21). Benzene and hydrogen chloride are the major peaks observed.

The total yield of benzene and HCl, as well as other aromatic pyrolyzates was calculated from the ion profile data for each of the sample pyrolyzed. The yields of each pyrolyzate was calculated relative to a yield of 100 for benzene produced from pure PVC. The total yield of all pyrolyzates between masses 91 and 131 accounted for most of the major aromatic pyrolyzates other than benzene that are produced during pyrolysis. Figure 8 shows the nature in which the production of benzene, HCl, as well as the other major aromatic pyrolyzates (m/z=91-131) was affected by the addition of MoO_3. It is known that the smoke suppression action of molybdenum oxide functions primarily by reducing the yield of benzene which is the principle cause of smoke formation (39,40). However, at the same time the amount of backbone char produced during the combustion of PVC was found to increase as

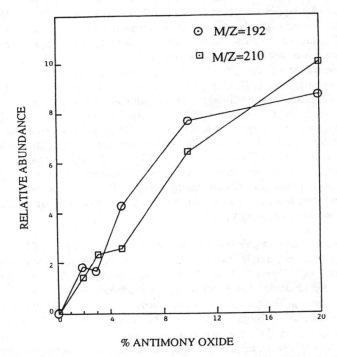

Figure 5. Relative Yield of PVC/Antimony Oxide Interaction Products.

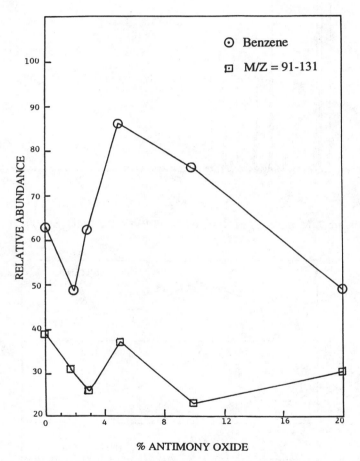

Figure 6. Relative Yield of Benzene and Aromatics in the Pyrolysis of PVC/Antimony Oxide.

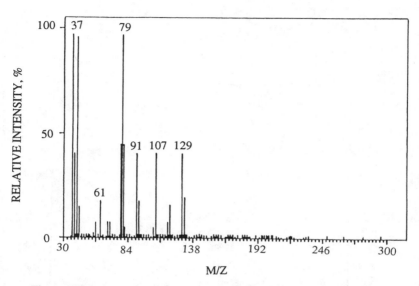

Figure 7. Methane-CI Mass Pyrogram of PVC/10% MoO_3 at 1000°C.

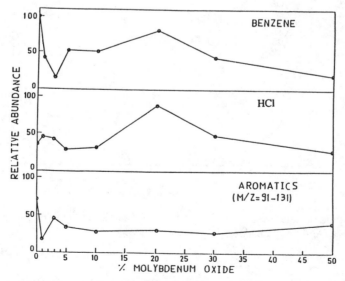

Figure 8. Relative Yield of Benzene, HCl and Aromatics in the Pyrolysis of PVC/Molybdenum Oxide.

the percentage of molybdenum oxide was increased (20). The plot of relative benzene production as a function of percent molybdenum oxide (Figure 8) in PVC indicates that the maximum amount of benzene was formed during the pyrolysis of pure PVC. The PVC sample containing 3% MoO_3 produces the least amount of benzene, and this concentration appears to be the optimum amount for maximum smoke suppression. A further increase in the MoO_3 concentration, from 4% to 20%, resulted in an increase in benzene formation, and a final decrease was observed when samples containing 30% and 50% MoO_3 were pyrolyzed.

HCl production was found to increase initially with the addition of up to 3% MoO_3. Greater amounts of the additive led to a decrease in the HCl production, but another maxima was reached on the plot at 20% MoO_3. Edelson, et al., in their investigation of molybdenum oxide smoke suppression have shown from the HCl ion profile data obtained from electron impact-mass spectrometry that molybdenum oxide catalyzes the dehydrochlorination process (21). Lum has also concluded that molybdenum oxide does catalyze the dehydrochlorination, making it occur at lower temperature (7,15). However, neither has reported a decrease in the HCl yield. In the present study of pyrolysis of PVC samples, a fast heating rate pyrolysis probe was used with a heating rate in excess of 20,000°C/sec. Thus the final pyrolysis temperature of 1000°C was reached in only 3 seconds. If the dehydrochlorination in the PVC samples containing molybdenum oxide occurred at lower temperatures than in the pure PVC, it would not be observed due to the fast temperature rise.

A comparison of the curves of benzene and HCl production reveal that the initial decrease in the yield of benzene from 0 to 3% molybdenum oxide content is accompanied by an increase in HCl production in the same concentration range. With concentrations of molybdenum oxide greater than 5%, the production of HCl and benzene appear to occur at the same rate, both reaching a maximum at 20% molybdenum oxide concentration.

The third curve in Figure 8 shows the nature in which the production of the other aromatic pyrolyzates besides benzene varies with the concentration of molybdenum oxide. The production of the other aromatic pyrolyzates represented toluene, dimethyl and ethyl benzene, indene and naphthalene. This curve again shows that pure PVC does produce the maximum amount of aromatic products. On the addition of 1% MoO_3 the aromatic yield decreases. Comparing this curve with that for benzene, it can be seen that decrease in the yield of benzene is compensated by an increase in the yield of other aromatics. With increasing molybdenum oxide concentration, the yield of aromatics levels off. However, at 50% MoO_3 concentration when the benzene yield further drops, the yield of other aromatics shows an increase. The production of higher mass fragments above m/z=131 was also monitored. The trend was very similar to that of the major aromatics in Figure 8 and the minimum amounts of each were produced in the 1% MoO_3 formulation.

Conclusions

1. The pyrolysis-CI mass spectra revealed that antimony oxide interacted with PVC in the gas phase resulting in the formation of volatile $SbCl_2$ as the major interaction product. No peak corresponding to volatile $SbCl_3$ was detected in any of the

PVC/antimony oxide pyrolysis spectra. The yield of the flame retardant species, $SbCl_2$, was found to increase with increasing amount of antimony oxide in the formulation. The present work can therefore be used to confirm that $SbCl_2$ is present in the mechanism of flame retardancy as has been previously suggested. It should be recognized that pyrolysis-chemical ionization-mass spectrometry is a technique to simulate the gas phase mechanism of flame retardancy of additives on different polymer systems and may not necessarily represent the species found in actual atmospheres.

2. The yield of benzene and other major aromatic pyrolyzates did not decrease due to the presence of increasing amounts of antimony oxide in the formulations. This shows that while antimony oxide performs well as a fire retardant additive in PVC, it is unsuccessful as a smoke suppressant as it cannot decrease the yield of aromatic compounds during thermal decomposition of PVC.

3. Molybdenum oxide smoke suppression appears to be a condensed phase mechanism as no volatile molybdenum containing species were detected in any of the spectra.

4. Minimum yield of benzene was observed for the PVC sample containing 3% molybdenum oxide and this appeared to be the optimum concentration of smoke suppressant amongst the compositions studied. The relative yield profiles of HCl and benzene showed that below 5% molybdenum oxide concentration, the benzene yield decreases sharply while there is an slight increase in HCl production. The combined aromatics yield in this concentration range drops initially but over the entire range studied it follows a trend opposite to the benzene yield.

Literature Cited

1. National Materials Advisory Board, NRC, *Fire Safety Aspects of Polymeric Materials*, Technomic, Westport, CT, 1977; *Vol. 1*.
2. Kuryla, W. C., In *Flame Retardancy of Polymeric Materials*, Kuryla, W. C. and Papa, A. J., Eds., Marcel Dekker, Inc., 1973; *Vol. 1*, 1.
3. Pitts, J. J., *J. Fire Flammability*, **1972**, *3*, 51.
4. Brauman, S. K., *J. Fire Retard. Chem.*, **1976**, *3*, 117-137.
5. *Encyclopedia of PVC*, Nass, L. I., Ed., Marcel Dekker, Inc., 1977, 818-824.
6. Lyons, J. W., *The Chemistry and Uses of Fire Retardants*, Wiley, N.Y., 1970.
7. Lum, R. M., *J. Polym. Sci. Polym. Chem. Ed.*, **1977**, *15*, 489-497.
8. Pitts, J. J.; Scott, P. H. and Powell, D. G., *J. Cellular Plastics*, **1970**, *6*, 35.
9. Hilado, C. J.; Cummings, H. J. and Machado, A. M., *Mod. Plast.*, **1978**, *55*, 61.
10. Kroenke, W. J., *J. Appl. Polym. Sci.*, **1981**, *26*, 1167.
11. Lawson, D. F., *J. Appl. Polym. Sci.*, **1976**, *20*, 2183.
12. Lecomte, L.; Bert, M.; Michel, A. and Guyot, A., *J. Macromol. Sci.-Chem.*, **1977**, *A11*, 1467.
13. Moore, F. W., *Soc. of Plast. Engr. Tech. Papers*, **1977**, *23*, 414-416.
14. Moore, F. W. and Church, D. A., *Intl. Symp. on Flamm. and Fire Retard.*, Technomic, Westport, CT, 1977, 216-227.

15. Lum, R. M., *J. Appl. Polym. Sci.*, **1979**, *23*, 1247.
16. Lum, R. M.; Siebles, L.; Edelson, D., and Starnes, Jr., W. H., *Org. Coat. Plast. Chem.*, **1980**, *43*, 176-180.
17. Ballistreri, A.; Montaudo G.; Puglisi, C.; Scamporrino, E. and D. Vitalini, *J. Polym. Sci. Polym. Chem., Ed.*, **1981**, *19*, 1397.
18. Ballistreri, A.; Foti, S.; Maravigna, P.; Montaudo, G. and Scamporrino, E., *J. Polym. Sci., Polym. Chem. Ed.*, **1980**, *18*, 3101.
19. Moore, F. W. and Tsigdinos, G. A., *J. Less-Common Met.*, **1977**, *54*, 297.
20. Lattimer, R. P. and Kroenke, W. J., *J. Appl. Polym. Sci.*, **1980**, *26*, 1191.
21. Edelson, D.; Kuck, V. J.; Lum, R. M.; Scalo, E. and Starnes, Jr., W. H., *Combust. and Flame*, **1980**, *38*, 271-283.
22. O'Mara, M. M., *J. Polym. Sci.*, **1970**, Part A1, *8*, 1887.
23. Wooley, W. D., *Br. Polym. J.*, **1971**, *3*, 186.
24. Starnes, Jr., W. H., *Adv. Chem. Ser.*, **1978**, *169*, 309.
25. Lattimer, R. P. and Kroenke, W. J., *J. Appl. Polym. Sci.*, **1980**, *25*, 101-110.
26. Starnes, Jr., W. H. and Edelson, D., *Macromolecules*, **1979**, *12(5)*, 797-802.
27. Starnes, Jr.,W. H.; Wescott, Jr., L. D.; Reents, Jr., W. D.; Cais, R. E.; Villacorta, G. M.; Plitz, I. M. and Anthony, L. J., *Org. Coat. Appl. Polym. Sci. Proc.*, **1982**, *46*, 556.
28. Starnes, Jr., W. H.; Chung, H.; Wojciechowski, B. J.; Skillicorn, D. E. and Benedikt, G. M., *Polymer Preprints*, **1993**, *34(2)*, 114.
29. Starnes, Jr., W. H. and Wojciechowski, B. J., *Makromol. Chem., Macromol. Symp.*, **1993**, *70-71*, 1.
30. Jeng, J. P.; Terranova, S. A.; Bonaplata, E.; Goldsmith, D. M.; Williams, K. and Starnes, Jr., W. H., *Poly. Mat. Sci. and Engr.*, **1994**, *71*, 299.
31. Munson, M. S. and Field, F. H., *J. Am. Chem. Soc.*, **1966**, *88*, 2621.
32. Israel, S. C., *Flame Retardant Polymeric Materials*, Plenum Press, NY, 1982; Vol.3.
33. Davé, V. and Israel, S. C., *Polymer Preprints*, **1989**, *30(2)*, 203.
34. Davé, V. and Israel, S. C., *Polymer Preprints*, **1990**, *31(1)*, 554.
35. Davé, V. and Israel, S. C., *J. Polym. Sci. Polym. Chem. Ed.*, (Submitted).
36. Israel, S. C.; Yang, W. C. and Bechard, M., *J. Macromol. Sci.-Chem.*, **1985**, *A22 (5-7)*, 779-801.
37. O'Mara, M. M., *Pure Appl. Chem.*, **1977**, *49*, 649.
38. Mayer, Z., *J. Macromol. Sci. Rev. Macromol. Chem.*, **1974**, *10*, 263.
39. Delman, A. D., *J. Macromol. Sci. Rev. Macromol. Chem.*, **1971**, *C3(2)*, 281.
40. Ritchie, P. D., *Soc. of Chem. Ind.*, London, **1961**, *Monograph No. 13*, 107.

RECEIVED January 4, 1995

SCIENCE-BASED REGULATION

Fire is a hazard involving loss of life and property. Regulations in various parts of the world are different, based upon different philosophies, tests, and standards. Cultural and emotional issues also play a role. Commercially, the success of products depends upon crossing lines in standards and regulations. Over the past 20 years fire science has advanced substantially, especially materials, test methods, and mathematical models. Fire regulations may take years, however, to benefit from that science. Because harmonization under the European Union is required, Europe may be closer than the United States in using science-based standards and other available tools.

Fire extends beyond the issues of ignitability, flame spread, heat release, and smoke. Corrosivity and "green" issues like the presence of halogen or the ability of a material to be recycled are also of interest, both from a regulatory as well as a commercial point of view.

Status reports on two major issues of controversy are presented in the first two chapters. Corrosivity of off gases in case of fire has been the focus of discussions and research by the plastics industry. Hirschler discusses world-wide test developments and their significance.

Plastics used for durables are increasingly facing requirements for recycling. In Europe the advent of "take back" programs may accelerate the trend for "green products" and materials. Nelson provides perspective on a series of issues involving the recycling of flame-retardant plastics.

In one-third of all fire deaths in the United States, residential furniture is the first item ignited. Hirschler discusses how advances in fire science and fire assessment tools are making their way into the design of furniture. Sundstrom presents a major European program designed to harmonize requirements and test methods for upholstered furniture.

Finally, Lyon discusses a program to develop technical parameters for commercial passenger aircraft interior materials that make a substantive difference in fire performance. Using current knowledge about post-crash fuel fires, studies are underway to determine which lightweight aircraft interior materials provide a truly "fire-resistant passenger cabin."

Chapter 34

Smoke Corrosivity: Technical Issues and Testing

Marcelo M. Hirschler

Safety Engineering Laboratories, 38 Oak Road, Rocky River, OH 44116

Until recently, it was believed that acid, mainly halogenated, gases were the only compounds causing corrosion. In reality, all smoke is corrosive, even in the absence of acid gases, although acid gases tend to be highly corrosive. Furthermore, acid gas emission in fires does not fully correlate with smoke corrosive potential, since even alkaline gases are corrosive. Corrosion is governed too by other factors, including smoke amount and composition, humidity and temperature.

Four bench scale performance tests for smoke corrosivity have been proposed and are described: CNET (ISO DIS 11907-2, static), ASTM E-5.21.70 draft radiant (static), ASTM D5485 (cone corrosimeter, dynamic) and DIN (ISO DIS 11907-3, tube furnace, dynamic). The cone corrosimeter is the only one of them which is a consensus standard. None of them have, as yet, been correlated with full scale fires.

The paper contains an analysis of the advantages and disadvantages of each one of the tests, based on concepts, on repeatability and on results supplied to date. Of particular interest in this respect was a study, by the Society of the Plastics Industry Polyolefins Fire Performance Council, which tested over 20 materials using three of the tests mentioned above. The analysis suggests ASTM D5485 is the most promising test, because: (1) it generates a broad range of smoke corrosivity data, (2) the data have good repeatability and (3) the apparatus is capable of yielding, with an excellent fire model, a whole range of fire properties other than smoke corrosivity. Moreover, the analysis also indicates that the CNET test is the one with the largest number of negative aspects.

Smoke corrosivity has been a subject of intense debate in recent years. The issue is, primarily, of commercial or marketing interest, while most other fire issues, are safety concerns. Thus, decisions on smoke corrosivity control are unlikely to appear in codes and regulations, but should surface in standards and specifications by powerful users. It is essential, thus, to examine tests being developed to assess smoke corrosivity.

Three types of corrosive effects of smoke on electrical or electronic circuitry can be identified: metal loss, bridging of conductor circuits and formation of non-conducting surfaces on contacts *(1)*. Metal loss results in an increase in resistance of the circuitry, so that electrical conduction is impaired. Bridging has the opposite effect: decrease in resistance by creating alternative simple paths for current flow. The formation of deposits can cause, just like metal loss, a loss of electrical conductivity and, thus, make an electrical contact unusable. It can also act mechanically, to render parts, such as ball bearings, ineffective because of it not being able to turn adequately. All main tests proposed to date address purely metal loss, although bridging can mask their results. In the remainder of this paper, thus, smoke corrosivity by metal loss will simply be labelled corrosion, for expediency's sake.

The combination of acid gases combine with water causes corrosion of metals. Originally, thus, acid gases were believed to be the **only** entities capable of causing corrosion. Thus, corrosive potential of smoke was determined based simply on rankings of acid gas emission following combustion of the materials in a hot tube furnace, under an air flow. The water soluble effluents were captured and the solutions titrated for their acid gas content (HCl, HBr, HF), acidity or conductivity. In practice, decisions were often taken based purely on chemical composition: i.e. halogen content. It has since been found that smoke corrosivity can occur with halogen-free smoke, and that it can, under certain conditions, be larger than that due to halogens *(1-3)*. Post-exposure treatments, such as cleaning, can retard (and can even fully stop) the corrosion process and save the equipment *(2-4)*.

INITIAL FUNDAMENTAL RESEARCH WORK

The first fundamental project to research smoke corrosivity in recent years looked at the effects of fuel composition, exposure conditions and post-exposure on smoke corrosivity of metallic samples in a 500 L chamber *(1)*. It used a range of combustible materials, and exposure conditions designed to simulate some of the wide range of conditions potentially found in an area affected by an accidental fire. The corrosion targets were simple: copper (mirrors, 25 x 6 mm, 500 Å thick) and steel (coupons 100 x 100 mm, 12 gauge thick). The effectiveness of some simplistic anti-corrosion treatments was also evaluated. Table 1 lists the materials used.

The targets (5 steel and 2 copper for each burn) were exposed for 1 h in a 500 L chamber (the NBS Smoke Density Chamber) fully lined with Teflon^R sheeting, to minimize HCl decay on the walls. Steel coupons were laid on the chamber floor while the copper mirrors were placed in a plastic holder. In each experiment 15 g of combustible were burnt, at a temperature of 600 °C, and water was injected, as steam, into the center of the chamber. In half the experiments the chamber air temperature was near ambient, while in the other half the chamber was heated to ca. 100-110 °C,

Table 1. Materials Used in First NBS Smoke Chamber Work

DFIR	Douglas fir wood
NPR	commercial polychloroprene (Neoprene® W)
NYL	commercial nylon (Du Pont Zytel® HSL 103)
PE	polyethylene non-halogen fire retarded commercial W&C (Union Carbide DEQD-1388, black)
PS	commercial polystyrene (Dow Styron® 6069)
PVC LH	low halogen (experimental) flexible poly(vinyl chloride) W&C
PVC WR	standard (non-commercial) flexible poly(vinyl chloride) W&C
WOOL	unbleached unwoven wool fibres (sample used for smoke toxicity testing)
NONE	no combustible material

with a ceramic plate. Each steel coupon was treated differently, to simulate a particular fire scenario. After exposure, all targets were kept in a controlled environment chamber (25 °C and 75 % relative humidity) for 28 days. The day after exposure, two of the steel coupons were taken out of the chamber, post-treated and returned to it. At the end of the 28 days, the steel coupons were mechanically stripped of corrosion debris and weighed to determine metal loss gravimetrically. Copper mirror corrosion was determined based on the metallic copper remaining on the mirror after exposure.

All the materials tested caused metal loss, as did even the simple presence of warm humidity. The highest overall corrosion was found on steel coupons placed, in the warm chamber, on the hot plate (Figure 1). The amount of metal lost in the warm chamber was fairly constant, irrespective of the material burnt, with no correlation with the chemical composition of the fuels. In particular, metal loss resulting from chlorinated material smoke did not correlate with acid gas emission (Figure 2), nor did it, as a whole, appear to be more corrosive than the smoke from the group of halogen free materials. Moreover, the most corrosive smoke was due to nylon combustion. The single set of conditions under which HCl emission directly correlated with corrosion level was: steel coupons on ice (Figure 2), an unrealistic exposure, but which is often simulated during testing. Immediately after exposure, these particular steel coupons were covered with abundant water droplets, while no other coupons were similarly wet, indicating that the condensed moisture promoted condensation of HCl. The corrosion of steel coupons at room temperature did not correlate well with HCl emission (Figure 2). In general, in the warm environment, the most corrosive smoke appeared to be that of nylon. There is, not yet, a clear explanation for some of these unusual results.

Rankings of the corrosive tendency of the smoke of all the materials tested in this series depend on the way in which smoke corrosivity is analyzed, as reflected by the cross-overs in Figure 3. The Figure shows (a) the acidity of the smokes ([H$^+$], in mol/dm^3), as measured by the coil test (a fast and accurate test in which a small sample of material is combusted in a closed environment and the effluent is dissolved in water

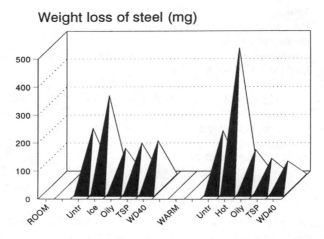

Figure 1: Effect of treatment on steel coupon corrosion. Legends: Untr, untreated; Ice, steel coupon on ice bag during exposure; Oily, treatment with machine oil pre-exposure; TSP, treatment with trisodium phosphate solution post-exposure; WD40, treatment with WD40 oil post-exposure; HOT: steel coupon lying on hot plate during exposure.

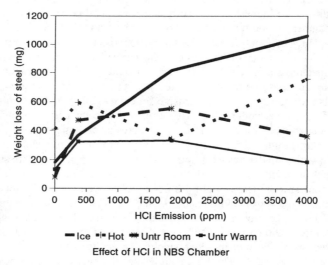

Figure 2: Effect of HCl concentration on steel coupon corrosion. Legends as in Figure 1.

(5), (b) the average corrosion of the steel coupons in all experiments, (c) the average corrosion of the steel coupons in the warm chamber experiments only and (d) the corrosion of the copper mirrors, as an average of the warm chamber experiments.

A follow-up study, in the same apparatus, used only copper targets: the same copper mirrors, and copper atmospheric corrosion probes, based on a Wheatstone bridge principle, with one arm of the bridge remaining unexposed to smoke but exposed to all thermal effects (commercially available Rohrback P610 TF50 C11000 targets, with a copper span of 2,500 Å). The additional current needed to reequilibrate the circuit was measured with a corrosometer (Rohrback CK3) and converted into equivalent metal loss (in Å). Copper mirror metal loss was assessed by resistance increase (at 1 h, end of exposure, and after 1, 3, 7, 14, 21 and 28 days) and by copper dissolution (as before). The targets were exposed in the warm chamber, with no treatments. The materials used in this study (Table 2) included 2 from the earlier work: NYL and PVC LH. Results in Figure 4 (comparing three measurement methods) show that (a) there is no simple correlation between the degree of acidity emitted and smoke corrosivity of copper and (b) there is a rough degree of correlation between the various targets, although they do not correlate exactly. The copper mirror resistance results at 28 days agree reasonably well with the copper mass loss data, as do the Rohrback probes at 1 day agree with the copper mirror resistance results at 1 day. Copper mirrors have severe reproducibility problems, partially because of uneven distribution of deposits on them and partially because of excessive sensitivity (at a 500 Å thickness), and cannot be used for standard testing methods. The Rohrback probes are much less sensitive, have a greater range before going off scale, can be used for continuous, or at least frequent, measurements, and thermal effects are minimized by the reference arm. Rohrback probes are also made for higher ranges, up to 45,000 Å, although the fabrication procedure is different.

Figure 5 shows a useful feature of keeping targets post-exposure: there are substantial crossovers between materials when smoke corrosivity is measured at different times after exposure, with copper targets. In Figure 6, the mass loss of copper is plotted in terms of the pH of the smoke (measured by the coil test *(5)*), for samples of very different chemical composition: containing Cl, Br, F, N, S and no heteroatom (other than oxygen). It is obvious that there is no direct correlation between pH of smoke and smoke corrosivity.

Table 2. Materials For Follow-up Work in NBS Smoke Chamber

ABS	Fire retarded acrylonitrile butadiene styrene (Borg Warner Cycolac® KJT)
F CB	Fluorinated cable jacket compound
HYP	Chlorosulphonated polyethylene compound (Hypalon®)
MEL	flexible polyurethane foam fire retarded with melamine (BASF Rest Easy®)
NYL	commercial nylon (Du Pont Zytel® HSL 103)
PU FM	ordinary commercial non fire retarded flexible polyurethane foam
PVC LH	low halogen (experimental) flexible poly(vinyl chloride) W&C

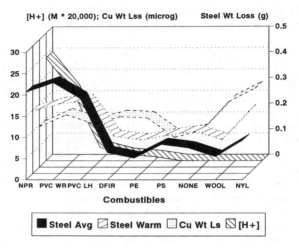

Figure 3: Effect of the combustible material on corrosion of metals; various reporting methods. Legends: Steel Avg: average corrosion of steel coupons in all experiments; Steel Warm: average corrosion of steel coupons in warm chamber experiments only; Cu Wt Ls: corrosion of copper mirrors, as an average of warm chamber experiments; [H+]: acidity of smokes in mol/dm^3 measured by the coil test.

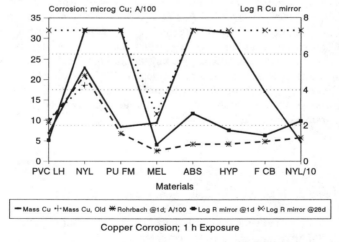

Figure 4: Corrosion of copper mirrors, after 1 h exposure. Legends: Mass Cu: mass of copper lost, at 28 days; Mass Cu, Old: mass of copper lost, at 28 days, in first series; Rohrback @ 1d; Å/100: resistance of Rohrback probe after 1 day post-exposure (X 0.01); Log R mirror @ 1d: log of the resistance of copper mirrors after 1 day post-exposure; Log R mirror @ 28d: log of the resistance of copper mirrors after 28 days post-exposure.

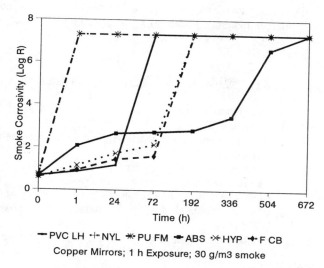

Figure 5:   Corrosion of copper mirrors, as resistance, after 1 h exposure in the NBS smoke chamber, and various post-exposure periods.

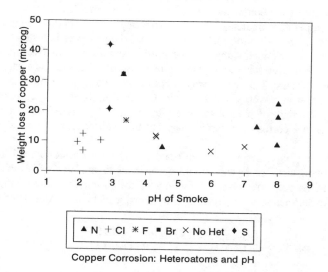

Figure 6:   Effect of heteroatoms in the fuel on the corrosion of copper mirrors, based on the pH of the smoke used.

LARGE SCALE TESTS

Large scale smoke corrosivity tests were conducted at the University of Edinburgh *(6)* using several combustible materials (Table 3) and copper mirror targets, in a 13.5 m^3 chamber.

Table 3. Materials Used for Large Scale Tests

ABS	Fire retarded acrylonitrile butadiene styrene (Borg Warner Cycolac® KJT)
NORYL	Polystyrene/polyphenylene oxide commercial compound (GE Noryl® N190)
NYL	commercial nylon (Du Pont Zytel® HSL 103)
PP FR	Fire retarded polypropylene compound
PVC CIM	Rigid commercial PVC custom injection moulding compound
PVC LF	PVC experimental flexible fire retarded and smoke suppressed W&C
PVC WR	standard (non-commercial) flexible poly(vinyl chloride) W&C

The most important results of the investigation were: (a) all materials tested caused corrosion (even those not containing halogens), (b) as the oxygen level increased in the proximity of the combustion zone, corrosion became less intense (Figure 7) and (c) as the temperature increased in the proximity of the exposure zone, corrosion became more intense (Figure 8). The effect of oxygen was investigated with three burning modes: unconfined samples (free burn), samples confined within an open fire box (confined burn) or samples within a fire box, with the opening restricted to a slit (closed burn). Figure 7 shows corrosion results at 3 days post-exposure. The effects of temperature were investigated on a single material, NYL: the corrosive potential increases by 8 orders of magnitude when the temperature is raised a total of 90 °C. This suggests that corrosive potential may change with environmental temperature.

Very few other series of large scale tests have been conducted to investigate corrosion. Any small scale test should correlate with the type of results found here. This issue is, as yet, undetermined for any of the proposed smoke corrosivity tests. One set of experiments, with personal computers, will be discussed later *(7)*.

FIRST MODERN APPROACH TO
SMOKE CORROSIVITY PERFORMANCE TESTING

Smoke corrosivity performance testing started with the French telecommunications industry test (CNET test, *(8)*), which first showed that smoke corrosivity could occur with materials containing no halogens. In the test the resistance of a copper printed circuit exposed to smoke is measured before and after exposure. The original version,

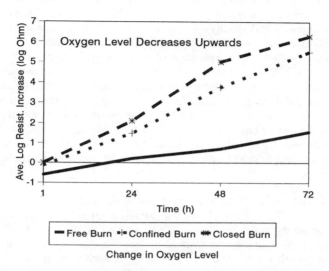

Figure 7: Overall average of corrosion results from changing oxygen availability in full scale burns at the University of Edinburgh, using copper mirrors, after three days post-exposure.

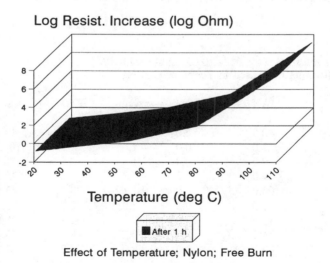

Figure 8: Effect of ambient temperature on corrosion of copper mirrors by nylon smoke, in full scale burns, using copper mirrors, after 1 h exposure, at free burn conditions.

proposes additional consideration of other fire properties: oxygen index (ASTM D2863) and ignitability (ASTM D1929). In the test a small material sample (600 mg) is mixed with polyethylene (100 mg) and placed within a 20 L chamber, conditioned to very high relative humidity and slightly elevated temperature. The target is a copper printed circuit board, at a total area of ca. 5.6 cm^2 and a copper thickness of 170,000 Å. Combustion results from rapid application of an intense flux, rapidly raising the sample temperature to ca. 800 °C. Water soluble combustion products are condensed onto the target, water-cooled to a temperature below that of the bulk of the chamber. The test apparatus is limited to testing of materials and is incapable of addressing corrosion by products. For chlorinated materials, CNET test results were found recently *(3)* to closely parallel those of two acid gas tests: hot tube furnace, as used by the Canadian Standards Association, *(9)* and the faster "coil" test *(5, 10)*. The CNET test, however, is much less sensitive to smoke corrosivity by other halogenated materials, particularly fluorinated ones.

The CNET test is on the way to becoming an international standard: ISO DIS 11907 part 2 *(11)*. Table 4 and Figure 9 show results of an interlaboratory evaluation carried out, under the auspices of ISO TC'61, on the CNET test. This illustrates that there are still some problems to overcome with the test. It is worth mentioning, however, that no other tests have, as yet, undergone such scrutiny.

The main overall problems with the CNET test are solvable: (a) the use of forced condensation, which means that the effects of non water-soluble products is lost, (b) the lack of post-exposure and (c) a very intense fire module. Some minor problems associated with the CNET test include the excessive time required to clean

Table 4. Results of CNET Round Robin 1991
(Results Expressed in Terms of COR%)

Material	MAX	MIN	Avg	STD	RSD
K	40.0	5.0	22.5	11.2	49.8
L	7.0	0.0	1.6	1.7	105.7
M	9.0	0.0	1.5	2.6	178.9
N	9.0	0.2	5.3	2.2	42.3
O	100.0	18.0	61.1	28.7	47.0
P	9.0	0.2	2.7	2.7	98.3
Q	12.9	3.0	7.1	2.7	38.9
K'	34.2	8.0	19.2	8.3	43.4
O'	100.0	33.0	69.0	28.2	40.8

Key: MAX: Maximum value reported by testing labs; MIN: Minimum value reported by testing labs; Avg: Average of all reported values; STD: standard deviation of all reported values; RSD: relative standard deviation, i.e. 100 times ratio of standard deviation to average. K-Q are the materials used; K' and O' are the same materials, under different conditions.

and condition (to the desired humidity) the chamber, the small chamber volume, and the type of sample chosen, which does not allow testing of finished products, or even mock-ups of finished products. The good fire performance of a tested material does not yield any advantages in the test itself, because of the forced combustion in the presence of polyethylene. Finally, test results can be difficult to interpret because the copper "lines" in the circuit board are so close that there often is a combination of resistance increase due to metal loss and resistance decrease due to bridging.

The most serious problem with the CNET test, however, was found when the results of tests on over 20 materials were compared with those using surrogate tests: no parallels with acid gas emission were found, *(12-13)*. The tests suggested that the only outliers were materials based on poly(vinyl chloride). On the other hand, materials containing bromine, and especially fluorine, seemed to be statistically almost indistinguishable from those containing no halogen.

THE NIST RADIANT SMOKE CORROSIVITY TEST (CLOSED SYSTEM)

As most cumulative fire properties, smoke corrosivity can be determined using closed systems (also called static) or open (flow-through) systems (also called dynamic). Some of the logic for using either one has recently been discussed *(3)*, and smoke corrosivity performance tests of both kinds are being considered for standardization.

In the United States, the ASTM E-5.21.70 task group developed a test method with a closed system, which is still under consideration. It has a plastic exposure chamber (ca. 200 L) used in the NBS cup furnace toxicity test and in the NIST radiant smoke toxicity test, associated with a radiant combustion chamber *(14)*. The sample (up to 50 x 50 mm, with up to 25 mm thickness) is subjected to radiant heat, produced by a set of quartz lamps, for 15 min. The lamps can vary the incident heat flux widely, but the recommended value is 50 kW/m^2. The test method contains all the smoke within the combustion/exposure chamber. However, it is not actually static, as the smoke circulates, impelled through a set of channels communicating the combustion and exposure chambers. There is a 1 h exposure in the test chamber, followed by a 24 h post-exposure, at room temperature and 75 % relative humidity. The apparatus is suitable to measure corrosive potential of products, if small representative samples can be placed in the holder.

The smoke corrosivity is measured with the copper circuit board targets described above, based on the principle of a Wheatstone bridge circuit, and manufactured by Rohrback Cosasco. The Rohrback targets used differ from the printed circuit boards used for the CNET test in that they have much larger gaps between copper runs, so that bridging is extremely unlikely. There is more experience with the more sensitive targets (2,500 Å span), but the less sensitive ones (45,000 Å span) show similar trends; results are, however, not interchangeable. In fact, the correlation between the two types of targets is unsatisfactory. The test apparatus can accommodate other targets, including full scale equipment, such as a small computer. This test appears to be promising, but no round robin series or other interlaboratory determination of reproducibility or reliability has yet been made.

TUBULAR COMBUSTION CHAMBER DYNAMIC CORROSION TEST

The tubular apparatus associated with the German smoke toxicity test (DIN 53436), either in a dynamic (flow-through) mode is being recommended for international standardization (ISO DIS 11907 - Part 3) *(15)*. This test is generally referred to as the DIN smoke corrosivity test. Some recent work has been published regarding this proposal, using steel coupons and CNET targets *(16)*. The sample is placed in a long combustion boat, and a 100 mm long tubular furnace, at a fixed temperature, is slowly (10 mm/min) slid over the sample. The smoke then exposes a variety of targets. There is, as yet, insufficient published information on the test procedure to be able to discuss its relevance to full-scale smoke corrosivity or its repeatability or reproducibility.

CONE CORROSIMETER

Most recently, work has focussed on the use of the cone calorimeter *(17)* for measuring smoke corrosivity. Researchers at Underwriters' Laboratories modified the sampling technique, in order to obtain high concentrations of corrosive species *(18)*. This concept has since been developed into the ASTM D5485 standard (also known as the cone corrosimeter), the first smoke corrosivity consensus standard fully approved, in 1994 *(19)*. The instrument, like the radiant apparatus described above, is now commercially available.

The cone corrosimeter can be built as an attachment to the cone calorimeter (ASTM E1354 *(20)*, ISO 5660 *(21)*) or as a separate unit. It consists of a conical heater (hence its name) which radiates heat input to a horizontal sample, on a load cell, an exhaust system and a corrosion sampling system. The corrosion sampling system takes an aliquot of the smoke from directly above the combustion zone and sends it to an 11 L polycarbonate exposure chamber. The samples are 0.1 x 0.1 m materials (as plaques) or mock-ups of components or products. Smoke passes through the exposure chamber for a period of time dependent on sample mass loss rate *(22)*. After this period, the exposure chamber is closed off for a total exposure time of 1 h, followed by 24 h post-exposure. The targets used are the Rohrback probes discussed above. No incident flux is specified, but fluxes of 25 and 50 kW/m^2 are recommended.

Figure 10 shows some results of this test, after 1 h and after 24 h, for a set of 19 materials. The Figure illustrates the type of behaviour encountered. Some materials may corrode the sample totally after the first hour, while other materials cause minimal

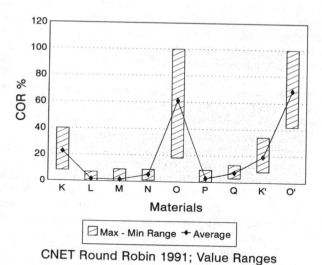

Figure 9: Results of preliminary ISO TC'61 1991 round robin evaluation of ISO DIS 11907-2. Max-Min Range: range of results; Average: average result.

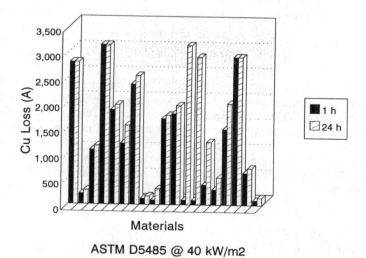

Figure 10: Results of smoke corrosivity tests using ASTM D5485, at 40 kW/m^2 incident flux. Exposure 1 h; post-exposure: 24 h.

corrosion even after 24 h. Of more interest are materials that fall between these two cases. The most noticeable fact is that, for some materials, particularly fluorinated ones, the amount of corrosion increases significantly between 1 h and 24 h, while the differences are small for others. No interlaboratory experience on the same materials exists as yet, but ASTM Committee D9 is planning a round-robin for 1995, and it is intended to become an international test, with the eventual designation ISO 11907-4.

A recent study, wherein actual personal computers were exposed to smoke from burning commercial cables, shows one of the problems with smoke corrosivity testing *(7)*. In the tests, personal computers were exposed, to smoke generated using the ASTM D5485 test as a source of smoke, in two ways: with a flow-through system (open) and enclosed so that all the smoke is trapped inside the computer. Simultaneously, Rohrback probes were also used for measurement, both inside the computer and inside the standard exposure chamber. The results are shown in Table 5, and they indicate that the bench-scale test seriously exaggerates the smoke corrosivity potential found in the real world, although it finds some of the same trends. Interestingly, the more sensitive probes (2,500 Å) are, as has been found elsewhere too, capable of determining the effect of halogen-free cables (ZERO HAL), while the least sensitive ones are not. A final trial conducted in this work was a qualitative check to determine whether the computer could still perform after exposure: in all cases where the "open" tests were conducted, the computer was still functional after the exposure. Only with the "closed" exposure were some cases found where the computer ceased to function after exposure. This work indicates that it is important to distinguish between the corrosive potential measured in small scale tests and real corrosion, which is much less.

SURROGATE TESTS

A number of surrogate tests have been proposed, which measure various properties, viz. halogen acid emission, overall acid gas emission (or pH) or conductivity, in aqueous solutions, instead of measuring smoke corrosivity directly. This has been described as measuring the "susceptibility of water to being corroded". The most common of these tests involve a hot tube furnace, with a small boat containing a specimen of the material to be tested, and undergoing a certain heating cycle. Such tests have been developed by the Canadian Standards Association *(8)*, IEC *(23-24)*, the German DIN organization *(25)* and the British Central Electricity Generating Board *(26)*, among others.

TEST COMPARISON STUDY CONDUCTED BY
THE SOCIETY OF THE PLASTICS INDUSTRY POLYOLEFINS COUNCIL

A comparative smoke corrosivity test study was conducted recently by the Society of the Plastics Industry Polyolefins Fire Performance Council (PFPC). It is a study of five tests with 20 materials (four additional materials were used for 4 of the tests) *(12, 13, 27, 28)*: three performance tests and two potential surrogates. The smoke corrosivity tests are the CNET test, NIST radiant test and cone corrosimeter, while the surrogate

Table 5. Exposure of Computers and Rohrback Targets to ASTM D5485-Generated Smoke

Cable Type	Material	Computer Exposure Open @ 1 h	Computer Exposure Closed @ 1 h	Computer Exposure Open @ 24 h	Computer Exposure Closed @ 24 h	ASTM D5485 Test Metal Loss @ 1 h	ASTM D5485 Test Metal Loss @ 24 h
		Target 45,000 Å	Target 45,000 Å	Target 45,000 Å	Target 45,000 Å	Target 45,000 Å	Target 45,000 Å
CMP	PVC/PVC	45	45	225	158	8911	12274
CMP	FEP/FEP	45	45	248	765	102	1463
CMR	PVC/PVC	90	450	338	585	15244	18833
CMP	ECTFE/ECTFE				45	4455	4568
IEC 332-1	ZERO HAL		68		293	60	158
CM	ZERO HAL				180	368	608
		Target 2,500 Å	Target 2,500 Å	Target 2,500 Å	Target 2,500 Å		
CMP	PVC/PVC	10	119	728	753		
CMP	FEP/FEP	175	650	1137	2500		
CMR	PVC/PVC	1	857	1059	1096		
CMP	ECTFE/ECTFE				508		
IEC 332-1	ZERO HAL		3		751		
CM	ZERO HAL		0		470		

Computer Function, after exposure, measured directly after test

Cable Type	Material	Open @ 1 h	Closed @ 1 h	Open @ 24 h	Closed @ 24 h
CMP	PVC/PVC	Yes	No	Yes	No
CMP	FEP/FEP	Yes	Yes	Yes	Yes
CMR	PVC/PVC	Yes	No	Yes	No
CMP	ECTFE/ECTFE	Yes	Yes	Yes	Yes
IEC 332-1	ZERO HAL	Yes	Yes	Yes	Yes
CM	ZERO HAL	Yes	Yes	Yes	Yes

tests measure acidity and conductivity of an aqueous solution, by the German protocol (25). The cone corrosimeter was run at an incident flux of 50 kW/m^2, using both Rohrback probes: 2,500 and 45,000 Å. Table 6 has a material number and a short description of the material tested and Table 7 shows the average results found on each test for all materials. Tables 8-9 show a standard deviation and a relative standard deviation (ratio of average and standard deviation) for all tests. Table 10 has a classification of materials proposed by the author, based on ranges of corrosivity. This classification is not intended to indicate that some level of corrosion is allowable, but simply intended to aid in the understanding of the data: no classifications have been issued which are known to correlate with corrosion in real fires. Table 11 shows how the materials tested in this series fare using this set of ranges.

Table 6. Materials in SPI Polyolefins Corrosion Study

Sample	Trade Name	#
XL Olefin elastomer + Me Hydrate	BP EXP839	1
HDPE + CPE elastomer blend	Dow 5435-30-11	2
CPE + fillers	Dow 5348-40-1	3
EVA polyolefin + ATH filler	Exxon EX-FR-100	4
Polyphenylene oxide/PS blend	GE Noryl® PX1766	5
Polyetherimide	GE Ultem® 1000	6
Polyetherimide/siloxane copolymer	GE Siltem® STM1500	7
Intumescent polypropylene	Himont EXP 127-32-6	8
Polyolefin copolymer + mineral filler	UCarb Unigard® RE DFDA 1735 NT	10
XL polyolefin copolymer + miner. filler	UCarb Unigard® RE HFDA 1393 BK	11
XL polyolefin copolymer + ATH filler	Quantum Petrothene® XL 7403	12
XL polyolefin copolymer + ATH filler	Quantum Petrothene® YR 19535	13
EVA polyolefin + mineral filler	Quantum Petrothene® YR 19543	14
Polyolefin + mineral filler	UCarb Ucarsil® FR-7920 NT	15
XL polyethylene copolymer + Cl addit.	BP Polycure® 798	16
Polyvinylidene fluoride	commercial PVDF	17
Polytetrafluoroethylene	commercial PTFE	18
Poly(vinyl chloride) material	commercial PVC	19
PVC building wire compound	commercial PVC compound	20
Polyethylene homopolymer	UCarb DGDK 3364 NT	21
Douglas fir	wood	22
EVA polyolefin copolymer	Quantum Ultrathene® UE631	23
Nylon 6,6	Nylet® P50	24
XL polyolefin copolymer + Br additive	UCarb Unigard® HP HFDA 6522 NT	25

Figure 11 shows the results of the surrogate tests and the CNET test, based on these ranges. All materials appear roughly similar in the CNET test, except for those based on PVC. The surrogate tests show high values for some other materials too, particularly materials 2, 3 and 17. The radiant test, Figure 12, and the cone

Table 7. Average Results of Corrosion tests by PFPC

PFPC No.	CNET COR %	Radiant test 60 min Å	24 h Å	Acid gas & Conductivity pH min -	Cond max Siemens	Cone corrosimeter 50 & 2,500 Å 60 min Å	24 h Å	50 & 45,000 Å 60 min Å	24 h Å
1	8.3	78	174	3.81	30.6	989	2406	240	600
2	13.7	2500	2500	2.21	3560.8	2675	2675	3608	4718
3	7.8	2500	2500	2.20	3378.7	2229	2675	1752	2148
4	2.6	125	307	3.91	13.1	105	1792	38	398
5	5.9	50	130	3.50	24.3	475	1639	180	480
6	5.4	20	46	3.69	212.1	517	790	180	203
7	5.0	40	123	3.56	55.6	37	1553	38	540
8	6.7	544	630	5.58	944.3	117	1672	60	540
10	5.1	716	941	3.60	16.0	85	1755	83	615
11	6.6	100	276	3.59	16.2	13	1993	23	615
12	5.5	51	64	4.10	15.0	257	331	165	75
13	3.2	134	186	3.70	16.8	286	1798	180	645
14	3.0	261	336	3.56	24.9	62	2052	15	495
15	4.5	86	398	3.73	13.5	9	2105	15	563
16	4.6	2500	2500	2.52	1435.8	2675	2675	1343	1965
17	9.6	2500	2500	1.99	4540.3	288	2675	53	2453
18	5.4	2500	2500	2.42	1952.5				
19	30.3	2500	2500	1.82	7207.7	2570	2675	3668	5205
20	47.8	2500	2500	1.75	10921.7				
21	7.4	67	146	3.58	35.2				
22	4.7	133	150	3.49	76.0	27	1913	75	555
23	6.8	61	82	3.27	114.4				
24	2.3	174	186	4.07	956.6	2075	2675	1343	1710
25	6.2	2268	2315	3.69	154.7	1257	1769	465	480
Average	8.7	934	1000	3.31	1488	731	1969	593	1151

Table 8. Standard Deviation of Results of Corrosion tests by PFPC

STD PFPC No.	CNET Test COR %	Radiant Test Å	Acid Gas & Conductivity		Cone Corrosimeter			
			pH min -	Cond max Siemens	50 & 2,500 Å		50 & 45,000 Å	
					60 min Å	24 h Å	60 min Å	24 h Å
1	0.5	105	0.14	0.0	677	380	193	120
2	6.1	96	0.10	140.9	0	0	743	727
3	4.5	35	0.05	31.8	631	0	1895	2006
4	1.1	551	0.17	0.5	104	281	28	59
5	2.6	114	0.03	1.9	325	779	97	106
6	2.5	73	0.09	3.1	324	584	120	150
7	2.3	189	0.06	1.8	30	150	28	115
8	2.0	196	1.05	30.2	32	192	42	115
10	2.7	895	0.14	2.8	23	223	59	46
11	2.7	164	0.33	3.1	5	43	18	142
12	1.7	56	0.08	3.1	40	50	46	56
13	1.7	269	0.16	1.8	217	427	150	101
14	2.3	452	0.06	6.2	69	476	21	18
15	1.5	443	0.06	2.3	6	409	21	80
16	1.8	85	0.09	187.7	0	0	355	449
17	2.3	53	0.09	323.4	319	0	46	361
18	1.7	72	0.05	84.0				
19	4.2	60	0.09	390.7	149	0	1059	1699
20	5.4	48	0.06	211.5				
21	2.7	349	0.09	3.9				
22	2.1	114	0.28	4.8	15	546	42	130
23	2.0	70	0.04	9.5				
24		179	0.33	93.2	848	0	1362	1177
25	2.4	74	0.09	35.6	105	98	153	148
Average	2.6	198	0.16	65.6	172	251	286	351

Table 9. Relative Standard Deviation of Results of Corrosion tests by PFPC

STD PFPC No.	CNET Test COR %	Radiant Test Å	Acid Gas & Conductivity		Cone Corrosimeter			
			pH min -	Cond max Siemens	50 & 2,500 Å 60 min Å	24 h Å	50 & 45,000 Å 60 min Å	24 h Å
1	5.8	51.1	3.7	0.0	68.4	15.8	80.4	19.9
2	44.2	3.8	4.7	4.0	0.0	0.0	20.6	15.4
3	57.3	1.4	2.2	0.9	28.3	0.0	108.2	93.4
4	43.2	87.9	4.4	3.7	99.0	15.7	75.0	14.9
5	44.5	70.3	1.0	7.6	68.3	47.5	54.0	22.1
6	46.0	110.4	2.4	1.5	62.7	73.9	66.7	74.2
7	46.8	102.7	1.7	3.3	81.0	9.7	75.0	21.2
8	29.3	29.6	18.8	3.2	27.4	11.5	70.7	21.2
10	53.8	87.5	4.0	17.7	27.1	12.7	71.5	7.5
11	41.0	27.9	9.2	19.3	35.0	2.1	81.1	23.2
12	31.5	73.7	2.0	20.5	15.7	15.0	27.8	74.8
13	53.6	122.6	4.3	10.8	75.7	23.7	83.4	15.6
14	77.9	108.8	1.8	24.8	112.1	23.2	141.4	3.7
15	33.7	66.9	1.7	17.0	62.1	19.4	141.4	14.2
16	39.7	3.4	3.4	13.1	0.0	0.0	26.5	22.8
17	24.3	2.1	4.3	7.1	110.6	0.0	88.2	14.7
18	31.2	2.9	2.0	4.3				
19	13.8	2.4	4.7	5.4	5.8	0.0	28.9	32.6
20	11.2	1.9	3.5	1.9				
21	36.2	120.0	2.4	11.0				
22	43.4	64.2	8.0	6.4	53.8	28.6	56.6	23.5
23	29.4	75.7	1.2	8.3				
24		90.2	8.0	9.7	40.9	0.0	101.4	68.8
25	39.4	3.1	2.5	23.0	8.4	5.5	32.9	30.9
Average	38.1	54.6	4.3	9.4	51.0	15.9	72.4	28.8

Table 10. Proposed Ranges of Corrosivity

Range	Rohrback Probes Å	CNET (%)	pH	Conductivity (Siemens)
8	> 2500	> 25	< 2	> 3,000
7	2000-2500	20-25	2.0 - 2.3	2,500-3,000
6	1700-2000	17-20	2.3 - 2.6	2,000-2,500
5	1400-1700	14-17	2.6 - 3.0	1,000-2,000
4	1000-1400	10-14	3.0 - 3.5	500-1,000
3	600-1000	6-10	3.5 - 4.0	300-500
2	200-600	2-6	4.0 - 5.0	100-300
1	< 200	< 2	> 5	< 100

Table 11. Ranges of Corrosivity of the Materials

STD	Cone corrosimeter				Radiant		CNET	pH	Cond.	Avg
	1 h 2,500 Å	24 h	1 h 45,000 Å	24 h	1 h 2,500 Å	24 h				
PFPC #	1	2	3	4	5	6	7	8	9	10
1	3	7	2	3	1	1	3	3	1	3
2	8	8	8	8	8	8	4	7	8	7
3	7	8	6	7	8	8	3	7	8	7
4	1	6	1	2	1	2	2	3	1	2
5	2	5	1	2	1	1	2	3	1	2
6	2	3	1	2	1	1	2	3	2	2
7	1	5	1	2	1	1	2	3	1	2
8	1	5	1	2	2	3	3	1	4	2
10	1	6	1	3	3	3	2	3	1	3
11	1	6	1	3	1	2	3	3	1	2
12	2	2	1	1	1	1	2	2	1	1
13	2	6	1	3	1	1	2	3	1	2
14	1	7	1	2	2	2	2	3	1	2
15	1	7	1	2	1	2	2	3	1	2
16	8	8	4	6	8	8	2	6	5	6
17	2	8	1	7	8	8	2	8	8	6
18					8	8	2	6	5	4
19	7	8	8	8	8	8	8	8	8	8
20					8	8	8	8	8	8
21					1	1	2	3	1	2
22	1	6	1	2	1	1	2	4	1	2
23					1	1	2	4	2	2
24	7	8	4	6	7	7	2	2	4	5
25	4	6	2	2	3	4	2	3	2	3
Avg	3	6	2	4	3	4	3	4	5	4

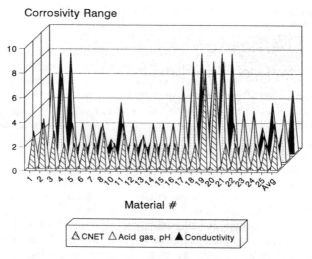

Figure 11: Corrosion ranges (Table 10) for the materials in the PFPC study, using the ISO 11907-2 test and surrogate tests (DIN 57472 (ph) and DIN 57472 (conductivity)).

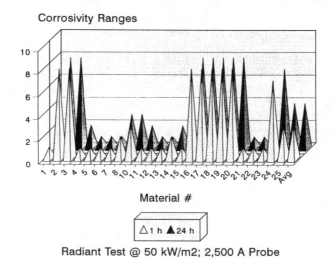

Figure 12: Corrosion ranges (Table 10) for the materials in the PFPC study, using the proposed ASTM E-5.21.70 radiant test, and with Rohrback (2,500 Å span) targets.

corrosimeter, Figure 13 show that the materials cover the whole set of smoke corrosivity potential ranges. Figure 14 shows a set of overall ranges of smoke corrosivity potential proposed, as the averages of the results of each test, with each test given equal weighting (which, although an inadequate way of handling the data, serves as a first approximation).

The first obvious conclusion of this analysis is that the materials generate different corrosivity results in the various tests. This conclusion is not based on individual material rankings, which are of little usefulness, but on the ranges proposed by the author. However, some materials always appear in the lower corrosion ranges and some others always appear in the higher corrosion ranges: this is frequently associated with the emission of acid gases, but not entirely. Other results were not predictable. For example, there is no correlation at all between pH and some of the corrosion results, since materials giving low pH values (high acidity), which would be expected always to generate high corrosion, generate low corrosion in some cases, and viceversa. The type of halogen evolved is clearly crucial, as is the detection system used: the only materials that look statistically different in the CNET test are based on PVC, while the results after 24 h (in the other tests) show them not to be that dissimilar from some other materials. It is encouraging that the overall average of all the ranges is exactly in the middle range, as are several of the tests. Interestingly, the cone corrosimeter at 1 h seems to give results which are in the lower ranges, many of which change considerably after 24 h. In fact, this surge in corrosion between 1 h and 24 h was found with several materials, particularly when they contain fluorine (Figures 12 and 13), as well as in the earlier series of experiments using the cone corrosimeter, as shown in Figure 8. The 45,000 Å probe gives results which cannot be compared with those using the 2,500 Å probe, and should be reserved for high corrosion.

The acid gas and conductivity tests have the lowest standard deviations, but such surrogate tests are unacceptable for other reasons, as discussed earlier. The relative standard deviations of the CNET test, the radiant test and the cone corrosimeter after 1 h are in the same range: 40-50. On the other hand, the cone corrosimeter after 24 h (with the 2,500 Å probes) gives a relative standard deviation of ca. 15, not much larger than that of the acid gas tests. This is an indication that this test shows great promise.

CONCLUSIONS

Many tests have now been proposed to determine the capability of materials to corrode various metallic surfaces after a fire. Real corrosion can be separated out into three zones, with two thresholds: (a) when corrosion is less than the lower threshold, the equipment can be recovered, with no interruption of service; (b) when the corrosion falls between the lower and upper thresholds, there is interruption of service, but the damaged equipment can be recovered; (c) if corrosion exceeds the upper threshold, the damaged products are unrecoverable. Unfortunately, no definitive values exists as yet for these thresholds, although some chloride deposit levels have been determined.

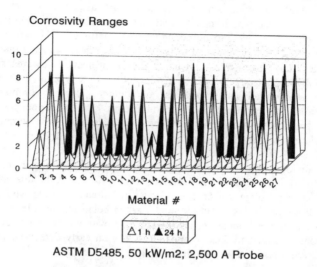

Figure 13: Corrosion ranges (Table 10) for the materials in the PFPC study, using ASTM D5485 at 50 kW/m² flux, and with Rohrback (2,500 Å span) targets.

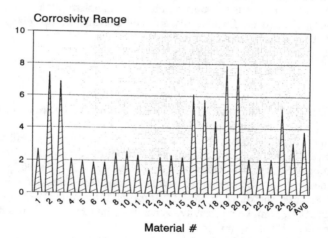

Figure 14: Average corrosion ranges (Table 10) for the materials in the PFPC study, using all tests, and assigning each test equal weight.

The interest in the issue of smoke corrosivity is extremely variable and is focussed in the electrical/electronic industry. Some manufacturers of materials have seen the issue as a potential marketing tool for advanced materials, as have some producers of the wires and cables into which the new materials are incorporated. On the other hand, users have been less inclined to be concerned. In particular, the computer industry, where equipment is already designed to resist corrosive environments (such as hot and humid climates) have been intent on using design protection features such as gold-plating, whereby copper targets lose their relevancy. This is particularly applicable in view of the work cited using personal computers *(7)*: the computer continued functioning even after exposures that generated over 1,000 Å copper loss. Thus, any classification based simply on results of the small scale tests must be considered with great care before being used for ranking actual materials or products.

Telephone operating companies have expressed some interest, in view of the potential for loss of service from a serious fire. However, in this industry many occupancies (such as telephone central offices) are already replete with cables of old vintage, which are rarely taken out ("mined") when being replaced by new, improved, cables. Thus, telephone operators are aware that material replacement represents a partial solution, at best, and these companies focus on early detection, careful design of occupancies and restoration and clean-up after the fire *(4)*.

The state of knowledge of smoke corrosivity achieved to date is sufficient to confirm that surrogate testing is unacceptable. Of the small scale tests proposed to date, the cone corrosimeter is the most promising, particularly after 24 h (exposure and post-exposure). However, final details are still unavailable, especially how well it would predict large scale smoke corrosivity test results.

LITERATURE CITED

1. Hirschler, M.M. and Smith, G.F., "Corrosive effects of smoke on metal surfaces", *Fire Safety J.*, **1989**, <u>15</u>, pp. 57-93.
2. Hirschler, M.M., "Update on smoke corrosivity", in Proc. 39th. Int. Wire & Cable Symp., US Army Communications-Electronics Command (CECOM), Fort Monmouth NJ, Ed. E.F. Godwin, Reno, NV, Nov. 13-15, 1990, p. 661-72.
3. Hirschler, M.M., "Discussion of Smoke Corrosivity Test Methods: Analysis of Existing Tests and of Their Results", *Fire and Materials*, **1993**, <u>17</u>, pp. 231-47.
4. Reagor, B.T., "Smoke corrosivity: generation, impact, detection and protection", *J. Fire Sci.*, **1992**, <u>10</u>, pp. 169-79.
5. Chandler, L.A., Hirschler, M.M., and Smith, G.F., "A heated tube furnace test for the emission of acid gas from PVC wire coating materials: effects of experimental procedures and mechanistic considerations", *Europ. Polymer J.*, **1987**, <u>23</u>, pp. 51-61.
6. Drysdale, D.D. and Macmillan, A.J.R., "The corrosivity of fire gases", *J. Fire Sci.*, **1992**, <u>10</u>, pp. 102-17.

7. Caudill, L.M., "The Corrosive Effect of Combustion Products on Personal Computers", in *19th. Int. Conf. Fire Safety*, Jan. 10-14, 1994, Hilado, C.J., Ed., Prod. Safety Corp., San Francisco, CA, 1994.
8. Fire Performance: Determination of the Corrosiveness of Effluents (158 CNET/LAB/SER/ENV). National Centre for Telecommunications Studies (CNET), France (1983).
9. Test to determine acid gas evolution, CSA Standard C22.2 No 0.3-M1985, February 1989, page 94, Canadian Standards Association, Rexdale, Ontario, Canada.
10. Smith, G.F., *J. Vinyl Technol.*, **1987**, $\underline{9}$(1) pp. 18.
11. ISO DIS 11907, Plastics - Burning Behaviour - Determination of the Corrosivity of Fire Effluents - Part 2: Static Test, International Standardization Organization, P.O. Box 56, CH-1211; Geneva 20, Switzerland, 1994.
12. Rogers, C.E., Bennett, J.G. and Kessel, S.L., "Corrosivity Test Methods for Polymeric Materials. Part 2 - CNET test method", *J. Fire Sci.*, **1994**, $\underline{12}$, pp. 134-54.
13. Bennett, J.G., Kessel, S.L. and Rogers, C.E., "Corrosivity Test Methods for Polymeric Materials. Part 3 - Modified DIN test method", *J. Fire Sci.*, **1994**, $\underline{12}$, pp. 155-74.
14. Hirschler, M.M., "Toxicity of the Smoke from PVC Materials: New Concepts", *Progress in Rubber & Plastics Technol.*, **1994**, pp. 154-69.
15. Barth, E., Müller. B., Prager, F.H. and Wittbecker, F.W., "Corrosive effects of smoke: decomposition with the DIN tube according to DIN 53436", *J. Fire Sci.*, **1992**, $\underline{10}$, pp. 432-54.
16. ISO DIS 11907, Plastics - Burning Behaviour - Determination of the Corrosivity of Fire Effluents - Part 3: Dynamic Test, International Standardization Organization, P.O. Box 56, CH-1211; Geneva 20, Switzerland, 1994.
17. Babrauskas, V., "Development of the Cone Calorimeter A Bench Scale Heat Release Rate Apparatus Based on Oxygen Consumption", *Fire and Materials*, **1984**, pp. 81-95.
18. Patton, J.S., "Fire and smoke corrosivity of metals," *J. Fire Sci.*, **1991**, $\underline{9}$, pp. 149-61.
19. ASTM D 5485, Standard Test Method for Determining the Corrosive Effect of Combustion Products Using the Cone Corrosimeter (ASTM D 5485), Annual Book of ASTM Standards, Volume 10.02, American Society for Testing and Materials, Philadelphia, PA, 19103.
20. ASTM E 1354, Standard Test Method for Heat and Visible Smoke Release Rates for Materials and Products Using an Oxygen Consumption Calorimeter (ASTM E 1354), Annual Book of ASTM Standards, Volume 04.07, American Society for Testing and Materials, Philadelphia, PA, 19103.
21. ISO 5660 - 1, Fire Tests - Reaction to Fire - Rate of Heat Release from Building Products (Cone Calorimeter Method), International Standardization Organization, P.O. Box 56, CH-1211; Geneva 20, Switzerland, 1993.
22. Hirschler, M.M., "Use of the Cone Calorimeter to Determine Smoke Corrosivity" in Fifth Ann. BCC Conf. Recent Advances in Flame Retardancy of Polymeric Materials, Stamford, CT, May 24-26 (1994).

23. International Electrotechnical Commission, Test on Gases Evolved During Combustion of Electric Cables, Part 1: Determination of the amount of halogen acid evolved during the combustion of polymeric materials taken from cables, IEC 754-1, Geneva, Switzerland, 1982.
24. International Electrotechnical Commission, Test on Gases Evolved During Combustion of Electric Cables, Part 2,, IEC 754-2, Geneva, Switzerland, 1982.
25. Prüfung an Kabeln und Isolierten Leitungen, Korrosivität von Brandgasen, Deutsche Elektrotechnische Kommission, Germany, DIN 57 472-Pt 813, 1983.
26. Central Electricity Generating Board, Leatherhead, UK, Test for Type Approval of Cables, E/TSS/EX5/8056, Pt. 3, 1984.
27. Kessel, S.L., Bennett J.G. and Rogers, C.E., "Corrosivity Test Methods for Polymeric Materials. Part 1 - Radiant Furnace Test Method", *J. Fire Sci.*, **1994**, 12, pp. 107-33.
28. Bennett,J. G., Kessel, S.L. and Rogers, C.E., "Corrosivity Test Methods for Polymeric Materials. Part 4 - Cone Corrosimeter Test Method", *J. Fire Sci.*, **1994**, 12, pp. 196-233.

RECEIVED January 23, 1995

Chapter 35

"Green Products," A Challenge to Flame-Retardant Plastics

Recycling, Marking, Ecolabeling, and Product Take-Back

Gordon L. Nelson

Florida Institute of Technology, 150 West University Boulevard, Melbourne, FL 32901–6988

Recycling will be an important factor in such high technology applications as automotive and electrical/electronic applications in the future. Recycling refers to the entire product, not just a few components such as plastics.
Elements of application design include:
- Ease of product disassembly
- Labeling for material identification
- Reuse and recycling of all components, including plastics
- Minimization of materials to maximize the recycle stream volume
- Compliance with "Eco-label" requirements as well as safety regulatory, and legal requirements.

In Europe there have been a number of challenges to the use of flame retardant plastics, particularly halogenated. Europeans also regard the use of flame retardant plastics in many electrical/electronic applications as unnecessary since electronics are designed to be "fail safe". Incineration and processing issues have raised questions about formation of brominated dioxan and furans from brominated flame retardants. Ecolabels such as TCO-95 have called for the absence of "bromided products." Material consolidation and reuse will change the economic equations. Indeed, The emergence of "green" products has added a new dimension to product design and is beginning to change the materials used in some high technology applications. The issue is more than simple material recyclability, but involves rethinking product life cycles to develop "environmentally conscious products."

Introduction

Materials are used in applications for a defined set of reasons, e.g., functionality, cost, regulation. For example plastics used in electrical/electric applications for enclosures must resist impact and heat in addition to being ignition resistant. The emergence of "green" products with the requirement for recycling has added a new dimension to product design. The issue is more than material recyclability, however simple or complex that may be, but involves rethinking the entire end product life cycle in an effort to develop products which in their totality have a lesser effect on human health and the environment than alternative products.

In the information and telecommunications products industries companies are much involved in life cycle analysis and in the development of environmentally conscious products (ECP). One company, IBM Corporation, for example, has released a policy letter (November, 1990) and a directive (January, 1991) and is continually tracking the progress being made by various IBM divisions in accomplishing ECP goals. In November, 1991, the Corporation established an Engineering Center for Environmentally Conscious Products at IBM, Research Triangle Park, N.C. The mission of the Engineering Center is "to provide guidance and leadership in the development and manufacturing of ECPs which are safe for their intended use, protective of the environment and that can be recycled, re-utilized, or disposed of safely." (1)

There is clear upsurge of interest in the U.S., Canada, Japan, and Europe in the environmental aspects of products, including eco-labeling of products. Internationally the International Standards Organization is developing ISO14000 on environmental management. This standard, which is expected to have an effect similar to ISO9000 on quality, will deal with both organization, evaluation and product evaluation; the first standards are expected in 1996. It is expected that documents on "Environmental Management System-Specification" and on "Guide to Environmental Management Principles, Systems and Supporting Techniques" will be approved as ISO Draft International Standards (DIS) in mid-1995 by ISO Technical Committee 207. Work in ISO TC 207 is divided into six areas:
- environmental management system
- environmental auditing
- environmental labeling
- environmental performance evaluation
- life cycle assessment, and
- environmental aspects of product standards

In the last area a "Guide for the Inclusion of Environmental Aspects in Product Standards" is also expected to be available as a DIS in 1995 (2-4). Guidance documents are being prepared on environmental labeling and on self-declaration of environmental claims. One view of the process to achieve ecologically sound manufacturing processes and products is shown in Figure 1. This is a process of continuous improvement.

In the United States President Clinton issued Executive Order 12873 on October 20, 1993, entitled Federal Acquisition, Recycling and Waste Prevention. This order encourages the Federal purchase of "environmentally preferable products." EPA

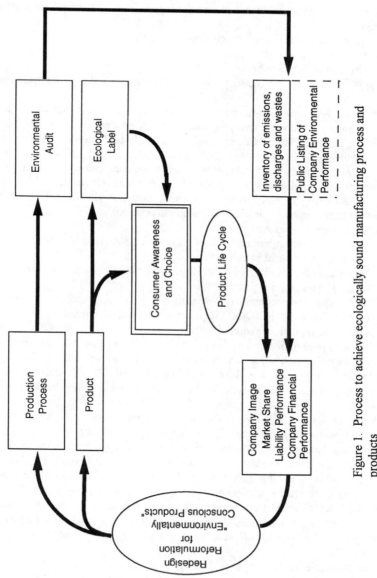

Figure 1. Process to achieve ecologically sound manufacturing process and products

is to issue guidance principles and procurement guidelines. Environmentally preferable products or services are those which "have a lesser or reduced effect on human health and the environment when compared with competing products or services that serve the same purpose." Product attributes among many as listed in the EPA concept paper include (5):
- Product contains higher recycled content
- Product is reusable or recyclable
- Safety (e.g. flammability)

Recycling is therefore in the broader context of environmentally preferable products (6,7).

Product life cycle assessment is of developing importance in such high technology applications as automotive and electrical/electronics. In the latter the International Electrotechnical Commission has a draft IEC Guide: Environmental Aspects: Inclusion in Electrotechnical Product Standards (8). Guidance includes design for disassembly and recyclability, and design for materials recyclability. Reuse, refurbishment, or reengineering are preferable to materials recycling. This is in the context of developing product take back requirements in Europe which may also substantiallly impact such applications as electrical /electronics.

Elements of materials recycling include:
- Ease of product disassembly - may require new materials
- Labeling for material identification
- Minimization of number of materials to maximize the recycle stream
- Ease of reuse or refurbishment
- Maintenance of properties on recycling
- Compliance with "Eco-label" requirements as well as applicable safety and regulatory requirements.

The need may be for fewer, more robust materials. One will choose the best compromise materials rather than a large number of tailored materials. There may be a larger number of applications in fewer materials. Indeed, GE Plastics for example announced in 1994 that it will be reducing the number of resin grades by 40% over the next three years.

How does all this affect the use of flame retardant flame plastics? Requirements for product take backs and for recycled content will cause Original Equipment Manufacturers (OEM's) to rethink product designs and the materials used. Are the materials used readily recyclable? By current technology? Are there environmental issues with the materials used? If the design is changed perhaps fire retardant materials are really not needed.

Indeed, recycling of products is an issue which is already here. For example, intense discussion in the electrical/electronics industry has occurred over the past three years. Design of products for disassembly, reuse and recycling is underway in many companies (6-7).

The remainder of this paper will focus on electrical/electronic products as the example. The reader should recognize that similar discussions are occurring in other industries.

Plastics Part Identification

The labeling of plastic parts has begun by many computer and automotive manufacturers using International Standard ISO11469 (9).

The chasing arrows symbol, which is mandated by law in 39 U.S. states, covers plastic bottles and rigid plastic containers (between 8 ounces and 5 gallons). This symbol has come under attack by some environmental groups on the basis that the implication to the consumer is that a product bearing the symbol is recyclable in an existing well defined collection route and recycling process. The Society of the Plastics Industry and the National Recycling Coalition had agreed to change the chasing arrows to a triangle and to use ISO acronyms for products not required to carry the chasing arrows. The ISO 11469 code was recommended for use for durable products. (10) For high density polyethylene the examples are as follows:

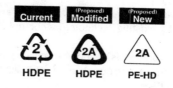

and the ISO 11469 mark

$$> PE - HD <$$

In May, 1994, this agreement broke down, with the National Recycling Coalition wanting a square or rectangle (11). On September 23, 1994, the SPI Board of Directors reaffirmed support for the "chasing arrows" identification code and endorsed efforts to encourage its proper use as an identifier for recyclers. (12-13).
With the breakdown of the SPI/NRC agreement the information processing industry is continuing its implementation of the ISO11469 marking system using ISO approved acronyms. It has been the recommendation of the Information Technology Industry Council (ITIC) Plastics Task Group that on the line below the ISO mark that a more complete resin designation be given either by trade name and number or a code (as is done in the case of IBM) (14). A 1992 survey noted that most companies are now using variants of this approach. Some within the flame retardant supplier community had recommended detailed notation of flame retardant type as part of part marking. Unless incorporated into the ISO standard, that is not likely to find acceptance as it is likely to be confusing and cumbersome. However "FR" for flame retardant has been accepted as an abbreviation in the ISO11469 Standard indicating that a material is flame retardant, e.g., >PC - FR< for flame retardant polycarbonate.

There was a copyright issue for use of the chasing arrows in Europe which was resolved in 1994. However, there is continuing discussion of the meaning of the symbol (15). Widening international acceptance of ISO11469 is leading to the general use of ISO11469 for plastics used in durable goods. Thus there is consensus for a molded-in identification of all but very small plastic parts in electrical/electronic applications.

Eco-labels

The second element of discussion is the advent of the ecological label. In September, 1991, the German environment agency launched an eco-label for copiers, "Blue Angel". By mid-1992 nearly every copier manufacturer in Germany had at least 1 copier with this new eco-label (16).

The plastics requirements for Blue Angel copiers include:
- Design for disassembly and recycling
- Use of a uniform plastic for large plastic housings and subassemblies, with the plastic recyclable by existing technologies for the manufacture of equivalent products
- No polybrominated biphenyls or polybrominated diphenyl ethers should be used and "flame retardant materials used must have no carcinogen or dioxin or furane-forming effect."
- Parts must be labeled using the ISO standard
- Plastics and flame retardants used must be specified.

"Blue Angel" published requirements for conditions for award of the environmental symbol for "environmental friendly personal computers "in late 1994 (17). The environment symbol is intended "to denote products in cases where the potentially long service life of the system and its components, a design which favors the recycling of parts, and possibilities for reusing and recycling used products or product components are all linked."

The requirements for plastics include specifications (17):
1) that large size-case parts and modules made of plastics consist of a uniform polymer (homopolymer or copolymer) to ensure reutilization on the basis of existing technologies for the production of high-quality and long lived industrial products. The plastic cases should consist of two separable polymers at the most.
2) that the plastics do not contain any cadmium-containing or lead-containing additives and that the plastics used are recyclable with the help of existing technologies. Excluded are small plastic parts (less than 25 g) which must be made of highly specialized plastics in order to achieve specific properties,
3) that no polybrominated biphenyls (PBB), polybrominated diphenyl ethers (PBDE) and short-chain chlorinated paraffins (chain length: 10-13 C-atoms, chlorine content $\leq$ 50%) are used for "the flame protection of the cases." The flame retardants used must not be carcinogenic according to Class III A or III B of the MAK-value-List ("carcinogenic working materials"), unless they are subject to the marking of carcinogenic substances and preparations according to section 5 of the Ordinance on Hazardous Substances, and must not form halogenated dioxins or furans, and

4) that plastic parts, except for ones weighing less than 25 g, are marked according to DIN (German Industrial Standard) 54840 or ISO 11469. Similar requirements are proposed for an environmental label for printers.

Draft requirements for environmental labeling of computers have also been issued in Sweden by TCO, the Swedish Confederation of Professional Employees in cooperation with the Swedish Society for Nature Conservation and NUTEK, the Swedish National Board for Industrial and Technical Development. Computer take back, recycling by a "professional computer recycler," and plastic component marking are required by "TCO-95". Plastic parts that weigh more than 25g "may not contain flame retardants that contain organically found chlorine or bromide." (18). This latter is a much broader specification than Blue Angel. The Dutch are also looking to develop eco-labels, through the "Stichting Milieukeur".

General features of eco-labels for equipment as they affect the use of plastics include design for disassembly, consolidation of resins, absence of contaminants, absence of certain heavy metals, absence of certain flame retardants, and labeling of components. In the United States, Green Seal has been formed and has discussed developing environmental standards on computer and computer peripherals, as well as copiers (19). Underwriters Laboratories is being used as the Green Seal test house. Green Seal already has standards or has standards under development for several products. A few products in other industries have been listed. Its draft standard on copiers simply mandates labeling of recyclable parts.

In Canada an "Environmental Choice Program" has been announced by the Ministry of the Environment, with a draft guideline for photocopiers issued. In this case, relative to materials, product information must state the parts or materials which have been recycled, reused, or remanufactured (20).

Eco-labels are proliferating, much to the displeasure of end product manufacturers. OEM's would much prefer an international eco-label under the auspicies of ISO. So far eco-labels leave a lot to be desired in terms of their technical basis. Yet if they became important in the market place then the technical quality behind the label becomes irrelevant. As shown above labels are already proposed or in place for some products in some countries. The importance of labels can be seen in the Underwriters Laboratories (UL) label for electrical/electronic products in the US. Its mark of evaluation for safety is widely expected despite limited legal requirement to do so. The use of the Underwriters Laboratories label has evolved in the US since 1898. One can expect eco-labeling to evolve as well. While the eco-label is on the final end-product the effect on the plastics used can be substantive as readily apparent from the above requirements.

Product Take Back

While many eco-labels include product "take back" requirements, more formal legal requirements have been proposed in Germany and the European Union for electronic products. Indeed electronic products have become a priority waste stream. Such requirements if and when implemented are nevertheless already providing a driving force for design for disassembly and for development of recycling schemes.

The proposed take back requirement in Germany is broadly based involving everything from small household appliances and office equipment, to large industrial items. After the date of the regulation the seller has the duty to take back used equipment parts from the end-user without charge. The manufacturer has the responsibility to reuse and recycle the returned equipment (21-22).

Published by the German Ministry for the Environment on July 11, 1991, the "Draft ordinance on the avoidance, reduction, and utilization of waste from used electrical and electronic appliances" (Electronic Scrap Ordinance) brought forth comments regarding practical collection and recycling schemes. A second draft was issued on October 15, 1992. In the Netherlands a similar proposal was developed.

Given that such requirements have broader implication the European Union accepted a proposal by Orgalime, a liaison group of European engineering industries, to develop a guideline for the waste stream "electronic waste". A working group under the auspices of the European Union and the Italian Government began meeting in 1993 (23). Finalization of legal take back requirements will likely await completion of these guidelines. Orgalime is scheduled to report back to the European Union in July, 1995. So far these initiatives will facilitate the recycling of electrical/electronic equipment and therefore support the recycling of flame retardant materials. In the Netherlands these recycle efforts may even be supported financially by raising a special fee on new electronic equipment or by the introduction of a return premium for old equipment.

The effect on plastics and on flame retardant plastics in particular is that take-back requirements will accelerate design for recycle, materials consolidation and the use of materials that do not have environmental issues of their own or complicate or are perceived to complicate the recycle stream.

Regulations Affecting Flame Retardants

Despite considerable regulatory activity particularly in Europe, over the past five years, as of this writing, there are no regulations in effect or actively under consideration which would directly ban the use of brominated flame retardants (other than PBB's) (24, 25).

In 1993, Germany proposed amending its existing Hazardous Substance Ordinance to include regulating brominated dibenzodioxins and furans in that country. The existing regulation specified maximum levels of chlorinated dibenzodioxins and furans that can be present in materials marketed in Germany. The amendment, known as the Dioxin Ordinance, added eight 2,3,7,8-substituted brominated dioxins and furans to the regulation. On July 15, 1994, the two Ordinances were combined and published in the official German journal as an amendment to the Chemicals Banning Ordinance. This regulation affects brominated flame retardants only indirectly and only those which contain greater than allowed limits of brominated dioxins or furans. DBDPO and TBBPA, the two largest volume brominated flame retardants, are reported to meet the requirements of the Ordinance. Other brominated flame retardants have also been tested and meet the requirements of the Ordinance. At least one flame retardant supplier is supplying compliance data on flame retardants in generic resins such as HIPS, PBT, ABS and PET (24).

In this ordinance a five year interim period is included where higher levels are permitted and components weighing less than 50 grams are excluded as follows (26):

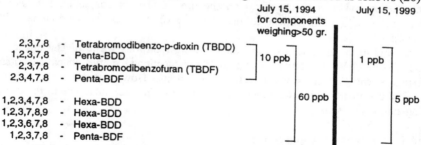

2,3,7,8	-	Tetrabromodibenzo-p-dioxin (TBDD)
1,2,3,7,8	-	Penta-BDD
2,3,7,8	-	Tetrabromodibenzofuran (TBDF)
2,3,4,7,8	-	Penta-BDF
1,2,3,4,7,8	-	Hexa-BDD
1,2,3,7,8,9	-	Hexa-BDD
1,2,3,6,7,8	-	Hexa-BDD
1,2,3,7,8	-	Penta-BDF

One published study on the formation of polybrominated dibenzodioxins and dibenzofurans (27) notes that when PBDE containing materials are processed the dioxin and furan levels exceed the final limits of 1 and 5 ppb for the various groupings as indicated in the "Dioxin Ordinance", but the levels are lower than the 10 and 60 ppb which are allowed during the 5 year interim period as described in the Ordinance, for some materials. This latter result is troublesome for equipment manufacturers. Workers at IBM have noted. (1): That the German Ordinances could :

1) "Hinder if not eliminate the recycling of plastics returning from the field at the end of life of the machine.... the machines already in the field have not been designed for the environment, i.e., the plastic parts have not been coded for material identification and some of the materials used may contain PBBs and/or PBDEs. Without the proper materials identification and lack of expedient and cost effective analytical techniques for determining the presence and levels of PBBs and PBDEs and the eight PBDDs and PBDFs, the recycler probably will have no choice but to landfill the materials as hazardous waste."

2) "Prevent the use of plastic resins fire retarded with brominated fire retardants. Without costly and time-consuming analytical testing of the finished part for levels of the regulated brominated dioxins and furans, it may not be possible for the users such as IBM to certify that their products meet the requirements of the German Ordinance, unless the German government states categorically that plastic resins free of PBBs and PBDEs are also considered free of PBDD and PBDF."

Currently the ordinance makes no requirement for testing, does not specify analytical methods to use, establish sampling protocols, or give a reporting mechanism for compliance (24). But the question is how is an equipment manufacturer to insure compliance?

Another indirect challenge is in government procurement (1). The Austrian government is requiring a declaration with each tender submitted for government procurement. Information is to be given if products or packaging materials offered include PVC, other halogenated plastics or halogenated hydrocarbons. For offered products the type, description, and expected life duration are to be given. The reasoning for use of PVC, halogenated plastics or halogenated hydrocarbons in the above products or their packaging material has to be documented in a separate attachment. The bidding manufacturer is to confirm by signature that statements are, to the best of knowledge, correct and confirms that all other offered equipment and packaging material, not specified, do not contain any substances as specified. The consequences of such requirements on the use of plastics containing halogens is significant should it be concluded that the use of halogenated material is being discouraged.

And in the European Union directive on hazardous waste of December 12, 1991, (91/689/EWC) antimony compounds are mentioned in the annex describing substances which classify a waste as being hazardous. How does this apply to flame retardant materials? While no direct regulations exist, indirect hindrance to the use of halogen is still present.

End-Product Safety Requirements

Plastics are extensively used for the fabrication of business machine parts. It is a common practice to use flame retarded plastics in compliance with safety standards such as International Electrotechnical Commission IEC-950, Underwriters Laboratories UL-1950 and Canadian Standards Association CSA 22.2 # 950-M89. These nearly identical standards define the minimum levels of flame retardancy required depending on the application of the part in a machine.

European electrical/electronics manufacturers have long believed that one can design fail-safe electronic components, and that therefore flame retardant enclosures are not required. IEC 950 on information processing equipment presently incorporates the UL approach to the use of flame retardant materials "to minimize the risk of ignition and the spread of flame both within the equipment and to the outside." Where required V-2 or HF-2 materials are specified as are V-1 or 5V fire enclosures. A detailed list of circumstances requiring a fire enclosure are specified. Often overlooked is the fact that secondary circuits supplied by a limited power source do not require a fire enclosure. It has always been permitted to exhaustively fault test all electrical components and thus avoid fire enclosures. To those exemptions are being added telecommunication circuits (to a maximum of 15VA). The addition of telecommunications, which traditionally has used only HB materials has invited a re-look at standards requirements. Indeed proposals have been received within IEC TC-74 which would potentially reduce further the circumstances requiring fire enclosures and rely more on fault tests. A few US manufacturers support this "freedom". Despite liability risks in this country there is some possibility that the approach being advanced by a Netherlands working group could further reduce requirements for flame retardant materials for information processing equipment. (28)

Plastics Recycling

The above considerations determine the context for reuse and recycling of flame retardant plastic parts, where flame retardant plastics are required, and where there may be issues. The information processing equipment segment is an ideal one for recycling of major plastic parts. There are only a few resins used: ABS, HIPS, polycarbonate, PVC, ABS/PC and PPE/HIPS, in only a few colors. It is a clean indoor application. The original product is expensive thus response to take back programs would likely be high. Design for disassembly would remove such issues as inserts and adhesives. Issues related to EMI and decorative paints are resolvable. Thus a clean resin stream is clearly possible.

Recycled flame retardant resins have been shown to largely retain properties of virgin resin from computer applications. Interesting case studies have been provided by IBM. In the case of companies like IBM plastics of course are only one item of concern. Many returned machines from the field are sent to IBM reutilization centers. The centers disassemble the machines and recover parts and subassemblies. The parts which are functional may be reused as "equivalent-to-new" or "certified service parts" or sold to brokers for resale as used parts. The identification of plastic parts for their chemistry is often difficult as the machines returning at the present time were not designed for the environment. However, plastics which can be identified by the generic and/or commercial grade are collected and where possible, recycled in a closed-loop system or sold to recyclers. One long term goal is closed-loop recycling. A number of plastic resins commonly used in IBM machine applications have been investigated for their mechanical, electrical, thermal and flammability performance upon their return from the field. The results have indicated that the extent of degradation of materials in the field, if any, is negligible. This important finding has enabled IBM to implement closed-loop plastic recycling activities for business machine applications. (1, 29). Indeed a number of papers have shown that flame retardant plastics can go through a number of molding cycles without serious property degradation (26,30,31). Of course, what we call recycling can encompass processes ranging from direct part reuse to energy recovery. (26)

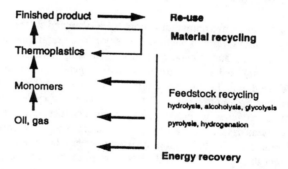

For materials recycling to be successful there must be proper collection of parts from end-of-life applications. A recycle stream must produce a quality feed stock, and there must be an appropriate application for the produced recycled material, with collection rate and application need in reasonable balance. That is a tall order. Significant research is underway by the American Plastics Council (APC) and plastics suppliers to develop practical recycle process lines. To date the main problem areas are associated with the level of contamination which may be present in the polymeric material and not degradation of the polymer during its initial application. Of course during reprocessing it is possible to restabilize or modify the recycle material to come to the required balance of properties. This applies to flame retardant materials, where in practice with polymers like polycarbonate, poly(phenylene oxide) and ABS it has been found that flame retardancy is little affected by recycling (26, 30, 31).

While part reuse and reengineering are not under the purview of plastics professionals, in the design of new equipment it is possible to select fewer more robust materials to simplify and increase the recycle stream, design for minimal contamination, and if needed to exclude materials which present environmental or other issues to a recycle stream including creating designs that don't require flame retardant plastics. Each of these have their own cost impact.

And technologies change. APC has shown interesting results that suggest that metal insert, fastener, or label free parts may not be essential in commercial recycle lines. Rapid instrumental part identification may also be possible, although part marking is still useful (APC reports to ITIC Plastics Task Group).

There are, of course, special questions for flame retardant resins, given acceptable reuse and recycle properties, i.e. incineration in inefficient incinerators (example - Netherlands) and collection in regions with insufficient recycle volume to permit single resin use. To the last point, in these areas (example - Scandinavia) mixed resin blends are being discussed. Flame retardant materials could significantly complicate those programs and may become a reason to design out of flame retardant resins (31).

Conclusion

A great deal of attention has been devoted by the fire retardant chemicals industry to environmental regulatory issues, particularly of brominated flame retardants in Europe. Those issues now appear to be resolved, with most brominated flame retardants remaining available.

At the same time environmental events have occurred in the user community which can also have a long term impact on the use of flame retardant materials. These involve take back, reuse and recycling, and eco-labeling. These applications may challenge certain additive flame retardant systems. Fewer different resins may be used, and those used will be broader in their property profiles. Limited volume tailored resins may be no longer used to increase volumes of recycle streams.

European approaches to electrical/electronic product safety may reduce the requirements for flame retardant materials. Added impetus may come from mixed resin recycle programs, again challenging the use of certain flame retardant materials.

While direct bans of PBDE's and PBB's are probably not going to occur, one sees reference to these flame retardants and brominated flame retardants in general in European eco-labeling programs, namely discouraging use of certain bromine compounds without in fact being regulation. The same is true for procurement requirements.

The issues discussed are not "over the horizon". OEM's are currently designing "environmentally preferable products", are preparing product take back programs, and are beginning recycle programs. Materials suppliers must always remember that an OEM has little choice but to design out materials or components if they become featured on customer negative lists. Such lists may have no basis in law or regulation. Such lists are not always rational or technically should. However, in the end the OEM's interest is to sell their product, not the material or component. While not always easy the key is to insure the technical soundness and rationality of such lists.

References

1. Kirby, Ray; Wadera, Inder L. Designing Business Machines for Environment with Flame Retarded Plastics, The Future of Fire Retarded Materials: Application and Regulations, 1994 Fall Conference Proceedings (Williamsburg, VA, October 9-12, 1994). The Fire Retardant Chemicals Association: Lancaster, PA, 1994; pp. 145-150.
2. Third Preliminary Draft of Environmental Management Systems- Specification, Working Paper SCI/WG1, ISO/TC 207/SC1/WG1 N62, 20 September 1994.
3. Guide to Environmental Management Principles, Systems, and Supporting Techniques, Committee Draft, ISO/TC207/SC1/WG2 N71, September 23, 1994.
4. Guide for the Inclusion of Environmental Aspects in Product Standards,. ISO/TC 207/WG1-N35, March 11, 1994.
5. Concept Paper for Development of Guidance for Determining "Environmentally Preferable "Products and Services, US EPA Office of Pollution Prevention and Toxins: Washington, D.C., January 28, 1994.
6. Berkman, Barbara N. "European Electronics Goes Green," *Electronics Business*, August 19, 1991, 50-53.
7. Cairnecross, Frances "How Europe's Companies Reposition to Recycle", *Harvard Business Review*, March - April, 1992, 34-45.
8. IEC Guide: Environmental Aspects - Inclusion in Electrotechnical Product Standards, 02 (Central Office) 594, International Electrotechnical Commission, November 11, 1994. (Draft)
9. Plastics - Generic Identification and Marking of Plastic Products, International Standard ISO 11469: 1993 (3)
10. White Paper #2: SPI Resin Identification Code, The National Recycling Coalition and the Society of the Plastics Industry, Inc: Washington, D.C., January 26, 1994.
11. Statement by Larry L. Thomas, President, the Society of the Plastics Industry, Inc., May 25, 1994.
12. *SPI Scope*, Vol. 2, No. 9, October, 1994.
13. Ford, Tom. "SPI Resolves To Keep Resin Code, Expand Use", *Plastics News*, October 3, 1994, 5.
14. Materials and Finishes, IBM Material Codes, Specifications Trade Names, Suppliers, Test Methods, Related Standards, IBM Materials Bulletin, M.B.5-0001-100, 1993-01, IBM Corporation: Armonk, N.Y., 1993; 414pp.
15. King, Roger. "Dutch Code Dispute Resolved, Work Continues on Global Code," *Plastics News*, November 14, 1994, 13.
16. Status Report on Environmental Achievement, Presented to IIIC Steering Committee in June 1992 in London by Facherband Informationstechnik (FVIT) within VDMA and ZVEI, International Information Industry Conference.
17. Basic Criteria for the Award of the Environmental Label, Environmentally Acceptable Workstation Computers, RAL-UZ 78, Berlin, July, 1994.
18. TCO'95 Certification Requirements for Environmental Labeling of Computers, The Swedish Confederation of Professional Employees TCO Development Unit: Stockholm, Sweden, Report No. 1, 1 December, 1994.

19. Summary, Meeting of Shareholders for Personal Computers and Printers, Green Seal: Washington, D.C., November 11, 1993.
20. Environmental Choice Program, Draft ECP 46-93, Environment Canada: Ottawa, Canada, January 14, 1995.
21. Draft Regulation on the Avoidance, Reduction and Utilization of Wastes from Used Electric and Electronic Equipment (Electronic Waste Regulation), dated ... 1991, The Federal Ministry for the Environment, Protection of the Environment Nature Conservation and Nuclear Safety, WA II 3-30 114-5.
22. Working Paper as of 15 October 1992, Regulation Regarding the Avoidance, Reduction and Recycling of the Waste/Used Electric and Electronic Equipment (Electronic Scrap Regulation), Federal Ministry for the Environment, Protection of the Environment and Reactor Safety, WA II 3-30 114/7.
23. Orgalime Guidelines for the Working Group "Priority Waste Stream on Electronic Waste", 24 January 1994.
24. Hardy, Marcia L. "Status of Regulations Affecting Brominated Flame Retardants in Europe and the United States," The Future of Fire Retarded Materials: Applications and Regulations, 1994 Fall Conference Proceedings (Williamsburg, VA, October 9 - 12, 1994) The Fire Retardant Chemicals Association: Lancaster, PA, 1994; pp123-128.
25. Risk Reduction Monograph No. 3: Selected Brominated Flame Retardants, OECD Environment Monograph Series No. 97, Environment Directorate, Organization for Economic Cooperation and Development: Paris, 1994; 150pp.
26. van Riel, Herman C.H.A. "Is There a Future in Flame Retardant Material Recycling: The European Perspective," The Future of Fire Retarded Materials: Applications and Regulations, 1994 Fall Conference Proceedings (Williamsburg, VA October 9-12 1994) The Fire Retardant Chemical Association: Lancaster, PA, 1994; pp167-174.
27. Meyer, H; Neupert, M; Pump, W. and Willenberg, B. "Flame Retardants and Recyclability", *Kunstoffe German Plastics* 83 (1993) 4, 253-257.
28. International Electrotechnical Commission, Technical Committee 74/WG8 (San Antonio/Ad Hoc 1 - Encl-2), May, 1995.
29. "Recycled Plastics for IBM computer Cover" *Appliance Manufacturer*, August, 1994, 43, 55.
30. Bopp, Richard C. "Recycling Post-Consumer FR Noryl Resins: A Case Study of Noryl Resin REN 814 in Roofing Applications," The Future of Fire Retarded Materials: Applications and Regulations, 1994 Fall Conference, Proceedings (Williamsburg, VA, October 9-12, 1994) The Fire Retardant Chemicals Association: Lancaster, PA, 1994.
31. Christy, Robert.; Richard Gavik," Recycling Feasibility Study: Ignition Resistant PC/ABS Blends," The Future of Fire Retarded Materials: Applications and Regulations, 1994 Fall Conference Proceedings (Williamsburg, VA, October 9 - 12, 1994) The Fire Retardant Chemicals Association: Lancaster, PA, 1994; pp151-166.
32. Carl Klason, "Recycling of Polymeric Materials," report, Chalmers University of Technology: Gothenburg Sweden, April, 1993.

RECEIVED March 6, 1995

Chapter 36

Tools Available To Predict Full-Scale Fire Performance of Furniture

Marcelo M. Hirschler

Safety Engineering Laboratories, 38 Oak Road, Rocky River, OH 44116

In recent years, it has been found that testing for the flaming ignition of upholstered furniture and mattresses is an essential means to help prevent serious fires. Tests, based on heat release rate, have been developed by California (CA TB 133), ASTM (ASTM E1537), UL (UL 1056) and NFPA (NFPA 266, draft). The equivalent tests for mattresses are CA TB 129, ASTM E1590, UL 1895. Some of them have now been written into the regulations of some states and into the Life Safety Code (NFPA 101). The reason for this emphasis is that heat release rate results can be used in realistic fire hazard assessment models. This has led to considerable work to try to predict the results, from small scale tests and from semi-empirical observations, without having to build furniture and test it with full instrumentation. The paper analyzes fundamental issues with the new test procedures, and some predictive techniques available, both theoretical and practical.

EARLY HISTORY OF FURNITURE TESTING

It was realized as far back as the mid 1970's that the fire performance of upholstered furniture is an important contribution to fire fatalities, at least in the United States *(1)*. Moreover, the single largest individual cause of fire fatalities has long been the ignition of upholstered furniture by smoking materials, specially cigarettes. This led, early on, to the development of two fire tests to determine the propensity of upholstered furniture or mattress components to ignition by cigarettes. The Upholstered Furniture Action Council (UFAC, which represents the manufacturers of residential upholstered furniture) adopted one of them, which became known as the UFAC test, as an industry-wide voluntary standard back in 1978. This fire test, on furniture components, together with another fire test, on furniture mock-ups, developed by the National Bureau of Standards (NBS) were eventually adopted as national standards by ASTM as ASTM E1352 *(2)* and ASTM E1353 *(3)* (the latter being the UFAC test), in 1989. As a consequence most new upholstered furniture sold in the United States is now unlikely to lead to flaming fires if the only ignition source is a cigarette. The UFAC test (or its ASTM version, ASTM E1353) requires testing of upholstered furniture components, including cover fabrics, interior fabrics, welt cords, fillings or paddings, decking materials or padding materials, to assess their ignitability by means of a cigarette. The materials pass the test (or are assessed to be Class A) as a function of the char length observed, and they are tested individually in conjunction with standard fabrics or foams.

By the mid to late 1980's, however, it had become clear to fire safety practitioners that protection against cigarette ignition, even if important, was insufficient. When upholstered furniture made with materials which did not ignite via cigarettes was exposed to flaming ignition sources, the result could be a very serious fire. The use of the UFAC test had succeeded in eliminating some of the worst actors, both as fabrics and as padding materials. However, some materials which did not ignite via smoldering sources, could ignite very easily when exposed to small flames. Moreover, it was found that a single upholstered chair could generate a fire large enough to engulf a whole room and take it to flashover.

This made it particularly important to develop a way of assessing the fire safety of upholstered furniture for use in high risk occupancies. The Boston Fire Department and the California Bureau of Home Furnishings (CBHF), independently, developed flaming ignition fire tests for full scale items of upholstered furniture, designed for high risk applications. The tests used by both regulatory bodies used sheets of newspaper as the ignition source, and had failure criteria based on temperature increases in a room. Moreover, both tests were designed with the concept of being simple "low-tech" tools for use by manufacturers, who could easily understand whether a chair burnt vigorously (and would likely fail the actual test criteria) or whether the fire never got out of control (and the chair would be likely to pass the test). In spite of these overall similarities, there were considerable differences between the tests. This caused concern, because manufacturers felt uncomfortable with the concept of potentially having to comply with two different tests and having to build furniture to two similar but not identical standards.

HEAT RELEASE AS A MEASURE OF FURNITURE FLAMMABILITY

By the 1980's, three further advances had taken place in fire science:

(1)   an empirical equation had been found showing that the heat release rate was proportional to the oxygen consumed *(4)*;

(2) very accurate analytical techniques had been developed, allowing measurement of oxygen concentrations with such precision and accuracy that oxygen concentration differences of 0.01 vol% could now be determined in fire tests *(5)* and

(3) it was understood that heat release rate is the most crucial property required for understanding fire safety, since the peak heat release rate was the numerical representation of the fire size, i.e. an answer to the question: "how big is the fire?" *(6)*.

As a consequence of a cooperative research program between CBHF and the National Institute of Standards and Technology (NIST, formerly the National Bureau of Standards), a square gas burner was developed *(7)* which, when used at a propane flow rate of 13 L/min for 80 s, had the same effect on ignition propensity as the original 5 sheets of newspaper used by CBHF, in California Technical Bulletin 133 (TB 133) *(8)*. Further research showed three additional important issues:

(1) the room temperature increase could be represented by a rate of heat release *(9)*;

(2) heat release rate measurements were more reliable than room temperature measurements and

(3) the rate of heat released was almost independent of room size (within certain limits: using the "ASTM" room (8 ft by 12 ft by 8 ft), the "California" room 12 ft by 10 ft by 8 ft) or the "furniture calorimeter") for low heat release rates (< 600 kW). It must be mentioned that the test is conducted under well-ventilated conditions.

CBHF amended its TB 133 test in order to incorporate the results of the joint study program. The newer versions of TB 133 recommended the use of the square gas burner developed in the CBHF/NIST program *(7)* (instead of the newspaper), allowed the use of different size rooms (or a furniture calorimeter) and allowed the use of heat release criteria instead of temperature increase criteria. By the fall of 1993, the only acceptance criteria were the ones based on heat release, specially a peak rate value of 80 kW, with the "old" criteria relegated to use for screening purposes. Underwriters Laboratories uses UL 1056, which has a wood crib as ignition source and measures heat release rate in a furniture calorimeter *(10)*.

HEAT RELEASE CONSENSUS CODES AND STANDARDS FOR FURNITURE

Both of the major organizations writing fire standards in the United States, namely ASTM and NFPA, set out to develop standards based on TB 133. ASTM wrote ASTM E1537 *(11)*, which was published in 1993, and NFPA is in the process of writing NFPA 266 *(12)*. There are some major differences between them: ASTM

E1537 explicitly allows three options (testing in: the ASTM room, the California room and the furniture calorimeter), while NFPA 266 requires testing in the furniture calorimeter, while stating that room testing may give similar results. Furthermore, the ASTM standard has no mention of any pass/fail criteria.

Once this was accomplished, both organizations picked up on a second, similar, development at CBHF: California Technical Bulletin 129, test for flaming ignition of mattresses in high risk applications *(13)*. This test used, from the start, a gas burner: in this case it was a T-shaped burner, for ease of application to the mattress. This test became ASTM E1590 *(14)*, published in 1994, and will become NFPA 267 *(15)*, to be published soon afterwards. The characteristics of these mattress standards are virtually identical to those of their corresponding "siblings" for upholstered furniture: ASTM E1537 and NFPA 266. Underwriters Laboratories has also developed a corresponding full scale test for mattresses (UL 1895) *(16)*.

The NFPA Life Safety Code, NFPA 101, incorporated the concept of heat release rate criteria for furniture in its 1991 edition. It recommended two cut-off points on peak heat release rate of a single item (whether an upholstered furniture item or a mattress): maximum heat release rate of 500 kW (which is waived if automatic sprinklers are present) or a maximum heat release rate of 250 kW (which is waived if either automatic sprinklers or smoke detectors are present). No test was specifically mentioned. In the 1994 edition of NFPA 101 *(17)*, the recommendation regarding the cut-off point at 500 kW and the waiver for smoke detectors were both eliminated, leaving the cut-off point for peak heat release rate of 250 kW (and 40 MJ of heat in the first 5 min of test), unless automatic sprinklers are present. The tests mentioned are ASTM E1537 (for upholstered furniture) and ASTM E1590 (or CA TB 129) (for mattresses). These requirements have only been adopted by NFPA 101 for a single type of occupancy: detection and correctional facilities. Separately, some states, including California, Illinois, Massachusetts and Minnesota, have adopted regulation based on this type of full scale heat release tests in different fashions.

Interestingly, the crucial importance of these tests is much broader than the limited application areas required by law: the fear of product liability has meant that most specifications for contract furniture in recent times call for furniture which passes the tests, even though that is beyond the legal minimum.

At ASTM, subcommittee E-5.15, on furnishings and contents, which developed ASTM E1537 and E1590, is also working on a parallel heat release test for stacking chairs, which has undergone a valid and successful ballot at subcommittee level in May 1994 and is expected to become a standard in 1995. Several studies have shown that a fire involving a stack of chairs can be significantly more severe than a fire involving a single chair of the same type *(18-22)*. In many public buildings, in facilities such as auditoriums, ballrooms or large meeting rooms, it is common to have hundreds of units of moveable furniture. In such facilities, which are often used for a multitude of purposes, stacking chairs are often present in a variety of configurations. When in use, the floor surface of a large ballroom, for example, may be covered by hundreds of stacking type chairs in an unstacked configuration. However, during cleaning, or when the facility is used for convention purposes, the chairs are usually stacked along the walls of the rooms, in passageways, or in storage areas. Often, the stacks may be 16 or more chairs high, in row after row. The potential fuel load presented by such an array of stacked chairs can be significant. Even lightly padded chairs, with less than 1 kg of combustibles each, may present a challenging fuel package when 5,000 or more chairs are stacked in close proximity. It is, therefore, of considerable importance to develop a test procedure that can determine the contribution to a fire of chairs in a stacked configuration. The potential for a serious fire from stacked chairs can also be affected by chair design. Many stacking chairs are designed to prevent the bottom of one chair from contacting the top of the chair below it when stacked. It is common for stacking chairs to be structurally designed with a gap of 12 to 100 mm (0.5 to 4 in.) between chair backs and adjacent seating surfaces, when in the stacked configuration. Under these conditions, and from a fire dynamics point of view, stacked stacking chairs create an ideal fuel array for promoting a rapidly developing fire, in a fashion not dissimilar to sticks placed in a fireplace in the form of a well-ventilated pile (or array). However, the potential fire risk resulting from a major stacking chair fire is quite small, due to its low likelihood *(1)*.

PREDICTION OF THE RESULTS OF TESTING FURNITURE ITEMS

Testing, in full scale, every item of upholstered furniture or mattress that is manufactured, is extremely expensive. The major expense results, actually, from manufacturing a full chair rather than from the test itself. This problem is enhanced by the proliferation of requests for "COM" fabrics. These are "customer owned materials", chosen by an architect or an interior designer, over which the furniture manufacturer has no control, and, often, even very little information. Thus, it is of major importance for such manufacturers (as well as for the manufacturers of fabrics or paddings) to be able to predict the heat release resulting from burning upholstered furniture or mattresses, without having to have every chair made up and sent to an official testing laboratory.

Backyard Test

The simplest way, albeit not necessarily the least expensive, of predicting full scale test results is by qualitative means. The use of a "backyard test" can be a useful screening method. In such a test, which is, in a way, a return to the original concept of a "low-tech" test for manufacturers, as opposed to real measurements useful for fire safety. The "backyard test" involves having an actual piece of furniture (or preferably a full scale mock-up) exposed to the same ignition source as that in ASTM E1537 (or even to the newspaper ignition source), in a relatively draft-free environment, together with visual observations of the results, but with no heat release measurements made. In this way it is possible to make reasonable predictions of the expected heat release rate results in the actual instrumented fire tests.

Figure 1 shows the results of "backyard" and fully instrumented tests on 30 chairs. In this case, the experienced observer suggested that the chairs be subdivided into two categories: (a) those that would give peak rates of heat release of under 40 kW (indicated in Figure 1 as 40 kW) and (b) those that would yield peak heat release rates of over 300 kW (indicated in Figure 1 as 300 kW). Fourteen of the systems tested were deemed to give off low heat release rates and 16 were deemed to give off high heat release rates. Of the 14 systems predicted to perform well, none actually exceeded peak values of 250 kW, and 4 exceeded 80 kW (the cut-off points of NFPA 101 and TB 133 respectively). Out of 16 systems tested and deemed to give off high heat release rate, none gave off peak values lower than 80 kW, and 3 gave off values between 100 and 250 kW. The experiments depicted in Figure 1 were all conducted with balanced, woven fabrics, of different types, but with the same foam and interliner barrier. This work indicates a definite usefulness of visual observations.

Fabric Weight

An attempt has been made to predict peak heat release rate values based on fabric weight. Figure 2 shows the results of two series of experiments. In the first case, shown at the left of the figure, a number of tests were carried out using identical chairs with the same specific fabric/interliner/foam/chair construction system. The only variable was the weight of the cover fabric, which was prepared by spinning. The tests indicate a clear trend with increase in fabric weight. However, the overlap between the error bars of the various fabric weights means that fabric weight is a very poor differentiator of resulting heat release rate.

In the second series of tests, a different type of fabric (Raschel knit) was used, and the results, shown on the right of Figure 2, were even more disappointing. For this series of tests, a 5-fold increase in fabric weight was not sufficient to significantly differentiate the fire performance of the chairs: all results were virtually undifferentiable. This is somewhat surprising, since other work has shown that in some systems the effect of the fabric is the dominant one on fire performance *(23)*.

Use of the Cone Calorimeter

It has been shown that the cone calorimeter, ASTM E1354 *(24)*, can be a very useful tool for predicting peak heat release rate data for ASTM E1537 *(25)*. Even better is the use of data from composites tested according to ASTM E1474 *(26)* (or NFPA 264A *(27)*), the cone calorimeter applications standard for upholstered furniture or mattress combinations. The initial work, done cooperatively between NIST and CBHF *(9)*, showed that the average (3 min) heat release rate was an excellent predictor of full scale peak heat release rate, when cone calorimeter tests are carried out at 35 kW/m^2 incident flux (Figure 3). In fact, the work showed that there was a cut-off at approximately 100 kW/m^2 average heat release rate, so that systems which generated values below that were likely not to develop self-propagating fires when they are made into actual furniture. Similarly, average heat release rate values above 200 kW/m^2 are likely to result in furniture that can cause self-propagating fires. Thus, three regions

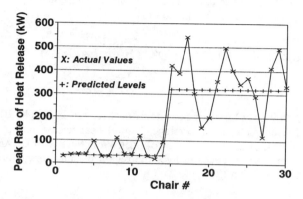

Figure 1. Comparison between the predicted peak rate of heat release levels, in kW, resulting from unistrumented full scale chair test burns with those obtained from using the ASTM E1537 standard test method.

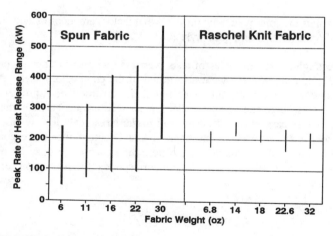

Figure 2. Effect of fabric weight (in ounces) on peak heat release rate (in kW) in the ASTM E1537 standard test method. Results are presented in terms of the range of results obtained, for two proprietary fabrics: a spun fabric and a Raschel-knit fabric.

can be identified: a non self-propagating region, up to an average bench-scale heat release rate of approximately 150 kW/m², a self-propagating region starting at an average bench-scale heat release rate of approximately 200 kW/m², and an intermediate region, where results are uncertain. The equation for the non self-propagating fires region is approximately shown as equation (1):

$$\dot{Q} \; (fullscale) = 0.75 \; x \; \dot{Q}'' \; (benchscale) \qquad (1)$$

where $\dot{Q}$ (full scale) is the peak rate of heat release in ASTM E1537 (in kW), and $\dot{Q}''$ (bench scale) is the average (3 min) heat release rate per unit area in ASTM E1474, at an incident flux of 35 kW/m² (in kW/m²). An approximate equation can also be developed for the region of self-propagating fires, and it is shown as equation (2):

$$\dot{Q} \; (fullscale) = 4 \; x \; \dot{Q}'' \; (benchscale) + 800 \qquad (2)$$

The results are improved even further when factors are incorporated representing the effects of total mass, frame and style *(28)*.

Interestingly, a recent series of tests was done wherein 9 chairs were tested in ASTM E1537, and the systems were also tested in the cone calorimeter. All systems were built using the same foam, interliner and chair construction, but different fabrics. As Figure 4 shows, there can be an excellent linear relationship between the peak (and not average) heat release rate in the cone and in the large scale test, with a regression correlation coefficient of 86%. The approximate equation that corresponds to this straight line is indicated in Equation (3) (where the bench-scale quantity measured is the peak rate of heat release, and which corresponds to the region of non self-propagating fires):

$$\dot{Q} \; (fullscale) = 0.85 \; x \; \dot{Q}'' \; (benchscale) - 2.7 \qquad (3)$$

Equation (3) cannot be compared to equation (1), since different magnitudes are being depicted. It is important to point out, however, that such correlations are heavily dependent on the systems tested. Figure 5 shows how, for three separate series of tests, the regressions found, although all linear, corresponded to different equations and had different (but similar) degrees of correlation, both based on bench-scale peak rate of heat release and both representing the non self-propagating fires region. Equations (4)

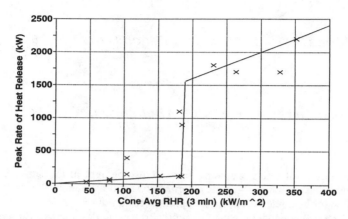

Figure 3. Comparison between the average rate of heat release (between ignition and ignition plus 3 minutes, in kW/m^2) test results obtained using the cone calorimeter applications (ASTM E1474) standard test method (at an incident heat flux of 35 kW/m^2, with horizontal samples) and the peak rate of heat release (in kW) obtained in the full scale standard test (ASTM E1537). Work conducted jointly by the National Institute of Standards and Technology (NIST) and the California Bureau of Home Furnishings and Thermal Insulation (CBHF).

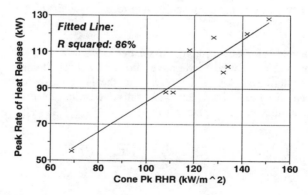

Figure 4. Comparison between the peak rate of heat release (in kW/m^2) test results obtained using the cone calorimeter applications (ASTM E1474) standard test method (at an incident heat flux of 35 kW/m^2, with horizontal samples) and the peak rate of heat release (in kW) obtained in the full scale standard test (ASTM E1537), for one series of material combinations.

and (5) present the best linear approximations for the other two series of experiments depicted in Figure 5:

$$\dot{Q} \ (fullscale) = 1.85 \times \dot{Q}'' \ (benchscale) - 280 \qquad (4)$$

$$\dot{Q} \ (fullscale) = 0.2 \times \dot{Q}'' \ (benchscale) - 14 \qquad (5)$$

The concept of self-propagating fire can also be applied to mattresses. In this case, the transition region in the cone calorimeter is still at roughly the same 3 minute average value for heat release rate as for upholstered furniture: 100-200 kW/m² average (3 min). The corresponding equation will also be similar to that for upholstered furniture *(29)*. Experience has shown that bedding (such as sheets and blankets) can substantially affect heat release from mattresses, particularly when the actual mattress has fairly poor fire performance. Thus, in general, tests with mattresses and bedding are of particular interest for systems with fairly high heat release rate values.

Relative Empirical Strategies

Two empirical strategies have been developed for predicting the fire performance of materials or systems, particularly with regard to their relative propensities to flashover.

The method of Ostman and Nussbaum *(30)* is a very simple empirical relationship for predicting time to flashover from room wall lining materials, in an international standard test, ISO 9705 *(31)*, with the ignition source set at 100 kW initially and then raised to 300 kW, and with walls and ceiling lined. It uses input data from the cone calorimeter and equation (6):

$$t_{fo} = k_a \times \frac{t_{ign} \sqrt{\rho}}{\sum \dot{Q}''_{pk}} + k_b \qquad (6)$$

where, t_{fo} is the predicted time to flashover in ISO 9705 (in s), t_{ign} is the time to ignition in the cone calorimeter at an incident flux of 25 kW/m² (in s), ρ is the density (in kg/m³), $\Sigma \dot{Q}''_{pk}$ is the total heat released per unit area during the peak period in the cone calorimeter at an incident heat flux of 50 kW/m², and

k_a and k_b are constants (2.76 x 10⁶ J (kg m)⁻⁰·⁵; -46.0 s, respectively). This method was applied with significant success to a full set of wall lining test data.

An even simpler method is a first order approximation for relative time to flashover in a room-corner scenario *(32)*, as shown in equation (7):

$$t_{fo} \propto FPI = \frac{t_{ign}}{Pk \, \dot{Q}''} \qquad (7)$$

where the t_{ign}, the time to ignition (in s), is measured, in the cone calorimeter, at an incident flux which is relevant to the scenario in question, Pk $\dot{Q}''$ is the peak heat release rate per unit area at that same incident flux (in kW/m²) and FPI is the fire performance index (in s m²/kW). If the material does not ignite, t_{ign} can be assigned a value of 10,000 s. This method is adequate simply as a relative indication of propensity to flashover, and cannot be used quantitatively.

It is interesting to note, however, that the method used on equation (7) has fairly broad applicability. Figure 6 shows its use for two very different sets of data, with very reasonable predictability. One of the sets of data, labelled FAA, refers to data obtained by burning aircraft panels in a fully furnished aircraft cabin. The other set, labelled EUREFIC, involves data from room-corner tests conducted using the ISO 9705 *(31)* test protocol, as indicated earlier. Both sets of materials were also tested in the cone calorimeter at the same incident flux, namely 50 kW/m². The linear regression correlation coefficients for both sets of data were in the range of 0.75-0.80.

Other strategies

Full scale tests on upholstered furniture or mattresses, such as ASTM E1537 or E1590, address the fire performance of the whole product. This means that it is not possible to develop any material that, when made into a product, is guaranteed to pass any of these tests. There are combinations of materials which may be antagonistic, so that the fire performance of the system is worse than that of the individual materials. Moreover, changes in construction can also result in rendering unfit a passing system or viceversa.

However, experience has shown that the use of certain combinations of materials is more likely to lead to a passing system. Thus, knowledge of the fire performance of materials and of the effect of furniture construction on fire performance is a helpful way of designing fire safe furniture.

The Ohio State University heat release rate calorimeter (OSU, ASTM E906 *(33)*) is an excellent test instrument for determining performance of composite systems *(23)*, in view of the good correlation of its results to those of the cone calorimeter.

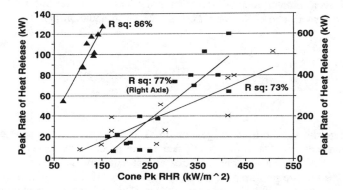

Figure 5. Comparison between the peak rate of heat release (in kW/m^2) test results obtained using the cone calorimeter applications (ASTM E1474) standard test method (at an incident heat flux of 35 kW/m^2, with horizontal samples) and the peak rate of heat release (in kW) obtained in the full scale standard test (ASTM E1537), for three series of material combinations.

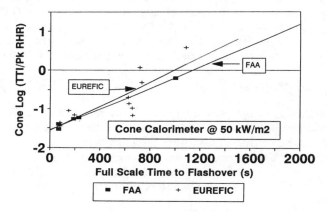

Figure 6. Comparison between the logarithm of the fire performance index (ratio of time to ignition, in s, TTI, and peak rate of heat release, in kW/m^2) test results obtained using the cone calorimeter (ASTM E1354) standard test method (at an incident heat flux of 50 kW/m^2, with horizontal samples) and the time to flashover (in s) obtained in the full scale standard test (ISO 9705, 100/300 kW, walls and ceiling), for two series of material combinations.

There is, usually, little advantage in using some of the more old-fashioned fire testing techniques. The only possible exception is when addressing a potential improvement in fire performance associated with the addition of fire retardants.

Effect of Vandalism

A common strategy for developing furniture which can pass full scale flaming ignition fire tests is the use of interliners (or barriers). However, it is often forgotten that such a strategy can be rendered useless by vandalism: if the barrier is penetrated and the padding material is exposed to the source of heat or flame, the interliner ceases being effective. This cannot be observed by carrying out standard tests, of course, but should be borne in mind, when fire safe furniture is constructed. The problem of vandalism is particularly frequent in three types of occupancies: correctional facilities, mental health institutions and urban transportation vehicles. Recently, a test method has been proposed to the ASTM committee on correctional facilities, F33, for standardization. It is a variant of ASTM E1474, cone calorimeter application to upholstered furniture or mattress composites, where the coverings (including any barriers) are vandalized in a "standard" fashion, in order to expose the padding material. The objective of this test is to observe the effect of exposure of the padding material on the fire properties of the composite.

CONCLUSIONS

In recent years, the fire performance of furniture, based on flaming ignition sources of full scale items, has become of great importance for fire safety. Fire tests, such as those developed by the California Bureau of Home Furnishings (TB 133 and TB 129), as well as their consensus standards counterparts, ASTM E1537 and ASTM E1590, are being demanded by specifiers even in occupancies where there is no legal requirements.

The fire performance of furniture depends on the materials used, the effects of their combinations and the construction of the final product. The use of any individual materials cannot ensure that the system will perform adequately. There are, however, a number of strategies that can be employed to understand whether the furniture item is likely to pass the test. They range from comparative fire tests on materials to qualitative tests of the furniture itself. The most effective means of predicting fire test results is the use of the cone calorimeter with combinations of materials, under the conditions described in ASTM E1474.

This paper presents information on a number of ways of predicting full scale furniture fire behavior. The use of expert observers, who have experience in constructing fire safe furniture can be extremely helpful when designing the furniture. However, it must be stressed that there are no guarantees in fire testing, and the safest way of constructing an adequate system is testing the actual product.

LITERATURE CITED

1. Hirschler, M.M., "Fire Tests and Interior Furnishings", in *Fire and Flammability of Furnishings and Contents of Buildings*, ASTM STP 1233, Fowell, A.J., Ed., Amer. Soc. Testing Materials, Philadelphia, PA, 1994; pp. 7-31.
2. ASTM E 1352-90, Test Method for Cigarette Ignition Resistance of Mock-Upholstered Furniture Assemblies, Annual Book of ASTM Standards, Volume 04.07, American Society for Testing and Materials, Philadelphia, PA, 19103.
3. ASTM E 1353-90, Test Methods for Cigarette Ignition Resistance of Components of Upholstered Furniture, Annual Book of ASTM Standards, Volume 04.07, American Society for Testing and Materials, Philadelphia, PA, 19103.
4. Huggett, C., "Estimation of Rate of Heat Release by Means of Oxygen Consumption Measurements", *Fire and Materials*, **1980**, $\underline{4}$, pp. 61-65.
5. *Heat Release in Fires*, Babrauskas, V. and Grayson, S.J. Eds, Elsevier, London, UK, 1992.
6. Babrauskas, V. and Peacock, R.D., "Heat release rate: The single most important variable in fire hazard", *Fire Safety J.*, **1992**, $\underline{18}$, pp. 255-72.
7. Ohlemiller, T.J. and Villa, K., "Furniture Flammability: an investigation of the California Technical Bulletin 133 test. Part II: Characterization of the Ignition Source and a Comparable Gas Burner", NISTIR 90-4348, Natl Inst. Stands Technology, Gaithersburg, MD, 1990.
8. CA TB 133, "Flammability Test Procedure for Seating Furniture for Use in Public Occupancies", State of California, Bureau of Home Furnishings and Thermal Insulation, 3485 Orange Grove Avenue, North Highlands, CA, 95660-5595, Technical Bulletin 133, January 1991.
9. Parker, W.J., Tu, K.-M., Nurbakhsh, S. and Damant, G.H., "Furniture Flammability: an investigation of the California Technical Bulletin 133 test. Part III: Full scale chair burns", NISTIR 90-4375, Natl Inst. Stands Technology, Gaithersburg, MD, 1990.
10. UL 1056, Standard for Fire Test of Upholstered Furniture, Underwriters Laboratories, Northbrook, IL (1989).
11. ASTM E 1537-93, Standard Method for Fire Testing of Real Scale Upholstered Furniture Items, Annual Book of ASTM Standards, Volume 04.07, American Society for Testing and Materials, Philadelphia, PA, 19103.
12. NFPA 266, Standard Method of Test for Fire Characteristics of Upholstered Furniture Exposed to Flaming Ignition Source (Draft), National Fire Protection Association, Quincy, MA, 1995.
13. CA TB 129, "Flammability Test Procedure for Mattresses for Use in Public Buildings", State of California, Bureau of Home Furnishings and Thermal Insulation, 3485 Orange Grove Avenue, North Highlands, CA, 95660-5595, Technical Bulletin 129, May 1992.
14. ASTM E 1590-94, Standard Method for Fire Testing of Real Scale Mattresses, Annual Book of ASTM Standards, Volume 04.07, American Society for Testing and Materials, Philadelphia, PA, 19103.

15. NFPA 267, Standard Method of Test for Fire Characteristics of Mattresses Exposed to Flaming Ignition Source (Draft), National Fire Protection Association, Quincy, MA, 1995.
16. UL 1895, Standard for Fire Test of Mattresses, Underwriters Laboratories, Northbrook, IL (1991).
17. NFPA 101, Life Safety Code 1994, Natl Fire Protection Assoc., Quincy, MA, 1994.
18. Woolley, W.D., Raftery, M.M., Ames, S.A. and Pitt, A.I., "Behaviour of stacking chair in fire tests", *Fire International*, **1979**, $\underline{6}$, No. 66, pp. 66-86.
19. Paul, K.T., "Demonstration of the effect of softening and fire resistance of materials on burning characteristics", *Fire and Materials*, **1980**, $\underline{4}$, pp. 83-86.
20. Babrauskas, V., "Will the second item ignite?", *Fire Safety Journal*, **1981/82**, $\underline{4}$, pp. 281-92.
21. Irjala, B.-L., "Rate of heat release of furniture", in *12th. Int. Conf. Fire Safety*, Jan. 12-16, 1987, Hilado, C.J., Ed., Prod. Safety Corp., San Francisco, CA, 1987, pp. 32-43.
22. Nurbakhsh, S. and Damant, G.H., "Development of a test method for the flammability of stacking chairs", in *19th. Int. Conf. Fire Safety*, Jan. 10-14, 1994, Hilado, C.J., Ed., Prod. Safety Corp., San Francisco, CA, 1994, pp. 32-58.
23. Hirschler, M.M. and Shakir, S., "Comparison of the fire performance of various upholstered furniture composite combinations (fabric/foam) in two rate of heat release calorimeters: cone and Ohio State University instruments", *J. Fire Sciences*, **1991**, $\underline{9}$, pp. 222-248.
24. Hirschler, M.M., "Heat release from plastic materials", Chapter 12 a, in *Heat Release in Fires*, Elsevier, London, UK, Babrauskas, V. and Grayson, S.J., Eds, 1992, pp. 375-422.
25. ASTM E 1354, Standard Test Method for Heat and Visible Smoke Release Rates for Materials and Products Using an Oxygen Consumption Calorimeter (ASTM E 1354), Annual Book of ASTM Standards, Volume 04.07, American Society for Testing and Materials, Philadelphia, PA, 19103.
26. ASTM E 1474, Standard Test Method for Determining the Heat Release Rate of Upholstered Furniture and Mattress Components or Composites Using a Bench Scale Oxygen Consumption Calorimeter, Annual Book of ASTM Standards, Volume 04.07, American Society for Testing and Materials, Philadelphia, PA, 19103.
27. NFPA 264A, Method of Test for Heat Release Rates for Upholstered Furniture Components of Composites and Mattresses Using an Oxygen Consumption Calorimeter, National Fire Protection Association, Quincy, MA, 1990.
28. Babrauskas, V., and Krasny, J.F., Fire Behavior of Upholstered Furniture (NBS Monograph 173). Natl. Bur. Standards, Gaithersburg, MD (1985).
29. Babrauskas, V., "Bench-scale predictions of mattress and upholstered chair fires - Similarities and differences", NISTIR 5152, NIST Internal Report, Natl Inst. Stands Technol., Gaithersburg, MD, March 1993.

30. Ostman, B.A.-L. and R.M. Nussbaum, "Correlation Between Small Scale Rate of Heat Release and Full Scale Room Flashover for Surface Linings", in *Fire Safety Science, Proc. 2nd. Int. Symp.*, Tokyo, Japan, 13-17 June, 1988, Wakamatsu, T., Hasemi, Y., Sekizawa, A., Seeger, P.G., Pagni, P.J. and Grant, C.E., Eds, Hemisphere, Washington, DC, 1989, pp. 823-32.
31. ISO 9705, Fire Tests - Full Scale Room Test for Surface Products, International Standardization Organization, P.O. Box 56, CH-1211; Geneva 20, Switzerland, 1992.
32. Hirschler, M.M., "Smoke and heat release and ignitability as measures of fire hazard from burning of carpet tiles", *Fire Safety J.*, **1992**, 18, pp. 305-24.
33. ASTM E 906, Standard Test Method for Heat and Visible Smoke Release Rates for Materials and Products (ASTM E 906), Annual Book of ASTM Standards, Volume 04.07, American Society for Testing and Materials, Philadelphia, PA, 19103.

RECEIVED January 4, 1995

Chapter 37

Combustion Behavior of Upholstered Furniture Tested in Europe
Overview of Activities and a Project Description

Björn Sundström

Sveriges Provnings-och Forskningsinstitut,
Swedish National Testing and Research Institute, Fire Technology,
Box 857, S-501, 15 Borås, Sweden

The European Commission has prepared a draft directive on the fire behaviour of upholstered furniture. Two essential requirements on fire properties are given. These are for ignition resistance and burning behaviour.

Four levels of ignition resistance are defined; ignition by cigarette, match flame, 20 g and 100 g of newspaper. CEN, the European Committee for Standardization, has almost finalized the standards for cigarette and match flame. Work is in progress to develop the needed test standards for the newspaper sources.

The second essential requirement addresses the problem when the furniture burns and people must be able to escape. However, required standards and evaluation techniques are not available. Therefore the Commission has funded the CBUF-programme to provide for technical solutions of testing and evaluation.

CBUF, Combustion Behaviour of Upholstered Furniture, started in 1993. The budget is approximately 2.8 million dollars. Fire experts from 8 European countries co-operate. The project manager is from DBI, Danish Institute of Fire Technology, Denmark, while SP, Swedish National Testing and Research Institute, is responsible for the technical co-ordination.

CBUF assumes that evaluation of the burning behaviour is carried out by the cone calorimeter (ISO 5660) for small scale testing and, as reference, the furniture calorimeter (NT FIRE 032) for large scale. Reliable test protocols are developed, hundreds of tests are performed and fire models that can predict the relevant hazard parameters are created.

The European Commission has prepared a draft directive on the fire behaviour of upholstered furniture[1]. The objective is to promote European harmonization of requirements and test methods that offer a high and consistent safety level. Surveys of European statistics show that a major cause of fire fatalities is associated with burning of upholstered furniture[2]. Ignition by smokers' materials are common events that start the fire. Once the furniture is burning, the growth rate of the fire may be such that people are not given sufficient time to escape. Consequently, the draft furniture directive includes requirements for ignition resistance of upholstered furniture, the so-called first essential requirement, as well as requirements for safe escape time from the burn room once the item is on fire, the so-called second essential requirement.

Ignition of Upholstered Furniture

The first essential requirement of the furniture directive deals solely with ignition resistance. Four levels are defined: Ignition by cigarette, by match flame, by 20 g of newspaper and by 100 g of newspaper. The cigarette and the match flame are intended for furniture in domestic use and the newspaper for public and high risk premises. CEN, the European Committee for Standardization, is producing European standards for this purpose, under the general designation of EN or "European norms".

CEN has so far met the task of producing standards for ignition resistance of upholstered furniture and mattresses against cigarette and match-like flame. The tests for upholstered furniture, EN 1021-1 and EN 1021-2 have already been accepted. A cigarette or a small gas flame is placed in a model sofa that is made of the test material. The occurrence of ignition is studied. The tests are very similar to the corresponding ISO standards with one important exception. The match flame test, 1021-2, is milder than the ISO version. The ignition flame exposure is only 15 s compared to 20 s in ISO. The tests for mattresses, EN 597-1 and EN 597-2 are about to be accepted. However, tests for simulation of newspaper sources have not advanced as far. In this case technical development work is required, so called prenormative research.

The Commission has sponsored a programme of work directed towards standards development. Eight laboratories are working together. The approach is to simulate the heat impact on upholstery when exposed to a fire from 20 g of newspaper and 100 g of newspaper, respectively. A square gas burner of a similar shape as the California TB 133 burner was chosen. However, the geometrical dimensions are smaller than the TB 133 burner; roughly half the size. At present no standardization work for the first essential requirement has started at CEN on these tests since the prenormative research is not yet finished.

SP is not involved in the work with the first essential requirement but we have nevertheless made some measurements of Heat Release Rate, HRR, from burning newspaper. Figure 1 shows data of HRR from different configurations of newspaper.

The paper was put on slabs of mineral wool in a furniture mock-up. The mineral wool had thermal properties similar to the padding in upholstered furniture.

The HRR data do not tell us which fraction of the heat was received by the furniture which would be a direct measure of ignition capacity. However, the geometrical configuration of the chair is very similar and therefore the size and duration (= HRR curve) of the source would give us a very good indication of the ignition capacity.

The different configurations gave different packing densities of the paper. Therefore the 100 g results give differences in peak HRR although the total energy release is the same as seen below in Table I. The 20 g newspaper consisted only of two sheets and therefore no difference occured. Paper size was 28 cm x 38.5 cm.

Table I. Summary Data on Experiments with Burning Newspaper

Source type	100 g paper pile	100 g paper ball	20 g paper pile	20 g paper ball
Maximum HRR (kW)	42	19	6.6	6.7
Total Heat Release (kJ)	1 688	1 720	301	356
Average HRR (kW)	14	12	2.4	2.7

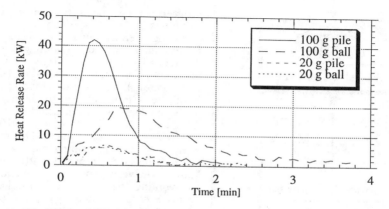

Figure 1 HRR measured on different configurations of newspaper. Sheets of paper were individually crumpled and laid up in a pile. Alternatively, a ball was formed where all sheets were crumpled together.

The TB 133 burner is designed to model the burning of 100 g of newspaper (when burning in a cage construction). The TB 133 burner, 20 kW for 80 s and THR 1600 kJ, is close to the 100 g paper data in the above experiments, although the HRR varies with paper configuration. This limited investigation (on HRR only) supports the TB 133 choice and the conclusions drawn by Ohlemiller and Villa[3]. They carried out a comprehensive investigation including testing a series of chairs and taking heat flux measurements on the samples, which are directly related to ignition propensity.

Combustion Behaviour of Upholstered Furniture-the CBUF programme

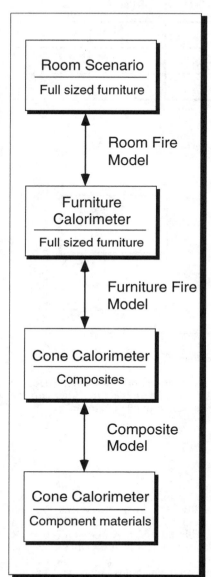

The second essential requirement in the draft directive states that *"the atmosphere in the room in which the upholstered furniture or related article are on fire should, despite the production of heat and smoke......., remain for a reasonable period of time after ignition such that it does not endanger the lives or physical well being of exposed persons. This will be achieved by controlling the rates of heat release, smoke and toxic gas production."*

A very large prenormative research programme has been funded by the Commission with the objective of providing scientific and technical support for the functional requirements given in the second essential requirement. This research programme is called CBUF; Combustion Behaviour of Upholstered Furniture. CBUF applies to domestic, public and high risk occupancies.

The CBUF-project started in 1993 and will last for two years. The total project budget is approximately 2.8 million dollars. Internationally well known and very experienced scientists are working together. Fire researchers, testing experts and industry experts from 11 organisations participate. They come from Sweden, UK, Italy, Finland, Denmark, Belgium, Germany and France. The project manager is from DBI, Denmark. The technical co-ordination is done by SP who also has the largest part in the project.

Relationships will be developed between (1) the conditions occurring in a full scale room scenario (ISO 9705 and larger rooms) due to the burning of a piece of furniture, and (2) the results of bench scale tests on composites consisting of various layers of fabric, interliner and padding used in the construction of the furniture. These relationships will be based on room fire models and furniture fire models insofar as possible. Rate of heat release, smoke and toxic gas production will be the main fire parameters considered.

The furniture calorimeter (NT FIRE 032)[4] and the cone calorimeter (ISO 5660)[5] are the tests used. Relationships will be developed between test data from the cone calorimeter on composites and the large scale burning behaviour on full sized furniture, as measured in the furniture calorimeter. An attempt will also be made to develop a model where data on component materials, i.e. fabric, interliner and padding, can be combined to predict the composite fire data as measured in the cone calorimeter.

Functional requirements will be examined by conducting room fire tests on furniture. Measured room fire conditions will be interpreted by doing fire hazard analysis and identifying tenability limits and associated times. These data will then be compared with predictions based on (1) furniture calorimeter data and existing room fire models, and (2) cone calorimeter furniture fire models and correlations.

Reliable test protocols have been developed and verified for the cone calorimeter and for large scale testing as well as procedures for toxic gas measurements. Hundreds of tests in small and large scale tests have been performed.

To determine time to tenability limits some hazard analysis needs to be carried out on the room scenarios which have been decided. Therefore room fire experiments are required. Then calculations are done with a room model, CFAST[6], on various geometries and conditions. In that way the choice of tenability limits and time can be validated.

The Furniture Calorimeter. Of particular importance are the test methods used in the CBUF-programme. They must be repeatable, reproducible, robust and yield relevant data for prediction of the burning behaviour. Let us examine some of their features.

The furniture calorimeter NT FIRE 032, see Figure 2, as defined by NORDTEST was first published in 1987. It was backed up with testing work on real furniture[7]. NT FIRE 032 was revised in 1991.

The combustion gases from the product are extracted into a 3 m x 3 m hood. In the exhaust duct, measurements of gas species concentration, gas flow rate and smoke optical density are performed. HRR is calculated according to the oxygen consumption technique.

A new technique for measuring a large variety of toxic gas species has been introduced, the so called FTIR (Fourier Transform InfraRed). This instrument makes infrared measurements simultaneously over the whole spectrum. Of special interest for furniture are for example HCN and NO_x. Quite recently, NORDTEST published a standard, NT FIRE 047[8] for FTIR measurements in fire testing. It was preceded by work done at VTT in 1990[9].

The FTIR technique is very promising. The concentration of a gas is measured continuously during a fire test. When combined with the mass burning rate, yield data can be determined. Older techniques often gave only the time average and are time-consuming and not very suitable for routine work. However, the FTIR used in fire testing is still a very new technique and therefore much care needs to be exercised when using it.

Smoke optical density is measured with a white light system having a detector with the same spectral responsivity as the human eye. The mass loss rate from the sample is measured simultaneously, therefore, yield data of smoke can also be determined.

The Furniture Calorimeter is reproducible. Prior to the CBUF-programme commencement a round robin was performed on NT FIRE 032. Six European laboratories participated. Good results were obtained for chair configurations, see Figure 3.

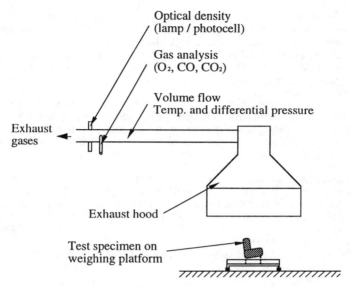

Figure 2 The furniture calorimeter. A full size, piece of furniture is placed on a scale under a hood. The test specimen burns without any restriction of air supply. Rates of heat release, smoke production, gas species production and mass loss are measured continuously.

It is clear that the results are very similar for the different laboratories. The best results were obtained with chairs, probably because the fire build-up does not rely on flame spread as for example with a three seat sofa.

The cone calorimeter test protocol. The cone calorimeter, see Figure 4, was chosen for small scale testing in the CBUF programme.

Both the composite product combinations and the individual components, fabric and padding, separately, are tested. All of the fire parameters that are measured in the furniture calorimeter are also measured in the cone calorimeter under controlled and defined conditions. Therefore it is assumed that cone calorimeter data can be used for predictions of the full-sized furniture items. A main problem is the prediction of the effects of furniture design. Therefore, a special test series of custom made chairs of similar materials but with different designs is included in CBUF to provide the necessary experimental data.

Much time has been devoted towards the development of a procedure for sample preparation which could give reliable test data. In addition a small video was produced that shows this procedure. The result is a very detailed description that has been verified in a comprehensive round robin. This work was done outside of CBUF by SP and is currently under publication[10]. An example of the good agreement between laboratories is given in Figure 5.

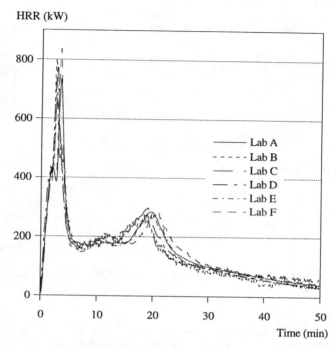

Figure 3   Round robin data from testing of a chair in the furniture calorimeter.

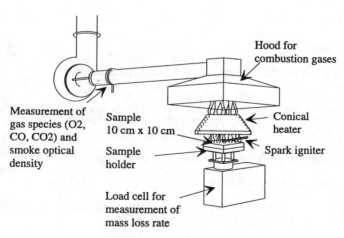

Figure 4   The cone calorimeter. A special procedure for sample preparation of furniture materials has been developed and verified.

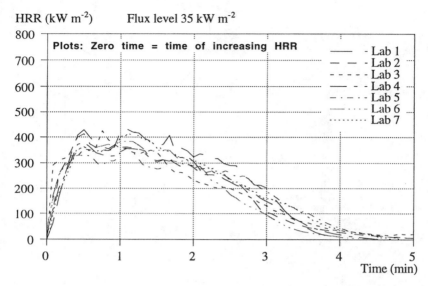

Figure 5 Round robin results from the cone calorimeter on a furniture composite.

CBUF will soon be finalized. The CBUF project is continuing and once the prenormative research for the furniture directive is finalized, which will be in 1995, there will be a massive amount of new data and knowledge available to be used by fire engineers and regulators for the benefit of fire safety and harmonization of requirements.

References

[1] Gravigny, L., "Draft European Directive on the Fire Behaviour of Upholstered Furniture", Flame Retardants '92., pp 238-242, Conference. Elsevier Science Publishers Ltd. ISBN 1-85166-758-x.

[2] de Boer, J.A., "Fire and Furnishing in Building and Transport. Statistical data on the Existing Situation in Europe.", Fire and Furnishing in Buildings and Transport Conference proceedings, pp 1-14, 6-8 November 1990, Luxembourg.

[3] Ohlemiller, T.J., Villa K., "Furniture Flammability: an Investigation of the California Bulletin 133 Test. Part II: Characterization of the Ignition Source and Comparable Gas Burner", NISTIR 4348, June 1990, USA.

[4] "Upholstered Furniture: Burning Behaviour - Full Scale Test", NORDTEST method NT FIRE 032, Helsinki, 1991.

5. International Standard-Fire Technology-Reaction to Fire-Part 1:Rate of Heat Release from Building Products (Cone Calorimeter method). ISO 5660-1:1993(E). International Organisation for Standardisation, Geneva, 1993.

6. Peacock, R.D.,Forney,G.P., Reneke, P., Portier, R., Jones, W.W., "CFAST, The Consolidated Model of Fire Growth and Smoke Transport", National Institute of Standards and Technology Technical Note, NIST TN 1299, 1993.

7. Sundström, B., "Full-Scale Fire Testing of Upholstered Furniture and the use of Test Data", Reprinted from "New Technology to Reduce Fire Losses and Costs", Elsevier, SP Technical Report SP-RAPP 1986:47.

8. "Combustible products: Smoke Gas Concentrations, Continuous FTIR Analysis", NORDTEST Method NT FIRE 047, Helsinki, 1993.

9. Kallonen. R., "Smoke Gas Analysis by FTIR Method. Preliminary Investigation", Journal of FIRE SCIENCES, VOL. 8-September/October 1990.

10. Babrauskas, V., Wetterlund, I., "Fire Testing of Furniture in the Cone Calorimeter - The CBUF Test Protocol", SP REPORT 1994:32.

RECEIVED March 23, 1995

Chapter 38

Fire-Safe Aircraft Cabin Materials

Richard E. Lyon

Fire Research Branch, Federal Aviation Administration Technical Center, Federal Aviation Administration, Atlantic City International Airport, Atlantic City, NJ 08405

Transient isothermal pyrolysis kinetics were used to relate polymer thermal stability to the fire-resistance of aircraft cabin materials in a post-crash fuel fire. The observed scaling of ignitability, time to ignition, and heat release rate with net surface heat flux was well predicted for polymeric solids using this simple approach.

Aircraft Cabin Materials

Fire Research is a long-range program within the Federal Aviation Administration (FAA) which includes advanced materials in a systems approach to improved aircraft cabin fire safety along with fire prevention, detection, and control. The objective of the Advanced Fire-Safe Materials portion of the Fire Research program is to discover the fundamental relationships between the composition and structure of lightweight aircraft interior materials and their behavior in fires to enable the design and manufacture of a totally fire-resistant passenger cabin for future commercial aircraft. This paper develops the technical requirements for fire-safe aircraft interior materials and discusses the importance of some material parameters related to fire performance.

The aircraft interior is the area within the pressure hull that includes the passenger compartment, cockpit, cargo compartments, and the various accessory spaces between the passenger compartment and pressure hull. Table I lists combustible cabin materials and their weight range in commercial passenger aircraft cabins (1). Multiplying the total weight of combustible textiles, foams, films, plastic parts, and, composites in the aircraft cabin by a typical heat of combustion of 25 MJ/kg for current cabin materials it is found that the fuel load from combustible materials in an aircraft cabin fire is on the order of 10^{11} Joules. While cabin materials represent only a few percent of the heat content of the aviation gasoline in the wing tanks, their involvement in post-crash aircraft fires is believed to be the most significant cause of death in impact-survivable accidents.

Thermoset composites form about eighty to ninety percent of the interior furnishings in today's commercial aircraft. Typically these composites are sandwich panels made of fiberglass-reinforced phenolic resin skins on NOMEX honeycomb core which are surfaced with an adhesively-bonded poly(vinyl fluoride) decorative film or painted to provide color, texture, and cleanability. These honeycomb decorative laminates are used as ceiling panels, interior wall panels, partitions, galley structures, large cabinet walls, structural flooring, and in the construction of overhead stowage bins. Until 1986 these

This chapter not subject to U.S. copyright
Published 1995 American Chemical Society

large-area component materials were only required to pass a vertical 60 second Bunsen burner ignitability test. Recently enacted regulations based on correlation of small-scale reaction-to-fire tests and full-scale aircraft cabin fire test data by the FAA specify maximum heat release values for large-area materials in an effort to delay the cabin flashover and provide increased escape time for passengers. The regulation also limits the maximum smoke release of these components.

Table I. Aircraft Cabin Materials

CABIN MATERIAL	Kilograms Weight per Aircraft	CABIN MATERIAL	Kilograms Weight per Aircraft
Acoustical Insulation	100-400	Paint	5
Blankets	20-250	Passenger service units	250-350
Cargo Liners	>50	Partitions and sidewalls	100-1000
Carpeting	100-400	Pillows	5-70
Ceiling	600	Thermoplastic parts	≈ 250
Curtains	0-100	Seat belts	5-160
Ducting	450	Seat cushions	175-900
Elastomers	250	Seat upholstery	80-430
Emergency slides	25-500	Seat trim	40-200
Floor panels	70-450	Wall covering	≈ 50
Floor coverings	10-100	Windows	200-350
Life rafts	160-530	Window shades	100
Life vests	50-250	Wire insulation	150-200
		TOTAL COMBUSTIBLES	**3300-8400**

The remaining twenty percent of aircraft cabin interior materials include floor coverings, textiles, draperies, upholstery, cushions, wall coverings, blankets, thermoacoustic insulation, cargo compartment liners, air ducting, trim strips, as well as molded and thermoformed plastic parts such as overhead passenger service units and seat components which are often painted to comply with aesthetic design requirements. These interior materials are not governed by the new heat release and smoke generation rules and are only required to pass a Bunsen burner ignitability test, or in the case of upholstered seat cushions and cargo liners, an oil-burner impingement test for ignitability.

Aircraft seats have been the primary fuel load in a cabin fire and are typically constructed of fire-retarded polyurethane foam encapsulated with a fire-blocking layer and covered with upholstery fabric. New exfoliated graphite-filled urethane foams pass the kerosene burner ignitability test without fire blocking layers and their use in aircraft passenger seating is increasing. Prior to 1987 seating materials were only required to pass a 12-second vertical Bunsen burner ignition test. Since then the FAA has established a kerosene burner test for seat back and bottom cushions in a chair configuration which more accurately simulates real fire conditions. The use of a fire-blocking layer material to encapsulate and delay ignition of the polyurethane foam was a practical alternative to inherently fire-resistant foam, acceptable versions of which were unavailable at the time the regulation was released. Aramid quilts or polybenzimidazole felt/fabric are now used as fire blocking layers over fire-retarded urethane foam in passenger aircraft. These seat fire-blocking layers prevent ignition of both fire-retarded and non-fire retarded urethane foams when subjected to small ignition sources such as cigarettes, newspapers, or a pint of gasoline. In simulated post crash cabin fires the seat fire-blocking layers slow fire growth and can provide 40-60 seconds of additional passenger escape time before full involvement of the seat cushions (2).

Aircraft Cabin Fires

Approximately forty percent of the fatalities in impact-survivable commercial aircraft accidents are due to fire. As more and more aircraft are added to the fleet, the number of aircraft accidents can be expected to increase as will the number of fatalities– including fire fatalities. A substantial increase in fire fatalities is an unacceptable prospect and the FAA has taken a bilateral approach to reduce the aircraft accident fatality rate. The first approach is to prevent new factors from increasing the accident rate through programs such as Aging Aircraft, Structural Airworthiness, Engine Reliability, and Catastrophic Failure Prevention. The second approach is to reduce the number of accidents of the type that have been occurring and to increase the survivability of such accidents through programs in Airplane Crashworthiness, Cabin Safety, Fire Safety, and Fire Research.

Compartment fires in aircraft, ships, ground vehicles, and buildings are the most severe from a fire safety perspective because enclosed spaces hold heat and combustion products which increase the severity of the fire and its impact on those exposed (3). Fires in aircraft, space vehicles, ships, and submarines are particularly hazardous because of the small size of the compartments and the difficulty or impossibility of escape. In aircraft, post-crash cabin fires ignited from spilled jet fuel become life-threatening when the cabin materials become involved and the fire propagates through the cabin generating heat, smoke, and toxic decomposition products. Hot combustion products rise from the fire entraining air and forming a distinct, hot, smoky layer just below the ceiling which deepens as the fire continues to burn. The availability of air influences the products of combustion as well as the intensity of a fire, and as oxygen is depleted during combustion the fraction of carbon monoxide in the smoke increases appreciably.

Cabin flashover is a non-survivable event characterized by localized ignition of the hot smoky layer containing incomplete combustion products and rapid fire growth through the cabin interior. Burning panels fall and ignite seats causing total involvement of the interior. Full-scale aircraft cabin measurements of fire hazards- temperature, smoke, oxygen deprivation, carbon dioxide, carbon monoxide, and irritant gases such as HCl and HF- indicate that these hazards increase markedly at flashover, exceeding individual and combined tolerance limits (4) at that time. Consequently the time required to reach flashover is a measure of the time available for escape from an aircraft cabin fire.

Aircraft cabin fires fall into three general categories: ramp, in-flight, and post-crash. Ramp fires occur when an aircraft is parked at the ramp during servicing. One past example was a smoldering cigarette in a trash bag which ignited an adjacent passenger seat in the unattended aircraft. To date ramp fires have resulted in the loss of property but not the loss of life. However, considering the current cost of a commercial aircraft ($\approx$ $100 million), ramp fires can be a significant problem.

In-flight fires most often occur in accessible areas such as the galley and are detected and extinguished promptly. On rare occasions in-flight fires originating in inaccessible areas have become uncontrollable leading to large loss of life, e.g. a cargo compartment fire claimed all 301 occupants when fire penetrated the cabin floor and ignited seats and other materials. In-flight fires are typically caused by electrical failures or overheated equipment.

The vast majority of fatalities attributable to fire have occurred in post-crash fire accidents (5). Fuel fires which penetrate the passenger cabin are the primary ignition source in these accidents. Newer regulations require a number of fire safety improvements in aircraft cabins including materials flammability upgrades in aircraft manufactured after 1988 which, depending on the accident scenario, may extend the passenger escape time by two or more minutes in a post-crash accident involving a fuel fire.

To characterize the fire hazards in a post-crash fuel fire, full-scale fire tests were conducted at the Federal Aviation Technical Center (6,7). One objective of the test program was to measure the heat flux and temperatures at various locations in the fuselage of a wide-body aircraft subjected to an external fuel fire entering an open cabin door to establish thermal exposure conditions for interior materials. Tests were conducted on a 5-meter diameter, C-133 fuselage, using a 2.4 x 3.0 x 0.1 meter pan to contain burning JP-4 jet fuel. The pan fire was elevated to the height of an open doorway to simulate fuselage fire conditions for a burning fuel spill following a landing accident. Some of the results of these tests are given in Figures 1 and 2. Figure 1 shows steady-state cabin heat flux measurements at the open doorway, along the cabin centerline, and along the far cabin wall at three vertical locations – floor (0.3 m), seat (1 m) and ceiling (2 meter) level – for quiescent wind conditions. Under quiescent (zero) wind conditions heat fluxes of approximately 50-60 kW/m² occur at the floor, seat, and ceiling levels in the vicinity of an open cabin door exposed to burning jet fuel. These heat fluxes decrease in proportion to the square of the distance away from the fire. Temperatures at the ceiling level are between 600 and 700C as seen in Figure 2.

The two principal conclusions from these studies and other full-scale fire tests relative to materials fire performance requirements for aircraft interiors are that 1) high heat fluxes (≥ 50 kW/m²) and temperatures (≥ 600 C) in and around the fire entry door are associated with a post-crash fuel fire, and, 2) smoke and heat from an external fuel fire are individually greater hazards than toxic gas production (carbon monoxide) although cabin conditions remain survivable if interior materials do not become involved. More recent fuel fire tests with furnished aircraft cabins (e.g., sidewalls, ceilings, seats installed) confirm that current interior materials ignite above the fire entry door where heat fluxes and temperatures are highest, and propagate to flashover within 3-4 minutes, after which cabin conditions become unsurvivable.

Figure 3 shows the equilibrium surface temperature for a 0.8 emissivity material as a function of incident heat flux according to (8)

$$\dot{Q}_i = \varepsilon\sigma(T_s^4 - T_o^4) + \bar{h}(T_s - T_o) \tag{1}$$

for a calculated surface convective heat transfer coefficient, $\bar{h} = 24$ W/m²-K. It is seen that the equilibrium surface temperature exceeds 600C for incident heat fluxes ≥ 50 kW/m², in qualitative agreement with the experimental fuel fire data in Figures 1 and 2. Consequently, a fire-safe material in a cabin door location should have as a minimum short term thermal stability (decomposition temperature) greater than about 600C or have negligible heat release under these conditions.

The relationship between material fire response characteristics (peak heat release rate, time-to-ignition) and the available escape time from an aircraft (time-to-flashover) is illustrated in Figure 4. Peak chemical heat release rate data is for decorative honeycomb aircraft panels used in sidewalls and ceilings which are designated by their surface film, polyvinylfluoride (PVF) or polyetherether ketone (PEEK), and the resin/fiber reinforcement used for the face skins on the 6 mm thick NOMEX honeycomb core. Results are for panel specimens tested in the vertical position at 50 kW/m² irradiance with the rate of heat release (RHR) calculated from oxygen consumption calorimetry. These bench-scale heat release data are plotted *versus* the reciprocal time to flashover ($1/t_{flashover}$) of these materials in full-scale, wide-body aircraft cabin fire tests. Also plotted in Figure 4 *versus* reciprocal time-to-flashover is the [peak rate of heat release]/[time-to-ignition] ratio, also known as the *flashover parameter*. Ratioing the peak rate of heat release with the ignition delay time seeks to remove the effect of sample thickness/thermal inertia on ignition time so as to provide a common basis for comparing the flashover potential of materials. It is seen that the correlation with reciprocal time-to-

List of Symbols

a_n	coefficients in transient conduction equation
α	thermal diffusivity = $\kappa/\rho C_p$
A	frequency or pre-exponential factor in Arrhenius equation
b	sample thickness
Bi	Biot modulus = $b/\bar{h}\kappa$
χ	combustion efficiency
C_p	heat capacity
E_a	activation energy for pyrolysis
Fo	Fourier number = $\kappa t/b^2$
ε	radiant emissivity
ΔH_c	chemical heat of complete combustion
ΔH_g	heat of gasification
H_d	enthalpy of primary bond dissociation
H_f	enthalpy of fusion
ΔH^*	activation enthalpy for pyrolysis
k_p	rate constant for pyrolysis
κ	thermal conductivity
λ	dimensionless parameter = $\Delta H_c / \Delta H_g$
m	mass
M_g	molecular weight of gaseous decomposition species
M_o	monomer molecular weight
$\dot{Q}_{net}$	net heat flux
$\dot{Q}_i$	incident heat flux
$\dot{Q}_{flame}$	flame heat flux
$\dot{Q}_{rerad}$	re-radiated heat flux
ρ	density
R	gas constant
t	time
T	temperature
T_p^{max}	temperature at maximum pyrolysis rate
T_s	surface temperature
T_s^∞	equilibrium surface temperature
ξ	time variable of integration

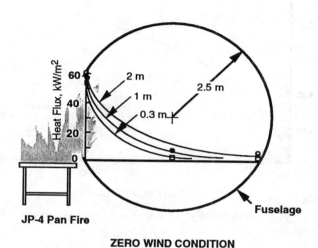

Figure 1. Heat fluxes measured at floor, seatback, and ceiling levels for three cabin locations during fuel fire at an open doorway.

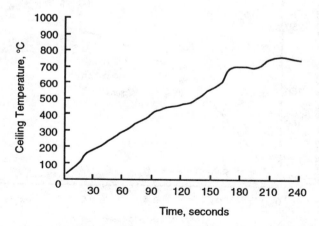

Figure 2. Cabin temperature above fire door during external fuel fire.

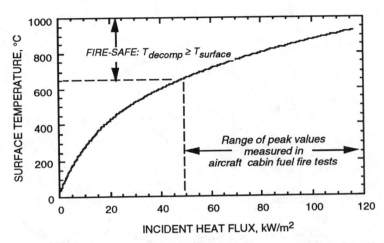

Figure 3. Equilibrium surface temperature of a polymer material as a function of incident heat flux.

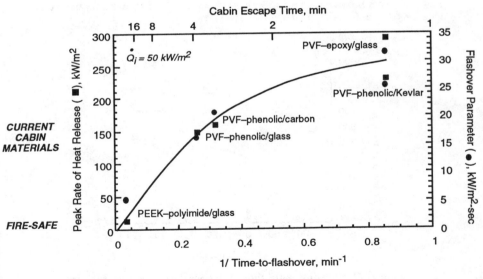

Figure 4. Peak rate of heat release and flashover parameter of panel materials tested at 50 kW/m^2 irradiance *versus* time-to-flashover for same materials in full-scale aircraft cabin fire tests.

flashover for both peak heat release rate and flashover parameter is equivalent, probably because of the nearly constant density and thickness of the aircraft panel specimens.

The above correlation between laboratory and full-scale fire test results indicate that incremental improvements in material fire-resistance will do little to increase passenger escape time (9). The high incident heat fluxes (50 kW/m²) and temperatures (600C) measured in the full-scale cabin fire tests require that essentially non-combustible aircraft cabin materials be developed to prevent ignition, flame spread, smoke, and toxic gas generation under post-crash fire conditions and achieve the FAA goal of a totally fire-resistant aircraft cabin (10).

Fire Response of Materials

To develop non-combustible aircraft materials it will be necessary to establish the relationship between the chemical structure of materials and the response parameters which determine the fire hazard of a material in an aircraft cabin. Material response parameters include ignition temperature, heat release and mass loss rate during combustion, flame spread rate, optical properties of smoke, toxicity of combustion products, response to suppressants, and fire endurance. In the following sections we will attempt to relate the kinetic and thermodynamic properties of materials to their behavior in aircraft cabin fires.

Solid material combustion can be separated into thermokinetic and thermodynamic processes as shown schematically in Figure 5. Plotted against a realistic energy scale is the reaction sequence by which solid reactants combust to gaseous products at various combustion efficiencies, χ. The combustion process requires that heat from an external source (heater or fire) be deposited in the solid until the net thermal energy into the material equals the activation energy for pyrolysis, E_a. At this point thermal decomposition of the solid begins at a rate which is determined by the heating rate and pyrolysis kinetics. It is often found that the activation energy for pyrolysis is roughly equal to the dissociation energy of the weakest primary chemical bond in the material. The chemical reactions in the solid and gaseous phase proceed to combustion products with the chemical heat of combustion, ΔH_c, released at a rate which depends on the temperature and fuel/oxygen stoichiometry in the combustion zone.

Thermokinetics: For a solid of mass, m, pyrolyzing via a single, first-order, isothermal process the mass loss per unit time, t, is

$$-\frac{dm}{dt} = -\dot{m} = k_p\, m \qquad (2)$$

where the reaction rate constant, k_p, is assumed to be of the Arrhenius form

$$k_p = A\, e^{-E_a/RT} \qquad (3)$$

and, A, E_a, are the pre-exponential factor and activation energy, respectively for the process. Under conditions where the temperature of the solid is uniform but increasing (e.g., thermogravimetric experiments) a pronounced peak in the pyrolysis rate will occur at a temperature, T_p^{max}, which depends on the rate of temperature rise. For a constant rate of temperature rise, $dT/dt = \dot{T}$, this peak pyrolysis temperature is obtained by setting the second time derivative of mass equal to zero at $T = T_p^{max}$, i.e., $-\ddot{m} = k_p\dot{m} + m\dot{k}_p = 0$ @ at $T = T_p^{max}$. The resulting non-dimensionalized equation in terms of T_p^{max} is

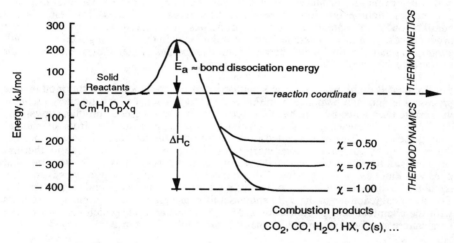

Figure 5. Energetics of solid combustion shown approximately to scale.

$$\ln\left[\frac{E_a}{R T_p^{max}}\right]^2 + \frac{E_a}{R T_p^{max}} + \ln\left[\frac{R \dot{T}}{A E_a}\right] = 0 \quad (4)$$

The appropriate $\dot{T}$ for surface heating is a constant, time-average, rate of temperature rise, $\langle dT/dt \rangle$, on the face of a slab of thickness, b, density, ρ, heat capacity, C_p, and thermal conductivity, κ, for a net heat flux to the surface, $\dot{Q}_{net}$. The semi-infinite slab result valid up to time, $\tau = \rho C_p b^2 / 4\kappa$, is

$$\dot{T} = \left\langle \frac{dT}{dt} \right\rangle = \frac{1}{\tau} \int_0^\tau \frac{\dot{Q}_{net}}{\sqrt{\pi k \rho C_p}} \frac{dt}{\sqrt{t}} = \frac{4}{\sqrt{\pi}} \frac{\dot{Q}_{net}}{\rho b C_p} \quad (5)$$

Table II lists values for the Arrhenius parameters A and E_a measured in our laboratory or obtained from the literature (*11-14*) for anaerobic pyrolysis of some polymeric materials along with the temperature of maximum pyrolysis rate, T_p^{max}, for $\dot{T} = 12$ K/s, calculated from Equation 5 for a 50 kW/m² net heat flux with b = 6 mm, ρ = 1200 kg/m³, and, C_p = 1300 J/kg-K (see Figure 9 for variable heating rate). Also tabulated is the calculated value of the reaction rate constant for pyrolysis, k_p, at T_p^{max}. Measured or calculated chemical heats of combustion, ΔH_c, for the polymeric materials are also listed in Table II for those polymers whose chemical structures are shown in Figure 6. Heat release rate data in Table II is a 300 second average obtained at 50 kW/m² irradiance for the indicated polymer in a glass or carbon fiber composite (*11,15,16*). Heat release rate values in parentheses were estimated by multiplying the heat release rate of the pure polymer at 50 kW/m² irradiance by 0.27 (approximately the resin mass fraction in the composite) as demonstrated experimentally for several resin composites (*17*).

The following simplifying assumptions are made with regard to the flaming combustion of polymeric materials: 1) the fuel generation process at a burning polymer surface is essentially an anaerobic, single-step pyrolysis which obeys first-order kinetics with rate-independent kinetic parameters, A, E_a; 2) the surface temperature of a polymer during flaming combustion is equal to its peak decomposition temperature, T_p^{max}, at a heating rate, $\dot{T}$, which depends on the thermal properties of the material and the net heat flux to the surface (e.g., Equation 5); 3) the pyrolyzing polymer surface and the gaseous decomposition product(s) in contact with the polymer surface are both at the same temperature, T_p^{max}, during flaming combustion; 4) the gaseous decomposition product (fuel) is reacting with oxygen in a first-order exothermic process with rate constant, k_g, and; 5) gaseous and thermal diffusion as well as char formation effects on ignitability and heat release rate can be neglected in this elementary treatment of flammability.

With these assumptions some conclusions can be drawn with regard to the effect of decomposition kinetics on polymer flammability. If T_p^{max} is greater than the equilibrium surface temperature, T_s^∞, for a particular heat flux, then the polymer should be difficult or impossible to ignite and have low heat release. Poor ignitability and zero-order heat release are in fact observed for the PEEK and PBO polymer composites which have maximum resin decomposition temperatures at 760C and 772C, respectively—significantly above the equilibrium surface temperature, T_s^∞ = 660C for $\dot{Q}_{net}$ = 50 kW/m². In contrast, polymers such as the epoxy and PMMA with $T_p^{max} \ll T_s^\infty$, would be expected to pyrolyze readily at 660C and have relatively high heat release rates, as observed. Polymers with $T_p^{max} \approx T_s^\infty$, such as the phenolic triazine and the polyimide should and do have moderate heat release rates when tested at 50 kW/m² irradiance.

At $T=T_p^{max}$ the calculated pyrolysis/gas phase reaction rate ratio, k_p/k_g, shown in Table II indicates that surface burning conditions are fuel-rich ($k_p/k_g > 1$) at low peak decomposition temperature/high heat release rates and fuel-lean ($k_p/k_g < 1$) at high peak

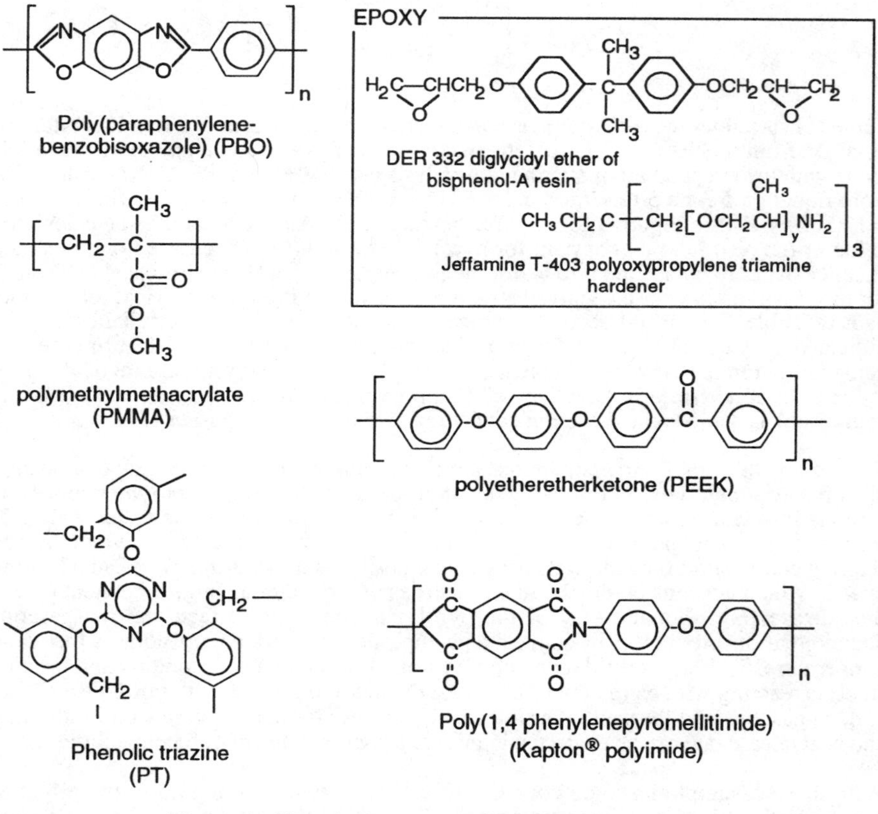

Figure 6. Chemical structure of polymeric materials.

decomposition temperature and low heat release rate. Qualitatively, the predicted fuel-rich gas phase for PMMA and epoxy burning should result in a low surface temperature due to incomplete combustion at the interface, a fuel-rich zone above the surface, and greater flame height as unburned fuel gases diffuse upward and mix with oxygen in the flame zone to complete the combustion process. Conversely, fuel-lean gas phases predicted for thermally-stable PEEK and PBO would be associated with a low flame in close proximity to the polymer surface and relatively low flame temperatures on the order of the pyrolysis temperature. Polymers with high thermal stability would therefor be expected to burn weakly with the potential for significant soot formation.

Since the heat of combustion of all of the polymers in Table II is relatively constant in the range, $\Delta H_c = -29 \pm 3$ kJ/g, the fuel value of the polymer has only a negligible effect on the total heat released and the heat release rate. More important in determining heat release and heat release rate of polymers are the combustion kinetics– which appear to be pyrolysis-limited in the case of thermally-stable materials. However, the validity of the contrasting phenomenological descriptions deduced from pyrolysis-limited combustion kinetics of low and high thermal-stability polymers remains to be established.

With regard to predicting the heat release rate of a polymer at a particular surface heat flux it is clear that for the short list of materials in Table II, no single kinetic parameter (A, E_a) correlates well with heat release rate. Moreover, the heat of combustion of these materials is relatively constant in the range, $\Delta H_c = -29 \pm 3$ kJ/g. Combined parameters such at, k_p (or k_p/k_g) and T_p^{max} evaluated at the equilibrium surface temperature for a particular heat flux appear to be qualitatively useful for the purpose of assessing the relative flammability of materials using only thermogravimetric data on pyrolysis kinetics.

Table II. Kinetic Parameters for Pyrolysis, Heat of Combustion, Heat Release Rate, and Peak Decomposition Temperature of Various Polymers

	MATERIAL	log A (sec^{-1})	E_a (kJ/mol)	k_p / k_g @ T_p^{max}	$-\Delta H_c$ (kJ/g)	Rate of Heat Release (kW/m^2)	T_p^{max} (C)
Pyrolysis reactions	PBO	14	290	10^{-4}	30.0	0	772
	PEEK	10	210	10^{-4}	32.5	8	760
							T_∞
	PI	19	340	10^{0}	26.7	27	646
	PT	18	309	10^{0}	29.5	(48)	608
	EPOXY	12	165	10^{2}	32.5	98	430
	PMMA	12	160	10^{2}	26.2	(167)	402
	Gas phase reactions	≈13	≈200				

Thermodynamics: The Arrhenius equation (Equation 3) was originally derived with the assumption that the reactants are at equilibrium with a high energy state and could proceed to products with no further energy requirements. This assumption of thermodynamic equilibrium between the reactants and activated complex allows us to attach a thermodynamic significance to the constants in the Arrhenius equation. From the definition of the free energy at equilibrium,

$$\ln K = \ln\frac{k_p}{k_b} = -\frac{\Delta G^*}{RT} = -\left[\frac{\Delta H^*}{RT} - \frac{\Delta S^*}{R}\right] \tag{6}$$

where k_p, k_b, are the forward and backward rates of pyrolysis and, ΔG^*, ΔH^*, and, ΔS^*, are the molar free energy, enthalpy, and entropy of pyrolysis, respectively. It follows that

$$k_p = Ae^{-(E_a/RT)} = \left[k_b e^{\Delta S^*/R}\right] e^{-(\Delta H^*/RT)} \tag{7}$$

where the pre-exponential factor now has the identity, $A = [k_b e^{\Delta S^*/R}]$, and, $E_a = \Delta H^*$, is the molar heat of pyrolysis,

$$E_a \equiv \Delta H^* = \Delta H_f + \Delta H_d + \Delta H_v \tag{8}$$

in terms of, ΔH_f the heat of fusion, ΔH_d the heat of dissociation of the primary chemical bonds in the molecules, and, ΔH_v the heat of vaporization of the low-molecular weight (liquid) decomposition products. In general, ΔH_f, ΔH_d, and, ΔH_v will be functions of temperature up to the decomposition temperature of the polymer, T_p, because of the temperature-dependent species distribution observed in pyrolysis reactions, so in general

$$\Delta H^* = \int_0^{T_p} C_p(T)dT + \int_0^{T_p} \frac{\partial H_d}{\partial T}dT + \int_0^{T_p} \frac{\partial H_v}{\partial T} dT \tag{9}$$

The molar heat of pyrolysis on a mass basis is conceptually equivalent to the heat of gasification, ΔH_g – a quantity used by fire scientists to relate the mass loss rate, $\dot{m}$, of a heated or burning specimen to the net heat flux to the surface, $\dot{Q}_{net}$, according to

$$\dot{m} = \frac{\left(\dot{Q}_i + \dot{Q}_{flame} - \dot{Q}_{reradiation}\right)}{\Delta H_g} = \frac{\dot{Q}_{net}}{\Delta H_g} \tag{10}$$

In Equation 10, $\dot{Q}_{net}$ is seen to be the difference between the incident heat from an external heater, $\dot{Q}_i$, or surface flame, $\dot{Q}_{flame}$, and the heat lost through reradiation to the surroundings, $\dot{Q}_{reradiation}$. In practice ΔH_g is determined experimentally as the slope of a plot of peak mass loss rate versus external radiant heat flux, $\dot{Q}_i$. If ΔH_g is related to the molar heat of pyrolysis determined in laboratory thermogravimetric experiments through the molecular weight of the decomposition products, M_g, then the following identity applies

$$\Delta H_g = \frac{\Delta H^*}{M_g} = \frac{E_a}{M_g} \tag{11}$$

Polymers which pyrolyze to monomer (depolymerise) at near-quantitative yield such as polymethylmethacrylate, polyoxymethylene, and polystyrene, should have M_g equal to the monomer molecular weight, M_o, i.e., $M_g/M_o \approx 1$. For polymers such as polyethylene and polypropylene which decompose by random scission to multi-monomer fragments, $M_g/M_o > 1$. In contrast, polymers with high molecular weight repeat units ($M_o \geq 200$ g/mol) such as nylon, cellulose, polycarbonate, or with good leaving groups (e.g., polyvinylchloride) are known to yield primarily low molecular weight species (water, carbon dioxide, alkanes, HCl) on pyrolysis and should have, $M_g/M_o < 1$. Table III shows the ratio of E_a/M_o to the heat of gasification, ΔH_g (*18*) as M_g/M_o for a variety of common polymers, according to the relation

$$\frac{E_d/M_o}{\Delta H_g} = \frac{M_g}{M_o} \tag{12}$$

The qualitative agreement between the observed modes of pyrolysis (random scission, depolymerization, solid-phase combustion/fragmentation) and the calculated fragment molecular weight using Equation 12 is strong support for the identity, $\Delta H_g = E_a/M_g$.

Table III. Heat of Gasification, Thermal Activation Energy, and Calculated Molecular Weight of Decomposition Products for Common Polymers with Known Pyrolysis Modes.

POLYMER	M_o (g/mol)	ΔH_g (kJ/g)	E_a (kJ/mol)	M_g / M_o	PYROLYSIS PRODUCTS
polypropylene	42	2.5	243	2.3	$C_2 - C_{90}$ saturated and unsaturated hydrocarbons
polyethylene	28	2.4	264	3.9	
polystyrene	104	2.2	230	1.0	40-60% monomer
polymethylmethacrylate	100	1.6	160	1.0	100% monomer
polyoxymethylene	30	2.7	84	1.0	100% monomer
nylon 6,6	226	2.6	160	0.3	H_2O, CO_2, C_5 HC's
cellulose	162	3.2	200	0.4	H_2O, CO_2, CO
polyvinylchloride	62	2.5	110	0.7	HCl, benzene, toluene

The activation energy for pyrolysis, E_a, can be related to the chemical heat release rate of a burning material, $\dot{Q}_c$, through the phenomenological definition (*19*)

$$\dot{Q}_c = \dot{m} \chi \Delta H_c = \frac{\chi \Delta H_c \dot{Q}_{net}}{\Delta H_g} \tag{13}$$

Combining Equations 11 and 13,

$$\dot{Q}_c = \lambda \chi \dot{Q}_{net}$$
$$= [\lambda \chi (\dot{Q}_{flame} - \dot{Q}_{reradiation})] + \lambda \chi \dot{Q}_i \tag{14}$$

where

$$\lambda = \frac{\Delta H_c}{\Delta H_g} = \left| \frac{M_g \Delta H_c}{E_a} \right| \tag{15}$$

The λ parameter is the ratio of the heat content to the thermal stability of a material and relates the combustion variables in Figure 5 to a measurable fire characteristic. Since ΔH_c and E_a are fundamental thermodynamic quantities which depend only on the chemical structure of the material and its additives, Equation 15 provides a physical basis for the hypothesis that the ratio $\Delta H_c/\Delta H_g$ is a characteristic material property (*20*).

It is now seen that the chemical heat release rate of a burning material is proportional to the net heat flux to the surface, and linearly related to the incident heat flux from a radiant heater, $\dot{Q}_i$, through a dimensionless material parameter, λ, which can be estimated from simple laboratory tests (bomb calorimetry, thermogravimetric analysis).

Heat Transfer: Relating the isothermal kinetics and thermodynamics of material combustion to the transient problem of ignition requires a heat transfer analysis. A simplified one-dimensional solution for transient heat conduction through a finite slab with a constant net heat flux to the surface and convective boundary conditions was used to couple the pyrolysis kinetics to the thermal radiation field in a fire. The geometry and boundary conditions for the problem of a horizontal 10 x 10 x 0.6 cm horizontal plate (cone calorimeter sample) with thickness b= 0.6 cm is shown schematically in Figure 7.

In the analysis the kinetic parameters in Table II were used to calculate the mass loss rate at the front surface (x=0) for constant net heat flux into the surface, $\dot{Q}_{net}$, and convection from the rear surface (x=b) into air at T_o =300K. The time-dependent surface temperature, $T_s(t) = T_{x=0}(t)$, was evaluated using the analytical result (21),

$$T_s(t) = T_o + \frac{b\dot{Q}_{net}}{\kappa Bi}[f_1(Bi,Fo) + Bi f_3(Bi,Fo)] \quad (16)$$

where, $Bi = b\bar{h}/\kappa$ is the Biot number for the dimensionless diffusion length in terms of the surface convective heat transfer coefficient, $\bar{h}$, sample thickness and thermal conductivity, κ. The Fourier number, $Fo = \kappa t/b^2$ is the dimensionless time in terms of the sample thermal diffusivity, α, and thickness. The functions, f_1, f_3 are

$$f_1(Bi,Fo) = 1 - \sum_{n=1}^{\infty}\left(\frac{2Bi}{Bi(Bi+1) + a_n^2}\right)e^{-a_n^2 Fo}$$

$$f_3(Bi,Fo) = 1 - \sum_{n=1}^{\infty}\left(\frac{2Bi^2}{a_n^2[Bi(Bi+1) + a_n^2]}\right)e^{-a_n^2 Fo}$$

with, a_n, the roots of the equation, $a \tan a = Bi$. For the analysis only the tabulated first six roots (n=1,2,...6) were used. Constants used in the transient heat transfer calculation were the surface convective heat transfer coefficient, $\bar{h} = 24$ W/m²-K, evaluated at a mean film temperature of 500C, and typical polymer thermal conductivity, $\kappa = 0.2$ W/m-K, and thermal diffusivity, $\alpha = \kappa/\rho C = 0.2$ W/m-K/(1200 kg/m³ x 1300 J/kg-K) = 1.3 x 10^{-7} m²/s.

It is worthwhile to compare the surface temperature history for a finite slab with a constant heat flux to the surface and convective rear boundary conditions (Equation 16) to commonly used analytic solution for the semi-infinite solid with constant surface heat flux (21)

$$T_s(t) = T_o + 2\dot{Q}_{net}\left[\frac{t}{\pi\kappa\rho C_p}\right]^{1/2} \quad (17)$$

with T_s the surface temperature at time t and, ρ, C_p, T_o, the density, heat capacity, and initial (ambient) temperature of the material, respectively. The thermal theory of ignition assumes that, $t = t_{ign}$, at, $T_s = T_{ign}$.

Figure 8 compares the finite (6 mm) and semi-infinite slab results for a typical polymer cone calorimeter specimen at an incident heat flux of 10 kW/m². For times, $t \leq \rho C_p b^2/4\kappa = 70$ seconds, the finite and semi-infinite results are equivalent. However, at longer times corresponding to the actual ignition times of PMMA ($t_{ign} \approx 1000$ seconds), Equation 17 under-predicts the actual surface temperature by 100-150C.

In calculating the fractional mass loss rate at the surface of the finite slab first-order pyrolysis kinetics were assumed. Combining Equations 2 and 3

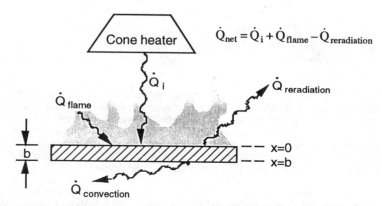

Figure 7. Geometry and boundary conditions for finite slab calculations.

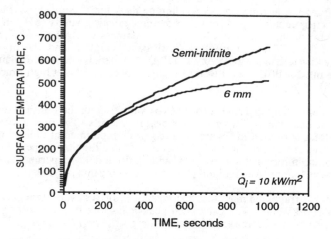

Figure 8. Comparison of semi-infinite and finite slab results for surface temperature of polymer prior to ignition at 10 kW/m^2 irradiance.

$$-\frac{dm}{dt} = m A \, e^{-[E_a/RT]} \tag{18}$$

from which the fractional mass loss history at the surface, m(t)/m$_o$, is obtained from

$$\ln\left[\frac{m}{m_o}\right] = -\int_0^t A \, e^{-[E_a/RT(\xi)]} d\xi \tag{19}$$

by numerical integration after substituting the surface temperature at each time step, $T_s(t)$, calculated with Equation 16 into the integrand of Equation 19. This is equivalent to transient heating of a thin, isothermal, polymer layer in contact with an inert substrate of identical thermal diffusivity.

An example of the results of these zero-dimensional calculations for a net external surface heat flux, $\dot{Q}_{net}$ = 50 kW/m^2, onto an infinitesimally thin polymer layer is shown in Figure 9. Plotted is the fractional surface mass loss rate due to pyrolysis adjusted for char yield, along with a surface temperature history which assumes no heat losses. In fact a constant surface temperature is seen to dominate the pyrolysis regime of real (finite thickness) polymer samples corresponding to a dynamic equilibrium between the net heat flux to the surface and heat removal by gasification of the polymer (22).

In terms of Figure 9 this would correspond to a leveling off of the surface temperature at the incipient mass loss temperature for the duration of each polymer mass loss event (peak width). Even though the surface temperature does not change significantly during the pyrolysis event under the assumed conditions (infinitesimal polymer thickness, no heat sink due to pyrolysis) it is seen that the calculated peak mass loss rates occur over a very broad range of heating time depending on the kinetic parameters. Moreover, no single kinetic parameter (e.g. A, or E_a) or thermodynamic quantity (ΔH_c) in Table II is a good predictor of the peak mass loss time or temperature. The peak mass loss rate temperature for PEEK and PBO is significantly higher, and the mass loss rate significantly lower, than for the epoxy and PMMA materials– perhaps explaining the low heat release rate for PEEK and PBO thermally stable materials at high irradiance levels (see Table II).

The kinetic parameters in Table II for PMMA were used in the transient heat transfer / mass loss rate calculations for various net heat flux levels. The time to peak mass loss rate for PMMA is plotted in Figure 10 along with experimental time-to-ignition data from cone calorimeter experiments (23). It is seen that the time-to-peak mass loss rate calculations are in reasonable agreement with ignition delay experiments suggesting a molecular basis for observed ignition phenomenon in organic polymers.

Figure 11 shows fractional mass loss histories for PMMA at various net heat flux levels from 20 to 100 kW/m^2. The calculated peak mass loss rate is a strong function of irradiance level as is observed experimentally for polymers (11,15,16,17).

Calculated surface temperatures at 5% mass loss are shown in Figure 12 for PMMA along with experimental data (24) for the equilibrium surface temperature of PMMA during anaerobic pyrolysis at 17 and 40 kW/m^2 irradiance. Figure 12 demonstrates that the decomposition temperature of a material (PMMA) depends strongly on the net heat flux to the surface and that this effect is reasonably well described by a heating rate effect on the incipient pyrolysis temperature acting through the decomposition kinetics.

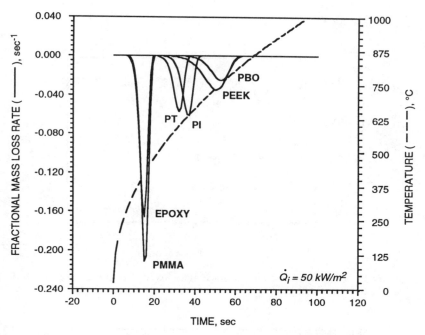

Figure 9. Fractional mass loss rate (———) for polymeric materials from first-order decomposition kinetics for indicated temperature history (– – – –).

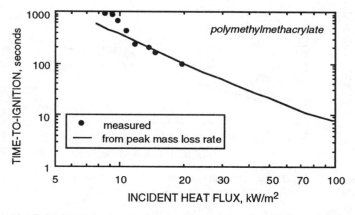

Figure 10. Calculated time-to-peak mass loss rate (———) at surface of 6 mm thick PMMA slab with real decomposition kinetics compared to measured time-to-ignition (●).

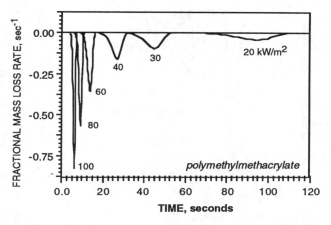

Figure 11. Calculated fractional surface mass loss rate histories for PMMA at various net heat flux levels.

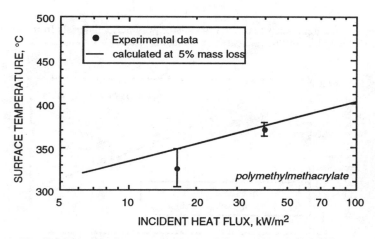

Figure 12. Calculated surface temperature at 5% mass loss for PMMA versus incident heat flux compared to experimental data of Kashiwagi and Ohlemiller [24].

Summary

Full-scale aircraft fuselage fire tests were conducted to establish the criteria for fire-safe cabin materials in a post-crash fuel fire. Heat fluxes of approximately 50-75 kW/m^2 and temperatures of 600–700C were recorded above fuselage openings exposed to jet fuel fires under quiescent conditions. The rate of heat release of cabin materials measured in bench-scale tests under these exposure conditions (e.g., 50 kW/m^2) was a reasonable fire hazard indicator. As a first step towards determining the effect of material composition on aircraft cabin fires correlations were developed between thermodynamic and kinetic variables, material properties, and the ignitability and heat release of polymeric materials. The limited data set analyzed indicates that the only kinetic parameter which correlates uniquely with heat release rate is the rate-dependent decomposition temperature of a material. No other single kinetic parameter such as activation energy, frequency factor, or heat of combustion was found to be a good predictor of fire-resistance nor were any dimensionless combination of these parameters. Inherently fire-resistant materials have an activation energy and frequency factor which satisfy the condition that the peak pyrolysis temperature of the material exceeds the equilibrium surface temperature at a specified heat flux so that ignition is delayed or suppressed. For fire-safe aircraft materials these requirements are that peak decomposition temperature is greater than about 700C for a 50 kW/m^2 incident heat flux. Once ignited the heat release rate of a burning polymer appears to scale with a dimensionless material parameter which is the ratio of the heat of combustion to the activation energy for pyrolysis per mole of degradation product. The coincidence of the measured ignition temperature and the calculated incipient decomposition temperature as function of incident heat flux suggests that pyrolysis kinetics govern ignition. Pyrolysis rates in comparison to gas phase reaction rates indicate that pyrolysis reactions in the solid-state are the rate-limiting step in the combustion process of low heat-release materials. Future efforts will extend this preliminary work to provide a methodology for relating chemical structure and properties to the fire-resistance of aircraft cabin materials.

Literature Cited

1. *Fire Safety Aspects of Polymeric Materials;* Aircraft: Civil and Military, NMAB 318-6; National Materials Advisory Board Publication: Washington, D.C., **1977**; Volume 6, pp. 68
2. Sarkos, C.P.; Hill, R.G. "Evaluation of Aircraft Interior Panels Under Full-Scale Cabin Fire Conditions," *Proc. AIAA 23rd Aerospace Sciences Mtg.*; AIAA: Reno, NV, Jan. 14-17, **1985**
3. *Fire & Smoke: Understanding the Hazards;* National Research Council Committee on Fire Toxicology; National Academy Press: Washington, D.C., **1986**, Chapter 1
4. Sarkos, C.P.;Hill, R.G.; "Effectiveness of Seat Cushion Blocking Layer Materials Against Cabin Fires," SAE Technical Paper 821484; Aerospace Congress and Exposition; Anaheim, CA October 25-28, **1992**
5. *Special Study: U.S. Air Carrier Accidents Involving Fire, 1965-1974 and Factors Affecting the Statistics;* Report NTSB-AAS-77-1; National Transportation Safety Board, **1977**
6. Brown, L.J.; *Cabin Hazards From a Large External Fuel Fire Adjacent to an Aircraft Fuselage*, FAA-RD-79-65; August **1979**
7. Hill, R.G.; Johnson, G.R.; and Sarkos, C.P.; "Postcrash Fuel Fire Hazard Measurements in a Wide-Body Aircraft Cabin," FAA-NA-79-42; December **1979**
8. Quintiere, J.G.; Harkelroad, M.T.; "New Concepts for Measuring Flame Spread Properties,"in *Fire Safety:Science and Engineering*, Harmathy, T.Z, ed; ASTM Special Technical Publication 882: Phila., PA **1985**; pp. 239-267

9. Hill, R.G. "The Future of Aircraft Cabin Fire Safety," *Proceedings of the International Conference for the Promotion of Advanced Fire Resistant Aircraft Materials*, Atlantic City, New Jersey, February 9-11, **1993**; p. 365
10. *Fire Research Plan*, U.S. Department of Transportation, Federal Aviation Administration, Atlantic City International Airport, NJ, January **1993**
11. Kim, P.K.; Pierini, P.; and Wessling, R.; "Thermal and Flammability Properties of Poly(p-phenylenebenzobisoxazole)," *J. Fire Science*, 1993, 11(4), pp. 296-307
12. Day, M.; Cooney, J.D.; Wiles, D.M.;"A Kinetic Study of the Thermal Decomposition of Poly(aryl-ether-ether-ketone)(PEEK) in Nitrogen," *Polym.Sci.Eng.*, **1989**; 29(1), pp.19-22
13. *Polymer Handbook*, Brandrup, J.; Immergut, E.H., Eds.; 2nd edition; Wiley-Interscience: NY **1975**
14. Van krevelen, D.W.; *Properties of Polymers*; Elsevier Scientific: NY, **1976**
15. Sorathia, U.; Rollhauser, C.M.; Hughes, W. A.; "Improved Fire Safety of Composites for Naval Applications," *Fire and Matls*. **1992**; 16, pp. 119-125
16. Sorathia, U.; Dapp, T.; Kerr, J.; "Flammability Characteristics of Composites for Shipboard and Submarine Internal Applications," *Proc. 36th Int'l SAMPE Symposium*, **1991**; 36(2), 1868-1877
17. Scudamore, M.J.; Biggs, P.J.; Prager, F.H.; "Cone Calorimetry–A Review of Tests Carried out on Plastics for the Association of Plastics Manufacturers in Europe," *Fire and Materials*, 1991; 15, pp. 65-84
18. Tewarson, A.; "Generation of Heat and Chemical Compounds in Fires," *SFPE Handbook of Fire Protection Engineering*; P.J. DiNenno, ed.; National Fire Protection Association Press: Quincy, MA, **1988**; Section 1, Chapter 13, pp. 1-179
19. Tewarson, A.; "Experimental Evaluation of Flammability Parameters of Polymeric Materials," in *Flame Retardant Polymeric Materials,* Lewin, M.; Atlas, S.M.; Pearce, E.M., eds.; Plenum Press: New York, NY, **1982**,Voume 3; Chapter 3, pp. 97-153
20. Tewarson, A., "Flammability Parameters of Materials: Ingition, Combustion, and Flame Spread," *Proceedings of the International Conference for the Promotion of Advanced Fire Resistant Aircraft Materials*, Atlantic City, New Jersey, Feb. 9-11, **1993**; pp. 263
21. Carslaw, H.S., JaegerJ.C.; *Conduction of Heat in Solids*, 2nd Edition, Clarendon Press: Oxford, **1976**; Chapter II and III, pp.50-193,
22. Quintiere, J.; Iqbal, N.; "A Burning Rate Model for Materials," in *The Science and Technology of Fire Resistant Materials–Proceedings of the Thirty-ninth Army Materials Research Conf;*, Macaione, D.P., ed.; Plymouth, MA, September 14-17,**1992**; Vol. 39, pp. 25-45
23. Kashiwagi , T.; Omori, A., *Twenty-Second Symposium (International) on Combustion*, The Combustion Institute: Pittsburg, PA, **1988**, pp. 1329-1338
24. Kashiwagi, T.; Ohlemiller, T.; "A Study of Oxygen Effects on Nonflaming Transient Gasification of PMMA and PE During Thermal Irradiation," *Ninteenth Symposium (International) on Combustion*, The Combustion Institute, Pittsburg, PA, **1982**; pp. 815-823

RECEIVED April 4, 1995

Author Index

Ahle, N. W., 312
Ali, Mohammad, 56
Aschbacher, D. G., 65
Benrashid, Ramazan, 217
Bonaplata, E., 118
Braun, Emil, 293
Brown, James E., 245
Caldwell, D. J., 366
Camino, G., 76
Chandrasiri, Jayakody A., 126
Christy, M. Robert, 498
Cleary, T. G., 422
Coad, Eric C., 256
Costa, L., 76
Davé, Vipul, 536
Deanin, Rudolph D., 56
Dietenberger, Mark A., 435
Dong, Xiaoxing, 236
Endo, Makoto, 91
Gilman, Jeffrey W., 161
Goldsmith, K., 118
Hirschler, Marcelo M., 553,593
Hurt, H. H., Jr., 312
Israel, Stanley C., 536
Janssens, Marc L., 409
Januszkiewicz, A. J., 323
Jeng, J. P., 118
Kaminskis, A., 312
Kampke-Thiel, Kathrin, 377
Kashiwagi, Takashi, 29,41,161
Keller, Teddy M., 267,280
Knauss, D. M., 41
Kuhlmann, K. J., 366
Kwei, T. K., 136
Lehnert, B. E., 323
Lenoir, Dieter, 377
Levin, Barbara C., 293
Lewin, Menachem, 91
Lomakin, S. M., 186,245
Luda, M. P., 76

Lyon, Richard E., 618
Macys, D. A., 344
Markezich, R. L., 65
Mayorga, M. A., 323
McGrath, J. E., 29,41
Moran, T. S., 312
Navarro, Magdalena, 293
Nelson, Gordon L., 1,217,579
Nyden, Marc R., 245
Oh, Sang Yeol, 136
Ohlemiller, T. J., 422
Paabo, Maya, 293
Pearce, Eli M., 136
Penkala, John J., 498
Petrella, Ronald V., 498
Rasmussen, Paul G., 256
Ritchie, G. D., 344
Roop, J. A., 366
Rossi, J., III, 344
Shah, Sunit, 536
Son, David Y., 280
Starnes, W. H., Jr., 118
Stemler, F. W., 312
Stotts, R. R., 312
Sundström, Björn, 609
Suzuki, Masanori, 236
Terranova, S. A., 118
Tewarson, A., 450
Tezak-Reid, T. M., 312
VanderHart, David L., 161
Voorhees, Kent J., 393
Wan, I-Yuan, 29
Wang, Jianqi, 518
Weil, Edward D., 199
Wilkie, Charles A., 126,236
Williams, D. M., 118
Wojciechowski, B. J., 118
Zaikov, G. E., 186
Zhu, Weiming, 199

Affiliation Index

Air Force Institute of Technology, 366
American Forest and Paper Association, 409
Beijing Institute of Technology, 518
College of William and Mary, 118
Colorado School of Mines, 393
Universitá di Torino, 76
Dow Chemical Company, 498
Factory Mutual Research Corporation, 450
Federal Aviation Administration, 618
Florida Institute of Technology, 1,217,579
Forest Service, 435
GBH International, 553,593
Geo Centers, Inc., 344
GSF Research Center for Environment and Health, 377
IBM Almaden Laboratories, 29
Los Alamos National Laboratory, 323
ManTech Environment, Inc., 366
Marquette University, 126,236
National Institute of Standards and Technology, 29,41,161,245,293,422
Naval Medical Research Institute Detachment (Toxicology), 344
Naval Research Laboratory, 267,280
Occidental Chemical Corporation, 65
Polytechnic University, 91,136,199
Russian Academy of Sciences, 186
Swedish National Testing and Research Institute, 609
U.S. Army Medical Research Detachment, 366
U.S. Army Medical Research Institute of Chemical Defense, 312
U.S. Department of Agriculture, 435
University of Massachusetts—Lowell, 56,536
University of Michigan, 256
Virginia Polytechnic Institute and State University, 29,41
Walter Reed Army Institute of Research, 323

Subject Index

A

Acetylene–linear siloxane polymers as precursors to high-temperature materials, *See* Linear siloxane–acetylene polymers as precursors to high-temperature materials
Acrylonitrile–butadiene–styrene, flame retardation, 68–72
Acrylonitrile–styrene copolymers, *See* Styrene–acrylonitrile copolymers
Additive(s), flame-retardant, *See* Flame-retardant additives
Additives, effects on thermal degradation of poly(methyl methacrylate)
  diphenyl disulfide, 131,133
  experimental procedure, 127–128
  mechanism for degradation, 133–134
Additives, effects on thermal degradation of poly(methyl methacrylate)—*Continued*
  oxygen indexes, 133
  phenyltin chloride, 129–130
  previous studies, 126–127
  pyrolysis of additives alone, 128–129
  tetraphenyltin, 131
Advanced composite materials
  applications, 366,367f
  potential risks, 366
  smoke production, 366–376
Aircraft cabin materials, fire-safe, *See* Fire-safe aircraft cabin materials
Alumina trihydrate, role in flame retardancy, 236–237
American Plastics Council, plastics recycling, 589–590

INDEX

American Society for Testing and Materials
ASTM D635, ignition test, 11
ASTM D2843, smoke test, 16
ASTM D2863, ease of extinction test, 14,16–20
ASTM D4100, smoke test, 1
ASTM E84 Steiner tunnel test, description, 499
ASTM E119, fire endurance test, 14,15f
ASTM E162 radiant panel test, flame spread test, 12
ASTM E662, smoke test, 16
ASTM E906, heat release rate test, 12,13f
ASTM E1317, heat release rate, 443–447
ASTM E1321
  ignitability, 436–440
  lateral flame spread, 440–443
ASTM E1354, heat release rate test, 12
ASTM E1537, description, 595–596
ASTM E1590, description, 596
Ammonium polyphosphate–poly(ethyleneurea formaldehyde) mixtures
  charring reaction, 84,87–88
  effect of polymer matrix on thermal behavior, 89
  experimental description, 77–78
  foaming behavior, 84–86f
  thermal degradation
    gas evolution, 79,82–84
    thermogravimetry, 78–81f
Ammonium polyphosphate–polypropylene mixtures
  additive amount vs. oxygen index, 92–95
  char morphology, 111–115
  comparison of synergisms, 97–98
  cone calorimetry, 104,107–111
  experimental description, 92t,93
  flame-retardant effectivity, 95,96f
  ingredients required, 91–93
  mechanism of synergism, 98,100
  relative importance of coadditives, 98,99f
  synergistic effectivity, 95,97
  thermogravimetric analysis, 100–106f

Animal models, use in behavioral toxicology, 347
Antimony oxide flame retardance, thermal analysis, 540–545
Antimony oxide–organic halogen compounds, use as flame-retardant additives, 7
Arapahoe smoke apparatus, smoke test, 16
Aromatic organic phosphate oligomers as flame retardants in plastics
  compounding, 58
  experimental description, 56
  flame-retardant elements vs. flame retardancy, 60–63
  halogen–antimony combinations in additives vs. flame retardancy, 59–60
  model reaction, 56–57
  molecular weight of phosphate vs. flame retardancy, 59
  oxygen index testing, 58–59
  polymers used, 58
  synthesis, 57–58
Austria, regulations affecting flame retardants, 587–588

B

Backbone incorporation of flame retardants
  advantages, 8
  examples, 8
Backyard test, full-scale fire performance prediction for furniture, 597–599f
Behavioral toxicity screening, 345
Behavioral toxicology, use of animal models, 347
Bench-scale radiant panel test, flame spread test, 12
[Bis(4-hydroxyphenyl)phenyl]phosphine oxide, synthesis, 43
Bismaleimide–poly(vinyl alcohol), pyrolysis, 171–184
Block copolymers, silicone moiety incorporation for flammability improvement, 217–233

Blood cyanide levels in miniature pigs
  animals used, 313
  blood cyanide concentrations, 315–317f,320
  blood gases, 315,320
  cardiopulmonary measurements and electrocardiograms, 315,318–319f,321
  clinical symptoms, 315
  experimental description, 313,314,316f
  pH, 315
  safety precautions, 314–315
  surgical procedure, 313
  validity of extrapolation from pigs to humans, 320–321
Blue Angel copiers, plastics requirements, 584–585
Brehob and Kulkarni's model, description, 424,429,431–433
Bridging, corrosive effect of smoke, 554
Brominated–chlorinated flame retardants, synergistic action, 65–66
Brominated flame retardants, polybrominated dibenzodioxin and dibenzofuran formation in laboratory combustion processes, 377–390
Buildings, hazard of fire, 1
Burning of polymers, model, 245
Butadiene–styrene–acrylonitrile, flame retardation, 68–72

C

^{13}C cross polarization–magic-angle spinning spectroscopy, honeycomb composites and phenol–formaldehyde resins, 246–247,252–254
^{13}C-NMR spectroscopy, effect of zinc chloride on thermal stability of styrene–acrylonitrile copolymers, 142
California Technical Bulletins 129 and 133, description, 596
Canadian Standards Association CSA 22.2, 950–M89, end-product safety requirements, 588
Carbon–carbon composites
  applications, 267
  oxidation problems, 267
  reason for interest, 267–268
Carbon dioxide, See CO_2–CO
Carbon dioxide generation calorimetry, determination of chemical heat release rate, 471,475–577,579
Carbon monoxide, See CO
Carbon–nitrogen materials
  2-(2-chloro-4,5-dicyano-1-imidazolyl)-4,5-dicyanoimidazole
    synthesis, 260,264
    thermogravimetric analysis, 264–266
    thermolysis structure, 264
  experimental description, 257–259
  isothermal thermogravimetric analysis, 260,263f
  thermogravimetric analysis
    carbon–nitrogen material from HTT under nitrogen and air, 260,262f
    carbonaceous material from HTT under nitrogen and air, 260,262f
    2-(2-chloro-4,5-dicyano-1-imidazolyl)-4,5-dicyanoimidazole, 264–266
    HTT under nitrogen and air, 260,261f
    1-iodo-2-chloro-4,5-dicyanoimidazole, 259–260,261f
  thermolysis of HTT, 260
Carbon oxidation protection method
  using additives, 268
  using external coating, 268
Carboxyhemoglobin levels, toxicity, 349
CFAST, description, 409–410
Char characterization, thermal decomposition of poly(vinyl alcohol), 161–184
Char formation
  advantages, 159
  reduction of polymer flammability, 161
Char-forming agents, incorporation for increased flame retardancy, 28
Char morphology, ammonium polyphosphate–polypropylene mixtures, 111–115
Characterization, carbon–nitrogen materials, 256–266
Charring process in condensed phase, X-ray photoelectron spectroscopy, 525–534
Charring reaction, ammonium polyphosphate–poly(ethyleneurea formaldehyde) mixtures, 84,87–88

INDEX

Chemical heat of combustion, 452
Chemical heat release rate
  definition, 471
  determination using carbon dioxide generation and oxygen consumption calorimetry, 471,475–477,479
Chemical ionization MS, analysis of pyrolysis products of polymers, 538
Chemistry, flame retardants, 2
Chlorinated–brominated flame retardants, synergistic action, 65–66
Chlorinated flame retardant use with other flame retardants
  experimental description, 66,67t
  flame retardant–acrylonitrile–butadiene–styrene
    antimony oxide content vs. retardancy, 68,70f,71–72t
    Cl:Br ratio vs. retardancy, 68,69f,71t
  flame retardant–polyolefin wire and cable
    alumina trihydrate vs. retardancy, 68,73f
    formulations, 68,72t
    MgOH content vs. retardancy, 68,70f
    smoke generation, 68,74t
    talc vs. retardancy, 68,73f,74t
  previous studies, 65–66,69f
  testing procedures, 66
2-(2-Chloro-4,5-dicyano-1-imidazolyl)-4,5-dicyanoimidazole, synthesis and characterization, 260,264–266
Cleary and Quintiere's model, description, 424,429–430,432–433
CO
  performance enhancement in operant behavior, 348
  toxicity, 24,291–292
CO–CO_2 exposures, effect on operant behavior, 348
CO–CO_2–NO_2–O_2–HCN mixture toxicity, N-gas model, 304,307,308t
CO-induced toxicity, effect of physiological irritancy, 348–349
CO_2–CO, measurement using controlled-atmosphere cone calorimeter, 513,515–516f

Coadditives, importance in ammonium phosphate–polypropylene mixtures, 98,99f
Codes and standards
  development, 21
  European harmonization, 21–24
  flame-retardant chemistry, 21–24
  ISO and IEC tests, 24
Combustion behavior of upholstered furniture study
  development, 609–610
  finalization, 616
  ignition
    cone calorimeter, 614,615f
    furniture calorimeter, 612,614f
  round-robin data
    cone calorimeter, 614,616f
    furniture calorimeter, 613,615f
Combustion of polymers
  analysis of soot produced, 393–406
  cycle, 2,3–4f
  interruption of cycle, 2,3–5f
  mechanism, 393
Combustion process, measurement, 451,464–474,486–490
Combustion test method
  generation rate of products, 468
  mass loss rate
    in flaming combustion process, 465–468,470
    in nonflaming combustion process, 464–465,466t
Combustion toxicology, application of NMRI/TD neurobehavioral screening battery, 345–362
Compartment fire models, examples, 409–410
COMPF2, description, 410
Complex combustion atmospheres, problems, 346–347
Complex combustion mixtures, prediction of toxic potency, 293–308
Composite materials
  advanced, smoke production, 366–376
  advantages, 422
  limitations of binder resin, 422–423
  upward flame spread, 423–434

Composite materials—*Continued*
  use for ship and submarine compartment constructions, 422
Conditioned position responding measurement
  higher cognitive decrement, 354
  sensory acuity, 352
  spatial and temporal discrimination, 353
Cone calorimetry
  ammonium polyphosphate–polypropylene mixtures, 104,107–111
  controlled atmosphere, *See* Controlled-atmosphere cone calorimeter
  development, 498
  ecologically safe flame-retardant polymer systems, 193,194t
  full-scale fire performance prediction for furniture, 598,600–602,604
  function, 498
  heat release parameter measurement, 476
  hydrolytically stable phosphine oxide comonomers, 54
  ignition test method, 460
  overventilation problem, 499
  procedure, 498
  smoke damage test method, 489
  surface modification of polymers for flame retardation, 242
  testing of upholstered furniture, 614–616
  thermal decomposition of poly(vinyl alcohol), 183
  triarylphosphine oxide containing nylon 6,6 copolymers, 35,36–38f
Cone corrosimetry
  comparison to other tests, 566,568–575
  problems, 566,567t
  procedure, 564
  results, 564–566
Controlled-atmosphere cone calorimeter
  background, 499
  burning characteristics of rigid polyisocyanurate foams, 502–503,508–509
  CO_2–CO for poly(methyl methacrylate), 513,515–516f
  combustion performance
    vs. oxygen concentration, 502,503–504t
    vs. ventilation rate, 502,505–506t

Controlled-atmosphere cone calorimeter—*Continued*
  description, 500,501f
  experimental description, 499,500,502
  heat release
    polyisocyanurate foams, 503,512f
    poly(methyl methacrylate), 498
  mass loss of poly(methyl methacrylate), 503,510f
  smoke
    polyisocyanurate foams, 513,514f
    poly(methyl methacrylate), 513,514f
Convective heat of combustion, definition, 452
Convective heat release rate
  definition, 471
  determination by gas temperature rise calorimetry, 471,475–477,479
Convective heat transfer, calculation using ROOMFIRE, 418
Copolycarbonates, phosphorus-containing, *See* Phosphorus-containing copolycarbonates
Copolymers from inorganic–organic hybrid polymer and multiethynylbenzene, high temperature, *See* High-temperature copolymers from inorganic–organic hybrid polymer and multiethynylbenzene
Copper compounds
  reductive coupling with poly(vinyl chloride), 118–124
  smoke suppressants for poly(vinyl chloride), 118
Copper targets, National Bureau of Standards smoke chamber work, 557–561
Corrosion, cause and description, 544
Corrosion index
  definition, 452,488
  measurements, 486–490
Corrosion test method
  measurements, 489,490t
  procedure, 488,490f
Corrosivity
  flammability tests, 16
  smoke, *See* Smoke corrosivity

INDEX

Critical heat flux
  definition, 451
  measurement, 460–464
Cross-linking, role in flame retardancy, 534–535
Cross polarization ^{13}C-NMR spectroscopy, thermal decomposition of poly(vinyl alcohol), 173,175–179
Cup burner test, Halon alternative screen, 492–494f
CVD coatings of silicon carbide, 268
Cyanide inhalation
  blood levels in miniature pigs, 313–321
  modeling problems, 312
  previous studies, 312–313

D

Deaths from fires, rate, 1
Decabromobiphenyl ether, thermal behavior, 383–386
Delayed matching to sample measurement
  higher cognitive decrement, 354
  sensory acuity, 352
Diacetylene groups, reactions, 280
Dibenzodioxin and dibenzofuran formation in laboratory combustion processes of brominated flame retardants, polybrominated, *See* Polybrominated dibenzodioxin and dibenzofuran formation in laboratory combustion processes of brominated flame retardants
Dichlorosiloxane compound, reaction with 1,4-dilithiobutadiyne to form high-temperature materials, 280–289
4,5-Dicyanoimidazole, use for synthesis and characterization of carbon–nitrogen materials, 256–266
Differential reinforcement of low rates of responding measurement
  higher cognitive decrement, 354
  spatial and temporal discrimination, 353

Differential scanning calorimetry
  effect of zinc chloride on thermal stability of styrene–acrylonitrile copolymers, 143–144
  high-temperature copolymers from inorganic–organic hybrid polymer and multiethynylbenzene, 272,273f
  hydrolytically stable phosphine oxide comonomers, 52,53f
  linear siloxane–acetylene polymers as precursors to high-temperature materials, 283,284f
  silicone moiety incorporation into block copolymers for flammability improvement, 225,226t
  1,2,4,5-tetrakis(phenylethynyl)benzene–poly[butadiyne-1,7-bis(tetramethyldisiloxyl)-*m*-carborane] mixture, 272,273f
  triarylphosphine oxide containing nylon 6,6 copolymers, 33–34
1,4-Dilithiobutadiyne, reaction with dichlorosiloxane compound to form high-temperature materials, 280–289
DIN smoke corrosivity test, 564
Diphenyl disulfide, role in thermal degradation of poly(methyl methacrylate), 131,133
Direct pyrolysis–GC–MS analysis, soot produced from polymeric material combustion, 404–406
Douglas fir plywood
  heat release rate using lateral flame spread using lateral ignition and flame spread test apparatus, 446–447
  ignitability using lateral ignition and flame spread test apparatus, 438–440
  lateral flame spread using lateral ignition and flame spread test apparatus, 442–443
Downward fire propagation, test method, 476,479–481f
Dynamic mechanical behavior, triarylphosphine oxide containing nylon 6,6 copolymers, 35

E

Ease of extinction
  fire property, 8–9
  flammability tests, 14,16–20
Ease of ignition, fire property, 8–9
Ecolabels, plastics requirements, 584–585
Ecological safeness of polymer flame retardants, problems, 186
Ecologically safe flame-retardant polymer systems
  cone calorimetry tests, 193,194t
  experimental description, 186,192
  heat release, 195,196f
  high-temperature polymer–organic char former, 187–192
  structure determination, 193,195
  synergistic carbonization, 195,197
Ecologically sound manufacturing processes and products, process, 580,581f
Electroencephalogram, measures of neurological activity, 351
Electron diffraction spectroscopy (EDS), silicone moiety incorporation into block copolymers for flammability improvement, 230,232f
Electron spectroscopy for chemical analysis, *See* X-ray photoelectron spectroscopy of flame retardancy of polymers
Emotionality, measures, 353
Emotionality rating scale measurement
  physiological irritation, 352
  social behavior and emotionality, 352
Empirical strategies, full-scale fire performance prediction for furniture, 602–604f
End-product safety, requirements, 588
Engineering application polymers, methods for flame retardancy improvement, 217–218
Engineering Center for Environmentally Conscious Products, IBM policy, 580
Environment, test apparatuses, 459
"Environmental Aspects: Inclusion in Electrotechnical Product Standards", description, 582
Environmental Choice Program, ecolabel requirements, 585
Environmental NO_2 exposure hazards
  bronchoalveolar lavage of NO_2 exposure in sheep
    procedure, 329–330
    results, 336,340–341f
  experimental description, 324
  extrapolation of sheep data to military personnel, 342
  lung lavage analysis
    procedure, 325
    results, 330,334f
  NO_2 exposure
    procedure, 324–328f
    results, 330–333,335–337
  nose vs. lung NO_2 exposure in sheep
    procedure, 329
    results, 336,338–339f
  reasons for interest, 323–324
Environmentally preferable products and services, description, 582
Escape conditioning measurement
  physiological irritation, 352
  social behavior and emotionality, 352
Ethylenevinyl acetate copolymers, charring process in condensed phase, 525–529,531
Europe
  codes and standards for flammability, 21–24
  end-product safety requirements, 588
  regulations affecting flame retardants, 588
  testing of upholstered furniture, 609–616
European Commission, draft directive on fire behavior of upholstered furniture, 609–610
Evoked potential, measures of neurological activity, 351
External heat flux capabilities, test apparatuses, 459

F

F33, description, 605
Fabric weight test, full-scale fire performance prediction for furniture, 598,599f

INDEX

FAR Part 25 tests for aircraft
  heat release rate test, 12
  ignition test, 11
Federal Acquisition, Recycling and Waste Prevention, function, 580,582
Ferrocene, use as smoke retarder in poly(vinyl chloride), 537
Fire
  causes in home, 1–2
  causes of death, 291,294
  death rate, 1
  history as building hazard, 1
  phases, 8,13f
  properties, 8–9
Fire chemistry
  investigation difficulties, 407
  mathematical models, 408
  techniques, 407–408
Fire endurance
  fire property, 8–9
  flammability tests, 14,15f
Fire extinction index
  definition, 452
  measurement, 489,491–495
Fire performance of furniture, prediction tools, 593–605
Fire propagation index
  calculation, 483–487
  definition, 452,482
  measurement, 476,479–490
Fire propagation test method
  classification of materials based on fire propagation behavior, 482–483
  downward fire propagation, 476,479–480,481f
  fire propagation index test, 483–486,487f
  upward fire propagation, 480–482
Fire properties of materials for model-based assessments for hazards and protection needs
  combustion process
    corrosion index, 452,486–490
    heat of combustion, 452,468–474
    heat of gasification, 451,464–468,470
    surface reradiation loss, 451,464–468,470
    yield of product, 452,468–474

Fire properties of materials for model-based assessments for hazards and protection needs—*Continued*
  fire extinction index, 452,489,491–495
  fire propagation index, 452,476,479–490
  ignition
    critical heat flux, 451,460–464
    thermal response parameter, 451,460,462–464
  reasons for interest, 450
  test apparatuses
    cone calorimetry, 453,457f
    environment, 459
    external heat flux capabilities, 459
    flammability apparatus, 453,455–456f
    measurements, 460
    Ohio State University heat release rate apparatus, 453,454f
    product flow, 460
    sample configuration and dimensions, 460
    test conditions, 453,458t
  test method development, 450
Fire regulations, *See* Regulations
Fire retardancy of polymers, methods, 186
Fire retardants, phosphorus compounds, 29
Fire-safe aircraft cabin materials
  aircraft cabin fires, 620–625
  aircraft cabin materials, 618–619
  fire response of materials
    thermodynamics, 629–631
    thermokinetics, 625–629
  future work, 637
  heat transfer, 632–636
Flame extinction test method
  cup burner test as Halon alternative screen, 492–494f
  description, 489
  Halon alternatives, 491–492
  procedure, 492,495
Flame heat flux
  definition, 451
  measurement, 464–468,470
Flame radiation scaling technique, procedure and validity, 467–468

Flame retardancy
  surface modification of polymers, 236–243
  X-ray photoelectron spectroscopy, 518–535
Flame retardant(s)
  brominated, polybrominated dibenzodioxin and dibenzofuran formation in laboratory combustion processes, 377–390
  developments, 25
  flammability tests, 8–20
  halogen, *See* Halogen flame retardants
  in plastics, aromatic organic phosphate oligomers, 56–63
  phosphorus, *See* Phosphorus flame retardants
  problems with ecological safeness, 186
  regulations, 586–588
  use with chlorinated flame retardants, 65–74
Flame-retardant additives
  examples, 7–8,236
  melamine, 199–214
Flame-retardant chemistry
  additives, 7–8
  backbone incorporation, 8
  codes and standards, 21–24
  flammability classification of polymers, 2,6t
  interruption of polymer combustion cycle, 2,5f
  intrinsically flame-retardant polymers, 7
  mechanism, polymers, 520–526
  polymer combustion cycle, 2,3–4f
  testing and regulation, 8–20
  toxicity, 24–25
Flame-retardant effectivity, ammonium polyphosphate–polypropylene mixtures, 95,96f
Flame-retardant moisture-resistant primary insulation with low power loss, 199
Flame-retardant polymer systems, ecologically safe, *See* Ecologically safe flame-retardant polymer systems
Flame spread
  fire property, 8–9
  flammability tests, 11–12
  models, 423–424,429–433

Flaming combustion process, 451
Flammability
  classification, 2,6t
  control by manipulation of condensed-phase chemistry, 161
  demand for control without use of halogenated additives, 161
  measurement using heat release, 594–595
Flammability apparatus
  heat release parameter measurement, 476
  ignition test method, 460–464
  measurement
    fire propagation, 476,479–483
    flame extinction, 492
  nonthermal damage measurement, 486–490
  smoke damage test method, 489
Flammability improvement of polyurethanes, silicone moiety incorporation into block copolymers, 217–233
Flammability properties of honeycomb composites and phenol–formaldehyde resins
  experimental description, 246–247
  honeycomb composites
    CH_4 produced from thermal degradation of resin vs. rate of heat released from composite, 247
    Fourier-transform IR–evolved gas analysis spectra of heat-treated and untreated composites, 247,249–250f
    rate of heat released, 247
    thermograms before and after heat treatment, 247,248f
  phenol–formaldehyde resins
    ^{13}C CP–MAS spectra of cured resins, 252–254f
    CO_2, CH_3OH, and CH_4 production during thermal degradation, 252,253t
    flammability properties, 252,253t
    Fourier-transform IR–evolved gas analysis spectra vs. temperature, 249,251f
    mechanism for CO and CO_2 formation during oxidative degradation, 252
    rate constants for CO_2 and CH_4 formation, 251
    structure vs. flammability, 255

INDEX

Flammability tests
  corrosivity, 16
  ease of extinction, 14,16–20
  fire endurance, 14,15f
  flame spread, 11–12
  heat release rate, 12–14
  ignition, 9–11
  large-scale tests, 16
  smoke, 16
Foaming behavior, ammonium polyphosphate–poly(ethyleneurea formaldehyde) mixtures, 84,85–86f
Forced swim test measurement
  motor system integrity and endurance, 351
  spatial and temporal discrimination, 353
Fourier-transform IR–evolved gas analysis spectroscopy, honeycomb composites and phenol–formaldehyde resins, 245–255
Forest Products Laboratory method, lateral flame spread using lateral ignition and flame spread test apparatus, 441–442
French telecommunications industry test
  comparison to other tests, 566,568–575
  interlaboratory evaluation, 562,565f
  problems, 562–563
  procedure, 562
  sensitivity, 562
Freon 12, effects on behavior and performance, 354–362
Full-scale fire performance prediction for furniture
  backyard test, 597–598,599f
  cone calorimeter use, 598,600–602,604
  early history, 593–594
  effect of vandalism, 605
  empirical strategies, 602–603,604f
  fabric weight test, 598,599f
  heat release as measure of flammability, 594–595
  heat release consensus codes and standards, 595–597
  Ohio State University heat release rate apparatus use, 603,605
Furniture
  fire performance prediction tools, 593–605

Furniture—*Continued*
  full-scale fire performance prediction, 593–605
  upholstered, testing in Europe, 609–616
Furniture calorimeter, testing of upholstered furniture, 612–615

G

Gas-layer conservation equations, ROOMFIRE, 411–414
Gas temperature rise calorimetry, determination of convective heat release rate, 471,475–477,479
Gasification, thermal decomposition of poly(vinyl alcohol), 183,184f
GC–MS analysis
  smoke production from advanced composite materials, 372–374
  soot produced from polymeric material combustion, 397–404
GC–MS–direct pyrolysis analysis, soot produced from polymeric material combustion, 404–406
Generation rate of products, measurement, 468–474
Germany, regulations affecting flame retardants, 586–587
Grafting percent, definition, 238
Green products
  ecolabels, 584–585
  end-product safety requirements, 588
  plastics part identification, 583–584
  plastics recycling, 588–590
  product take back, 585–586
  regulations affecting flame retardants, 586–588
  role in product design, 580
Grip strength response, measures of motor system integrity and endurance, 351
Group D–1 materials, description, 482
Group N–1 materials, nonpropagating, description, 482
Group P–2 materials, propagating, description, 482
Group P–3 materials, description, 483

H

^{1}H-NMR analysis
　effect of zinc chloride on thermal stability of styrene–acrylonitrile copolymers, 142
　hydrolytically stable phosphine oxide comonomers, 45–46,48–49,52
Halogen(s), role in flame retardancy, 236–237
Halogen flame retardants
　backbone incorporation, 27
　property improvements, 27–28
Halon 1211, effects on behavior and performance, 354–362
Hazard
　definition, 2
　model-based assessments of fire properties, 450–495
Hazard identification and prevention in oxidative thermal degradation of polymers, importance, 377
Hazardous Substance Ordinance, description, 586–587
HCl, effect of physiological irritancy, 348–349
HCN, toxicity, 292
Heat conduction, calculation using ROOMFIRE, 418–419
Heat of combustion
　definition, 452
　measurement, 468–474
Heat of gasification
　definition, 451
　measurement, 464–468,470
Heat release
　flammability measurement, 594–595
　measurement using controlled-atmosphere cone calorimeter, 503,511–512f
Heat release calorimetry, use for fire chemistry analysis, 407–408
Heat release consensus codes and standards, full-scale fire performance prediction for furniture, 595–597
Heat release parameter
　definition, 475
　determination, 475–476,477f,479t
Heat release rate
　definition, 12–14
　measurement, 471,475–477f,479t
　using lateral ignition and flame spread test apparatus
　　description, 443–444
　　Douglas fir plywood example, 446–447
　　Forest Products Laboratory modification, 444–446
Heat transfer, fire-safe aircraft cabin materials, 632–636
Heterocyclic materials exhibiting enhanced thermal stability, examples, 256
Hexacarbonitrile tris(imidazo)triazine, 260
High-temperature copolymers from inorganic–organic hybrid polymer and multiethynylbenzene
　aging studies in air, 274,276–278
　DSC, 272,273f
　experimental description, 268–271
　future work, 279
　previous studies, 268
　pyrolysis, 272,274–276f
　synthesis of copolymer, 271–272
High-temperature IR spectroscopy, effect of zinc chloride on thermal stability of styrene–acrylonitrile copolymers, 139–141,152–157
High-temperature materials, linear siloxane–acetylene polymers as precursors, 280–289
High-temperature polymer–organic char former, preparation, 187
Higher cognitive decrement, measures, 354
Homopolycarbonate
　^{1}H-NMR analysis, 45,46f
　^{31}P-NMR analysis, 45,47f
　synthesis, 45
　thermogravimetric analysis, 45–47f
Honeycomb composites, flammability properties, 245–250
Houben–Hoesch synthesis, styrene–acrylonitrile copolymers, 151–152
Hue discrimination, measures of sensory acuity, 352
(4-Hydroxyphenyl)diphenylphosphine oxide, synthesis, 44

INDEX

Hydrogen cyanide, toxicity, 24
Hydrolytically stable phosphine oxide comonomers
  experimental description, 41–43
  future work, 55
  homopolycarbonate
    ^{1}H-NMR analysis, 45,46f
    ^{31}P-NMR analysis, 45,47f
    synthesis, 45
    thermogravimetric analysis, 45–46,47f
  model reactions, 44–45
  monomer synthesis, 43–44
  phosphorus-containing copolycarbonates
    cone calorimetry, 54
    DSC, 52,53f
    ^{1}H-NMR analysis, 52
    properties, 52
    synthesis, 51
    thermogravimetric analysis, 52,53f
  phosphorus-containing polyarylate
    ^{1}H-NMR analysis, 48,49f
    ^{31}P-NMR analysis, 48,50f
    synthesis, 48
    thermogravimetric analysis, 48,50–51
  previous studies, 41

I

Ignitability using lateral ignition and flame spread test apparatus
  calibrations, 436–438
  Douglas fir plywood example, 438–440
Ignition
  flammability tests, 9–11
  measurement, 451,460,462–464
Inorganic elements, advantages of combining with organic polymers, 280–281
Inorganic–organic hybrid polymer and multiethynylbenzene, high-temperature copolymer formation, 267–279
International Electrotechnical Commission (IEC) tests
  codes and standards for flammability, 24
  IEC-950, end-product safety requirements, 588

International Standards Organization (ISO)
  codes and standards for flammability, 24
  ISO 9000, description of quality, 580
  ISO 13344, toxicity standard, 24
  ISO 14000, description of environmental management, 580
  Technical Committee 207, areas of work, 580
Intracranial self-stimulation thresholds, measures of motivational level and frustration, 353
Intrinsically flame-retardant polymers, 7
Intumescent fire retardants
  advantages, 76
  ammonium polyphosphate–poly(ethyleneurea formaldehyde) mixtures, 76–89
  applications, 77
  improvement of effectiveness, 76
Intumescent flame retardants
  ammonium polyphosphate–polypropylene mixtures, 92–115
  ingredients required, 92–93
Ion chromatographic analysis, soot produced from polymeric material combustion, 396–397
IR spectroscopy, high-temperature, effect of zinc chloride on thermal stability of styrene–acrylonitrile copolymers, 139–141,152–157

K

Kaolin
  component loading level vs. flame retardancy, 202–207
  experimental procedure, 200–201
  flammability, 201
  mode of action, 207,209,210f
  morphology of burning residues, 211,213–214f
  processability, 207–208
  water absorption, 207
Kashiwagi mechanism, poly(methyl methacrylate) degradation, 133–134
$KMnO_4$, role in production of ecologically safe flame-retardant systems, 186–197

L

Laboratory combustion processes of brominated flame retardants, polybrominated dibenzodioxin and dibenzofuran formation, 377–390
Laboratory flammability tests, 499
Ladder-type polymers, examples, 256
Large-scale tests, flammability, 16
Lateral flame spread using lateral ignition and flame spread test apparatus
  description, 440–441
  Douglas fir plywood example, 442–443
  Forest Products Laboratory method, 441–442
Lateral ignition and flame spread test apparatus
  experimental description, 435–436
  heat release rate, 443–447
  ignitability, 436–440
  lateral flame spread, 440–443
Limiting oxygen index, description, 14
Linear siloxane–acetylene polymers as precursors to high-temperature materials
  cure studies, 286–289
  experimental description, 281–282
  future work, 289
  previous studies, 281
  synthesis, 282–284,286
  thermal behavior
    DSC, 283,284f
    thermogravimetric analysis, 283,285f,287f
Live Fire Testing Program, testing, 324

M

Mass flow entrained in room fire, calculation using ROOMFIRE, 417
Mass flows through doorway, calculation using ROOMFIRE, 414–417
Mass loss of poly(methyl methacrylate), measurement using controlled-atmosphere cone calorimeter, 503,510f

Mass loss rate measurement
  in flaming combustion process, 465–468,470f
  in nonflaming combustion process, 464–465,466t
Materials, reasons for use, 580
Materials recycling, elements, 582
Mathematical models, fire chemistry, 408
Melamine as flame-retardant additive
  applications, 199
  component loading level vs. flame retardancy, 202–207
  experimental procedure, 200–201
  flammability, 201
  mode of action, 207,209,210f
  morphology of burning residues, 211,213–214f
  previous studies, 199
  processability, 207–208
  water absorption, 207
Melem, structures, 524–525,527t
Melon, structures, 524–525,527t
Metal(s)
  studies, 117
  use as flame-retardant additives, 7
Metal-containing additives for poly(vinyl chloride), role as Lewis acid catalysts for cross-linking reactions, 118
Metal loss, corrosive effect of smoke, 554
Metal oxides, use as flame retardants in polymer systems, 537
Methacrylic acid, surface modification of polymers for flame retardation, 236–243
Miniature pigs, blood cyanide levels, 312–321
Mitler's model, description, 424,429,431–433
Modified forced swim test, measures of higher cognitive decrement, 354
Molybdenum additives, reactions with poly(vinyl chloride), 119,120f
Molybdenum-based compounds, use as smoke retarder in poly(vinyl chloride), 537–538
Molybdenum oxide smoke suppression poly(vinyl chloride), 537–538
  thermal analysis, 543,546–547

INDEX

Motivational level and frustration, 353
Motor system integrity and endurance, measures, 351
Multiethynylbenzene and inorganic-organic hybrid polymer, high-temperature copolymer formation, 267–279

N

N-gas method, 296
N-gas model to toxicity potency using seven gases
 animals, 297–298
 $CO-CO_2-NO_2-O_2-HCN$ mixture toxicity, 304,307,308t
 description, 294–295
 development into N-gas methods, 296–308
 development objectives, 294
 empirical mathematical relationship, 295
 gas analytical procedure, 297
 NO_2-CO_2-HCN ternary mixture toxicity, 304–306f
 NO_2 plus CO toxicity, 299–301f
 NO_2 plus CO_2 toxicity, 298
 NO_2 plus HCN toxicity, 299,302–304,306
 NO_2 plus O_2 toxicity, 299,301–302f
 NO_2 toxicity, 298,300f
 toxicity determination procedure, 298
 validity, 295–296
National Bureau of Standards smoke chamber
 combustible materials vs. metal corrosion, 555,557,558f
 description, 554
 experimental description, 554–555
 follow-up work with copper targets, 557–561
 HCl concentration vs. steel coupon corrosion, 555,556f
 smoke test, 16
 treatment vs. steel coupon corrosion, 555,556f
National Fire Protection Association
 NFPA 101, description, 596
 NFPA 266, description, 595–596
 NFPA 267, description, 596

National Institute of Standards and Technology (NIST) radiant smoke corrosivity test
 comparison to other tests, 566,568–575
 description, 563–564,595
Natural and synthetic polymers, problem of flammability, 76
Naval Medical Research Institute Toxicology Detachment (NMRI/TD) neurobehavioral screening battery
 application to combustion technology, 354–362
 description, 350–351
 experimental description, 344
 function, 350
 measurement
  higher cognitive decrement, 354
  motivational level and frustration, 353
  motor system integrity and endurance, 351
  neurological activity, 351
  physiological irritation, 352
  sensory acuity, 32
  social behavior and emotionality, 352
  spatial and temporal discrimination, 353
 previous studies, 348–350
 problems with complex combustion atmospheres, 346–347
 reasons for development, 345
 use of animal models, 347
Negative reinforcement paradigms, measures of motivational level and frustration, 353
Net heat of complete combustion, 452
Net rate of heat released, measurement of fire size, 245–246
Neurobehavioral screening battery
 Naval Medical Research Institute Toxicology Detachment (NMRI/TD), See Naval Medical Research Institute Toxicology Detachment (NMRI/TD) neurobehavioral screening battery
Neurobehavioral tests and batteries, use in combustion toxicology, 348–350
Neurobehavioral toxicity evaluation, 345
Neurological activity, 354
Nitrogen-containing polymeric materials, oxygen indexes, 256,257t

NO_2 exposure hazards, environmental, *See* Environmental NO_2 exposure hazards
NO_2–CO_2–HCN ternary mixture toxicity, N-gas model, 304,305t,306f
NO_2 plus CO toxicity, N-gas model, 299,300–301f
NO_2 plus CO_2 toxicity, N-gas model, 298
NO_2 plus HCN toxicity, N-gas model, 299,302f–304,306f
NO_2 plus O_2 toxicity, N-gas model, 299,301–302f
NO_2 toxicity, N-gas model, 298,300f
Nonconducting surface formation on contacts, corrosive effects of smoke, 554
Nonflaming combustion process, 451
Nonpropagating group N–1 materials, description, 482
Nonthermal damage test method
 classification of equipment contamination levels, 486–488
 corrosion test, 488–490f,t
 smoke damage test, 489
North American wood industry, single-compartment mathematical fire model, 409
Nylon 6,6, production of ecologically safe flame-retardant systems, 186–197
Nylon 6,6 copolymers, triarylphosphine oxide containing, *See* Triarylphosphine oxide containing nylon 6,6 copolymers

O

Ohio State University heat release rate apparatus
 full-scale fire performance prediction for furniture, 603,605
 heat release parameter measurement, 476
 ignition test method, 460
 smoke damage test method, 489
Oligomer, synthesis for increased flame retardancy, 27
Operant progressive ratio measurement
 higher cognitive decrement, 354
 motivational level and frustration, 353
Organic halogen–antimony oxide compounds, use as flame-retardant additives, 7

Organic halogen compounds, use as flame-retardant additives, 7
Organic phosphate(s)
 reduction of flammability of plastics, 56
 use as flame-retardant additives, 7
Organic phosphate oligomers as flame retardants in plastics, aromatic, *See* Aromatic organic phosphate oligomers as flame retardants in plastics
Organic polymers, advantages of combining with organic polymers, 280–281
Original Equipment Manufactures, product design questions, 582
Oxidative thermal degradation
 comparison to pyrolysis, 379
 importance of hazard identification, 377
Oxygen concentration, measurement using controlled-atmosphere cone calorimeter, 502, 504–505t
Oxygen consumption calorimetry, determination of chemical heat release rate, 471,475–477,479
Oxygen depletion calorimetry, heat release rate test, 14
Oxygen index
 comparison with UL 94 test ratings, 16,20f
 description, 14
 effect of C:H ratio, 256,257t
 effect of flame retardants, 14,16–19
 factors affecting sensitivity, 14
 flammability measurement, 256
 nitrogen-containing polymeric materials, 256
 silicone moiety incorporation into block copolymers for flammability improvement, 225,228–229,230–231
 surface modification of polymers for flame retardation, 242–243
 test (ASTM D2863) as measure of ease of extinction, 14
Ozone-depleting substance replacement candidate, R–134a, effect of exposure on behavior and performance
 apparatus, 356
 behavioral incapacitation, 358–359,362
 effects of oxygen replacement, 357–358,362
 epileptic seizures, 359–360

INDEX

Ozone-depleting substance replacement candidate, R–134a, effect of exposure on behavior and performance—*Continued*
  experimental description, 354–355
  gas exposure groups, 355–356
  neurobehavioral end points, 356–357
  operant chamber, 357
  operant conditioning, 360–361
  physiological symptoms, 359
  recovery from incapacitation, 359,362
  subjects, 355
  toxicity comparison to Freon 12 and Halon 1211, 361–362
  training procedure, 356

P

P–300 wave, measures of neurological activity, 351
^{31}P-NMR analysis, hydrolytically stable phosphine oxide comonomers, 45–48,50
Pattern vision, measures of sensory acuity, 352
Percent of grafting, definition, 238
Phenol–formaldehyde resins, flammability properties, 245–247,249–255
Phenyltin chloride, role in thermal degradation of poly(methyl methacrylate), 129–130
Phosphate oligomers as flame retardants in plastics, aromatic organic, *See* Aromatic organic phosphate oligomers as flame retardants in plastics
Phosphine oxide comonomers, hydrolytically stable, *See* Hydrolytically stable phosphine oxide comonomers
Phosphorus
  advantages of polymer backbone incorporation, 41
  role in flame retardancy, 236–237
Phosphorus compounds, advantages of polymer backbone incorporation, 29
Phosphorus-containing copolycarbonates
  cone calorimetry, 54
  DSC, 52,53*f*

Phosphorus-containing copolycarbonates—*Continued*
  ^{1}H-NMR analysis, 52
  properties, 52
  synthesis, 51
  thermogravimetric analysis, 52,53*f*
Phosphorus-containing polyarylate
  ^{1}H-NMR analysis, 48,49*f*
  ^{31}P-NMR analysis, 48,50*f*
  synthesis, 48
  thermogravimetric analysis, 48,50–51
Phosphorus flame retardants
  backbone incorporation, 27
  property improvements, 27–28
Physiological irritation, measures, 352
Plastics
  aromatic organic phosphate oligomers as flame retardants, 56–63
  Blue Angel requirements, 584–585
  part identification, 583–584
  recycling, 588–590
Plastics use in building and construction, flammability considerations, 56
Polyaryl phosphate oligomers, flame retardants in plastics, 56–63
Polyarylate, phosphorus containing, *See* Phosphorus-containing polyarylate
Polybrominated dibenzodioxin and dibenzofuran formation in laboratory combustion processes of brominated flame retardants
  effect of temperature, 390
  experimental description, 377–379, 382–383
  furnaces, 379,380–381*f*
  hazard studies, 390
  single electron transfer mechanism, 386,388*f*
  thermal behavior of decabromobiphenyl ether
    metal oxide vs. product formation, 386,389*f*
    metal species vs. product formation, 386,387*f*
    product yields vs. temperature, 383–386

Poly[butadiyne-1,7-bis(tetramethyl-disiloxyl)-*m*-carborane], pyrolysis and synthesis, 270

Poly[butadiyne-1,7-bis(tetramethyl-disiloxyl)-*m*-carborane]–1,2,4,5-tetrakis(phenylethynyl)benzene mixture, *See* 1,2,4,5-Tetrakis(phenylethynyl)-benzene–poly[butadiyne-1,7-bis-(tetramethyldisiloxyl)-*m*-carborane] mixture

Poly(2,6-dimethylphenylene oxide), component loading level vs. flame retardancy, 202–207
experimental procedure, 200–201
flammability, 201
mode of action, 209,211,212f
morphology of burning residues, 211,213–214f
processability, 207–208
water absorption, 207

Polyethylene–inorganic phosphorus, charring process in condensed phase, 530–535

Poly(ethyleneurea formaldehyde)–ammonium polyphosphate mixtures, *See* Ammonium polyphosphate–poly(ethyleneurea formaldehyde) mixtures

Polyfunctional arylacetylenes, use as precursors to carbon, 267–279

Polyisocyanurate foams, property measurement using controlled-atmosphere cone calorimeter, 502–503,508–509,512–514

Polymer(s)
burn risk level, 2
chemical ionization MS of pyrolysis products, 538
flammability classification, 2,6t
surface modification for flame retardation, 236–243
X-ray photoelectron spectroscopy of flame retardancy, 518–535

Polymer combustion, *See* Combustion of polymers

Polymer matrix, effect on thermal behavior of ammonium polyphosphate–poly(ethyleneurea formaldehyde) mixtures, 89

Polymeric materials, oxygen index vs. C:H ratios, 256,257t

Poly(methyl methacrylate), property measurement using controlled-atmosphere cone calorimeter, 503, 510–516

Polyolefin wire and cable, flame retardation, 68,70,72–74

Polyphenylene, example of intrinsically flame-retardant polymers, 7

Polypropylene–ammonium polyphosphate mixtures, *See* Ammonium polyphosphate–polypropylene mixtures

Polyurethanes, silicone moiety incorporation for flammability improvement, 217–233

Poly(vinyl alcohol), pyrolysis, 164–171

Poly(vinyl alcohol)–bismaleimide, pyrolysis, 171–184

Poly(vinyl alcohol) system, preparation of high-temperature polymer–organic char former, 187–192

Poly(vinyl chloride)
advantages, 536
consumption, 536
flammability, 536
smoke production on combustion, 537
smoke suppression, 118–124
thermal analysis using pyrolysis–chemical ionization MS, 536–547

Potential toxicants, screening, 344–345

Processable thermosetting resins with high thermal and oxidative stability, demand, 280

Processing window, definition, 268

Product flow, test apparatuses, 460

Product generation parameter
definition, 469
measurement, 469–474

Product life cycle assessment, 582

Product take back, requirements, 585–586

Propagating group P-2 materials, 482

Propagating group P-3 materials, 483

Protection needs, model-based assessments of fire properties, 450–495

Pyrolysis
comparison to oxidative thermal degradation, 379
definition, 451

INDEX

Pyrolysis—*Continued*
  high-temperature copolymers from inorganic–organic hybrid polymer and multiethynylbenzene, 272,274–276*f*
  poly(vinyl chloride), 539–540,541*f*
Pyrolysis–chemical ionization MS
  applications, 538
  thermal analysis of fire-retardant poly(vinyl chloride), 536–547
  use for fire chemistry analysis, 407
Pyrolysis–GC–MS analysis, soot produced from polymeric material combustion, 404–406
Pyrolysis front, definition, 476
Pyrolysis products of polymers, chemical ionization MS, 538

R

R–134a, *See* Ozone-depleting substance replacement candidate, R–134a
Radiant smoke corrosivity test, *See* National Institute of Standards and Technology (NIST) radiant smoke corrosivity test
Radiative heat of combustion, 452
Radiative heat release rate, 471
Radiative heat transfer, calculation using ROOMFIRE, 417–418
Rat complex operant discrimination measurement
  higher cognitive decrement, 354
  spatial and temporal discrimination, 353
Rate of heat release
  fire property, 8–9
  measurement using lateral ignition and flame spread test apparatus, *See* Heat release rate using lateral ignition and flame spread test apparatus
Rating of convulsive behaviors, measurement of neurological activity, 351
Recycling
  elements, 582
  plastics, 588–590

Reductive coupling of poly(vinyl chloride) promoted by zero-valent copper
  activated copper–organic chloride reactions, 121–122
  activated copper slurry–poly(vinyl chloride) solution reaction, 122–123
  experimental description, 119–120
  reactions, 119
  solid-state copper additive–poly(vinyl chloride) reaction, 123,124*f*
Regulations
  flame-retardant chemistry, 8–20,586–588
  influencing factors, 551
  studies, 551–552
Relaxation studies, thermal decomposition of poly(vinyl alcohol), 179–181
Residue morphology, thermal decomposition of poly(vinyl alcohol), 181–183
Risk, definition, 2
ROOMFIRE
  applications, 420
  auxiliary equations
    convective heat transfer, 418
    heat conduction, 418–419
    mass flow entrained in fire, 417
    mass flows through doorway, 414–417
    radiative heat transfer, 417–418
  description, 410
  enthalpy flows, 413–414
  gas-layer conservation equations, 411–414
  heat flows, 413–414
  mass flows in room fire, 411,412*f*
  overview, 410–411
  sequence of calculations, 419
  wall sections for heat-transfer calculations, 410,411*f*
Rotowheel performance, measures of motor system integrity and endurance, 351

S

Sample configuration and dimensions, test apparatuses, 460

Schedule-induced polydipsia, measures of motivational level and frustration, 353
Screening of potential toxicants, importance, 344–345
Self-driven treadmill, measures of motor system integrity and endurance, 351
Scanning electron microscopy (SEM), silicone moiety incorporation into block copolymers for flammability improvement, 230
Semiladder-type polymers, examples, 256
Sensory acuity, measures, 352
Sensory and respiratory system irritancy, measures of physiological irritation, 352
Setchkin apparatus, ignition test, 9
Silicon carbide, CVD coatings, 268
Silicone moiety incorporation into block copolymers for flammability improvement
  advantages, 218
  backbone, 218
  DSC, 225,226t
  EDS, 230,232f
  experimental description, 218,220
  measurement procedures, 222,225
  oxygen index, 225,228–229,230–231
  previous studies, 218
  SEM, 230
  synthesis, 218–224
  thermogravimetric analysis, 225,226t,227f
  XPS, 230,232–233
Siloxane–acetylene polymers as precursors to high-temperature materials, *See* Linear siloxane–acetylene polymers as precursors to high-temperature materials
Siloxyl groups, advantages of combining with organic polymers, 280–281
Single-compartment mathematical fire model
  applications, 409
  ROOMFIRE, 410–420
Single behavioral end points, assessment of behavioral toxicity, 348
Small-scale flammability tests, 499
Smoke
  definition, 16
  evaluation of effects, 292
  flammability tests, 16

Smoke—*Continued*
  measurement using controlled-atmosphere cone calorimeter, 513,514f
Smoke corrosivity
  cause, 554
  commercial and marketing interest, 553–554
  initial fundamental research work, 554–561
  types of effects, 554
Smoke corrosivity performance testing
  applications, 576
  cone calorimeter, 564
  French telecommunications industry test, 560,562–563,565f
  NIST radiant smoke corrosivity test, 563–564
  surrogate tests, 566
  test comparison study conducted by Society of Plastics Industry Polyolefins Council, 566,568–575
  tubular combustion chamber dynamic corrosion test, 564
Smoke damage test method, procedure, 489
Smoke production from advanced composite materials
  aerosol characterization, 371–372
  apparatus, 367–368
  comparison of UPITT II with thermogravimetric analysis, 375
  experimental description, 367–368
  future work, 376
  identification and quantification of compounds by GC–MS, 372–374
  mass loss rate prediction, 374–375
  mass loss rate vs. air flow, 369,370f
  mass loss rate vs. heat flux, 368–370f
  smoke characterization, 371–372
  thermogravimetric analysis, 369,371t,375
  time to ignition vs. heat flux, 368–369
  toxicity prediction, 376
Smoke release, fire property, 8–9
Smoke retarders, use in poly(vinyl chloride), 537–538
Smoke suppression for vinyl chloride polymers, reductive coupling of poly(vinyl chloride) promoted by zero-valent copper, 118–124

INDEX

Smoldering, definition, 451
Social contact behavior measurement
  sensory acuity, 352
  social behavior and emotionality, 352
Society of Plastics Industry Polyolefins
  Council, test comparison study of
  smoke corrosivity performance testing,
  566,568–575
Solid-phase inhibition, interruption of
  polymer combustion cycle, 2,5f
Soot, composition, 393
Soot produced from polymeric material
  combustion
  direct pyrolysis–GC–MS analysis,
    404–406
  experimental description, 394–395
  GC–MS analysis, 397–404
  ion chromatographic analysis, 396–397
  previous studies, 394
Spatial discrimination, measures, 353
Standards, flame-retardant chemistry, 21–24
Standards and codes for flammability, *See*
  Codes and standards
Startle response measurement
  neurological activity, 351
  sensory acuity, 352
Stichting Milieukeur, ecolabel
  requirements, 585
Styrene–acrylonitrile–butadiene, flame
  retardation, 68–72
Styrene–acrylonitrile copolymers
  effect of zinc chloride on thermal
    stability, 136–157
  properties, 136
Surface chemistry, role in ignitability
  and flame spread, 159–160
Surface modification of polymers for
  flame retardation
  char yield in thermogravimetric
    analysis, 239–243
  cone calorimetry, 242
  experimental description, 238–239
  oxygen index, 242–243
  previous studies, 237
Surface reradiation loss
  definition, 451
  measurement, 464,–468,470
Surrogate tests, description, 566

Synergistic effectivity
  ammonium polyphosphate–polypropylene
    mixtures, 95,97
  calculation, 95
Synthesis
  carbon–nitrogen materials, 256–266
  high-temperature copolymers from
    inorganic–organic hybrid polymer and
    multiethynylbenzene, 271–272
  hydrolytically stable phosphine oxide
    comonomers, 45,48,51

T

Tail flick measurement
  neurological activity, 351
  sensory acuity, 352
TCO–95, ecolabel requirements, 585
Temporal discrimination, measurement, 353
Tertiary phosphine dichloride, formation
  and hydrolysis, 44–45
Testing, flame-retardant chemistry, 8–20
Test(s)
  evaluation of effect of toxic substance on
    behavior, 345
  flammability, *See* Flammability tests
1,2,4,5-Tetrakis(phenylethynyl)benzene,
  oxidation of carbon residue formed, 270
  polymerization and carbonization,
    268–269
  synthesis, 269
1,2,4,5-Tetrakis(phenylethynyl)benzene–
  poly[butadiyne-1,7-bis(tetramethyldisiloxyl)-*m*-carborane] mixture
  aging studies in air, 274,276–278
  differential scanning calorimetry, 272,273f
  future work, 279
  oxidative stability of chars, 271
  polymerization, 271–272
  pyrolysis, 271–272,274–276
Tetraphenyltin, role in thermal degradation
  of poly(methyl methacrylate), 131
Thermal analysis
  fire-retardant poly(vinyl chloride) using
    pyrolysis–chemical ionization MS
    antimony oxide flame retardance,
    540–545
  experimental description, 536,539

Thermal analysis—*Continued*
  fire-retardant poly(vinyl chloride) using pyrolysis–chemical ionization MS—*Continued*
    molybdenum oxide smoke suppression, 543,546–547
    previous studies, 537–538
    pyrolysis, 539–540,541f
  ignition test, 9
Thermal damage, influencing factors, 486
Thermal decomposition of poly(vinyl alcohol)
  characterization procedure, 164
  experimental description, 161–162
  poly(vinyl alcohol)–bismaleimide pyrolysis
    ^{13}C-NMR cross polarization spectroscopy, 173,175–179
    cone calorimetry, 183
    gasification, 183,184f
    poly(vinyl alcohol), 164–171
    procedure, 162,164
    pyrolysis at 300 °C, 171,173,174f
    reactions, 171,172f
    relaxation studies, 179–181
    residue morphology, 181–183
    thermogravimetric analysis, 162,163f
Thermal degradation
  ammonium polyphosphate–poly(ethyleneurea formaldehyde) mixtures, 78–84
  poly(methyl methacrylate)
    effect of tin and sulfur additives, 126–134
    monomer as product, 126
    products, examples, 393
  styrene–acrylonitrile copolymers, 147,149–151
Thermal hazard, definition, 469
Thermal response parameter
  definition, 451
  measurement, 460,462–464
Thermal response parameter test, fire propagation index test, 483
Thermal stability of styrene–acrylonitrile copolymers, effect of zinc chloride, 136–157
Thermodynamics, testing of fire-safe aircraft cabin materials, 629–631

Thermogravimetric analysis
  ammonium polyphosphate–polypropylene mixtures, 100–106
  carbon–nitrogen materials, 256–266
  effect of zinc chloride on thermal stability of styrene–acrylonitrile copolymers, 145–147,148f
  hydrolytically stable phosphine oxide comonomers, 45–48,50–53
  linear siloxane–acetylene polymers as precursors to high-temperature materials, 283,285–286f
  silicone moiety incorporation into block copolymers for flammability improvement, 225,226t,227f
  smoke production from advanced composite materials, 369,371t,375
  surface modification of polymers for flame retardation, 239–243
  thermal decomposition of poly(vinyl alcohol), 162,163f
  triarylphosphine oxide containing nylon 6,6 copolymers, 34
Thermokinetics, testing of fire-safe aircraft cabin materials, 625–629
Thermolysis of monomers and dimers of 4,5-dicyanoimidazole, synthesis and characterization of carbon–nitrogen materials, 256–266
Thermophysical processes, description, 423–424
Thermoplastics, thermal properties, 9,10t
Tin tetrachloride, role in thermal degradation of poly(methyl methacrylate), 131–132
Toxic gas evolution, fire property, 8–9
Toxic potency
  measurement, 291
  prediction using N-gas model, 293–308
Toxic potency tests in air, reasons for development, 292
Toxicity, flame-retardant chemistry, 24–25
Triarylphosphine oxide containing nylon 6,6 copolymers
  cone calorimetry, 35,36–38f
  DSC thermograms, 33–34
  dynamic mechanical behavior, 35

INDEX

Triarylphosphine oxide containing nylon 6,6 copolymers—*Continued*
 experimental description, 29–31
 previous studies, 30
 properties, 33
 synthesis of copolymers and monomers, 32
 thermogravimetric analysis, 34
 XPS, 35,38,39*f*
Tris(imidazo)[1,2-*a*:1,2-*c*:1,2-*e*]-1,3,5-triazine-2,3,5,6,8,9-hexacarbonitrile, 260
Tubular combustion chamber dynamic corrosion test, description, 564
Tunnel tests, flame spread, 11–12

U

Underwriters Laboratories (UL) 1950, end-product safety requirements, 588
United States, end-product safety requirements, 588
Upholstered furniture, testing in Europe, 609–616
UPITT II combustion toxicity method, smoke production from advanced composite materials, 366–376
Upward fire propagation test
 fire propagation index test, 483
 method, 480–482
Upward flame spread on composite materials
 comparison of intermediate-scale data with model predictions, 429–433
 cone calorimetry, 427,429,430*f*
 experimental description, 423
 experimental procedure
  flame spread, 425–426,428*f*
  heat release, 426
 flame heat fluxes, 426–427,428*f*
 future work, 434
 models
  application to composites, 425
  Brehob and Kulkarni's model, 424,429,431–433
  Cleary and Quintiere's model, 424,429–430,432–433
  Mitler's model, 424,429,431–433
 thermophysical processes, 423–424

U.S. Army, environmental NO_2 exposure hazards, 323–342
U.S. Army Health Hazard Assessment Program, function, 324
U.S. Environmental Protection Agency Office of Toxic Substances, guidelines for conducting neurobehavioral tests of motor activity, schedule-controlled operant performance, and functional observational battery, 349
U.S. Navy, use of composite materials for ship and submarine compartment construction, 422

V

Vandalism, role in full-scale fire performance prediction for furniture, 605
Vapor-phase inhibition, interruption of polymer combustion cycle, 2,5*f*
Ventilation rate, measurement using controlled-atmosphere cone calorimeter, 502,506–507*t*
Vinyl chloride polymers, smoke suppression, 118–124

W

Wool, thermal degradation on heating, 519–520

X

X-ray photoelectron spectroscopy (XPS)
 flame retardancy
  applications, 518
  charring process in condensed phase
   effect of oxygen, 528,530
   ethylenevinyl acetate copolymers as model system, 525–529,531
   polyethylene–inorganic phosphorus, 530–534
  cross-linking vs. flame retardancy
   decontamination, 534
   strong cross-linking, 535
   weak cross-linking, 534–535
  development, 518
  experimental description, 518,519

X-ray photoelectron spectroscopy (XPS)—*Continued*
  flame retardancy—*Continued*
    flame-retardant mechanism
      in condensed phase, 521–524,526
      in gas phase, 520–521,522f
    structural analysis of intractable intermediates, 524–525,527t
    thermal degradation on heating for wool, 519–520
  silicone moiety incorporation into block copolymers for flammability improvement, 230,232–233
  triarylphosphine oxide containing nylon 6,6 copolymers, 35,38,39f
  use for fire chemistry analysis, 407

Y

Yield of product
  definition, 452
  measurement, 468–474

Z

Zinc chloride, effect on thermal stability of styrene–acrylonitrile copolymers
  char formation chemistry by high-temperature IR spectroscopy, 152–157
  characterization procedures, 138
  experimental description, 137
  Houben–Hoesch synthesis application, 151–152
  kinetic analysis of thermal degradation, 147,149–151
  material preparation, 137–138
  polymer structure characterization
    ^{13}C-NMR spectroscopy, 142
    ^{1}H-NMR spectroscopy, 142
    Houben–Hoesch synthesis application, 151–152
    IR spectroscopy, 139–141
  previous studies, 136
  thermal stability by thermogravimetric analysis, 145–147,148f
  thermal transition by DSC, 143–144

Production: Susan Antigone
Indexing: Deborah H. Steiner
Acquisition: Barbara Pralle
Cover design: Amy Hayes

Printed and bound by Maple Press, York, PA

RETURN TO: **CHEMISTRY LIBRARY**
100 Hildebrand Hall • 510-642-3753

LOAN PERIOD	1	2	3
4		5	6

1-MONTH USE

ALL BOOKS MAY BE RECALLED AFTER 7 DAYS.
Renewals may be requested by phone or, using GLADIS, type **inv** followed by your patron ID number.

DUE AS STAMPED BELOW.

DEC 1 9 2002

DEC 2 0 2005

FORM NO. DD 10
2M 5-01

UNIVERSITY OF CALIFORNIA, BERKELEY
Berkeley, California 94720–6000